# Advanced Models and Tools for Effective Decision Making Under Uncertainty and Risk Contexts

Vicente González-Prida
*University of Seville, Spain & UNED, Spain*

María Carmen Carnero
*University of Castilla-La Mancha, Spain*

IGI Global
PUBLISHER of TIMELY KNOWLEDGE

A volume in the Advances in Business Strategy and Competitive Advantage (ABSCA) Book Series

Published in the United States of America by
IGI Global
Business Science Reference (an imprint of IGI Global)
701 E. Chocolate Avenue
Hershey PA, USA 17033
Tel: 717-533-8845
Fax: 717-533-8661
E-mail: cust@igi-global.com
Web site: http://www.igi-global.com



Library of Congress Cataloging-in-Publication Data

Names: Gonzalez-Prida, Vicente, 1975- editor. | Carnero, Maria Carmen, 1970- editor.
Title: Advanced models and tools for effective decision making under uncertainty and risk contexts / Vicente González-Prida and María Carmen Carnero, editors.
Description: Hershey, PA : Business Science Reference, [2020] | Includes bibliographical references and index. | Summary: “”This book explores Industry 4.0 and its effect on decision-making, particularly how this digital revolution may reduce the uncertainty and risk associated with choosing between alternatives”--Provided by publisher”-- Provided by publisher.
Identifiers: LCCN 2019049631 (print) | LCCN 2019049632 (ebook) | ISBN 9781799832461 (hardcover) | ISBN 9781799832478 (paperback) | ISBN 9781799832485 (ebook)
Subjects: LCSH: Decision making--Mathematical models. | Risk--Mathematical models.
Classification: LCC T57.95 .A38 2020 (print) | LCC T57.95 (ebook) | DDC 658.4/0301--dc23
LC record available at https://lccn.loc.gov/2019049631
LC ebook record available at https://lccn.loc.gov/2019049632

This book is published in the IGI Global book series Advances in Business Strategy and Competitive Advantage (ABSCA) (ISSN: 2327-3429; eISSN: 2327-3437)

British Cataloguing in Publication Data
A Cataloguing in Publication record for this book is available from the British Library.



For electronic access to this publication, please contact: eresources@igi-global.com.

# Advances in Business Strategy and Competitive Advantage (ABSCA) Book Series

Patricia Ordóñez de Pablos
Universidad de Oviedo, Spain

ISSN:2327-3429
EISSN:2327-3437

## Mission

Business entities are constantly seeking new ways through which to gain advantage over their competitors and strengthen their position within the business environment. With competition at an all-time high due to technological advancements allowing for competition on a global scale, firms continue to seek new ways through which to improve and strengthen their business processes, procedures, and profitability.

The **Advances in Business Strategy and Competitive Advantage (ABSCA) Book Series** is a timely series responding to the high demand for state-of-the-art research on how business strategies are created, implemented and re-designed to meet the demands of globalized competitive markets. With a focus on local and global challenges, business opportunities and the needs of society, the **ABSCA** encourages scientific discourse on doing business and managing information technologies for the creation of sustainable competitive advantage.

## Coverage

- Tacit Knowledge
- Value Creation
- Ethics and Business Strategy
- Core Competencies
- International Business Strategy
- Differentiation Strategy
- Resource-Based Competition
- Outsourcing
- Balanced Scorecard
- Co-operative Strategies



The Advances in Business Strategy and Competitive Advantage (ABSCA) Book Series (ISSN 2327-3429) is published by IGI Global, 701 E. Chocolate Avenue, Hershey, PA 17033-1240, USA, www.igi-global.com. This series is composed of titles available for purchase individually; each title is edited to be contextually exclusive from any other title within the series. For pricing and ordering information please visit http://www.igi-global.com/book-series/advances-business-strategy-competitive-advantage/73672. Postmaster: Send all address changes to above address. 

## Titles in this Series

*For a list of additional titles in this series, please visit:*
http://www.igi-global.com/book-series/advances-business-strategy-competitive-advantage/73672

***Ethical Standards and Practice in Managerial and Financial Accounting***
Ionica Oncioiu (Titu Maiorescu University, Romania)
Business Science Reference • © 2021 • 300pp • H/C (ISBN: 9781799840121) • US $195.00

***Sustainability Reporting, Ethics, and Strategic Management Strategies for Modern Organizations***
Ionica Oncioiu (Titu Maiorescu University, Romania) Sorinel Căpuşneanu (Titu Maiorescu University, Romania) Dan Ioan Topor (1 Decembrie 1918 University, Romania) and Dana Maria (Oprea) Constantin (University of Bucharest, Romania)
Business Science Reference • © 2021 • 320pp • H/C (ISBN: 9781799846376) • US $195.00

***International Perspectives on the Youth Labor Market Emerging Research and Opportunities***
Samir Amine (Université du Québec en Outaouais, Canada)
Business Science Reference • © 2020 • 217pp • H/C (ISBN: 9781799827795) • US $170.00

***Open and Innovative Trade Opportunities for Latin America and the Caribbean***
Pablo Alberto Baisotti (University of Costa Rica, Costa Rica)
Business Science Reference • © 2020 • 255pp • H/C (ISBN: 9781799835035) • US $215.00

***Cases on Performance Improvement Innovation***
Darlene M. Van Tiem (Capella University, USA & University of Michigan – Dearborn, USA) and Nancy Crain Burns (Crain Burns Associates LLC, USA)
Business Science Reference • © 2020 • 377pp • H/C (ISBN: 9781799836735) • US $195.00

***Handbook of Research on Entrepreneurship Development and Opportunities in Circular Economy***
Neeta Baporikar (Namibia University of Science and Technology, Namibia & University of Pune, India)
Business Science Reference • © 2020 • 696pp • H/C (ISBN: 9781799851165) • US $265.00

***Digital Transformation and Innovative Services for Business and Learning***
Kamaljeet Sandhu (University of New England, Australia)
Business Science Reference • © 2020 • 282pp • H/C (ISBN: 9781799851752) • US $235.00

***Development and Implementation of Adaptive Trade Policy in the Era of Digital Globalization***
Vardan Mkrttchian (HHH University, Australia) and Yulia Vertakova (Financial University Under the Government of the Russian Federation, Kursk, Russia)
Business Science Reference • © 2020 • 300pp • H/C (ISBN: 9781799831341) • US $215.00

IGI Global
PUBLISHER of TIMELY KNOWLEDGE

701 East Chocolate Avenue, Hershey, PA 17033, USA
Tel: 717-533-8845 x100 • Fax: 717-533-8661
E-Mail: cust@igi-global.com • www.igi-global.com

# Editorial Advisory Board

# Table of Contents

# Detailed Table of Contents

Risk management is a core element from a strategic and operational point of view. Conditions of the environment, financial scandals like Enron and Worldcom, globalization, and new technologies make risk a priority for companies that want to keep a sustainable competitive advantage under uncertainty. In line with these, there are several frameworks, initiatives, and ways to approach risk under a structure organization and integrated approach, including the development of internal control standards like Sarbanes-Oxley (SOX). One of the elements to be considered when taking optimum decisions is a proper risk management. In this chapter, the key elements of assets risk management will be exposed, starting from main concepts, constructing the elements that are applicable to build a risk management system, and developing a practical example under SOX framework with a management and system approach.

**Section 2**



This chapter is intended to deal with the topic of standardization as a solution for risk control and decision making, mainly in high technologized context. For that purpose, this document starts introducing the reason of controlling risk as well as the uncertainty background that lays in different sectors. After that, the risk management evolution is outlined, defining briefly the risk concept, mentioning some current management models and enumerating the more relevant norms, standards, and guidelines for risk management. From this point, an overview on the standard ISO 31000 about risk management is provided in order to finally summarize some possible future research lines and some conclusions about this groundwork on risk and uncertainty.



The overall perspective of this study is related to the concept of risk and uncertainty in the world today. In this sense, it considers Popper's contributions together with the deductive method, contrasted Bayes' contributions with the inductive method. On one hand, induction allows to generate results considered probabilistically true. This is basically the method used by supervised predictive methods of machine learning, where a general rule is inferred from particular examples in which solutions are known, inducing consequently to possible results for new inputs. On the other hand, deduction is a process in which general hypotheses are proposed, and from them, particular statements are obtained. These particular statements can obviously generate the rejection of those initial hypotheses. Under these considerations, Bayes' and Popper's postulates should not be understood as opposed methods. With this, the specific objectives of this chapter states on an overview about technology and its relationship with science, being analyzed from the Popperian and Bayesian perspective.



The essential aim of Industry 4.0 is to enable industries to be more productive, efficient, and flexible. A predictive maintenance strategy can make a positive contribution to all these things, as it uses industrial IoT technologies to monitor asset health, optimise maintenance schedules, provide real-time alerts about operational risks, and maximise uptime, and can provide digital services to customers based on data from its machines. It improves productivity, improves customer satisfaction, and therefore gives the company a competitive advantage. Nevertheless, decision making in relation to a predictive maintenance strategy is not systematised, and this may lead to some inappropriate decisions, which do not achieve the goal sought. This chapter describes a multicriteria model, designed with the analytic hierarchy process, to systematise decision making with respect to a predictive maintenance strategy.



Information security has become one of the most important responsibilities of all organizations due to increasing cyber threats. Attackers take advantage of systems vulnerabilities; therefore, system administrators should be aware of potential threats to take necessary actions to protect their organizations and stakeholders. At this point, a risk assessment is needed to discover possible threats for vulnerable systems of the organization and to implement strategies for the business goals. This study proposes a hybrid risk management framework using both qualitative and quantitative methods to analyze risk within organizations and reduce them with practical countermeasures. Based on this framework, case studies have been carried out considering three hypothetical companies identifying possible information security risks, and these risks have been reduced to an acceptable level by applying the proposed risk analysis methodology.



In recent years, there has been a growth of interest in the development of systematic search methods for solving problems in operational research and artificial intelligence. This chapter introduces a new idea for the integration of approaches for hard combinatorial optimisation problems. The proposed methodology evaluates objects in a way that combines fuzzy reasoning with a greedy mechanism. In other words, a fuzzy solution space is exploited using greedy methods. This seems to be superior to the standard greedy version. The chapter consists of two main parts. The first part focuses on description of the theory and mathematics of the so-called fuzzy greedy evaluation concept. The second part demonstrates through computational experiments the effectiveness and efficiency of the proposed concept for hard combinatorial optimisation problems.

## Section 3



The main objective of this chapter is to suggest a methodology for the application of ISO 55000 for a water utility in order to obtain all the benefits of asset management from the starting point of application. In addition, risk is an important part of this methodology, so it also complies with the clauses of ISO 31000. The methodology consists of six steps to be carried out: plan (1), implement (2), risk management (3), monitor (4), analyze (5) and make decisions and improvements (6). The application of this methodology is an iterative process in which the information obtained is going to be used in the previous and subsequent steps so that its benefits are greater as time goes by. At each point in the guide, the actions to be taken in compliance with the clauses and sub-clauses of ISO 55001 will be proposed along with some documents so that companies have a clearer idea of how to proceed.



High works-in-process (WIP) levels usually involve significant losses for companies as they trigger dispatching delays, product shortages, increases in immobilized capital, and greater storage costs, among others. This study provides a strategy for gradually decreasing the WIP levels associated with the production line of a pharmaceutical company located in Latin America. The methodology used is based on Lean practices and, more specifically, organized into phases: the diagnosis of the current situation, information processing, mapping out the current situation through value stream mapping, analysis and determination of key factors for uncertainty control, and improvement proposals. The main tools used are value stream mapping, 5-whys, single-minute exchange of dies, and total productive manufacturing.

This chapter aims to propose an improvement in decision making in the planning sector and production control (PPC) with application of a mathematical model. In the methodology, the qualitative approach was used because the linguistic codifications are interpreted and characterized by a case study applying a questionnaire to the managers of the company of the metal mechanic sector. In this context, six constructs were structured as a proposal for performance improvement, being composed of costs, management, inspection, processes, and capacity. The chapter reports the main results achieved during fuzzy sets application, obtaining a better result compared to FAHP in which there were certain oscillations between the percentage of constructs. The construct prioritized by managers and specialists was the cost construct, reaching 38.60%, being advantageous for the industry when the cost is placed in order of manufacture (subconstruct), followed by the prioritized management construct with 28.50%.



This chapter provides an implementation of an integral methodology for the simulation of renewable energy scenarios applied to an isolated village situated in Democratic Republic of Congo. A specific area in Democratic Republic of Congo is selected as a case study in order to validate the methodology. In addition, this chapter includes the experimental validation of the HRES configuration, operating a scalable prototype in LabDER at UPV (Universitat Politècnica de València, Spain). Experimental tests conducted in the laboratory are described and main results and conclusions are discussed.



An investigation on cellular automata (CA)-based validation of PageRank with reference to the power-law distribution is presented in this chapter to enhance the trustworthiness and safety of Clouds. Web traffic data for several Clouds were analyzed in view of power-law distribution to explore whether they are natural or manmade. Results obtained for CA-based PageRank validation were compared with Alexa®, which further supported the power-law distribution with some cut-offs, which ensured effectiveness and accuracy for CA-based validation of PageRank. Hence, this exploration helps to enhance the trustworthiness

# Preface

## OVERVIEW

In 2017, the book *Optimum Decision Making in Asset Management* was published reaching an outstanding success. Now, with the view in the present year 2020, the editorial house has considered to enhance that publication with the release of a second part. While this new book is derived mostly from the themes and concepts presented in the previous publication, each chapter provides now new research directions, updated state of the art, new conclusions, etc. In this sense, besides contributors from the previous publication, this book adds new authors with entirely new content, ensuring all together the highest quality.

In the same way as the former publication, the new book *Advanced Models and Tools for Effective Decision Making Under Uncertainty and Risk Contexts* is also focused on an effective decision making based not only on mathematical techniques and models, but also in all those technological tools that are available nowadays from the so-called 4$^{th}$ Industrial Revolution. These advanced models and tools shall provide an optimal and objective decision-making, which guarantees improving competitiveness reducing risk and uncertainty. Basically, this new book keeps on contributing towards the research about the new term 'Industry 4.0' and its effect on decision-making, particularly how this digital revolution may reduce the uncertainty and risk associated with choosing between alternatives.

Therefore, the objective of this book is to continue the first part called *Optimum Decision Making in Asset Management*, providing a complementary vision on digital transformation that identifies methodologies related to decision making, risk and uncertainty, as well as design and development on scientific and industrial topics. Discussions about risk and uncertainty, together with a humanistic view of technology, particularly in the fields of logic, philosophy and history of science, and complemented with case studies related to the phenomenon of the Fourth Industrial Revolution, provide a high valuable publication that can be identified as an essential reference source in this area.

## SUMMARY OF TOPICS

The topics gathered in this book show diverse understandings of the new high-tech advances, what they means to the control of risk and uncertainty, pros and cons, including the challenges and benefits from the application of machine learning, big data, virtual reality, cloud computing, IoT platforms, disruptive technologies, or even additive 3D manufacturing. This book analyses these factors, which are divided into the following categories:

- Risk and Uncertainty
- Decision making
- Digitalization
- Science policies
- Advanced and disrupted technologies
- Industrial assets management
- Globalization of science
- Condition based maintenance and predictive maintenance
- Asset management benchmarking
- Performance evaluation and key performance indicators
- Case studies on decision making in asset management
- Asset management best practices in manufacturing organizations
- Forecasting of life cycle
- Decision support systems on asset management
- Asset management standard: ISO 55000
- Integrated Logistic Support
- Aftersales Management and Customer Service
- Social and industrial implications of technology disruption

## TARGET AUDIENCE

*Advanced Models and Tools for Effective Decision Making Under Uncertainty and Risk Contexts* is aimed at a worldwide target audience consisting of industrial and manufacturing engineers, managers and plant supervisor, academicians, researchers, advanced-level students (both postgraduate and doctoral), logic philosophers, technology developers, and managers who take decisions exposed to contexts under innovative disrupted technologies.

Due to the chapters contained in the book and the variety of the subjects analysed, it provides an in-depth look at current global concerns.

## IMPORTANCE OF EACH CHAPTER

The book is structured in three parts. The first section consists of five chapters which include updated and improved contributions of the previous book *Optimum Decision Making in Asset Management.* Section 2 consists of totally new chapters from 6 to 10 with mostly theorical content. The third section includes also new chapters, from 11 to 15, with practical content as applications to various fields.

Of more than fifty members of the editorial board and authors who have participated in this book, most of them have carried out activities related to asset management on risk contexts in different commercial organizations, bringing practical vision, but also motivating improvements, advances and new techniques to be applied to companies. This favours orientation of research towards real application in organizations, which is part of the value added by this book. There is also a significant flow of ideas and experience through collaboration as reviewers with a number of authors who have contributed to the book. The suggestions, comments and ideas provided by these colleagues have been enriching for all of us, seeding meaningful developments in knowledge. For all these reasons, we wish to thank each of them.

A brief description of the fifteen chapters, with regard to their research material and the conclusions reached, are collected and summarized as follows:

Chapter 1 describes how to select the most suitable combination of maintenance policies in the different systems that make up an operating theatre: air conditioning, sterile water, power supply, medicinal gases and operating theatre lighting. To do so a multicriteria model will be developed using the Measuring Attractiveness by a Categorical Based Evaluation Technique (MACBETH) approach considering multiple decision centres.

Chapter 2 presents how in industrial production environments, maintenance and exploitation activities are little studied in reference to the mechanisms of transmission of knowledge that generate, based on the difficulty of their capture, because is fundamentally based on tacit knowledge acquired by operators for years from their work experiences in the industrial plant. In this article, is to approach the transcendence of the tacit knowledge among staff, as well as mark the conditions for transmission as a basis for the learning of other partners of the Organization.

Chapter 3 focuses on implementation of Industry 4.0 in energy management of district heating system. The present objectives for implementation of digital technologies in district heating are analyzed. The barriers for sustainable and flexible district heating systems and challenges to overcome them are discussed. The objectives of energy management in district heating are presented. This chapter describes the decision support system "HeatCAM" as a tool of energy management in district heating.

Chapter 4 describes how high strength low alloy steels (HSLA) are currently used in large quantities for constructions in the North Sea and elsewhere. For success operation under such rigorous conditions it is essential to ensure adequate weld metal toughness at temperature down -10ºC or even lower is used. To comply with this high requirement the heat input control it is needed to avoid forming martensitic microstructure in High strength low alloy steels (HAZ) near to the fusion line during weldment cooling. This work carried out CTOD tests to ensure that, under welding conditions, the fatigue crack, toughness, hardness and resilience in the HAZ are acceptable with the standards.

Chapter 5 analyses how there are several frameworks, initiatives and ways to approach risk under a structure organization and integrated approach, including the development of internal control standards like Sarbanes-Oxley (SOX). One of the elements to be considered when taking optimum decisions is a proper risk management. In this chapter, the key elements of assets risk management will be exposed, starting from main concepts, constructing the elements which are applicable to build a risk management system and developing a practical example under SOX framework with a management and system approach.

Chapter 6 shows how to deal with the topic of standardization as a solution for risk control and decision making, mainly in high technologized context. For that purpose, this document starts introducing the reason of controlling risk as well as the uncertainty background that lays in different sectors. After that, the risk management evolution is outlined, mentioning some current management models and enumerating the more relevant norms, standards and guidelines for risk management. From this point, an overview on the standard ISO 31000 about risk management is provided.

Chapter 7 considers Popper's contributions together with the deductive method, contrasted with Bayes' contributions and the inductive method. Bayes' and Popper's postulates should not be understood as opposed methods. With this, the specific objectives of this chapter states on an overview about technology and its relationship with science, being analyzed from the Popperian and Bayesian perspective.

Chapter 8 describes how decision making in relation to a Predictive Maintenance strategy is not systematised, and this may lead to some inappropriate decisions, which do not achieve the goal sought. Therefore, this Chapter describes a multicriteria model, designed with the Analytic Hierarchy Process, to systematise decision making with respect to a Predictive Maintenance strategy.

Chapter 9 proposes a hybrid risk management framework using both qualitative and quantitative methods to analyze risk within organizations and reduce them with practical countermeasures. Based on this framework, case studies have been carried out considering three hypothetical companies identifying possible information security risks, and these risks have been reduced to an acceptable level by applying the proposed risk analysis methodology.

Chapter 10 introduces a new idea for the integration of approaches for hard combinatorial optimisation problems. The proposed methodology evaluates objects in a way that combines fuzzy reasoning with a greedy mechanism. The monograph consists of two main parts. The first part focuses on description of the theory and mathematics of the so-called fuzzy greedy evaluation concept. The second part demonstrates through computational experiments, the effectiveness and efficiency of the proposed concept within search, optimisation and learning systems for hard combinatorial optimisation problems.

Chapter 11 suggests a methodology for the application of ISO 55000 for a water utility, in order to obtain all the benefits of asset management from the starting point of application. The application of this methodology is an iterative process in which the information obtained is going to be used in the previous and subsequent steps so that its benefits are greater as time goes by. At each point in the guide, the actions to be taken in compliance with the clauses and sub-clauses of ISO 55001 will be proposed along with some documents so that companies have a clearer idea of how to proceed.

Chapter 12 provides a strategy for gradually decreasing the WIP levels associated with the production line of a pharmaceutical company located in Latin America. The methodology used is based on Lean practices and, more specifically, organized into five phases: the diagnosis of the current situation, information processing, mapping out the current situation through Value Stream Mapping, analysis and determination of key factors for uncertainty control and improvement proposals. The main tools used are: Value Stream Mapping, 5-Whys, Single-Minute Exchange of Dies and Total Productive Manufacturing.

Chapter 13 proposes an improvement in decision making in the planning sector and production control (PPC) with application of a mathematical model. Six constructs were structured as a proposal for performance improvement. The article reports the main results achieved during Fuzzy Sets application, obtaining a better result compared to FAHP. In which, there were certain oscillations between the percentage of constructs. The construct prioritized by managers and specialists was the cost construct, reaching 38.60%, being advantageous for the industry when the cost is placed in order of manufacture (subconstruct), followed by the prioritized management construct with 28.50%.

Chapter 14 provides an implementation of an integral methodology for the simulation of renewable energy scenarios applied to an isolated village situated in Republic Democratic of Congo. In addition, this chapter includes the experimental validation of the HRES configuration, operating a scalable prototype in LabDER at UPV (Universitat Politècnica de València, Spain). Experimental tests conducted in the laboratory are described and main results and conclusions are discussed.

Chapter 15 explores Cellular Automata (CA)-based validation of PageRank with reference to the Power-Law distribution to enhance the trustworthiness and safety of Clouds. Web traffic data for several Clouds were analyzed in view of Power-Law distribution to explore whether they are natural or man-made. Results obtained for CA-based PageRank validation were compared with Alexa®, which further supported the Power-Law distribution with some cut-offs; which ensured effectiveness and accuracy for CA-based validation of PageRank.

## CONCLUSION

This book stimulates the idea that decisions related to the management of physical assets should cease to be based purely on empirical evidence from experience and understanding of maintenance engineers and managers. By contrast, it should begin to incorporate techniques, tools and models using integrated mathematical techniques, together with better data and analytical and clarification techniques, the judgements of experts, and common sense. This should lead not only to satisfactory or good decisions, as optimal decisions do not generally exist due to frequently conflicting criteria, but it will also lead to efficient and effective maintenance and asset management systems which separate world class companies from the rest.

This book includes contributions developed by authors from different countries including: Chile, Portugal, Ukraine, Canada, Turkey, India, Brazil and Spain. This gives a more complete view of the state of asset management and decision under uncertainty and risk contexts around the world.

The interplay of techniques required achieving this, the application to ever more complex and exclusive areas, and the inclusion of risk and uncertainty in decision-making are questions whose analysis this book attempts to address, and they represent a challenge for future research in in the industry 4.0 environment.

*Vicente González-Prida*
*University of Seville, Spain & UNED, Spain*

*María Carmen Carnero*
*University of Castilla-La Mancha, Spain*

# Acknowledgement

To my parents, Juan Antonio (in memoriam) and Pilar.To my wife, Margarita, and my daughters, Pilar and Paloma.

Vicente González-Prida

To my parents, Manuel and Carmen and to my husband, José Luis.

María Carmen Carnero

# Section 1

# Chapter 1
# Selection of Maintenance Strategies in an Operating Theatre

**María Carmen Carnero**
*University of Castilla-La Mancha, Spain*

**Andrés Gómez**
*University of Castilla-La Mancha, Spain*

## ABSTRACT

*The aim of this chapter is to select the most suitable combination of maintenance policies in the different systems that make up an operating theatre: air conditioning, sterile water, power supply, medicinal gases, and operating theatre lighting. To do so, a multicriteria model will be developed using the Measuring Attractiveness by a Categorical Based Evaluation Technique (MACBETH) approach considering multiple decision centres. The model uses functional, safety, and technical-economic criteria, amongst which is availability. Mean availability for repairable systems has been measured to assess this criterion, using Markov chains from the data obtained over three years from the subsystems of a hospital operating theatre. The alternatives considered are corrective maintenance; preventive maintenance together with corrective maintenance by means of daily, weekly, monthly, and yearly programmes; periodical predictive maintenance together with corrective maintenance; and corrective together with preventive and predictive maintenance.*

## INTRODUCTION

The most up-to-date hospitals are experiencing increasing automation and monitoring of installation and equipment and, in general, greater complexity in the assets to be maintained (Wolf, 2008). To adjust to this new environment, a hospital's technical services (including maintenance engineering, safety and environmental services) should adapt to modern strategies and management tools that have, in fact, already been operating for some time in the manufacturing, energy, etc. industries. Currently, due to

DOI: 10.4018/978-1-7998-3246-1.ch001

the lack of structured criteria in hospital activity, it is normal to find different strategies and approaches to achieve similar ends with a great variety of costs and results. However, the activities carried out are essentially corrective and technical-legal (Fernández & Canals, 2006).

It is essential to analyse, from the qualities perceived by patients and healthcare professionals, the aspects that have a bearing on a hospital as a building answering to customer needs and increased safety. It is in these areas that the technical services have a predominant role, although there are deficiencies in recognizing the importance of maintenance management in hospitals (Rani, Baharum, Akber, & Nawawi, 2015). In general, users of the Spanish health system believe that the scientific-technical quality of its health professionals is excellent (SESCAM, 2007). They do not think the same about the other services, and particularly those regarding buildings, facilities, cleaning and those necessary for minimum comfort standards.

The application of a methodology such as that described in this research should therefore lead to a very significant change in the appreciation that people in general have of the technical services of the hospital responsible for carrying out maintenance.

Furthermore, health care professionals should be made aware of the situation of the facilities and equipment they use as internal users, as they need greater safety in their working environment. To do so, procedures should be established so that the condition of the equipment, facilities and working environment is precisely determined, and the staff need to be included in the assessment process.

The aim of this article is to select the most suitable maintenance policies in the different subsystems that make up an operating theatre. To do so, a multicriteria model is designed, using the Measuring Attractiveness by a Categorical Based Evaluation Technique (MACBETH) (Bana e Costa & Vansnick, 1994), considering multiple decision centres: managers from the Hospital's technical services, environmental and occupational risk prevention sections, healthcare managers (operating theatre and health activity programming), healthcare staff, technicians, purchasing managers and Hospital executives. The model uses safety, functional and technical-economic criteria, among which is availability. Mean availability for repairable systems has been measured to assess this criterion, using Markov chains from the data obtained over three years from the subsystems of a Hospital operating theatre. All this is aimed at increasing the operating theatre's availability and, consequently, increasing physical safety in patient operations and reducing the amount of delayed operations due to technical malfunctions.

Thus, the application of this research could lead to a significant increase in care quality, when carried out successfully in a state general hospital.

To choose the type of Multi-Criteria Decision Making (MCDM) to apply in this case study, it was felt that, given the multi-faceted profile of the decision-making groups, it should be easy to handle, intuitive, and respond to qualitative stimuli, without the need for the decision-making group to have to establish a numerical scale for subsystems. The choice of MACBETH is justified latter.

The originality of this chapter rests on the following points:

- The inclusion of the multicriteria technique MACBETH in group decision-making and its use with Markov chains in repairable systems to assess the availability of subsystems in the Hospital.
- It is the first research to analyse the selection of the best combination of maintenance policies, assessing all the systems which make up the operating theatre system. To this end the interrelations between the different systems of the Hospital were analysed.
- Although practical research on choice of maintenance policy in manufacturing companies has been done, this is one of the first studies that analyses this matter in a public service company

such as a Hospital, and specifically an operating theatre. The environment, the facilities and the need for availability and quality of service in those facilities or systems may be very different from other industrial settings, due to the fact that its efficiency has a very important impact on the way in which health services are provided and on the life of patients. This is expressed in the decision criteria used, which are adapted to the context; for example, Risk to users (patients).

The rest of this chapter is structured as follows. First, there is a review of the literature about maintenance policy selection; next it describes the characteristics of the MACBETH multicriteria technique and the Markov theory for stable repairable systems; then the multicriteria model for the selection of maintenance policies developed for an operating theatre's subsystems is presented, followed by a sensitivity analysis. The paper ends with future lines of research and conclusions.

## BACKGROUND

Different types of maintenance policy can be identified:

- Corrective maintenance, carried out after a failure with the aim of returning the machine to a state in which it can perform its desired function (Kelly, 2006).
- Preventive maintenance, carried out at set intervals throughout the working life of the machine (Knezevic, 1996), with the aim of reducing the probability of problems or wear in the machine (Kelly, 2006).
- Condition-based maintenance or predictive maintenance, based on the control of physical parameters of a working machine that can be registered, periodically or continuously, by a set of sensors, to analyse and compare data which allow decisions to be made about the working and maintenance of the machine (Rao, 1996).
- Opportunity-based maintenance, which involves using the stoppage of a system due to a fault in a component to carry out preventive maintenance in other components which are working normally (Pintelon & Gleders, 1992).
- Proactive maintenance, which detects and removes problems before they come about, hence avoiding their negative effects (Gómez & Crespo, 2009).
- Perfective maintenance, based on continuous improvement. The aim is to improve the behaviour, ease of maintenance and the service which the system can provide (Gómez & Crespo, 2009).
- Total Productive Maintenance (TPM) is defined by Nakajima (1988) as productive maintenance involving total participation in addition to maximizing equipment effectiveness and establishing a thorough system of planned preventive maintenance.
- Reliability Centred Maintenance (RCM) is a systematic methodology for the allocation of efficient predictive and preventive maintenance aimed at preventing the dominant causes of failure of critical equipment, and, in turn, towards achieving acceptable levels of equipment availability and costs by reducing corrective maintenance (Martorell et al., 1999).

There is little research that analyses the choice of maintenance policy despite its importance in industry (Bertolini & Bevilacqua, 2006; Wang, Chu, & Wu, 2007); those worth mentioning include Al-Najjar

(1999), Azadivar and Shu (1999), Sharma et al. (2005) Bashiri et al. (2011), Zaim et al., (2012), Chan and Prakash (2012), Shahin, Shirouyehzad and Pourjavad (2012).

In most of these previous papers, cost is the key determinant. However, the maintenance of medical equipment or installations should bear in mind other factors in addition to cost, such as the safety of the patient and professionals, the impact on healthcare activity and the environment. MCDM constitutes a rational and objective tool to improve understanding of underlying decision processes in the systemic processes and, to help the decision-making groups to compare the alternatives (Romero, 1996). The application of an MCDM for optimising maintenance policies provides the advantages of analysing each criterion from the point of view of the different actors that have decision-making capacity on the equipment analysed. Thus, Almeida and Bohoris (1995) show that many maintenance decisions need to include criteria such as ease of maintenance, reliability, availability and safety in industrial settings, such as a Hospital, due to the high technology of the facilities and devices and the need for availability (Martorell et al., 2005).

The use of multicriteria techniques in choosing optimal maintenance policies can be found in the following work. Vatn (1997) uses utility functions and influence diagrams for the choice of maintenance actions (scheduled failure-finding task, scheduled on-condition task, scheduled restoration task and scheduled discard task) within an RCM programme. Labib et al. (1998) develop a maintenance decision model using the Analytic Hierarchy Process (AHP) together with the available data for equipment failure. Ramadhan et al. (1999) use AHP to decide priority factor weights which in order to develop a pavement maintenance priority ranking procedure for a road network. Bevilacqua and Braglia (2000) use AHP to select the best maintenance strategy in an integrated gasification and combined cycle factory. Kodali and Chandra (2001) use AHP to justify the application of TPM in Indian industries; the alternatives considered in this model are the application of a TPM system or a traditional maintenance system. Emblemsvag and Tonning (2003) determine appropriate maintenance policy for a weapon of the Norwegian Army by means of AHP; in this case, four maintenance echelons are considered and Monte Carlo simulations are used to evaluate the robustness of the decision. Al-Najjar and Alsyouf (2003) use fuzzy multicriteria decision making taking as criteria the causes of failure and as alternative corrective, preventive, and condition monitoring maintenance based on vibration analysis with continuous and periodic data collection, RCM, TPM and Total Quality Maintenance. Labib (2004) analyses the application of maintenance policy to equipment with different level of criticality, using fuzzy logic to obtain criteria in the form of a set of functions. In Carnero (2005), AHP and factorial analysis are used to choose the most suitable diagnostic techniques and instrumentation for conditional maintenance programmes based on analysis of lubricants and vibrations, taking into account different levels of technology, categorized by cost and diagnostic capacity. Bertolini and Bevilacqua (2006) set out a model combining goal programming and AHP to define the best maintenance policy to be applied to critical centrifugal pumps in an oil refinery, giving a priority vector of the possible policies: corrective, preventive and predictive for each type of failure analysed. By the use of multicriteria classification for critical equipment, Gómez de León and Ruiz (2006) give indices of criticality for each subsystem and the contribution of each device or piece of equipment to the criticality value obtained. Thus, it is possible to quantify the degree of importance of each subsystem and apply the most suitable maintenance policy depending on the resources available. In Carnero (2006), AHP is used to obtain a strategy for introducing predictive maintenance (internal, subcontracting, carrying out improvements in advance and not introducing any strategy) in a petrochemical plant, in a mixed form with Bayesian techniques to minimize uncertainty, where there is insufficient information. Carnero (2007) describes a model using AHP to choose the most suitable predictive maintenance techniques by vibration analysis, lubricant analysis and the integration of both techniques depending on the different technological levels. Ishizaka and Nemery (2014) apply ELECTRE-SORT, to consider an unlimited number of criteria in order to assign machines to non-comparable strategies, because maintenance strategies cannot always be completely ordered. Cavalcante and Lopes (2015) use a multi-attribute value function to select the maintenance policy for a cogeneration system in an ethanol and sugar plant. Goossens and Basten (2015) use AHP to model the choice of failure-based, time/use-based, and condition-based maintenance strategies in naval ships. From an initial study of 187 criteria, 29 criteria were finally used in building the model. Emovon et al. (2018) applied Delphi-AHP and Delphi-AHP-PROMETHEE to choose between corrective, preventive and condition-based maintenance for ship machinery systems. Cost, (spare parts inventories, maintenance cost, crew training cost and equipment damage cost), safety (personnel, equipment, environment), added value (minimisation of operational loss and equipment reliability) and applicability (system failure characteristics, available monetary resource, equipment risk level) are the criteria used. Condition based maintenance is the alternative chosen for application in the high-pressure pump of the fuel oil system of the marine diesel engine. Animah and Shafiee (2019) combine AHP and PROMETHEE to choose the maintenance strategies in critical machinery in the engine rooms of ships, and whose results will be applied to 17 different shipboard machinery systems. Shafiee et al. (2019) describe the use of Analytic Network Process (ANP) together with a cost risk criticality analysis model to choose a maintenance strategy between: failure-based, time-based, risk-based and condition-based maintenance that is cost effective and low risk. The main criteria used in the study are maintenance implementation cost and failure criticality. This is all applied to the design stage of a wind turbine.

The literature over the current decade show an increase in contributions applying fuzzy logic to the choice of maintenance policies through MCDM techniques. Thus, Wang et al. (2007) select an optimal combination of maintenance policies which increase availability and reliability of productive facilities without significantly increasing investment, using the criteria cost, safety, feasibility and added value. This is achieved via a fuzzy modification of AHP so that the uncertain and imprecise judgements of decision makers are translated into fuzzy numbers. Ghosh and Roy (2009) apply fuzzy AHP to select the optimal mix of maintenance approaches, including corrective maintenance, time-based preventive maintenance and condition-based predictive maintenance in a safety process plant. Also, a goal-programming model is included to optimize risk reduction and cost minimization, which are subject to resource constraints. Ilangkumaran and Kumanan (2009) combine fuzzy AHP and TOPSIS to select an optimum maintenance strategy for a textile plant. Chan and Prakash (2012) describe a model using TOPSIS in the fuzzy environment. The criteria used in the model are: capital cost, running cost, maintenance downtime, reliability, capability, repair load, operator skills, flexibility, efficiency, facilities utilization and resource availability. The result is a ranking of the alternatives analysed, condition-based maintenance, TPM and total quality maintenance. Lazakis et al. (2012) set out a model for choosing a maintenance strategy for the diesel generator of a cruise ship, using the fuzzy TOPSIS technique. Maintenance cost, efficiency/effectiveness, system reliability, management commitment, crew training, company investment, spare parts inventories and operation loss are the criteria considered; the alternatives analysed are corrective maintenance, preventive maintenance and predictive maintenance or condition based maintenance. The results of the model suggest that this last maintenance strategy is the best. Subsequently, Lazakis and Olcer (2015) combine this fuzzy TOPSIS methodology with AHP to obtain the weightings of the criteria. The results show that preventive maintenance is the best maintenance strategy for a ship diesel generator. Kirubakaran and Ilangkumaran (2016) use fuzzy AHP to calculate the weightings of the criteria safety, cost, added value, and feasibility, and Grey Relational Analysis (GRA) –TOPSIS to determine the ranking from among the alternatives: corrective maintenance, predictive maintenance, time-based preventive maintenance, and condition-based maintenance. Srivastava et al. (2017) apply fuzzy AHP and fuzzy TOPSIS together with graphical approaches to select the optimum maintenance strategy for a steam generating unit in a coal fired thermal power plant in India. The alternatives of the study included the following: corrective maintenance, preventive maintenance, condition based maintenance, predictive maintenance and opportunistic maintenance. The criteria used for the MCDM techniques are: safety (including the subcriteria environment, personnel and equipment safety), cost (with the subcriteria hardware, software and personnel training), value added (including the criteria spare parts inventories, production loss, fault identification) and capability of execution (comprising the criteria human resources, technology). Predictive maintenance is the optimal type of maintenance proposed by the research. Hemmati et al. (2018) produce an Analytic Network Process (ANP) model using Chang's extended analysis to choose a maintenance policy for the sulphur fuel furnace, absorption tower, molten sulphur ponds, cooling towers, converter, boiler and heat exchanger of an acid manufacturing company; the maintenance strategies considered are: corrective maintenance, time-based maintenance, condition-based maintenance and shutdown maintenance. The results returned by the model are that condition-based maintenance should be chosen to be applied in the cooling tower and absorption tower, as these are elements with respectively high-risk and high-added-value equipment. Time-based maintenance was chosen to be applied to boilers and converters while shutdown maintenance is appropriate for molten sulphur ponds. Other apparatus with high-cost, low-risk and low-added-value equipment will be assigned corrective maintenance strategy. Ighravwe and Oke (2019) describe a methodology that combines stepwise weight assessment ratio analysis (SWARA), weighted additive sum product assessment (WASPAS), fuzzy axiomatic design (FAD) and the additive ratio assessment (ARAS) to choose the best maintenance strategy for public buildings taking sustainable practices into account. The alternatives included in the methodology are corrective maintenance, preventive maintenance, predictive maintenance and condition-based maintenance. The criteria used are: environmental (composed of the subcriteria: energy efficiency, emission reduction, waste reduction and water usage reduction), economic (with the subcriteria: improved building value, better rental returns, reduced operational cost and extended building life-span), social (with the subcriteria: improved productivity, occupant health and safety, human traffic, occupant satisfaction) and technical (with the subcriteria: maintenance failure and downtime, spare parts quality and materials, staff expertise and business operations). Asuquo et al. (2019) use fuzzy TOPSIS to choose run-to-failure, preventive maintenance, condition-based maintenance, and RCM for marine and offshore machines. Reliability, cost, safety, equipment downtime and availability are the criteria used in the model; Condition-based maintenance is the maintenance strategy proposed by the model.

In this sense, as with other fuzzy techniques, the MACBETH approach allows ambiguities and uncertainties to be included in the judgements given by the decision group or centre, even though the analysis of the information is done in a crisp environment. This is because M-MACBETH software allows the giving of judgements from ranks within semantic categories: very weak, weak, moderate, strong, very strong and extreme; so, for example, a judgement may be between very strong and extreme or be only given as positive, showing that an element is higher than another, but without giving the strength of the domination.

Furthermore, the needs and characteristics of the processing or manufacturing plants or ships and the machinery and devices they have available, analysed in the literature, are very different from those of healthcare organizations, and so, in many cases, the criteria and models described cannot be extended to hospital system.

A further literature review on selection or formulation of maintenance strategies has been carried out. for example, Velmurugan and Dhingra (2015) identify the frameworks that exist for making a choice of maintenance strategies and their introduction into industry. They therefore develop their own framework for the formulation of maintenance strategies, and the introduction of the selected strategies. They next analyse the impact of introducing maintenance strategies into the maintenance function. Shafiee (2015), on the other hand, carries out a review of contributions applying MCDM in the choice of maintenance policies. Chong et al. (2019) review and discuss the literature on maintenance prioritization in asset management over the period 1990 to 2016. AHP is seen to be the most commonly-used method, together with failure mode and effect analysis (FMEA) and the priority criterion and matrix. The aspects most frequently considered in the research are technical, financial, social and political.

However, there is a significant lack of literature related to the application of multicriteria techniques in the selection of the most suitable maintenance strategy for a Hospital. Such literature as exists begins to appear only in the last decade. Contributions worthy of note include Taghipour et al. (2011) which uses AHP to prioritise medical devices. The multicriteria model uses the following criteria in its construction: function, mission criticality, age, risk, recall and hazard alerts and maintenance requirements. In the case of medical equipment with high criticality, the possibility of developing user training or redesigning the equipment is considered, and subsequently guidelines are assigned to select appropriate maintenance strategies or policies depending on the typology of the equipment and the level of criticality. Jamshidi et al. (2015) describe a risk-based prioritization framework for choosing the most suitable maintenance strategy. Firstly, fuzzy failure mode and effect analysis must be applied to calculate a risk priority index; this considers the criteria severity, occurrence and detection, and eight subcriteria. An additive weighting model is then used comprising these criteria: number of available identical devices, usage-related hazards, age, function, utilization, recalls and hazard alerts and maintenance requirements; this model allows a total intensity score to be obtained, in such a way that other aspects of hazards, additional to those in fuzzy failure mode and effect analysis first developed, can be considered. Finally, a maintenance planning diagram was produced from the scores previously obtained. Ben Houria et al. (2015) and Ben Houria et al. (2015) also prioritize medical equipment according to its criticality; this was done using the criteria degree of complexity of the maintenance, function, risk, degree of the importance of the mission, age, recalls and user errors and class of equipment in a model built with fuzzy AHP; this allows weightings to be obtained for the criteria. They then apply fuzzy TOPSIS to choose the most appropriate maintenance policy via the closeness coefficient. The choice of maintenance strategies is made from: corrective maintenance, time-based or preventive maintenance and condition-based maintenance. The study proposes the use of time-based maintenance for high-criticality medical equipment. Subsequently, Ben Houria et al. (2016) produce a model using AHP and TOPSIS and mixed integer problems MILP to determine the criticality of medical equipment and choose the optimal maintenance strategy for the devices, but with the same maintenance budget. Mahfoud et al. (2016) describe a model that combines AHP and Preference Ranking Organization METHod for Enrichment of Evaluations (PROMETHEE) for the prioritization of medical devices. These are the criteria used: function, recalls and hazard alerts, utilization, redundancy, age, technological obsolescence, maintenance requirements and a risk priority index using multiple clinical expert judgements. Carnero and Gómez (2016) set out a multicriteria methodology for choosing a combination of maintenance policies in dialysis subsystems associated with acute or chronic patients infected with hepatitis B or C. The model requires the use of continuous-time

Markov chains to estimate the availability of each alternative. The model was built with the Measuring Attractiveness by a Categorical Based Evaluation Technique (MACBETH) approach; as a result it is seen that the optimal decision is to apply corrective and preventive maintenance plus two reserve devices; the implications of this change of maintenance policies as against those currently applied in the hospital are described from the viewpoint of quality of care. Ighravwe and Oke (2019) apply stepwise weight assessment ratio analysis (SWARA), weighted additive sum product assessment (WASPAS), fuzzy axiomatic design (FAD) and the additive ratio assessment (ARAS) to choose the best maintenance strategy in public buildings (hospitals, research institutions, universities and polytechnics), and the result was that the most suitable order is predictive maintenance, followed by preventive maintenance, condition-based maintenance and corrective maintenance. With the aim of increasing availability and reduce costs, Hutagalung and Hasibuan (2019) determined a priority level for medical device maintenance based on the criticality scores of the medical devices using AHP. The criteria used are maintenance complexity, function, risk (comprising detectability of failure, frequency of failure, downtime, safety, repair costs), degree of importance of the mission (with the subcriteria utilization rate and availability), age of medical equipment, recall and user errors and medical equipment class. In this case the alternatives are the various medical devices analysed. Carnero and Gómez (2018) use the MACBETH approach to build a model optimizing the maintenance applied to production and storage systems for domestic water in a hospital. Maintenance costs, safety of maintenance workers, suitability of maintenance actions, impact on the environment and impact on hospital activity are the criteria used in building the model. In this case the following alternatives are used: corrective maintenance and monthly preventive maintenance, corrective maintenance and daily preventive maintenance and corrective maintenance and continuous preventive maintenance. Also, Mahfoud et al. (2018) review maintenance strategies within the healthcare environment and, and set out the practical needs and gaps between research and practice, as well as quantitative and qualitative criteria which to be taken into account, and data collection techniques and applied approaches to study the optimization of maintenance in healthcare.

Generally, the previous contributions build models which they apply to a diverse set of medical devices with very different characteristics, use, redundancy, technological obsolescence and risks. Nevertheless, in the research presented here, the model is developed exclusively for the systems which are part of the operating theatre, that is: air conditioning, sterile water, power supply, medicinal gases and operating theatre lighting.

Moreover, most of the previous research chooses the most suitable maintenance strategy, but does not take into account that maintenance strategies are usually applied in combination; this is because corrective maintenance is always present, even when technologically more advance maintenance strategies, such as condition based maintenance, are also being used

This research may serve as an example to other hospitals that can develop their own decision groups and make assessments depending on the specific conditions, the combination of maintenance policies best applied based on objective mathematical tools, instead of taking decisions based solely on experience. This is the only research which integrates MCDM techniques and Markov chains. Furthermore, it should be noted that it has been successfully applied to a state-run general hospital.

Among the decision criteria used in the multicriteria model presented in this paper is availability. To assess them in the different subsystems that exist in an operating theatre, it was necessary to assess the mean availability of repairable systems with Markov chains. Markov chains in stable repairable systems are frequently used to predict the behaviour of complex repairable elements in the pipework of nuclear power plants with constant failure and repair rates (Fleming, 2004; Gosselin & Fleming, 1997) so that periodical revisions can be programmed for the cost of the shutdowns to be minimum, according to Crespo and Sánchez (2002), and decide on when to intervene depending on the condition of the system (Castanier et al. 2005).

## MACBETH INTRODUCTION

MACBETH (Measuring Attractiveness by a Categorical Based Evaluation Technique) (Bana e Costa & Vasnick, 1999) builds a cardinal scale of the alternatives, based on the information provided by the decision centre, for each pair of criteria submitted for judgement, so that this information is a group of attractiveness differences between the two stimuli. For a decision-making group without a great deal of technical knowledge on the information to be assessed, this method is very useful, as it builds a scale of preferences from the indirect information obtained from the answers from simple and direct questions, without this group needing mentally to make a numerical scale, for which extensive knowledge of the material is required.

MACBETH is applied in numerous real cases (see Bana e Costa, Nunes da Silva, & Vansnick, 2001; Bana e Costa, Da Costa-Lobo, Ramos, & Vansnick, 2002; Bana e Costa, Oliveira, & Vieira, 2008; Carvalho & Lourenço, 2002; Dhouib, 2014; Rodrigues, 2014).

The procedure by which MACBETH builds a scale of quantitative values is the following. A facilitator asks the decision-making group to judge the attractiveness difference between the two stimuli *(x, y) Î S* in 6 semantic attractiveness categories, $C_k$, $(k=1, 2, \ldots, 6)$; these categories are: very weak, weak, moderate, strong, very strong and extreme. Furthermore, the decision-making group may consider *x* and *y* equally attractive, or consider that they are not comparable in attractiveness terms.

The application of the method is carried out in three stages:

(1) Obtaining the cardinal scale for each criterion. In each criterion the impact levels that constitute the overall judgements are established by *J* so that between the elements $S = \{s_1, \ldots s_n\}$ there is a strict weak order, so that $s_1$ is more attractive than $s_2$, this than $s_3$ and successively, regarding the considered criterion.

Two reference levels "good" and "neutral" are chosen from the *S* group.

M-MACBETH assesses the degree of inconsistency of the judgement matrix and tells the assessor the corrections to make to solve this. M-MACBETH determines, from all the possible scales, a special numerical scale μ in S, with linear programming.

The numerical scale presented by M-MACBETH, called the basic scale, is represented in a standard way or with the thermometer tool, so that the new scale is obtained making the good and neutral levels coincide with 100 and 0 respectively.

In Figure 1, the numerical scale and thermometer are represented as well as the margin of values offered by M-MACBETH, due to the scale obtained not being rigid, but rather there being an infinite number of scales inside the preference margin given by the decision-making group and, therefore, that do not alter the consistency of the judgement matrix. Thus, the decision-making group can choose the values to adopt within these margins, according to their judgement after observing the thermometer scale and comparing margins between the different attractivenesses.

(2) Obtaining the MACBETH scale for the criteria. To measure the attractiveness of each stimulus, the weight of each criterion has to be determined. To do so, the previous process is applied, comparing the attractiveness difference between pairs of criteria. M-MACBETH analyses the consistency of the judgement matrix, as it is filled in and allows the decision-making group to correct them at the same time as creating the weight scale in ascending order and from greater to lesser weighting.

The scale offered by M-MACBETH is, therefore, linear. To obtain the standardised MACBETH scale, $\sum_{i=1.n} K = 1$; $K_i$ is applied as the specific standardised weight of each criterion and $n$ the number of criteria.

As in the previous case, M-MACBETH offers the possibility of modifying the values obtained within a range.

(3) Impact of the alternatives on the criteria. The third judgement expression level by the decision-making group is considering what impact value each alternative has on each criterion. To do so, the decision-making group considers a criterion and assigns it a given impact level.

For a detailed explanation of the MACBETH technique, see Bana e Costa and Vansnick (1994), Bana e Costa and Vasnick (1999), Bana e Costa et al. (2006) and Bana e Costa et al. (2011).

## MARKOV CHAINS FOR REPAIRABLE SYSTEMS

In a Markov process, "the occurrence of a future state depends on the immediately previous state and only on it" (Taha, 2004). If the system's different states are considered, so that they are exhaustive and mutually exclusive, a special case is obtained for a Markov process, called a Markov Chain. To describe a Markov chain, we have to know the current state and the transition possibilities. If transition probabilities between states are maintained constant throughout time, they are called Homogenous Markov Chains. The Homogeneous Markov chains' most important properties are (Prawda, 1999):

- For a high n number of transitions and considering infinite time, the rows of the transition matrix tend to become equal to the value of the system in each state and at a constant value Õ.

$$\lim_{t\to\infty} P_t(t) = \prod_i \quad (1)$$

- The sum of probable values of the system in each state is equal to 1.

$$\sum_{i-0}^{m} \prod_i = 1 \quad (2)$$

- The transition between states is considered a Markov chain, and the change in state occurs in a sufficiently small Δt so that only a fault or a repair occurs.

*Figure 1. MACBETH, thermometer and variation margin in the M-MACBETH thermometer.*

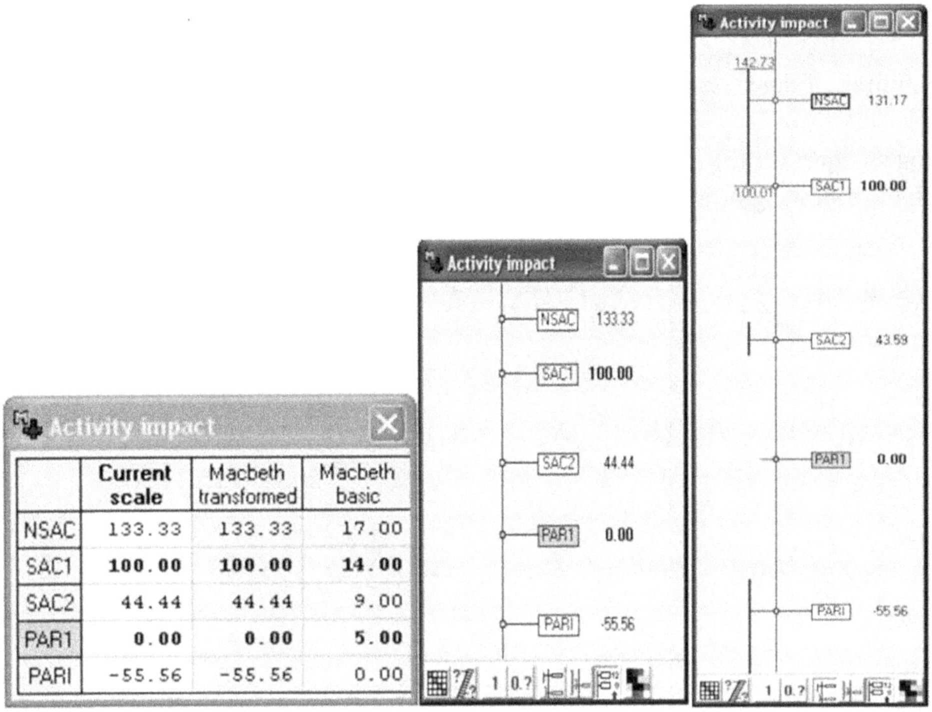


According to international standard IEC 61165 "Application of Markov techniques", the probability of a system being in state *i*, after applying a $\Delta t$ is called $P_i\ (t + \Delta t)$. The system passing from any state to *i* may occur by means of a fault or a repair, not simultaneously. Therefore, two transit situations are considered: states *j (0* to *i-l)* for states in which the transit is carried out by means of a fault and states *h (i+1* to *n)* for those which the transit is carried out by means of a repair.

$$P(t+\Delta t)=\sum_{j=0}^{i-1}P_j(t)P_{ji}(\Delta t)+P_i(t)\left(1-\left(\sum_{j=i+1}^{n}P_{ij}(\Delta t)+\sum_{g=0}^{i-1}P_{ig}(\Delta t)\right)\right)+\sum_{h=i+1}^{n}P_h(t)P_{hi}(\Delta t) \qquad (3)$$

where:

$P_i\,(t + \Delta t)$ is the probability of the system being in state *i*, after applying a $\Delta t$.
$P_i\,(t)$, $P_j\,(t)$, $P_h$ are the probabilities that the system is in the *i, j, h* states in *t*.
$P_{ij}$, $P_{ji}$, $P_{ig}$ are the probabilities that the system changes state in a $\Delta t$.

The probabilities of a change of state to another imply the non-reliability or the maintainability of the system in this state. They are called:

- $R_0(\Delta t)$, $R_1(\Delta t)$, ..., $R_n(\Delta t)$: Reliability of the system in each state between *t* and $t+\Delta t$ (probabilities of the system being in state *i* in $t+\Delta t$).

- $M_1(\Delta t)$, $M_2(\Delta t)$, ..., $M_n(\Delta t)$: maintainability function of the system in each state between $t$ and $t+\Delta t$ (probabilities of the system passing from state $i+1$ to state $i$ in $t+\Delta t$).
- $F_1(\Delta t)$, $F_2(\Delta t)$, ..., $F_n(\Delta t)$: unreliability function of the system in each state between $t$ and $t+\Delta t$ (probabilities of the system passing from state $i$ to state $i+1$ in $t+\Delta t$). $F_i(\Delta t) = 1 - R_0(\Delta t)$.

Replacing in equation (3) and considering that fault and repair density functions are taken to be exponential, the following is obtained for each state:

$$P_0(t+\Delta t) = P_0(t)(1 - \sum_{j=1}^{n} \lambda_{oj}\Delta t + \sum_{h=1}^{n} P_h(t)\mu_{h0}(\Delta t)$$

$$P_1(t+\Delta t) = P_0(t)\lambda_{01}\Delta t + P_1(t)(1 - (\sum_{j=2}^{n} \lambda_{1j}\Delta t + \mu_{10}\Delta t)) + \sum_{h=2}^{n} P_h(t)\mu_{h1}(\Delta t)$$

$$P_{n-1}(t+\Delta t) = \sum_{j=0}^{n-2} P_j(t)\lambda_{j,n-1}\Delta t + P_{n-1}(t)(1 - (\lambda_{n-1}\Delta t + \sum_{g=0}^{n-2} \mu_{n-1,g}\Delta t)) + P_n(t)\mu_{n,n-1}(\Delta t)\,.$$

$$P_n(t+\Delta t) = \sum_{j=0}^{n-1} P_j(t)\lambda_{j,n}\Delta t + P_n(t)(1 - \sum_{g=0}^{n-2} \mu_{n,g}\Delta t) \tag{4}$$

Operating on equation (4) and letting $\Delta t$ ®$0$, the derivatives in each state can be represented in vector form; the system of equations to be considered is that equivalent to:

$$P' = A * P \tag{5}$$

applying transposition, the system ends up as shown in equation (6).

$$\begin{pmatrix} P_0' \\ P_1'' \\ \cdot \\ \cdot \\ P_{n-1}' \\ P_n' \end{pmatrix} = \begin{pmatrix} -\sum_{j=1}^{n} \lambda_{oj} & \mu_{10} & \cdot & \cdot & \mu_{n-1,0} & \mu_{no} \\ \lambda_{01} & -(\sum_{j=2}^{n} \lambda_{1j} + \mu_{10}) & \mu_{21} & \cdot & \cdot & \mu_{n1} \\ \cdot & \cdot & \cdot & \cdot & \cdot & \cdot \\ \cdot & \cdot & \cdot & \cdot & \cdot & \cdot \\ \lambda_{0,n-1} & \lambda_{1,n-1} & \cdot & \cdot & -(\lambda_{n-1,n} + \sum_{j=0}^{n-2} \mu_{n-1,g}) & \mu_{n,n-1} \\ \lambda_{0,n} & \lambda_{1,n} & \cdot & \cdot & \lambda_{n-1,n} & -\sum_{g=0}^{n-1} \mu_{n,g} \end{pmatrix} \begin{pmatrix} P_0 \\ P_1 \\ \cdot \\ \cdot \\ P_{n-1} \\ P_n \end{pmatrix} \tag{6}$$

In the matrix in Equation (6), Equation (2) must be satisfied.

For stable Markov processes, the probability at each instant $P_i$ tends to be a constant for infinite $t$, where, according to Equation (1):

$$\lim_{t\to\infty} P_i(t) = 0 \tag{7}$$

and in algebraic form, the equation system is:

$$A\bar{\prod}\bar{0} \tag{8}$$

where $\bar{\Pi} = (\Pi_0, \ldots, \Pi_n)^T$. The sum of all probabilities is, according to (3), equal to 1. Replacing equation (6) a new, non-homogeneous equation system is obtained with a unique solution:

$$A_1\vec{\prod}\vec{0} \tag{9}$$

where:

$$A_1 = \begin{pmatrix} -\sum_{j-1}^{n}\lambda oj & \mu 10 & . & . & \mu n-1,0 & \mu 10 \\ \lambda 01 & -\left(\sum_{j-2}^{n}\lambda_1 j+\mu 10\right) & & & . & \mu n1 \\ . & . & & & . & . \\ . & . & & & & \\ \lambda 0_1 n-1 & \lambda 1, n-1 & . & . & -\left(\lambda n-1, n+\sum_{j-0}^{n-2}\mu n-1, g\right) & \mu n, n-1 \\ 1 & 1 & . & . & 1 & 1 \end{pmatrix} A_1= \tag{10}$$

Thus, the mean availability $D_M$ for $k$ minimum operating states is:

$$D_M = \Pi_0 + \Pi_1 + \ldots + \Pi_k \tag{11}$$

## MULTICRITERIA MODEL FOR THE SELECTION OF MAINTENANCE POLICIES IN OPERATING THEATRE SUBSYSTEMS

The choice of MACBETH for studying maintenance policies in medical installations and equipment has been made by the technical services manager of the Hospital, acting as decision analyst. He has extensive knowledge of multicriteria techniques and has previously used ELECTRE, Analytic Hierarchy Process (AHP) and MACBETH. The choice of MACBETH as opposed to other techniques was due to the possibility of obtaining a quantitative scale value from the qualitative judgements based on good and neutral reference levels and of making a comparison between criteria based on those references rather than subjectively. The analytical ability of the M-MACBETH software was another important factor since it allowed an assessment of the quantitative ranges associated with each level on the scale of a criterion or possible ranges in the weighting of criteria, according to the judgements given. The existence of detailed procedures for decision making; for example, how to construct the judgement matrix between levels of a descriptor associated to a criterion, how to fill out the judgement matrix over the criteria (Bana e Costa & Chagas, 2002) has also been considered. Techniques like the multi-attribute utility theory could have introduced probabilistic information into the decision problem; however, in the case of the availability criterion, the calculation of mean availability for each subsystem has been made by the analysis of Markov chains. There therefore exist data for the mean availabilities associated with each subsystem. For this reason, the decision makers have preferred to work with deterministic mean availability values for each subsystem, which are also those typically used in a maintenance department, and also in the Hospital, rather than using utility functions, which were, in any case, not understood by some decision makers.

In the case of the criterion Risk to employees, although the risk has a certain associated probability, the maintenance activities to be carried out for each action have been performed countless times and have been deeply analysed, and so the association of a probability to these actions is close to the value 1. Thus, it was decided to define criteria and descriptors with deterministic scale levels.

For the application of the MACBETH method in this paper, a decision-making group has been established made up of the managers of the following areas of the General Hospital of Ciudad Real: installations, safety, environment and medical equipment, and technical services. The decision-making group has been coordinated by the technical services manager, who acted as facilitator, helping the decision makers in the decision process.

The group decision-making process took into account, among others, the following factors, described in León (1993), Munda (2005) and Zahir (1999): the solution obtained is more easily put into practice if the members who will be affected by the decision have participated in making it; by Pareto's principle, the decision-making process has not imposed the preference of any individual on the group and if, after the group has given a classification of alternatives, the alternative of lowest preference is eliminated, the classification should subsist as made. To collate the judgements given by the members of the group the models set out in Forman and Peniwati (1998) have been considered: joining together individual judgements and individual preferences. The first of these has been used, since the individuals form a homogeneous group.

The structuring phase includes the following activities:

1. Identifying the criteria and organizing them into a hierarchical structure;
2. Constructing an impact descriptor for each criterion, to make the evaluation of the alternatives operational;

Carrying out an impact estimate of the different alternatives on each criterion.

## Markov Chain Analysis

In this case study, the subsystems included in the operating theatre system are: power supply, air conditioning, operating theatre lighting, sterile water and medicinal gases. All of them are critical. These subsystems include connections with the IT subsystem of the Hospital. This subsystem, though very important in a hospital, is not considered an independent subsystem of the operating theatre; that is, the connections or devices are considered parts of the power supply, air conditioning, operating theatre lighting, sterile water and medicinal gases subsystems and therefore their failures are analysed as failures of the subsystem they form part of.

To be able to study the behaviour of each subsystem, they must be modelled according to the operating states, bearing in mind different eventualities that might appear. This section shows an example of application to the power supply subsystem. Similar developments have been carried out on the rest of the subsystems in an operating theatre in the University General Hospital of Ciudad Real.

The power supply subsystem includes the elements that provide uninterrupted power supply to the vital attendance and operating theatre lighting, when there is a general power failure. The elements that make it up are as follows: insulation transformers, automatic circuit breakers, uninterrupted power supply system, electric leakage detection system and operating theatre lamp.

The electricity supply comes from an Uninterrupted Power Supply (UPS), powered by an insulation transformer. That of the subsystem has the following states:

1. In this state, the power supply is provided via the corresponding transformer centre by the electricity supplier company. This supply passes through the operating theatre's protection panel and insulation and through the UPS, before powering the operating theatre equipment.
2. If the insulation fails, an audible alarm goes off in the operating theatre, indicating the existence of an electrical leakage in a phase. If the electrical connection has a malfunction, all the charge is powered by the UPS batteries. Ten minutes later, all the power socket power circuits are disconnected, except for the vital attendance and operating theatre lamp. The latter, respirator and monitor are disconnected one hundred and twenty minutes later.
3. This state occurs when a second derivation occurs in another phase, with the risk of electrocution of patients or staff, if an electrical leakage occurs. The UPS may automatically go into bypass, if any internal anomaly is detected, disconnecting the accumulators of the charge, and these will not have any backup supply.
4. Fault in the main supply. All the charge is supplied from the UPS batteries.
5. UPS fault. The UPS goes into by-pass and, therefore, the equipment is not powered if a fault occurs in the main supply.

The subsystem is considered to fail when it is in states 2 and 4. Figure 2 represents the Markov state diagram. The system is considered operative when it is in states 0, 1 or 2. The mean availability $D_M$ is,

$$D_M = D_0 + D_1 + D_2$$

Matrix $A_1$ is that shown in equation (12). The row of ones allows the constraint of a unique solution to be included.

$$\begin{pmatrix} -(\lambda_1+\lambda_2+\lambda_3) & \mu_1 & \mu_2 & 0 & \mu_4 \\ \lambda_1 & -(\lambda_2+\mu_1) & 0 & 0 & 0 \\ \lambda_3 & 0 & -(\lambda_4+\mu_2) & 0 & 0 \\ \lambda_2 & 0 & 0 & -\lambda_5 & 0 \\ 1 & 1 & 1 & 1 & 1 \end{pmatrix} \quad (12)$$

## Description of Criteria and Impact Levels

*Figure 2. Markov availability diagram.*

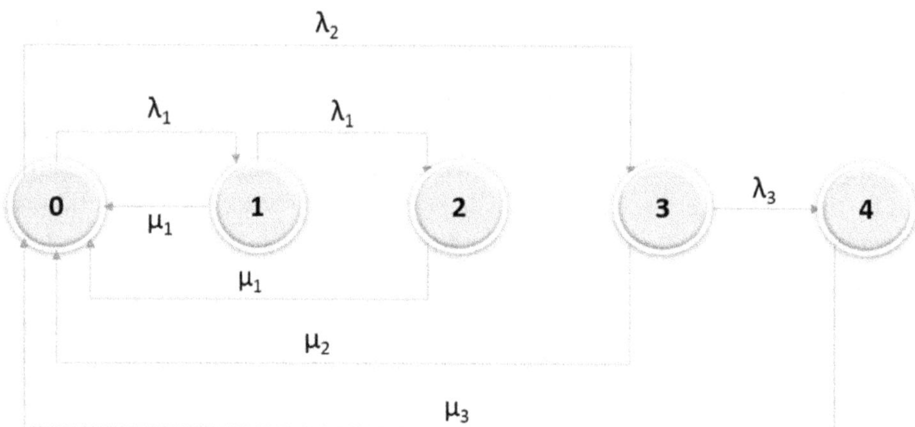


In the first stage of the structuring phase, the decision criteria were chosen by consensus.

The technical services manager acted as facilitator, helping the decision makers in this decision process. If a disagreement occurred among the decision makers, the facilitator asked each member to state and discuss his opinion until a final consensus was reached. Nevertheless, the initially established criteria do not differ greatly from those chosen. A hierarchical structure was obtained made up of the following three areas and six decision criteria:

1. Technical-economic: availability, failure detection time and cost.
2. Safety: risk to employees and risk to users.
3. Functional: activity impact.

Any criterion must be associated to a descriptor in order to make it operational. A descriptor is a set of impact levels that can measure, either quantitatively or qualitatively, a criterion's level of compliance (Bana e Costa & Vasnick, 1999).

There exist the following types of descriptor: natural, when the impact levels reflect the effects directly; proxy, when the impact levels mainly reflect the causes rather than the effects; and constructed, when the impact levels are defined as a finite combination of reference levels.

In order to define each descriptor a meeting of the group was arranged in which each member set out, anonymously, five scale levels per descriptor, in order of importance, and the normal and good reference levels. During this stage, changes were made to the definition of the criterion to make it intelligible to all members of the group. The final scale levels were obtained by consensus of all members of the group, following the procedure of the Delphi method. Fulfilment of the properties of completeness, mutual independence preference, non-redundancy, concision, degree of consensus, measurability, operationality and non-ambiguity was checked during the process.

The different decision criteria used and impact levels of the associated descriptors are defined below, with an indication of the neutral (N) and good (G) levels.

The considered decision criteria are:

- Availability (Avail). The mean availability foreseen throughout its operating life is considered, bearing in mind the historical data obtained and applying the Markovian model, according to section 3. Therefore, the descriptor is the mean value obtained for the availability using the repairable system Markovian model. The scale levels considered in the mean availability indicator are:

  0: $99 \leq D_M < 100$
  1: $95 \leq D_M < 99$ (G)
  2: $85 \leq D_M < 95$
  3: $60 \leq D_M < 85$ (N)
  4: $D_M < 60$

The mean availability descriptor and its scale levels are defined based on the results of a Markov chain analysis of the more than 700 subsystems that exist in the Hospital. Therefore, the mean availabilities associated with each subsystem are known, which also happens to be the information most commonly used in a maintenance department, and so in the Hospital.

The normal and good values have been set by the decision-making group such that the range corresponding to the normal value is that most often obtained among the values calculated in the Markov chain analysis, and the good value corresponds to the range of the best availabilities of the 700 subsystems analysed, until an availability of 100% is reached. The other values, which correspond to levels 2, 3 and 4, have been distributed in proportion up to a value 0.85 below the value at which the least favourable level is reached. This is established as the quality standard of the Hospital.

- Activity impact (Activ). Defined as how much it affects the last operating function, which in this example is surgery. The descriptor is the foreseen suspension percentage of the operating theatre activity as a result of a fault in the subsystem, considering as scale levels:
  0: Never suspends activity.
  1: Suspends activity in less than 10% of the stoppages (G).
  2: Suspends activity from 10 to 25% of the stoppages.
  3: Suspends activity from 25 to 50% of the stoppages (N).
  4: Always suspends activity.

- Failure detection time (Tfailur). It is the opportunity of the moment when the fault is detected, due to the implications it may have. The descriptor is the estimate of the time when the fault is detected progressively, where the scale margin is between before the start of the activity and after, with its intermediate values varying in percentage. The following scale levels have been established, according to the accumulated experience:
  0: Always before affecting the service.
  1: Before affecting the service at least in 90% of cases (G).
  2: Before affecting the service at least in 70% of cases.
  3: At the time of the fault (N).
  4: Indirectly, after the fault.
- Cost (Cost). The necessary resources for applying any of the maintenance policies considered, are considered generally as in practice, in-house and subcontracted resources intervene. Therefore, the decision-making group has estimated that the associated indicator is the type of complexity of the required resources:
  0: Specific non-programmed resources.
  1: Standard programmed resources (G).
  2: Specific non-programmed resources and specialised programmed resources.
  3: Non-programmed, programmed and specialised services (N).
  4: A change to the existing technology.
- Risk to employees (Risk1). This is defined as the risk to which the workers located in the operating theatre area are subjected. The descriptor measures the electrical hazard in the operating theatre. The associated scale levels are:
  Does not involve a risk for the professional in the event of a fault (G).
  The professional always detects the risk before the fault occurs.
  There may be a risk of electric shock, without the professional having time to detect it (N).
  There is a risk of an electric shock before the professional detects it.
  High probability of electric shock.
- Risk to users (Risk2). This is defined as the risk the user is subjected to, where the user is understood to be the medical service requester. The associated descriptor is the level of proximity and severity of the possible damage suffered by the patient. The established scale levels are:
  0: Does not have implications for the patient (G).
  1: Has indirect implications for the patient, without causing damage: deprogrammings.
  2: The patient is subjected to direct action without damage: temporary suspension of the operation (N).
  3: The patient is subjected to direct action with mild damage: mild burns.
  4: Has direct, very serious effects on the patient with the possibility of death: electrocution, respiratory failure, etc.

All the criteria included in the model are independent. This should be especially noted for the Availability and Activity impact criteria due to the fact that in the case of a failure of one of the subsystems, it will have a different level of influence in the surgery, depending on the type of subsystem and independently of the time that it is available. For example, a failure of the electrical installation subsystem will have different levels of severity or impact on an activity, depending on the nature of the failure and independently of the time that it is not operational (availability). In the same way an electrical fault, above the permitted values, will mean the cessation of activity, and it may be traced and repaired immediately or it may require a time-consuming, thorough review of the electrical system.

To determine the impact level in the availability criterion, stable system Markov chains have been used, with a 3-year breakdown and repair log for each operating theatre subsystem.

## Criteria Weighting Process and Degree of Attractiveness Between Levels in Each Criterion

To perform the weighting process between levels of a single criterion and to establish the degree of attractiveness between criteria, a decision-making group was created made up of the maintenance manager of the Hospital installations area, the manager of the medical equipment, and the person in charge of the management of the operation of the different equipment and installations, from a healthcare point of view (operating theatre supervisor). The person in charge of the technical services of the Hospital acted as facilitator, without making any judgements, performing the following actions: explaining the procedure to all members of the group, implementing the judgements given by each member of the group in M-MACBETH and analysing the results.

Each member of the decision-making group has filled out, anonymously, the opinion forms shown in Table 1 in the Appendix. In most of the judgements given, the results were consensual, even though no one was aware of the judgements made by the other decision makers. Only when comparing the availability criterion with the criteria failure detection time, risk to employees and risk to users, were there discrepancies in the judgements (see Appendix A); in these cases, the final judgements have been obtained by applying the most voted judgement method. Thus, if inconsistencies appear, the following most voted judgement is chosen following a descending order. The results obtained from the decision-making group are shown in Appendix A.

The third stage of the structuring phase considers the impact of each criterion, with the managers of each area involved intervening. Thus, the experts were the technical managers for the technical-economic criteria, the environmental and occupational risk prevention managers for safety criteria and finally, the healthcare managers (operating theatres and activity programming) for functional criteria.

The results of the judgements expressed regarding the weighting of the availability criterion levels corresponding to the electrical installation subsystem are shown in figure 3. The same process has been applied to the different criteria of each subsystem. Figure 4 shows the results of the judgements made for comparing criteria, as well as the weightings of the criteria obtained for all subsystems. In this case, the same result is obtained for all subsystems, as the study is carried out in an operating theatre environment, where the importance of each criterion is the same regardless of the subsystem considered.

## Definition of Actions

The actions or maintenance policies considered are:

- Corrective maintenance (Corr).
- Preventive maintenance together with corrective maintenance by means of daily, weekly, monthly and yearly plans (Prev+Corr).
- Periodical predictive maintenance together with corrective maintenance (Pre+Corr).
- Corrective together with preventive and predictive maintenance (C+Prev+Pred).

*Figure 3. Judgement matrix between levels of availability criteria of the power supply subsystem.*

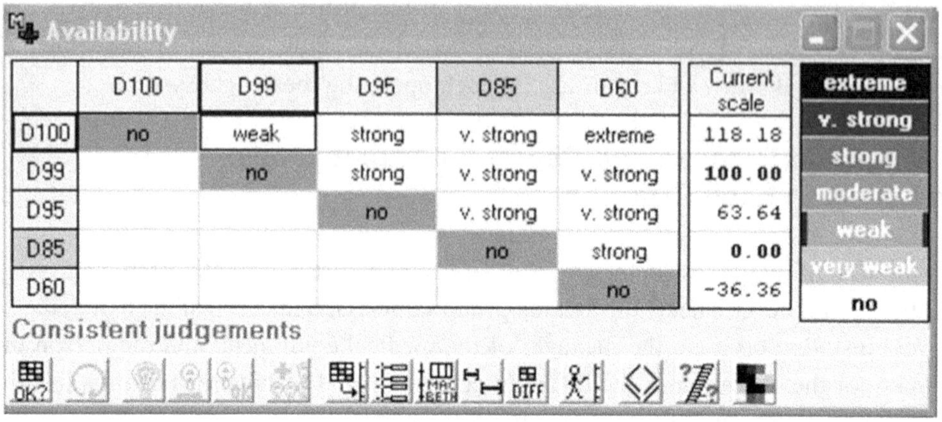


*Figure 4. Judgement matrix between criteria via M-MACBETH and bar chart of the weights obtained for each criterion.*

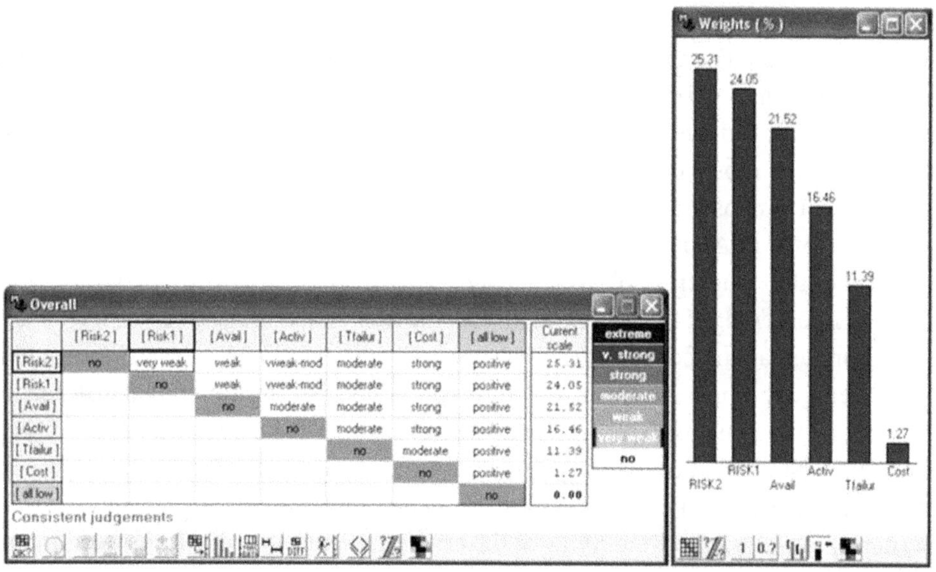


The possibility of applying proactive, perfective and total productive maintenance among the actions was considered, however, after assessing the Hospital resources they were not held to be viable in the short term, although the possibility of including them in future analyses was not completely rejected.

## RESULTS

Figure 5 shows the ranking obtained from the different maintenance policies in the different subsystems analysed in the operating theatre system.

*Figure 5. Maintenance policy classification in the Hospital's different subsystems.*

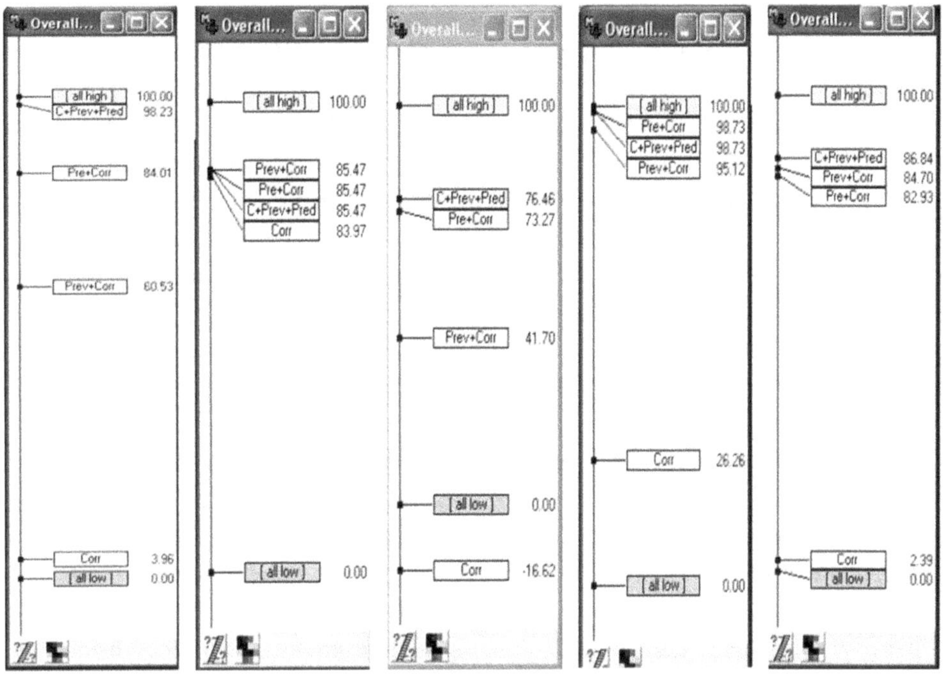


The maintenance policies considered have a distinct preference in each subsystem, as different impact levels have been considered for a single criterion. Thus, in the Climate Control subsystem, the impact on the availability criterion is clearly different according to the maintenance policy, while for the other criteria it is practically the same, except for the corrective maintenance policy, where the acceptance level is very low.

The availability criterion, therefore, has a higher relative weight than the others, which means that the variations within it are highly significant in defining the maintenance policy preference graph. A 99% mean availability value corresponds to a C+Prev+Pred combination of maintenance policies, giving an overall valuation index of 98.23%; for the Pre+Corr action, with a mean availability of 97%, this index is 84.01%. For the Prev+Corr, and Corr actions, with a mean availability of 95% and 75%, the overall valuations are 60.52% and 3.96% respectively.

In the Electrical Supply subsystem the situation is almost the same, with identical mean availability values to the Climate Control subsystem, although the C+Prev+Pred and Pre+Corr actions give practically the same preferences, 76.46% and 73.27% respectively. This is because the Risk to users and Risk to employees criteria assume the same impact in both policies.

In the Supply of Medicinal Gases and Lighting of the Operating Area subsystems the C+Prev+Pred, Pre+Corr and Prev+Corr combinations of maintenance policies give practically the same preferences; since the impact of each action and the mean availability on the criteria varies little they may be taken as more or less the same for each one. This is because predictive maintenance is based on a study of the evolution of the parameters of gas leaks and the electrical consumption of the surgical lamps respectively. This study is carried out using manually acquired data from preventive reviews, and so the difference in the results obtained by using Pre+Corr and Prev+Corr combinations of maintenance policies is virtually non-existent.

In the Supply of Sterilized Water subsystem, although the mean availability is slightly lower for the corrective maintenance policy, 85% as opposed to 90% for the other combinations of policies, the remaining criteria give almost the same impact for each policy combination; therefore, the degree of preference is slightly lower, 83.97% as against 85.47% for the other combinations of maintenance policies.

Currently, maintenance policies applied to the subsystems considered in this paper are daily corrective and preventive maintenance. In the study carried out, a considerable increase in the preferences on applying corrective plus preventive plus predictive maintenance policies is observed in the power supply and air conditioning subsystems, compared to the currently applied policy. In the rest of the subsystems, the relevance of predictive maintenance is practically insignificant. Furthermore, it can be observed that in the sterile water subsystem, including corrective maintenance has a preference practically the same as the rest of the policies.

In the analysis of the results obtained, technical services have opted for assessing the introduction of predictive maintenance in the power supply and air conditioning subsystems and only assigning corrective maintenance to the sterile water subsystem. In the first case, the introduction of a control system and charge measuring of the UPS batteries is considered, with an indication on the centralised control panel, that allows the operator to monitor them at all times, and be able to take the necessary actions sufficiently in advance: battery replacement, revision of environmental work conditions, etc. The cost of introduction per operating theatre is estimated at 6,000 euros.

In the air conditioning subsystem, the centralisation of alarms was studied due to filter dirt and wear of the fan transmission system and air extractor in the centralised control panel. The cost of introduction per operating theatre is estimated at 10,000 euros.

In the sterile water subsystem preventive maintenance has been eliminated with a saving of 0.5 euros/day.

Investment in predictive maintenance represents a single initial investment of 16,000 euros, with a break-even period of 10 years, including maintenance costs. Furthermore, with predictive maintenance, an increase in availability of 95 to 99% is foreseen in the power supply and air conditioning subsystems. This implies an increase in productivity per operating theatre, bearing in mind that the usage rate under normal conditions is 75%, of 0.4*0.75 = 3%, only in these subsystems, also obtaining an increase in the safety of operation.

The results obtained for the operating theatre system are considered satisfactory by the Hospital Management. The model described has given results which allow decisions to be taken and quantified and which provide a rational approach to the choice of combination of maintenance policies, changing the starting position and the decision-making process considerably from now on.

## SENSITIVITY ANALYSIS

Sensitivity analysis allows us to examine how the ranking of actions might change under different weighting.

M-MACBETH allows us to analyse the effect of the variations in the weights of the criteria in the considered combination of maintenance policies or actions. The technical services manager considered assessing possible modifications in the weightings of the cost and availability criteria of the power supply and operating theatre lighting subsystems. These subsystems were chosen because they were the subject of longer discussions during the meeting of the decision-making group.

Figure 6 shows the evolution of the actions when the weighting of the cost criterion in the operating theatre lighting and power supply subsystems is adjusted. The original weighting assigned by the decision makers to the cost criterion are shown on the vertical axis. The value of this weighting is 1.27. It can be seen that a very high increase in the weighting of the cost criteria will be necessary, to change the classification of actions in both subsystems. In the case of lighting subsystem, the cost should reach 9, while for the power supply subsystem it should reach 6.5. The weighting change would be so high that it would not be considered a possible value, and there is therefore no modification in the classification of alternatives in the two considered subsystems.

Figure 7 shows the changes the classification of actions would undergo when the weighting of the criterion availability in the operating theatre lighting and power supply subsystems is adjusted. For the availability criterion, the original weightings set by the decision makers 21.52 (identified by a vertical line). It can be seen that the weighting should be changed significantly, over 100%, in both subsystems to change the classification of alternatives.

The results obtained from the sensitivity analysis have been considered satisfactory, since, as can be seen, there are no modifications in the action classification in any subsystem when the weightings are modified, considering logical modification ranges in the weights in the different criteria.

## FUTURE RESEARCH DIRECTIONS

New maintenance policies could be applied to some or all of the subsystems of the operating theatre, for example TPM. In this case, new alternatives, in isolation or in combination with others should be proposed.

Additionally, judgements made by the decision centres could be brought together by geometric means, rather than considering the consensus. The results obtained by each procedure could be compared. In addition, the judgements given by the decision group should be reviewed periodically to consider a new economic situation for the healthcare organisation, new requirements in quality of care or legislation, as well as possible changes in those in charge of the health services who took part in this research.

The application of other multicriteria techniques, other than MACBETH, and the comparison of the results achieved by both techniques, would confirm the reliability of the results.

## CONCLUSION

This paper sets out a case study of integrating Markov chain techniques and multicriteria techniques for the selection of the most suitable maintenance policy in subsystems with direct interaction with hospital patients. Therefore, this is an innovative application on integrating these tools, as well as being applied to a health environment, in this case to an operating theatre's subsystems.

The analyses carried out are currently being applied in the University General Hospital of Ciudad Real, and it is notable as one of the first cases of applying advanced maintenance tools, such as MCDM techniques and Markov chains in hospitals. There is only one precedent in the literature where multicriteria decision techniques for the choice of the most suitable maintenance policy in a health environment,

*Figure 6. a) (top) Sensitivity analysis in the cost criterion of the operating theatre lighting subsystem. b) (bottom). Sensitivity analysis in the cost criterion of the power supply subsystem.*

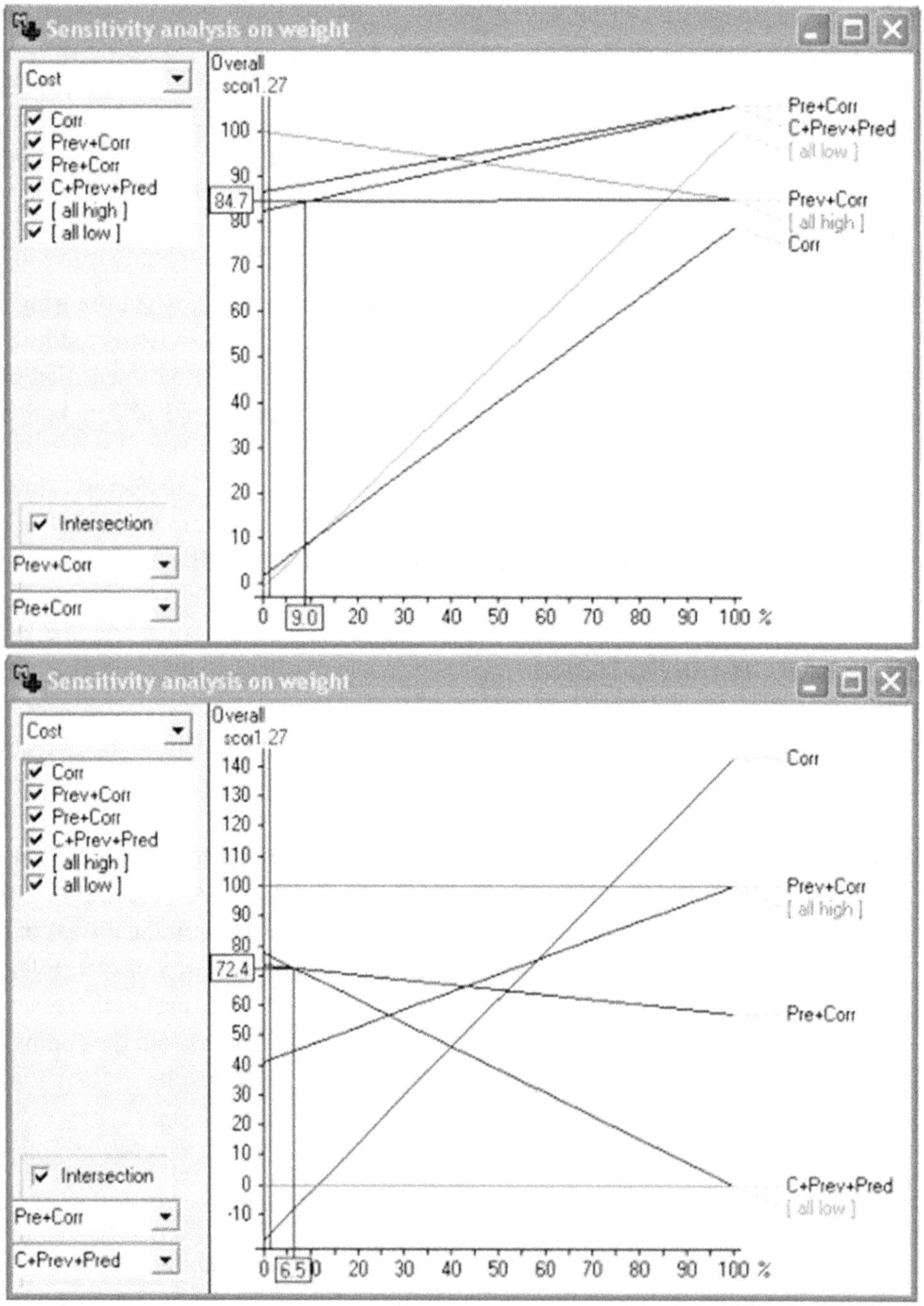


in spite of the large repercussions subsystems with direct interaction with the patient can have on their treatment and/or on their lives. Thus, the research could lead to a change in the mindset of technical services responsible for maintenance. Until now they have been focused on traditional maintenance systems, however, this research shows that small improvements in designing maintenance policies may

*Figure 7 a) (top) Sensitivity analysis in the availability criterion of the operating theatre lighting subsystem. b) (bottom). Sensitivity analysis in the availability criterion of the power supply subsystem.*

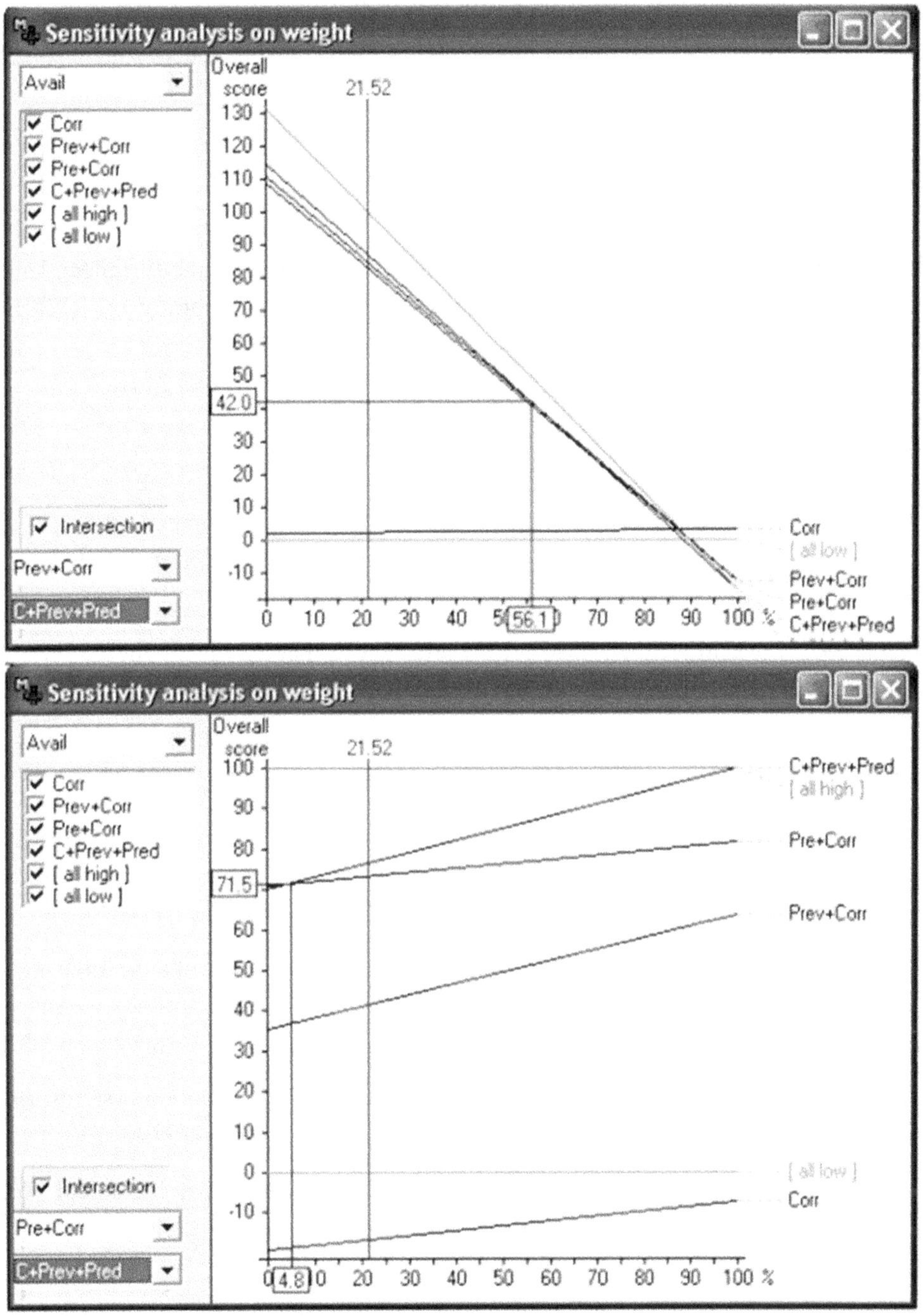


have significant repercussions not only on the programming and cost of maintenance, but also on the satisfaction of care staff and especially on patients.

Finally, this methodology can mean a significant increase in quality of care. An important point to be noted is the successful application of the research to a general hospital.

The creation of a decision-making group made up of different functional managers in the hospital has allowed the development of an awareness of the importance these subsystems can have on the treatment given to the patient. This aspect which has traditionally been known to the Hospital's technical sector has now also been acknowledged by the medical sector. Thus the importance of the judgements expressed by the decision-making group.

## ACKNOWLEDGMENT

This research was supported by the University of Castilla-La Mancha and the European Union through the European Regional Development Fund to the Predictive Analysis Laboratory (PREDILAB) group [2019-GRIN-27187].

## REFERENCES

Al-Najjar, B., & Alsyouf, I. (2003). Selecting the most efficient maintenance approach using fuzzy multiple criteria decision making. *International Journal of Production Economics*, *84*(1), 85–100. doi:10.1016/S0925-5273(02)00380-8

Al-Najjar, G. (1999). Economic criteria to select a cost-effective maintenance policy. *Journal of Quality in Maintenance Engineering*, *5*(3), 236–247. doi:10.1108/13552519910282692

Almeida, A. T., & Bohoris, G. A. (1995). Decision theory in maintenance decision making. *Journal of Quality in Maintenance Engineering*, *1*(1), 39–45. doi:10.1108/13552519510083138

Animah, I., & Shafiee, M. (2019). Maintenance strategy selection for critical shipboard machinery systems using a hybrid AHP-PROMETHEE and cost benefit analysis: A case study. *Journal of Marine Engineering and Technology*, 1–12. doi:10.1080/20464177.2019.1572705

Asuquo, M. P., Wang, J., Zhang, L., & Phylip-Jones, G. (2019). Application of a multiple attribute group decision making (MAGDM) model for selecting appropriate maintenance strategy for marine and offshore machinery operations. *Ocean Engineering*, *179*, 246–260. doi:10.1016/j.oceaneng.2019.02.065

Azadivar, F., & Shu, V. (1999). Maintenance policy selection for JIT production systems. *International Journal of Production Research*, *37*(16), 3725–3738. doi:10.1080/002075499190013

Bana e Costa, C. A., De Corte, J. M., & Vansnick, J. C. (2006). On the Mathematical Foundation of MACBETH. In J. Figueira, S. Greco & M. Ehrgott (Eds.), Multicriteria Decision analysis: State of the art surveys (pp. 409-444). New York, NY: Springer.

Bana e Costa, C. A., & Chagas, M. P. (2002). *A career choice problem: An example of how to use MACBETH to build a quantitative value model based on qualitative value judgments*. Working Paper LSEOR 02.53 ISBN 0 7530 1517, The London School of Economics and Political Science.

Bana e Costa, C. A., Da Costa-Lobo, M. L., Ramos, I. A., & Vansnick, J. C. (2002). Multicriteria Approach for Strategic Town Planning: The Case of Barcelos. *International Journal of Management Sciences*, *30*(3), 227–242.

Bana e Costa, C. A., De Corte, J. M., & Vansnick, J. C. (2011). MACBETH (Measuring Attractiveness by a Categorical Based Evaluation Technique). In J. J. Cochran, L. A. Cox Jr, P. Keskinocak, J. P. Kharoufeh, & J. C. Smith (Eds.), *Encyclopedia of Operations Research and Management Science*. John Wiley & Sons. doi:10.1002/9780470400531.eorms0970

Bana e Costa, C. A., Nunes da Silva, F., & Vansnick, J. C. (2001). Conflict dissolution in the public sector: A case-study. *European Journal of Operational Research*, *130*(2), 388–401. doi:10.1016/S0377-2217(00)00037-0

Bana e Costa, C. A., Oliveira, C. S., & Vieira, V. (2008). Priorization of bridges and tunnels in earthquake risk mitigation using multicriteria decision analysis: Application to Lisbon. *Omega*, *36*(3), 442–450. doi:10.1016/j.omega.2006.05.008

Bana e Costa, C. A., & Vansnick, J. C. (1994). MACBETH - An interactive path towards the construction of cardinal value functions. *International Transactions in Operational Research*, *1*(4), 489–500. doi:10.1016/0969-6016(94)90010-8

Bana e Costa, C. A., & Vansnick, J. C. (1999). The MACBETH approach: basic ideas, software and an application. In N. Meskens & M. Roubens (Eds.), *Advances in Decision Analysis* (pp. 131–157). Kluwer Academic Publishers. doi:10.1007/978-94-017-0647-6_9

Bashiri, M., Badri, H., & Hejazi, T. H. (2011). Selecting optimum maintenance strategy by fuzzy interactive linear assignment method. *Applied Mathematical Modelling*, *35*(1), 152–164. doi:10.1016/j.apm.2010.05.014

Ben Houria, Z., Besbes, M., Elaoud, B., Masmoudi, M., & Masmoudi, F. (2015). Maintenance strategy selection for medical equipments using fuzzy multiple criteria decision making approach. *Proceedings - CIE 45: 2015 International Conference on Computers and Industrial Engineering.*

Ben Houria, Z., Masmoudi, M., Al Hanbali, A., Khatrouch, I., & Masmoudi, F. (2016). Quantitative techniques for medical equipment maintenance management. *European Journal of Industrial Engineering*, *10*(6), 703–723. doi:10.1504/EJIE.2016.081017

Bertolini, M., & Bevilacqua, M. (2006). A combined goal programming-AHP approach to maintenance selection problem. *Reliability Engineering & System Safety*, *91*(7), 839–848. doi:10.1016/j.ress.2005.08.006

Bevilacqua, M., & Braglia, M. (2000). The analytic hierarchy process applied to maintenance strategy selection. *Reliability Engineering & System Safety*, *70*(1), 71–83. doi:10.1016/S0951-8320(00)00047-8

Carnero, M. C. (2005). Selection of diagnostic techniques and instrumentation in a predictive maintenance program. A case study. *Decision Support Systems*, *38*(4), 539–555. doi:10.1016/j.dss.2003.09.003

Carnero, M. C. (2006). An evaluation system of the setting up of predictive maintenance programmes. *Reliability Engineering & System Safety*, *91*(8), 945–963. doi:10.1016/j.ress.2005.09.003

Carnero, M. C. (2007). Model for the Selection of Predictive Maintenance Techniques. *INFOR*, *45*(2), 83–94. doi:10.3138/infor.45.2.83

Carnero, M. C., & Gómez, A. (2016). A multicriteria decision making approach applied to improving maintenance policies in healthcare organizations. *BMC Medical Informatics and Decision Making*, *16*(1), 47. Advance online publication. doi:10.118612911-016-0282-7 PMID:27108234

Carnero, M. C., & Gómez, A. (2018). Optimization of Maintenance in Production and Storage Systems for Domestic Water. *Water Resources Management*, *32*(1), 359–380. doi:10.100711269-017-1815-2

Carvalho, R., & Lourenço, J. C. (2002). A multicriteria model for assigning new orders to service suppliers. *European Journal of Operational Research*, *139*(2), 390–399. doi:10.1016/S0377-2217(01)00367-8

Castanier, B., Grall, A., & Bérenguer, C. (2005). A condition-based maintenance policy with non-periodic inspections for a two-unit series system. *Reliability Engineering & System Safety*, *87*(1), 109–120. doi:10.1016/j.ress.2004.04.013

Cavalcante, C. A., & Lopes, R. S. (2015). V. R. S. Multi-criteria model to support the definition of opportunistic maintenance policy: A study in a cogeneration system. *Energy*, *80*, 32–40. doi:10.1016/j.energy.2014.11.039

Chan, F. T. S., & Prakash, A. (2012). Maintenance policy selection in manufacturing firms using the fuzzy MCDM approach. *International Journal of Production Research*, *50*(23), 7044–7056. doi:10.1080/00207543.2011.653451

Chong, A. K. W., Mohammed, A. H., Abdullah, M. N., & Rahman, M. S. A. (2019). Maintenance prioritization – a review on factors and methods. *Journal of Facilities Management*, *17*(1), 18–39. doi:10.1108/JFM-11-2017-0058

Crespo, A., & Sánchez, A. (2002). Models for maintenance optimization: A study for repairable system and finite time periods. *Reliability Engineering & System Safety*, *75*(3), 367–377. doi:10.1016/S0951-8320(01)00131-4

Dhouib, D. (2014). An extension of MACBETH method for a fuzzy environment to analyze alternatives in reverse logistics for automobile tire wastes. *Omega*, *42*(1), 25–32. doi:10.1016/j.omega.2013.02.003

Emblemsvag, J., & Tonning, L. (2003). Decision support in selecting maintenance organization. *Journal of Quality in Maintenance Engineering*, *9*(1), 1–24. doi:10.1108/13552510310466765

Emovon, I., Norman, R. A., & Murphy, A. J. (2018). Hybrid MCDM based methodology for selecting the optimum maintenance strategy for ship machinery systems. *Journal of Intelligent Manufacturing*, *29*(3), 519–531. doi:10.100710845-015-1133-6

Fernández, J. C., & Canals, X. (2006). Quality standards for medical devices. In Proceedings of the XV Winter Course Standards in Telemedicine CATAI (pp. 133-140). Tenerife.

Fleming, K. N. (2004). Markov models for evaluating risk-informed in-service inspection strategies for nuclear power plant piping systems. *Reliability Engineering & System Safety*, *83*(1), 27–45. doi:10.1016/j.ress.2003.08.009

Forman, E., & Peniwati, K. (1998). Aggregating individual judgments and priorities with the Analytic Hierarchy Process. *European Journal of Operational Research*, *108*(1), 165–169. doi:10.1016/S0377-2217(97)00244-0

Ghosh, D., & Roy, S. (2009). A decision-making framework for process plant maintenance. *European Journal of Industrial Engineering*, *4*(1), 78–98. doi:10.1504/EJIE.2010.029571

Gómez, J. F., & Crespo, A. (2009). Framework for implementation of maintenance management in distribution network service providers. *Reliability Engineering & System Safety*, *94*(10), 1639–1649. doi:10.1016/j.ress.2009.04.003

Gómez de León, F. C., & Ruiz, J. J. (2006). Maintenance strategy based on a multicriterion classification of equipments. *Reliability Engineering & System Safety*, *91*(4), 444–451. doi:10.1016/j.ress.2005.03.001

Goossens, A. J. M., & Basten, R. J. I. (2015). Exploring maintenance policy selection using the Analytic Hierarchy Process; An application for naval ships. *Reliability Engineering & System Safety*, *142*, 31–41. doi:10.1016/j.ress.2015.04.014

Gosselin, S. R., & Fleming, K. N. (1997). Evaluation of pipe failure potential via degradation mechanism assessment. *Proceedings of the 5th International Conference on Nuclear Engineering (INCONE).*

Hemmati, N., Galankashi, M. R., Imani, D. M., & Farughi, H. (2018). Maintenance policy selection: A fuzzy-ANP approach. *Journal of Manufacturing Technology Management*, *29*(7), 1253–1268. doi:10.1108/JMTM-06-2017-0109

Hutagalung, A.O., & Hasibuan, S. (2019). Determining the priority of medical equipment maintenance with analytical hierarchy process. *International Journal of Online and Biomedical Engineering, 15*(10), 107-120.

Ighravwe, D. E., & Oke, S. A. (2019). A multi-criteria decision-making framework for selecting a suitable maintenance strategy for public buildings using sustainability criteria. *Journal of Building Engineering*, *24*, 100753. doi:10.1016/j.jobe.2019.100753

Ilangkumaran, M., & Kumanan, S. (2009). Selection of maintenance policy for textile industry using hybrid multi-criteria decision making approach. *Journal of Manufacturing Technology Management*, *20*(7), 1009–1022. doi:10.1108/17410380910984258

Ishizaka, A., & Nemery, P. (2014). Assigning machines to incomparable maintenance strategies with ELECTRE-SORT. *Omega*, *47*, 45–59. doi:10.1016/j.omega.2014.03.006

Jamshidi, A., Rahimi, S. A., Ait-kadi, D., & Ruiz, A. (2015). A comprehensive fuzzy risk-based maintenance framework for prioritization of medical devices. *Applied Soft Computing*, *32*, 322–334. doi:10.1016/j.asoc.2015.03.054

Kelly, A. (2006). *Strategic Maintenance Planning*. Butterwoth-Heinemann.

Kirubakaran, B., & Ilangkumaran, M. (2016). Selection of optimum maintenance strategy based on FAHP integrated with GRA–TOPSIS. *Annals of Operations Research*, *245*(1), 285–313. doi:10.100710479-014-1775-3

Knezevic, J. (1996). *Mantenimiento*. Isdefe.

Kodali, R., & Chandra, S. (2001). Analytical Hierarchy Process for justification of total productive maintenance. *Production Planning and Control*, *12*(7), 695–705. doi:10.1080/09537280010024045

Labib, W. A. (2004). Towards an Intelligent Holonic Maintenance System. *Maintenance Journal, 17*(2).

Labib, W. A., O'Connor, F. R., & Glyn, B. W. (1998). An effective maintenance system using the analytic hierarchy process. *Integrated Manufacturing Systems*, *9*(2), 87–98. doi:10.1108/09576069810202005

Lazakis, I., & Olcer, A. I. (2015). Selection of the best maintenance approach in the maritime industry under fuzzy multiple attribute group decision-making environment. *Proceedings of IMechE part M: Journal of Engineering for Maritime Environment*, 1-13. doi:10.1177/1475090215569819

Lazakis, I., Turan, O., & Olcer, A. I. (2012). Determination of the optimum ship maintenance strategy through multi attribute decision making. *11th International marine design conference*, 473-487.

León, O. (1993). *Análisis de decisiones: técnicas y situaciones aplicables a directivos y profesionales.* McGraw-Hill.

Mahfoud, H., Abdellah, E. B., & El Biyaali, A. (2018). Dependability-based maintenance optimization in healthcare domain. *Journal of Quality in Maintenance Engineering*, *24*(2), 200–223. doi:10.1108/JQME-07-2016-0029

Mahfoud, H., El Barkany, A., & El Biyaali, A. (2016). A Hybrid Decision-Making Model for Maintenance Prioritization in Health Care Systems. *American Journal of Applied Sciences*, *13*(4), 439–450. doi:10.3844/ajassp.2016.439.450

Martorell, S., Sánchez, A., Muñoz, A., Pitarch, J. L., Serradell, V., & Roldan, J. (1999). The use of maintenance indicators to evaluate the effects of maintenance programs on NPP performance and safety. *Reliability Engineering & System Safety*, *65*(2), 85–94. doi:10.1016/S0951-8320(98)00088-X

Martorell, S., Villanueva, J. F., Carlos, S., Nebot, Y., Sanchez, A., Pitarch, J. L., & Serradell, V. (2005). RAMS+C informed decision-making with application to multiobjective optimization of technical specifications and maintenance using genetic algorithms. *Reliability Engineering & System Safety*, *87*(1), 65–75. doi:10.1016/j.ress.2004.04.009

Munda, G. (2005). Multiple Criteria Decision analysis and sustainable development. In J. Figueira, S. Greco, & M. Ehrgott (Eds.), *Multicriteria Decision analysis. State of the Art Surveys* (pp. 953–988). Springer Verlag. doi:10.1007/0-387-23081-5_23

Nakajima, S. (1988). *TPM: Introduction to total productive maintenance*. Productivity Press Inc.

Pintelon, L., & Gelders, L. (1992). Maintenance management decision-making. *European Journal of Operational Research*, *58*(3), 301–317. doi:10.1016/0377-2217(92)90062-E

Prawda, J. (1999). *Métodos y modelos de investigación de operaciones. Modelos estocásticos*. Limusa S.A.

Ramadhan, R. H., Hamad, I., Wahhab, A., & Duffuaa, S. O. (1999). The use of an analytical hierarchy process in pavement maintenance priority ranking. *Journal of Quality in Maintenance Engineering*, *5*(1), 25–39. doi:10.1108/13552519910257041

Rani, N. A. A., Baharum, M. R., Akbar, A. R. N., & Nawawi, A. H. (2015). Perception of Maintenance Management Strategy on Healthcare Facilities. *Procedia: Social and Behavioral Sciences*, *170*(27), 272–281. doi:10.1016/j.sbspro.2015.01.037

Rao, B. K. N. (1996). *Handbook of Condition Monitoring*. Elsevier.

Rodrigues, T. C. (2014). The MACBETH Approach to Health Value Measurement: Building a Population Health Index in Group Processes. *Procedia Technology*, *16*, 1361–1366. doi:10.1016/j.protcy.2014.10.153

Romero, C. (1996). *Análisis de Decisiones Multicriterio*. Isdefe.

SESCAM. (2007). *Valoración de la calidad de la atención sanitaria pública en hospitales*. Memory of activities of SESCAM.

Shafiee, M. (2015). Maintenance strategy selection problem: An MCDM overview. *Journal of Quality in Maintenance Engineering*, *21*(4), 378–402. doi:10.1108/JQME-09-2013-0063

Shafiee, M., Labib, A., Maiti, J., & Starr, A. (2019). Maintenance strategy selection for multi-component systems using a combined analytic network process and cost-risk criticality mode. Proceedings of the Institution of Mechanical Engineers, Part O. *Journal of Risk and Reliability*, *233*(2), 89–104.

Shahin, A., Shirouyehzad, H., & Pourjavad, E. (2012). Optimum maintenance strategy: A case study in the mining industry. *International Journal of Services and Operations Management*, *12*(3), 368–386. doi:10.1504/IJSOM.2012.047626

Sharma, R. K., Kumar, D., & Kumar, P. (2005). FLM to select suitable maintenance strategy in process industries using MISO model. *Journal of Quality in Maintenance Engineering*, *11*(4), 359–374. doi:10.1108/13552510510626981

Srivastava, P., Khanduja, D., & Agrawal, V. (2017). A framework of fuzzy integrated MADM and GMA for maintenance strategy selection based on agile enabler attributes. *Mathematics in Industry Case Studies*, *8*(1), 5. doi:10.118640929-017-0014-y

Taghipour, S., Banjevic, D., & Jardine, A. K. S. (2011). Prioritization of medical equipment for maintenance decisions. *The Journal of the Operational Research Society*, *62*(9), 1666–1687. doi:10.1057/jors.2010.106

Taha, H. A. (2011). *Operations research. An introduction*. Prentice Hall.

Vatn, J. (1997). Maintenance optimisation from a decision theoretical point of view. *Reliability Engineering & System Safety*, *58*(2), 119–126. doi:10.1016/S0951-8320(97)00025-2

Velmurugan, R., & Dhingra, T. (2015). Maintenance strategy selection and its impact in maintenance function. *International Journal of Operations & Production Management*, *35*(12), 1622–1661. doi:10.1108/IJOPM-01-2014-0028

Wang, L., Chu, J., & Wu, J. (2007). Selection of optimum maintenance strategies based on a fuzzy analytic hierarchy process. *International Journal of Production Economics*, *107*(1), 151–163. doi:10.1016/j.ijpe.2006.08.005

Wolf, M. (2008). Strategic management of medical equipment. Starting and operating a TSC. *Proceedings of the 20th Congress of the International Federation of Hospital Engineering*.

Zahir, S. (1999). Geometry of decision making and the vector space formulation of the analytic hierarchy process. *European Journal of Operational Research, 112*(2), 373–396. doi:10.1016/S0377-2217(98)00020-4

Zaim, S., Turkyílmaz, A., Acar, M. F., Al-Turki, U., & Demirel, O. F. (2012). Maintenance strategy selection using AHP and ANP algorithms: A case study. *Journal of Quality in Maintenance Engineering, 18*(1), 16–29. doi:10.1108/13552511211226166

## ADDITIONAL READING

Al-Najjar, B., & Alsyouf, I. (2000). Improving effectiveness of manufacturing systems using total quality maintenance. *Integrated Manufacturing Systems, 11*(4), 267–276. doi:10.1108/09576060010326393

Burton, J., & Hubaeck, K. (2007). Is small beautiful? A multicriteria assessment of small-scale energy technology applications in local governments. *Energy Policy, 35*(12), 6402–6412. doi:10.1016/j.enpol.2007.08.002

Carnero, M. C. (2014). Multicriteria model for maintenance benchmarking. *Journal of Manufacturing Systems, 33*(2), 303–321. doi:10.1016/j.jmsy.2013.12.006

Chen, F. L., & Chen, Y. C. (2010). Evaluating the maintenance performance of the semiconductor factories based on the analytical hierarchy process and grey relational analysis. *American Journal of Applied Sciences, 7*(4), 567–574. doi:10.3844/ajassp.2010.568.574

Crespo, A. (2007). *The maintenance management framework. Models and methods for complex systems maintenance*. Springer.

De Jonge, B., Klingenberg, W., Teunter, R., & Tinga, T. (2015). Optimum maintenance strategy under uncertainty in the lifetime distribution. *Reliability Engineering & System Safety, 133*, 59–67. doi:10.1016/j.ress.2014.09.013

Iung, B., Véron, M., Suhner, M. C., & Muller A. (2005). Integration of maintenance strategies into prognosis process to decision-making aid on system operation. *CIRP Annals - Manufacturing Technology, 54*(1), *5-8, doi:* . doi:10.1016/S0007-8506(07)60037-7

Macchi, M., & Garetti, M. (2006). Information requirements for e-maintenance strategic planning: A benchmark study in complex production systems. *Computers in Industry, 57*(6), 581–594. doi:10.1016/j.compind.2006.02.015

Mobley, R. K. (2004). *Maintenance fundamentals*. Elsevier Butterworth-Heinemann.

Moore, W. J., & Starr, A. G. (2006). An intelligent maintenance system for continuous cost-based prioritisation of maintenance activities. *Computers in Industry, 57*(6), 595–606. doi:10.1016/j.compind.2006.02.008

Pintelon, L., Gelders, L., & Van Puyvelde, F. (1997). *Maintenance Management*. Acco Leuven Amersfoort.

Regattieri, A. A., Gamberi, G. M., & Gamberini, R. (2015). An innovative method to optimize the maintenance policies in an aircraft: General framework and case study. *Journal of Air Transport Management*, *44-45*, 8–20. doi:10.1016/j.jairtraman.2015.02.001

Taghipour, S., Banjevic, D., & Jardine, A. K. S. (2011). Prioritization of medical equipment for maintenance decisions. *The Journal of the Operational Research Society*, *62*(9), 1666–1687. doi:10.1057/jors.2010.106

Teixeira de Almeida, A. (2001). Multicriteria decision making on maintenance: Spares and contracts planning. *European Journal of Operational Research*, *129*(2), 235–241. doi:10.1016/S0377-2217(00)00220-4

Vineyard, M., Amoako-Gyampah, K., & Meredith, J. R. (2000). An evaluation of maintenance policies for flexible manufacturing system. A case study. *International Journal of Operations & Production Management*, *20*(4), 409–426. doi:10.1108/01443570010319156

Wang, H. (2002). A survey of maintenance policies of deteriorating systems. *European Journal of Operational Research*, *139*(3), 469–489. doi:10.1016/S0377-2217(01)00197-7

## KEY TERMS AND DEFINITIONS

**Descriptor:** Is an ordered set of impact levels which allows the degree of fulfilment of a criterion by an alternative to be assessed quantitatively and qualitatively.

**MACBETH:** Measuring Attractiveness by a Categorical Based Evaluation Techniques is a multicriteria decision-making technique based on the additive value model. The methodological background to the technique includes the structuring of value trees, the building of descriptors associated with each criterion, the weighting of criteria, the building of value functions from the judgements of the decision-maker by linear programming, a sensitivity and robustness analysis and finally the considering of alternatives.

**Maintenance Policy:** Organization of the human and material resources to address faults, wear or breakdowns in machines, devices or facilities, to guide maintenance management decision making, with the aim of having facilities which work continuously in conditions of optimal availability, safety and quality, minimizing the cost of maintenance.

**Multicriteria Decision Making:** Consists of structuring a complex problem and considering the multiple criteria by which it is affected, which are frequently in conflict, to determine a preferred alternative, choose from a small group of the best alternatives or group alternatives into different preference sets. This is because there is no optimal alternative and so the solution to the problem will be found from a set of non-dominated solutions. In the solution process, the judgements and preferences of the decision maker or decision-making group are taken into account.

**Preventive Maintenance:** This is a maintenance policy consisting in periodic, systematic revision and repair of machines and facilities, to prevent faults and breakdowns before they happen, so as to guarantee optimal working and to increase the life cycle of assets.

**Subsystem:** Systems from the decomposition of a group of elements made up of equipment and installations and their interconnections, intended to provide a product or service, so that they can be analysed and assessed more easily.

## APPENDIX

*Table 1. Judgement attractiveness criteria for intermediate subsystems.*

| EXPRESSION OF JUDGEMENTS OF ATTRACTION BETWEEN CRITERIA | | | | | | | | | | | | |
|---|---|---|---|---|---|---|---|---|---|---|---|---|
| | Very weak | Weak | Moderate | Strong | Very Strong | Extreme | Very Strong | Strong | Moderate | Weak | Very weak | |
| Availability | | | | | | | | | ✓ * | | | Activity impact |
| Availability | ✓ | ✓ * | ✓ | | | | | | | | | Failure detection time |
| Availability | | | ✓ * | | | | | | | | | Cost |
| Availability | | | | | | | | ✓ | ✓ * | | | Risk to employees |
| Availability | | | | | | | | ✓ | ✓ * | | | Risk to users |
| Activity impact | | | ✓ * | | | | | | | | | Failure detection time |
| Activity impact | | | | ✓ * | | | | | | | | Cost |
| Activity impact | | | | | | | | | | | ✓ * | Risk to employees |
| Activity impact | | | | | | | | | | | ✓ * | Risk to users |
| Failure detection time | | | ✓ * | | | | | | | | | Cost |
| Failure detection time | | | | | | | | | ✓ * | | | Risk to employees |
| Failure detection time | | | | | | | | | ✓ * | | | Risk to users |
| Cost | | | | | | | | ✓ * | | | | Risk to employees |
| Cost | | | | | | | | ✓ * | | | | Risk to users |
| Risk to employees | | | | | | | | | | | ✓ * | Risk to users |

✓ Outcome of the judgement expressed by the decision-making group made up of each manager of the following Hospital areas: installations, safety, environment and medical equipment, coordinated by the technical services manager.

* Semantic value included in M-MACBETH.

This table shows how each decision maker, when comparing one criterion with another, gave a judgement from the semantic values provided by MACBETH: very weak, weak, moderate, strong, very strong and extreme. These judgements are shown in the table with a square symbol. Where the criterion in the left-hand column is more attractive than that in the right-hand column, the values are included in the left side of the table.

On the other hand, if the criterion from the column on the right is more attractive than the one on the left (see Risk to employees against Availability), the judgement given by the decision makers is shown on the right of the table. In some cases, the group as a whole reached a consensus on the judgement given, for example in the case of Cost against Risk to employees; All the members of the decision-making group considered that Risk to employees was more strongly attractive than cost; thus, it only takes up a single square of the table.

Nonetheless, when Availability is compared to Failure detection time, it is seen that there are squares in the very weak, weak and moderate columns. This means that there was no consensus and that the members of the decision group considered that the three vales were appropriate. Finally, the model created with M-MACBETH includes the most voted judgement, identified in the table by *.

# Chapter 2
# An Analysis of the Relationship Between Maintenance Engineering and Knowledge Management

**Francisco Javier Cárcel-Carrasco**
*Universitat Politècnica de València, Spain*

**Elisa Peñalvo-López**
*Universitat Politècnica de València, Spain*

**Maria Carmen Carnero**
*University of Castilla-La Mancha, Spain*

**Vicente López-Mateu**
*Universitat Politècnica de València, Spain*

## ABSTRACT

*Knowledge is vital in everyday life, but from the viewpoint of the companies, it is a powerful intangible capital that can check the efficiency and existence of the organizations themselves. However, in industrial production environments, maintenance and exploitation activities are little studied in reference to the mechanisms of transmission of knowledge that are generated based on the difficulty of their capture, because it is fundamentally based on tacit knowledge (intrinsic fond professions operating personnel) acquired by operators for years from their work experiences in the industrial plant. These highly qualified professionals, accustomed to solve technical problems and maintain the required state of production with a high load of stress, traditionally function as islands of knowledge, and the company loses an important asset when one of them leaves. This chapter approaches the transcendence of the tacit knowledge among staff and marks the conditions for transmission as a basis for the learning of other partners of the organization.*

DOI: 10.4018/978-1-7998-3246-1.ch002

## 1. INTRODUCTION

In the resources of management engineering, it is required a measuring equipment and a human component with high specialization. All this implies a strong knowledge that must be managed in its creation phase, dissemination and use phase. The indicators (normally used) of maintenance performance are based on various pillars such as reliability, availability, maintainability, security, operational costs, and a fundamental one for operability such as the human factor (Cárcel, 2013a), given the high tactical knowledge that is generated in the performance of the assigned functions.

The knowledge is the ability to act, process and interpret the information to generate more knowledge or solve a particular problem. In this new century there has been a fundamental change, in which the growth of many sectors and companies is driven by the knowledge and ideas, more than just taking into account the traditional resources (Del Moral, 2007). This drives us to a strategic management of knowledge-driven companies, where the active assets are losing value in favor of intangibles (Sánchez, 1999; Peña, 2001). This is why the knowledge can be considered as the main intangible ingredient both in companies and in the economy as a whole (OCDE, 2004), and in the evolution of maintenance, where the importance of the human component takes a large incidence, is a value which has taken great importance with the of techniques and the resources destined to the improvement.

When planning a maintenance service, it is vitally important to have a deeper knowledge of the facilities, to transform the tactical strategic knowledge of the operational experiences of the maintenance operators in explicit, which will undoubtedly deepen the study of the measures of energy efficiency and value the reliability of the facilities, with the knowledge failure process, which improves the productivity of the company (Alsyouf, 2007; López, 2006), identifying the relevant data and information to improve the service.

With a shift towards a model based in the knowledge and learning, the organization is focuses on the ability to innovate and learn, to more efficiently solve their daily work, as well as to solve new or non-routine actions, creating a value of the intangible based on knowledge and its quick update in the working environment of the maintenance organization. It must be assumed as a long-term strategy, visualizing knowledge as a strategic factor, as well as a tool for problem solving and decision making (Peluffo, 2002).

There is an abundant literature on knowledge management in various industrial activities and services, and the effects of their application (Bahoque, Gómez & Pietrosemoli, 2007; Colino, Martinez & Martinez, 2010; Colino & Riquelme, 2000; Chua & Gu, 2008; Ferrada & Serpal, 2009; Rivas & Flores, 2007; Ventura & Ordoñez, 2007; Yang, 2006;), but normally these studies focus on global management (especially in the more explicit part), with an emphasis on trade, accounting or internal administrative management, or development activities, but dealing sparingly with the tactical actions of industrial trades, normally considered as an "expense for the company", but having the effect, however, of substantially reducing costs involved (often taken by management itself). Therefore, to manage knowledge in these areas of work means, in itself, not only an improvement in the efficiency of the processes within industrial trades, but also a reduction in the spending caused to the company (by stoppage of production, loss of energy, loss of efficiency or reliability of systems and facilities and longer adaptation time of new technicians).

It is necessary to analyze the personal knowledge to develop organizational knowledge (Martin, 2008; Pauleen, 2009; Volkel & Haller, 2009), that allows to make an analysis of cost-benefit on its application (Volkel & Abeckar, 2008). The internal activities of the industrial enterprise, maintenance needs deep technical knowledge, highly experienced staff and has traditionally been the structure within the company where there is major component of tacit knowledge. Since their duties directly affect the reliability of systems and installations (Sols, 2000), eliminating unwanted stops and performance to critical processes, the need for the appropriate management of such information/knowledge is given that you can have a great strategic value for the company.

Personnel policies, so prevalent today, that they include subcontracting of industrial services, the objectives of the staff, serving only for reasons of age, without submitting to the recently incorporated into a loyalty preprocessing to reduce the external rotation, it is possible to achieve, in some cases, successes in the short term but certainly the lack of sufficient time to create and transfer knowledge in a controlled way, and the lack of sedimentation cultural to assimilate it and implement it, will be the implementation of any model of knowledge management in the medium term (Muñoz, 1999), and to a large extent, the loss of control of the reliability and efficiency of production processes or industrial internal services.

It should be borne in mind that the operational impact of maintenance actions has an impact on the company, affecting most of the fundamental tactical actions, and given that this action can be strategic, it is necessary to check conditions for the capture of that knowledge (Cárcel, 2014b).

This chapter aims to make a brief review of the maintenance evolution to define the basic elements that define its nature and to mark its relation to processes linked to the knowledge, to thereby understand future trends in the management of such an important intangible in these departments, whose operation exponentially effects the productivity of any company.

## 2. THE KNOWLEDGE AND THE MAINTENANCE ENGINEERING

In numerous existing literature it is defined the knowledge is power, but as Nonaka affirms (Nonaka, 1998), the importance of knowledge in organizations depends on what can be done with it within a business environment. Which means, the knowledge by itself is not relevant, it can not be used for bringing out actions of value creation (Xiomara, 2009).

The knowledge management implies, therefore, the use of practices that are difficult to observe and manipulate and which are sometimes even unknown for those who possess them. This presents a problem for the companies, more familiar with the management and accounting of fixed capital. Among the various categories of investment related to knowledge (education, training, software, R&D,etc), knowledge management is one of the least known both qualitatively and quantitatively, as well as in terms of costs and economic returns(OCDE, 2004).

There are numerous factors of knowledge experience that affect maintenance (Figure 1), which oscillate from the mission itself to create and achieve the required goals, innovate or improve the future of the activity.

In the generation of knowledge there is a transformation of tactical knowledge of individuals into explicit knowledge at the group and organizational level (Nonaka, 1999) and each of the members of such groups internalize it, turning into tactical again. This process generates four phases, which are: socialization, outsourcing, combination and internalization (Figure 2).

1. *Socialization (from tactical to tactical):* It is a process in which tactical knowledge of others is acquired, sharing experiences and thoughts with them, and both communicating, so that whoever receives them increases their knowledge and reaches to the level close to the issuer. This is a process of great incidence in industrial trades, specially in industrial maintenance: there is a coupling phase in which the new operator must assimilate all the equipment and facilities of the industrial

*Figure 1. Knowledge factors in relation to maintenance.*

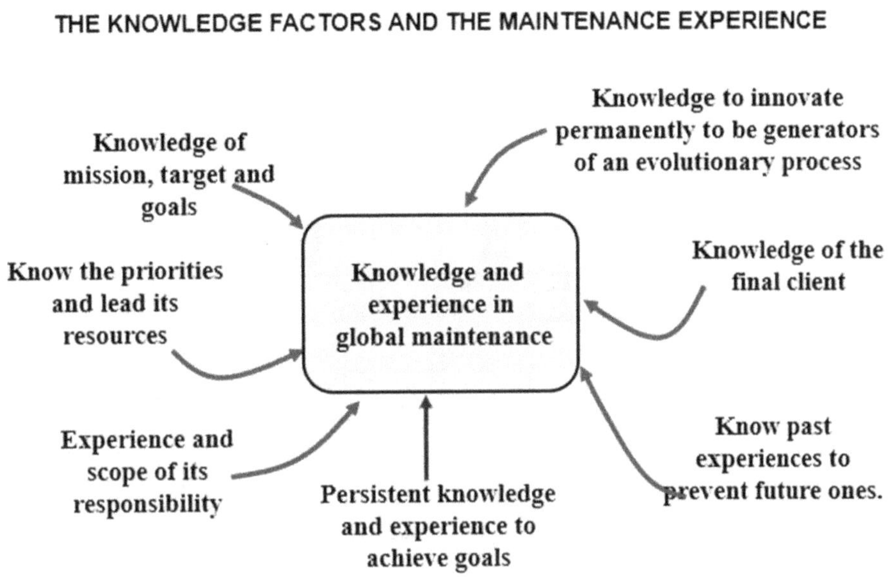


plant, usually through comments and explanations of experienced colleagues. It is a process that in large industrial plants can take months or years to fully operate to operator.

2. *Outsourcing (from tactical to explicit):* stage in which tactical knowledge is transformed into explicit or understandable concepts for the organization or for any individual, through its own articulation and its transfer to quickly understandable supports (Nonaka, 1999a) b. In industrial trades, this stage uses to be partially biased. Parts of the work are specified, times of execution or periods of work accomplishment, but normally, the most valuable experiences (description of the repair of a critical breakdown, emergency maneuvers, reassembly, etc.), are explained with brief parts of work with indication of experience.
3. *Combination (from explicit to explicit):* it is a part of the process that synthesizes the explicit concepts and transfers them into a knowledge base, through following procedures (Nonaka, 1998): capturing and integrating new essential explicit knowledge, through the compilation, reflection and synthesis; dissemination of it using the transfer processes normally used in the organization, such as presentations, meetings or emails; and processing in documents, layouts, reports and market data. It is produced in industrial trades through the technical information of the plant itself (assembly drawings, equipment manuals, review of regulation, etc.) as well as consultation of other processed information (maintenance history, maneuvers of facilities, etc)
4. *Internalization (from explicit to tactical):* it is a stage of the process in which the tactical knowledge of the individuals is expanded from the explicit knowledge of the organization, after purified and converted into each person's own knowledge. Such internalization requires, on the one hand, the updating of the explicit concepts or methods and, on the other hand, the inclusion of such explicit

*Figure 2. Industrial knowledge according to Nonaka model. Source: own elaboration from (Nonaka, 1998)*

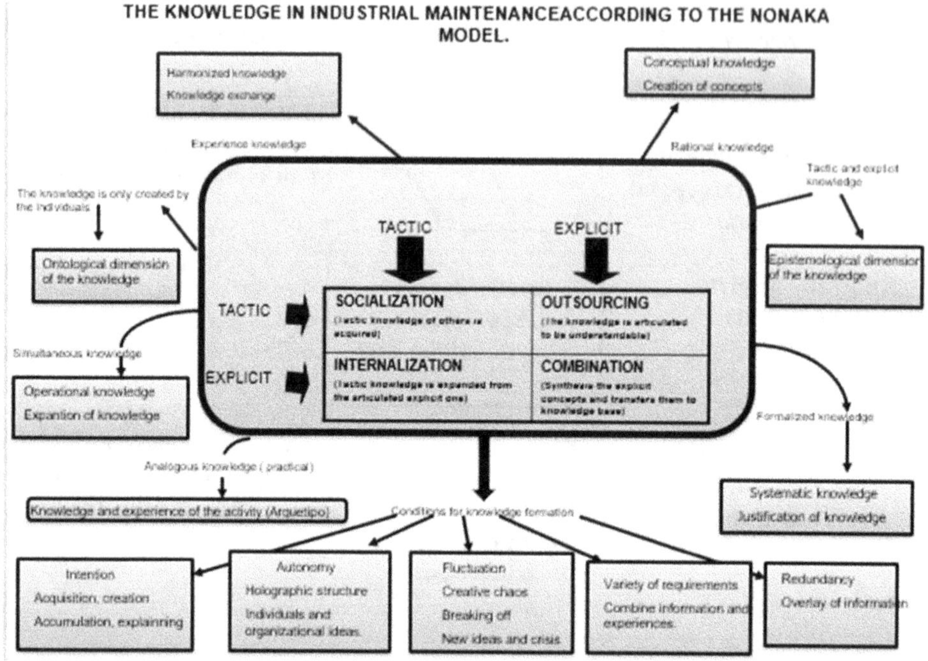


knowledge in tactical (Nonaka, 1998). It is necessary that explicit knowledge be experienced, either personally doing the activity or through participation, so that the individual internalizes it according to their own tactical activities of maintenance or exploitation, it is produced through the technical information of the plant, which operators must assimilate to perform the specified work.

## 3. HISTORICAL EVOLUTION AND ITS RELATION WITH THE KNOWLEDGE

In a very clear way, a brief review of the evolution of maintenance is carried out below, in order to determine the basic line of this evolution, extracting the common elements to it and certifying the definitive need for processes that are linked to the knowledge, as something essential and valuable maintenance activity.

The maintenance in its optimal performance has to go through three basic operations, from inspection to conservation actions as well as repair work and experiences, which implies states of knowledge that must be synthesized. Althought it is true mentioned operations are common in the daily basis of a plant, it is also true that the fact that they are carried out depends on the maintenance policy set by the corresponding responsible. In addition, the fact of carrying them out does not imply that maintenance is optimal or that the knowledge generated is properly collected.

Although the origin of maintenance is undoubtedly as old as first machines used by man, industrial maintenance, as we understand it, made its appearance, as a systematically organized activity, at the dawn of the twentieth century. Apparently, the first known case took place, in foundries in the United States, and in the military sector: in the airplanes and submarines during World War 1. In 1920, the mechanical maintenance was already practiced in industrial plants, transport activities, etc. From 19228 to 1930, the first consulting companies in this area appear. It is not by chance that consultants whose product is a service based on their experience and knowledge appear so early.

After this initial stage, the maintenance, like other fields of Industrial Organization, undergoes a remarkable development during the World War II and in the postwar period, in different applications of military interest. Preventive maintenance programs are developed consisting of inspectioning the aircraft before each flight, checking its condition and replacing components after a certain number of operating hours (technique, also used nowadays, known as "hard time"). Thus, knowledge about condition of the equipment and the cost-effectiveness of the replacement are revealed, as something basic to efficiently proceed to the conservation of the mentioned one.

From the 1945, the redundancy of components in the design is used, reducing the "criticality of the breakdown". Test techniques and physical measurements are also formalized, in order to know the probability of breakdown of each component. The maintenance activity in relation to tests and trials begins to be "protocolized", thus obtaining a registered and formalized information, object, therefore, of explicit knowledge. The measurement introduces the rigorous scientific method aimed to the knowledge of the most valuable and substantial element of the target of maintenance: the breakdown and its characteristics.

At the end of the 60s, the systematic application of reliability techniques begins, allowing the prediction of costs derived from failures and the calculation of the profitability of the maintenance activity. Even so, the behavior of certain components did not present easy prediction, escaping from models to use; something that still exists today, as it happens, from the classic model of the bathtub failure rate curve, to six different models in reliability based maintenance (RCM), and research continues to try to model those behaviors that do not adapt sufficientemente to the proposed. The knowledge about the behavior of the breakdown, throughout the life cycle of the equipment, is revealed as complex and specific and it is expected that, from now on, multidisciplinary models will be raised that address the cause-effect relationship of the breakdown in all its complexity, that is, understanding the machine or the equipment as a whole.

In addition to the attention to the analysis and study of the causes and effects of the incidences of the industrial equipment, basic target of the maintenance, also its management was already from the beginning of the 60s element of the study and analysis, the utility of having historical breakdowns statistics for the analysis and planning of maintenance (Greer, 1960; Hanks, 1961), as well as the description of various methods of optimizaztion of maintenance policies (Barlow, 1963) is highlighted. Along these lines, but aiming at increasing productivity, Darnell and Bert in 1978(Darnell, 1978), publish a contribution advocating scheduled maintenance as a means of increasing productivity (Darnell, 1978). Christer, in 1981 and Boland and Proscan in 1982, insis on the same line, highlighting maintenance attitude, as opposed to passive and uncontrolled attitudes (Christer, 1981; Boland, 1982)

The already complex and specific information, in the analysis of causes, is now overflowing because of the many aspects of interest and their relationship. The treatment of the the information requires refined computer and operational research techniques that give meaning to that information, guide it to the goals proposed in the planning and direct its use to the optimization of the maintenance activity.

Although it is at the end of the 60s when the maintenance associated promoted the studies, corresponding to this discipline, especially in England (British Standards, 1964), it is in the 70s when the two lines of analysis appear that make up two schools of though about maintenance.

An analysis line, the Soviet School, contemplates the parameters of machine maintenance from a constructive and biological point of view, introducing concepts of health and aging. Artobolevski and Gorvachkin highlight the relationship of the state parameters in the design of machines: the last one referring, in its analysis, to the modeling of machine wear,and incorporatinf specifivations oriented to the ease with which the repaid is carried out. Also, to hightlight the works of Artemieb and Raiman on obtaining and analyzing data in the process of tractor waer. Finally, Selivanov introduced in 1972 the notion of utility as an objective of variable or characteristic of machine service.

Since taht time when it begins to be seen with greater need to classify the types of maintenance, introducing the greater measure the planned maintenance, with a proactivity that would give rise to the different organizational techniques of industrial maintenance that evolve until today, marcando the need for a greater degree of transfer and knowledge management in all the activities involved and that deeply involves the human component, so intensive in this activity (Figure 3). A further definition of the maintenance classification can be found in the UNE EN 13306 standard where generic definition and terms for technical, administrative and management maintenance areas are specified

*Figure 3. General classification of maintenance according to UNE-EN 13306.Source UNE EN 13306*

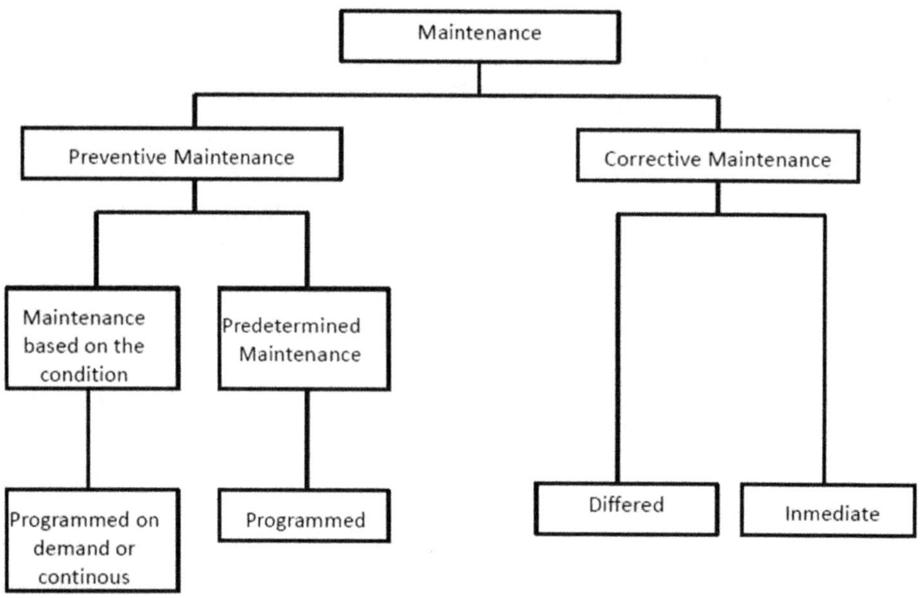


On the other hand, the western school contemplates, within the study of maintenance, the basic economic concepts of the Theory of the Firm, to determine the profitability of the repair, and consequently the setting of maintenance policies, in relation, for example, with the renovation of the machinery. To this approach that can be described as classic, from 1975 the terotechnology is introduced, which contemplates the study of the life cost cycle (LCC or "life cycle costing"), as necessary when concluding strategies for investment, operation and disposal in maintenance. (Kelly,1978)

Both approaches are moderatly integrated into a systems enfoque in which conservation can not be demarcated from design and logistics throughout the life of the equipment and the plant (Figure 4). The knowledge in this field is interrelated with the rest of the information and experience of the production and distribution processes. The integrated knowledge takes a value and it appears as essential, if what is sought are objectives of sustainability and shared productivity in industrial projects.

Also the maintenance activity has been considered from a macroeconomic or sectoral point of view, thus, in 1970, the United Kingdom Ministry of Technology publishes a report on the incidence of maintenance in the national economy and even, subsequently, in 1977, presents forecasts for the end of the decade (Butlery, 1978). The forecast predicted a notable increase in maintenance activity and in parallel and increase in costs, in line with data from the american economy, in the sense that more modern plants have a more expensive maintenance. This report from the British labor party about the situation in the industry, already contemplated the maintenance activity separated from the production

*Figure 4. General scheme of the evolution of maintenance. Source: self-made.*

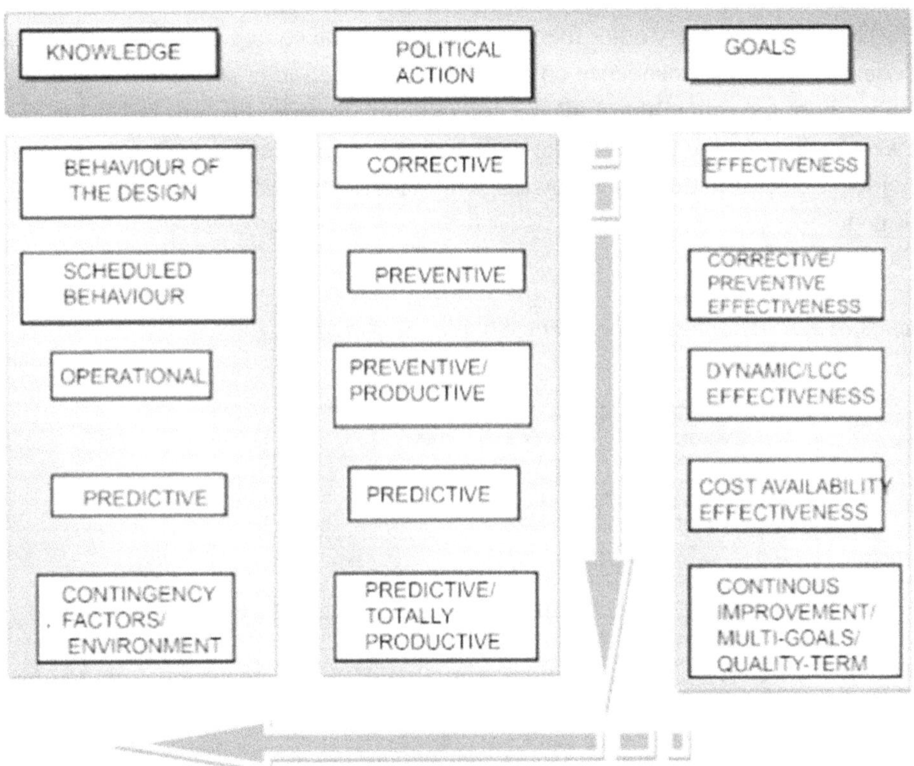


This systematic approach is reflected in the document "Maintenance Steering Group", also known as MSG-2, the result of the deliberations of a committee promoted by Boeing with Pratt and Whitney. In the document the aircraft systems were identified, within them, subsystems, and so on until reaching the critical units, using criteria of security or economy (Matteson, 1985; Ackert, 2010). From the critical units the individual functions were defined for each of them. Finally, for each function, the possible failure modes were specified and from them the maintenance policies were established.

With the expanstion of new informational technologies, the treatment of knowledge in general, and specially the one obtained in the industrial plant, has taken a new dimension, appearing as the most valuable asset in future scenarios (Crespo, 2004; Muller, 2008).

The effects of automation are enabling real time monitoring of maintenance (Reiner, 2005). Thus, for example, certain systems, such as some aircrafts, that already incorporate status monitoring that analysis the data of the units in service and specifies the actions to be taken. Strong synergies are predictable in the combined use of new information and communication technologies and maintenance technologies (predictive, self diagnosis, remote maintenance, maintenance systems through the internet, etc).

It is specially from the decade of the 70s, 80s and 90s of the 20th century, when the use of organizational maintenance strategies is intensified, although normally in the environment of large companies, such as based in reliability (RCM) (Moubray, 1991; Smith, 1992), total productive maintenance (TPM) (Nikajima, 1988), the effective maintenance (Conde, 1999; Cárcel, 2012c), proactive (Oiltech,1995; Pirret 1999), reactive (Idhamman, 1997), world class WCM (Hiatt, 1999), risk centered maintenance (Serratella, 2007), as well as many other theoretical models. The strategic level of this activity must be taken into account, with great dependence on the areas of production or services (Crespo, 2004), as well as an in-depth study of the reliability of all the components involved (Turan, 2011; Lazakis, 2007). In the figure 5 we can observe the trends in the management of maintenance services in which the need for greater incidence of information and knowledge for an efficient and economic practice of the service is also mentioned.

*Figure 5. Trends in maintenance management. Source: Self-made.*

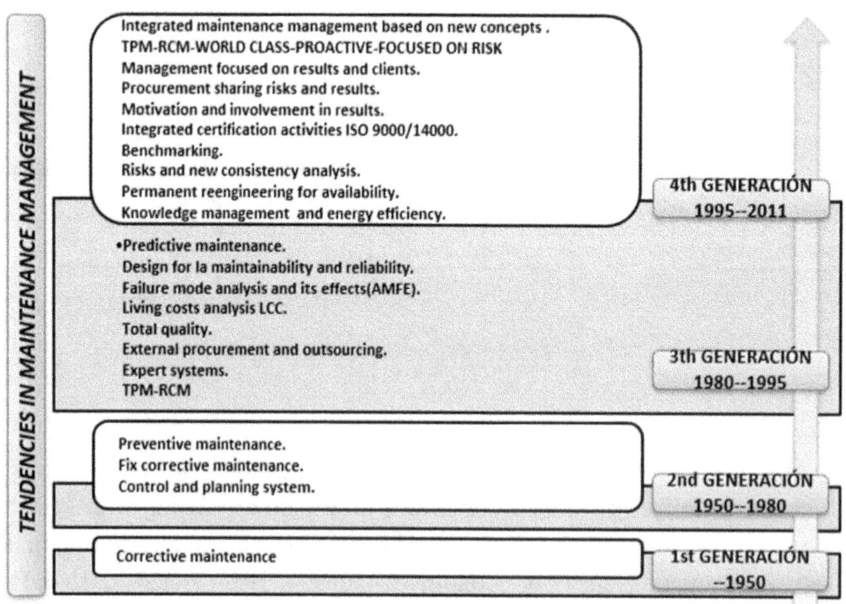

It is important to clarify that not all companies evolve historically as they go through each of the tactics sequentially, the simply adopt their own tactic that brings together the best practices of several of them, to emphasize that the TPM is the most basic of all.

The increasing importance of computer aided maintenance management, condition based maintenance and the application of remote devices and functions such as e-maintenance, have opened up great possibilities for maintenance optimization (Jonsson, 1997), and the improvement of knowledge management (Cárcel, 2014d).

As already described, once it has been verified how, throughout the history and evolution of industrial maintenance, the essential elements thereof are outlined, and that they need given their intimate relationship, with the factors that allow a proper management of the information and knowledge.

## 4. KNOWLEDGE MANAGEMENT IN PROFESSIONS OF INDUSTRIAL ENVIRONMENTS.

Knowledge comes up when a person considers, interprets and uses the information combined with their own experience and ability (Zapata, 2011). Within organizations it is necessary to study the knowledge to mark the incidence of its application (Ajmal & Koskinen, 2008; Chua & Goh, 2008; Davenport, 1997; Geisler, 1997; Lin, 2007; Nonaka, 1994), and to mark the productivity from a human perspective based on knowledge (Marvel & Rodríguez, 2011), which can be used in numerous industrial and service activi-

*Figure 6. Knowledge understanding framework in the maintenance activity.*

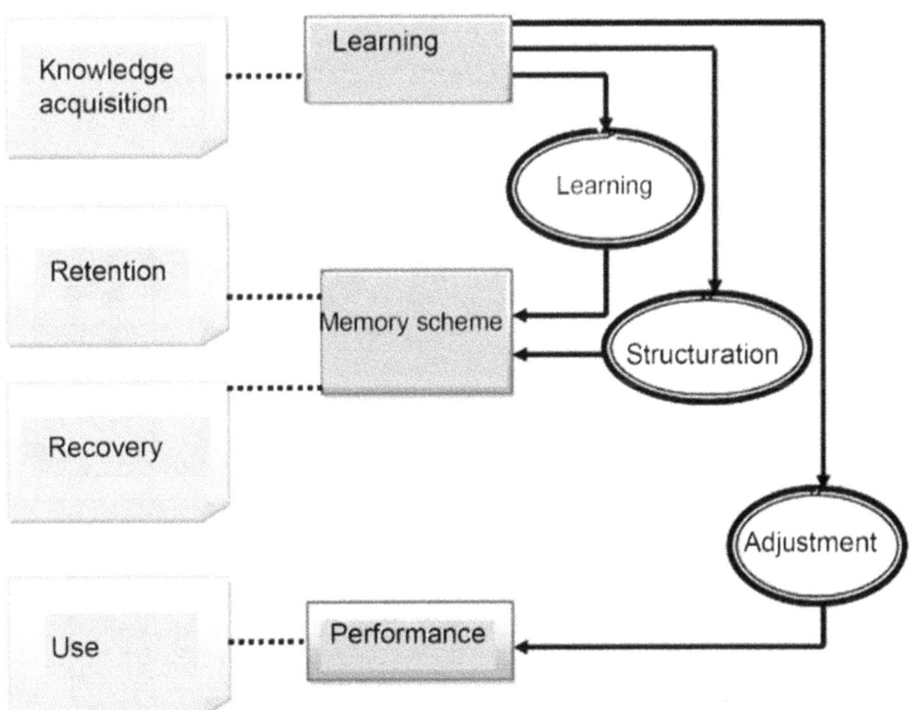

ties, such as industrial activities (Cárcel el al, 2016a; 2016b; 2019a; 2019b), renewable energy systems (Cárcel et al., 2018; Gómez et al 2018; Peñalvo et al., 2017; 2019a; 2019b), etc.

The basic principles on which a knowledge management model should be centered and that is the basis as a learning system in its application to industrial maintenance should be based on the mechanisms observed in how knowledge acquisition occurs, how retention occurs, recovery and its use (Figure 6). This will lead to the study of how learning occurs and its aggregation and structuting to memory schemes for its retention and recovery and the relevant adjustments that must be taken into account for the use of strategic and tactical knowledge that improves the efficiency of thereof service. The proposed system must try to integrate concepts and techniques of application to the maintenance, in order to respond to the problem of loss of experience, reduce the time of action and increase the efficiency of the maintenance service (in light of operation, reliability and improvement of energy efficiency).

People acquire an active and central role, because the knowledge is born, developed and changed from them.

*Figure 7. Kantian approach to maintenance activity in relation to the G.C.*

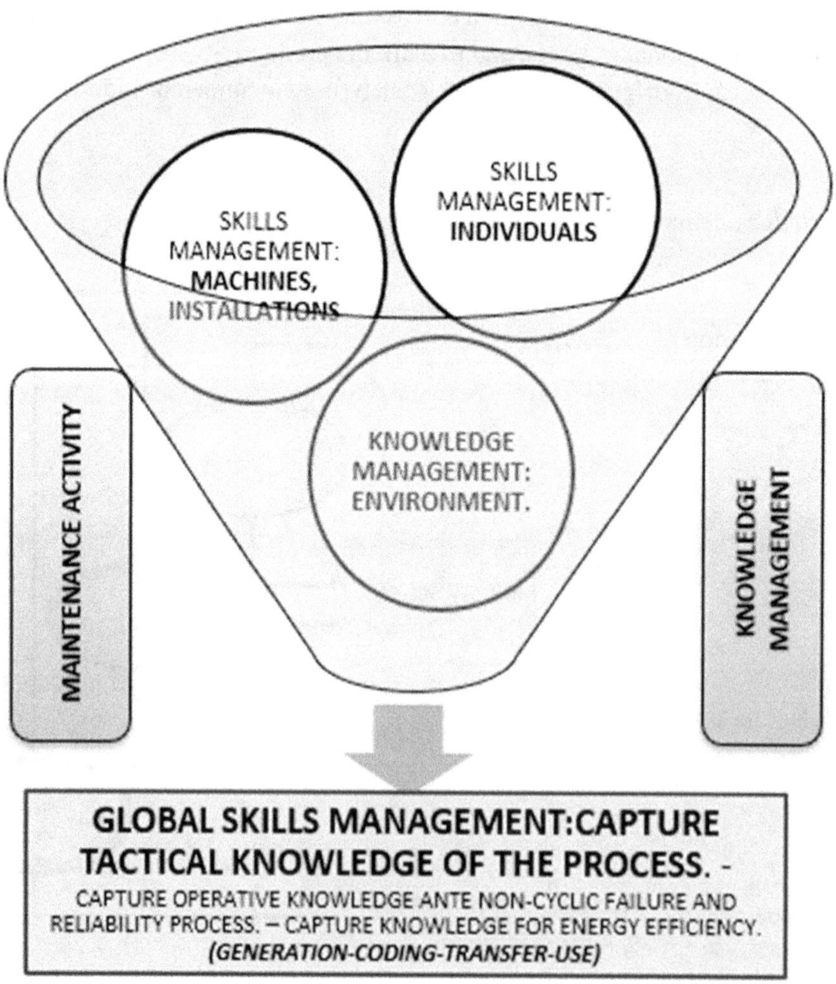

It should seek to strengthen the spaces so that the agents obtain better results in the actions of strategic knowledge management, among which we can mention:

a) The mechanisms necessary to obtain the information and knowledge required by a person must be marked (Barragán, 2009; González, 2010; Kim & Breslin, 2009), and enhance the ability to respond to the ideas obtained from the information and the tactical knowledge that they possess. All this must be integrated based on the fact that the operational staff of industrial maintenance can operate in Kantian environment where they must combine their own skills with the knowledge of the environment where they work (large areas), as well as the knowledge required to operate and maintain machined and facilities that are not found in the traditional training (figure 7)

b) Manage knowledge and organizational learning in order to promote medium and long-term development strategies.
c) Define the strategic knowledge (O'dell & Grayson, 1998; Pauleen, 2009) that will give efficiency and security to the process in the maintenance organization, and that can achieve a vision of utility and economic results or efficiency in processes.
d) Create a simple technological base where the managed knowledge resides and its transfer to the various users for its use, taking advantage of the most successful experiences and the ways in which, the most frequent errors were solved. This makes it possible to solve problems faster and adapt more flexibility (Dale & Boaden, 1997; Sher & Lee, 2004). In this technological base, all operative experiences carried out by the operators must be collected based on their best experiences (figure 8), collected in files easily understandable by the rest of the operating organization staff.

e) Define the agents that will pursue the proper management during all the processes that are manifested (generation, production, transfer and use).

The knowledge management is faced with a series of difficulties that come from the same environment especially from cultural factors (individualisms, the lack of a culture based in knowledge, isolation from the environment and the members of that environment, the short-term orientation, etc) (Peluffo et al, 2002).

The principles on which the foundations for learning must be based have been divided into three aspects, in relation to people, the environment and the means and tools for G.C. These are:

In relation to people:

- In the field of industrial maintenance, people are usually evaluated based on their tactical knowledge. We must highlight and reward the contribution of people to the creation of strategic knowledge, and that serve to learn the group.
- Integrate and coordinate individual and group knowledge this being one of the main objectives that ensure the success of GC system.
- To create strategic knowledge for the maintenance organization and therefore to the company, individual proposals and experiences must be socialize and a common base of tactical knowledge

*Figure 8. File card to perform a task (greasing points of a machine) that serves to learn new operators.*

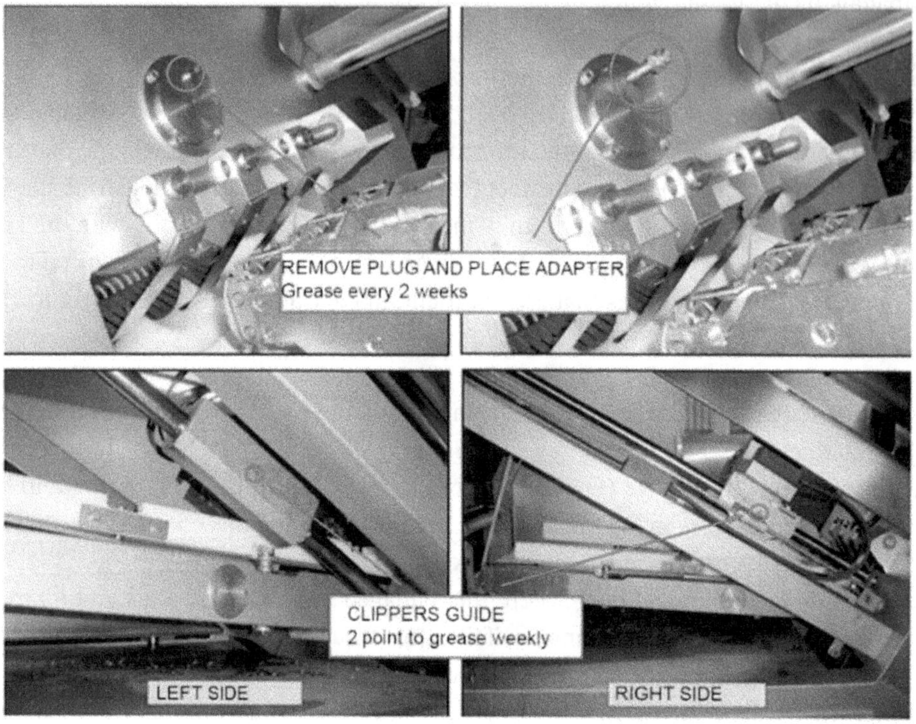


must be achieved that will allow for externalizing ideas, making them explicit and to be understandable and shared by all members.

- Informal means of information capture are usually the common rule among maintenance members (informal conversations, hall meetings, etc.) The situations where they are presented must be transformed into productive and generate informal work networks from them.
- The synergy with the users of the maintenance activity (usually other departments of the company), generates new procedures and improves the transmission of strategic knowledge of the company.
- The capture of personal experiences in the workplace (successes and errors fore various ordinary or extraordinary situations), is a fundamental base as a generator engine in the learning and appropriation of knowledge, being able to be used in the self-learning of other staff of the organization.

In relation to the work environment:

- Training and motivation towards GC in the organization's environment is essential in the initial phase for the success of the processes of knowledge acquisition, generation and transfer.
- The knowledge of various areas of specialization or functional maintenance must be integrated and combined. Through this combination, fundamental tactical knowledge is extracted, deleting islands of knowledge.
- The current explicit information (Layouts, projects, manuals, data, etc.), must be organized and "lightened" in order to increase efficiency for the capture of useful knowledge.

- The incorporation of knowledge in the maintenance organization must have a dynamic effect in a cycle of continuous improvement.
- The GC in the work environment can induce innovation in the creation of new capabilities.
- A knowledge of crises and emergencies is essential to anticipate threats and opportunities.
- Pro-activity should be encouraged in the searcg for tactical knowledge.

In relation to the means and tools for GC:

- The activity of managing strategic knowledge in maintenance must be structured as an operational action within the organization's own work, for which it must be equipped with the necessary means (human and technological)
- The technological and organizational tools must be simple in their use and oriented towards the operational staff, normally not so used to the use of computer means, and that introduce a synergy in the GC system.
- The technological base or tool where strategic knowledge is collected, must be available in a simple way to all the members of the maintenance organization, promoting its circulation, new acquisition of experiences, and promoting self-learning among all members.
- Complex computer tools are not necessary, but in their creation, operational technicians must participate by marking their opinions and inherent wisdom in their expertise in maintenance tactics(at first with common office tools in all work environment, photo and video cameras, etc., is enough for a uniform start).
- The fundamental thing with the tools that has been used, is to guide them towards the gather of tactical knowledge and know-how (introducing expertise), introducing the best practices and experiences that can be useful to the rest of the people in the maintenance organization that have not experienced in first person.

## 5. THE INCIDENCE OF LEARNING IN PROFESSIONS IN INDUSTRIAL ENVIRONMENTS.

With a shift towards a model based on knowledge and learning, the organization focuses on the ability to innovate and learn, to more efficicently solve their daily work, as well as solve new or non-routine actions, creating a value of the intangible based on knowledge and its rapid update in the scope of the work environment of the maintenance organization. It must be assumed as a long-term development strategy, visualizing knowledge as a strategic factor, so problem solving, and decision making must have a support based on (Peluffo et al, 2002)

- The availability of key information and knowledge in all members of the organization, based on the key tactical actions of industrial maintenance.
- The ability to analyze, classify, model and systematically relate data and information on key values for the mentioned company.
- The ability to build the future of that society in an integral and equitable manner (directionality to goals)

- It must be accompanied by key tranformations in the administration and development of the organization, which focus on:
- The way in which things are done (it tends to be managed by competences rather that by job),
- The ways to address the combination of the use of technology with accumulated individual and organizational knowledge (emphasis of thinking skills, active knowledge search, communities of practice, etc)
- Training and self-learning, for the achievement of skills.
- New ways of communicating knowledge and to build it (tactical stored knowledge, techniques for information analysis, idea banks, knowledge, best practices).
- The cultural change experienced by the acceptance of the benefits of the new model over the traditional one among others (new ways of valuing work, the role of the human factor, the greater autonomy to develope tasks, the alignment between individual and organizational interests).

Within the tactical maintenance context, if we define the management of knowledge as a process to be taken into account within this activity, an approach could be integrated basically, by the generation, coding, transfer and use of knowledge (Nonaka, 1995; Wiig, 1997; Bueno, 2002)

Knowledge generation: study the processes of acquiring external knowledge and creating it in the organization itself, putting into action the knowledge possessed by people.

- *Coding, storage or integration of knowledge:* make available to everyone organizational knowledge, either in writing or by location the person who concentrates it.
- *Transfer of knowledge*: it analysis the spaces of knowledge exchange and the technical processes or platfrom that make it possible. This phase can be carried out through formal or informal communication mechanisms.
- *Use of knowledge:* application of knowledge recently acquired in the routine activities of the company.
- However, on numerous ocations the knowledge gap that usually exists in the maintenance functions is mainly due to the following causes:
    - There is no strong culture of writing and conserving knowledge.
    - It has not been appreciated that a breakdown can be a source of knowledge and that this experience should be capitalized by recording causes, phenomena and actions taken, and normally, due to the inertia of the work performed in an impulsive manner and under strong stress and on numerous occasions before acritical actions.
    - Information is not normally used to gain knowledge. Statistics are not understood as diagnostic tools. Experience, technical skill, and therefore strong tactical knowledge prevail.
    - The management of the company does not take it serious and neither stimulate the work with data.
    - Reliability and maintainability techniques may have some degree of difficulty for the maintenance staff with less experience in industrial etatistics, and who normally performs manual labor.

At the operational level of the maintenance organization itself and after the analysis and review of any formalized sector studies (AEM, 2010 ; Sepi, 2009; INE, 2008), the possible incidence of the use of knowledge management techniques can be considered (table 1) that would helo to soften or minimize

*Table 1. Tactical aspects of maintenance and their impact on knowledge and learning management actions.*

| TACTIC ASPECTS OF MAINTENANCE | POSSIBLE INCIDENCE BY THE ACTION OF KNOWLEDGE MANAGEMENT |
|---|---|
| Reliability, availability in the production / exploitation in the company | The storage, transmission and management of knowledge increases the overall productivity of the company (lower non-programmed stops) |
| Life cycle of equipment and the facilities | Operational information of the equipment that affects its durability and good practices |
| Repairs and recovery | Capturing knowledge of what has been done eliminates unwanted stoppages. Knowledge transmission to other operators. |
| Staff | Recruitment of tactical knowledge of staff based on operational experience. Reduction of coupling times of new person. Recycling aid for existing personnel. |
| Staff qualification and teaching | The training must have an important component on the management of operational experiences in the plant itself creation of self-learning systems. |
| Organizative maintenance tactics | They must be implemented, and capture and transmit the knowledge generated. They must be implemented by the staff themselves. Analysis of data obtained. |
| Preventive / corrective maintenance | Management of experience and knowledge in carrying out maintenance activities. |
| Critics or emergency works | Any experience of urgency or criticism must be recorded. It should serve to learn before future performances. |
| Use of information and its management | Information management must be agile and useful. The records should show the experiences and concerns of the maintenance |
| Energy management and its efficciency | Capture of experiences and good practices. Analysis by maintenance members. Bidirectional |

the negative points observed or mark new lines of action that can make maintenance activities more efficient and, consequiently, greater productivity, efficiency and cost reduction of the entire company.

Taking into account the environment of the maintenance activity itself, the following conclusions can be drawn based on the relevance, characteristics and experience that are implicit in the performance itself:

***Relevance of knowledge:*** knwoledge affects the strategic and tactical elements in maintenance management systems. Maintenance requires very special technical and organizational knowledge, skills and experience; not only theoretical knowledge, but mainly the practical knowledge. An important part of the plant experiences configures tactical knowledge and in general, maintenance management systems do not contain tools to make it explicit. Nor is the transmission of explicit knowledge satisfactory.

***Characteristics of knowledge***: The strategic knowledge (process and failure chain, availability, etc.) and tactics (tactical options and organizational systems) have two fundamental origins: experiences of the industrial plant (a large part of which derives in tactic knowledge) and theoretical approaches (failure modes, reliability theory, replacement of equipment, life cycle, etc.) There would be at least three types of experience: The experience provided by the operational life in the plant, that derived from controlled experiments and explicit or recorded historical knowledge.

***Unregistered experience:*** The knowledge based on experience (tactical) is difficult to extract and formalize, as it is fragmented, complex knowledge, has few regularities, confusing, collected from unforeseen events, guided by emergency, with impositions of time, space, activity not that regulated, and barely "protocolizable", incomplete, isolated, infrequent, local (applicable to specific spaces and situations), and contingent (located in little repeatable scenarios).

***Historical data.*** Historical data is not usually stored selected or filtered much less goal-oriented or on a relational basis, the information they contain is fragmented and is usually unreliable, so its effective utility is usually weak and difficult to transmit.

***Learning and training***. Maintenancec organization systems decisively promote that the acquisition and transmission of knowledge, and its updating, be efficiently and effectively achieved through training, learing and training of human resources, a keystone, especially of the TPM (total productive maintenance)

Learining strategies can be understood as the organized, conscious and intentional set of what the learner does to effectively achieve a learning objective in each context (Gargallo et al, 2009).

## 6. PROCESSES, DIMENSIONS AND TOOLS IN KNOWLEDGE MANAGEMENT

Barragan analyzes the taxonomies of models of knowledge management (Barragan, 2009) where you can find points in common that allow summarize them and reassemble them to homogenize criteria in areas where the study and development of knowledge management have had an important development; These criteria include theoretical, conceptual, philosophical, technical, scientific, cognitive, intellectual capital, knowledge management and social aspects (Cárcel et al, 2016c):.

- The dimensions in the management of the knowledge according to empirical studies and theoretical (Tari & Garcia, 2009), that are directly related to the quality control, can be enumerated in the following:
- Creation of knowledge (organizational learning): acquisition of information, dissemination of information and shared interpretation.
- Transfer and storage of knowledge (organizational knowledge): store knowledge and knowledge transfer.
- Application and use of knowledge (learning organization): teamwork, promote dialogue, establish systems to capture and share learning, relationship between different departments or functional areas and commitment to learning.

Within the context of a company, if we define knowledge as a process management, an approach to this process could be composed basically of the generation, coding, transfer and use of knowledge (Wiig, 1997).

The dimensions in the management of the knowledge according to empirical studies and theoretical (Tari & Garcia, 2009), that are directly related to the quality control, can be enumerated in the following:

- Creation of knowledge (organizational learning): acquisition of information, dissemination of information and shared interpretation.
- Transfer and storage of knowledge (organizational knowledge): store knowledge and knowledge transfer.

- Application and use of knowledge (learning organization): teamwork, promote dialogue, establish systems to capture and share learning, relationship between different departments or functional areas and commitment to learning.

Within the context of a company, if we define knowledge as a process management, an approach to this process could be composed basically of the generation, coding, transfer and use of knowledge (Wiig, 1997).

The generation and transfer of knowledge are processes with a greater amount of tacit knowledge. Both the stage of coding in the use stage, tacit knowledge is converted into explicit knowledge to understand and available to all members of the company itself. (Bueno, 2002) points out that the fundamental aspects of the knowledge management are the creation and distribution of knowledge, although there are qualifications expressed by other authors (Cárcel et al, 2016c) surplus processes involved in the management of knowledge (Quintana, 2006).

## 7. FACTORS IN THE MANAGEMENT OF KNOWLEDGE IN INDUSTRIAL MAINTENANCE

Reviewing the bibliography on analysis of the situation in industrial sectors (Cárcel 2014a, 2014b), the main factors involved in knowledge management in maintenance have been identified, and they have an exponential influence on the overall efficiency of the company, given that they affect operational reliability and have economic repercussions. Although the data shown suggest great similarity between what is seen by technicians and maintenance managers, there are some differences due to the specific nature of the jobs, since maintenance managers are normally educated to University level, are used to using computer-based tools, and have a higher level of organization in their role, as compared to technicians.

The following categories or dimensions to describe the phenomena of the knowledge management in maintenance personnel can be assigned (Cárcel et al, 2016c):

- Acquisition and generation of internal and external knowledge: activities from the outside or inside that support organizational processes and provide new knowledge on the maintenance activity.
- Important elements of the knowledge generation: individual and organisational actions that support or limit the generation of knowledge.
- Transfer of knowledge through formal and informal mechanisms: formal mechanisms encoded or stored knowledge in databases, documents or software tools, which you can access and easily used by any member of the Organization (Hansen, Nohria & Tierney, 1999). Informal mechanisms that help transfer knowledge through direct contact from person to person.
- Important elements of knowledge transfer: individual and organisational actions that support or limit the transfer of knowledge.
- Use of knowledge: the application of newly acquired knowledge in routine activities in the company.
- Tools for knowledge management: use of audits and knowledge maps that improve the processes in which the maintenance activity is involved.

## 8. CONCLUSION

This chapter focuses on the ways that maintenance organizations of companies generate, transfer and use their knowledge, and the impact it can produce on the organization as a whole. These processes are characterized by a Kantian process (people, means and environment).

Both the acquisition of external knowledge and internal knowledge creation are important activities to generate that knowledge to critical actions (breakdowns, emergencies, etc.) not cyclical, can assume an important strategic value. All of them are aware that having a proper knowledge management, positively affects the following actions performed by maintenance:

- Capture of strategic tacit knowledge of technical maintenance operations.
- Resolution of critical faults in shortest time (especially the non-cyclic).
- Reduction of the times of operational maneuvers.
- Facilitate area change or replacement of staff.
- Decrease of coupling again personal times.
- Capture of information and transfer of subcontractors.
- Share knowledge of employees who can be used by others that can detect new opportunities for improvement.
- Improvement of the knowledge of the reliability of the equipment and facilities.
- Improvement of the knowledge for the detection and improvement of energy efficiency actions.
- Optimization of the time, which is again in the knowledge management and the reduction of the maintenance costs.

Self-learning is key in this kind of activity in a technological environment and with demand for rapid and efficient action. The knowledge management is enhanced by a proactive and participatory management style that promotes the emergence of new ideas and work processes. Likewise, this organizational culture must be opened, allowing the address encourage employees to share their knowledge and to facilitate communication between the members of the company. These findings are supported by studies (O'Dell & Grayson, 1998; Ruggles, 1998) where it was noted that companies with an open culture that motivates to generating and sharing knowledge will have more success in the implementation of these processes.

A greater use of informal meetings as a means of generation and transfer of knowledge, especially among the groups of technical operations, with a minor organizational culture that managers or heads of maintenance is detected.

Audits (maintenance, knowledge, energy, etc.), tend not to be used, what shows that the application of these techniques would enhance a first process in the elaboration of a global strategy of knowledge management. To the maintenance organization perform successfully the replica of his know-how, by technical operations, require agile and simple mechanisms that allow them to share their experiences, which generate knowledge quickly and efficiently.

Similarly, has been detected that on numerous occasions, the documentation for use in their activities, tends to be disaggregated, and often not up to date.

The key barriers identified for the proper management of knowledge have been the availability of time to properly document important actions, cultural barriers with a culture based on "own knowledge", not shared, especially in technical operations (with a high component of knowledge, tacit and therefore not registered), as well as getting the total involvement of the staff they are barriers in the present study.

The need for the figure of a "Knowledge Manager", as an important facilitator is confirming in the catchment of the transfer and use of knowledge within maintenance departments. This figure should be a person with training technical, organisational and notions of knowledge management, with extensive experience in the operational area of maintenance (who knows first hand deep in the factors influencing their work), and which brings together all the efforts of the maintenance organization to manage a strategic knowledge that can be used throughout the Organization not being an accidental Manager (Dorfman, 2001), but with a partial or total dedication that mark the continuity of the project. Since innovation is a knowledge intensive process, this new concept will be necessary to rethink many of the management decisions associated with him (Gonzalez & Garcia, 2010).

The foregoing suggests that the knowledge that you want to transfer needs to be a priority in the activity of the industrial maintenance, i.e. must be planned and included in the strategic planning of the company. The power of tacit knowledge is combined with explicit knowledge and objective continuously to go to generating new knowledge (Arbonies & Calzada, 2004).

There is a high tactic component among all the human components of the maintenance activity, a value that still measured from the subjective view of the participants cab be considered to range between 50 and 90%, depending on the area of activity of the company and the extension of the work area. The tactic knowledge rate is given to a greater extent when the area of performance of the maintenance activity is greater. Operations experiences, which is an internal knowledge that is accumulated by all the members that operate in maintenance, and hardly transmitted, causes that in the periods of coupling and that person is really operative. Apart from the operational and economic loss that involves the coupling of a new person, experienced staff must use part of their time in the training of new members, that latter must be assessed from an economic or lack of effectiveness during these processes. These repercussions of operability and economic value suppose a great value in companies with high cariation of the staff, and it is seen the need to look for learning systems based in this knowledge, which motivates their monitoring by the staff itself, and improves their work processed.

## ACKNOWLEDGMENT

This work has been conducted within the framework of the two research project: "Monitoring, control and discrimination of electrical loads connected to renewable generation hybrid systems in configuration on island", within the programme for the promotion of the scientific research, technological development and innovation in the Comunitat Valenciana [2017/647] funded by the Conselleria de Educación, Investigación, Cultura y Deporte of the GeneralitatValenciana and reference number: GV/2017/023; and research project "Investigación experimental de la integración de sistemas renovables híbridos para un desarrollo sostenible de países emergentes" funded by Universidad Politècnica de València, under grant Primeros Proyectos de Investigación (PAID-06-18), Vicerrectorado de Investigación, Innovación y Transferencia de la Universitat Politècnica de València (UPV).

## REFERENCES

Ackert, S. P. (2010). Basics of aircraft maintenance programs for financiers—evaluation & insights of commercial aircraft maintenance programs. www. aircraftmonitor. com/uploads/1/5/9/9/15993320/ basics_of_aircraft_maintenance_programs_for_ financiers___v1. pdf

Al-Najjar, B., & Kans, M. (2006). A model to identify relevant data for problem tracing and maintenance cost-effective decisions. A case study. *International Journal of Productivity and Performance Management, 55*(8), 616–637. doi:10.1108/17410400610710170

Alsyouf, I. (2007). The role of maintenance in improving companies, productivity and profitability. *International Journal of Production Economics, 105*(1), 70–78. doi:10.1016/j.ijpe.2004.06.057

Arbonies, A., & Calzada, I. (2004). El poder del conocimiento tácito: Por encima del aprendizaje organizacional. *Intangible Capital, 6*(0), 1–17.

Baez, J. (2007). *Investigación cualitativa.* Esic Editorial.

Bahoque, E., Gómez, O., & Pietrosemoli, L. (2007). Gestión del Conocimiento en la Industria de la Construcción: Estudio de un caso. Revista Venezolana de Gerencia (RVG). *Año, 12*(39), 393–409.

Barlow, R. E., Hunter, L. C., & Proschan, F. (1963). Optimum checking procedures. *Journal of the Society for Industrial and Applied Mathematics, 11*(4), 1078–1095. doi:10.1137/0111080

Barragán, A. (2009). Aproximación a una taxonomía de modelos de gestión del conocimiento. *Intangible Capital, 5*(1), 65–101. doi:10.3926/ic.2009.v5n1.p65-101

Boland, P. J., & Proscan, F. (1982). Periodic Replacement with Increasing Minimal Repair Cost at Failure. *Operations Research, 30*(6), 1083–1089. doi:10.1287/opre.30.6.1183

Bueno, E. (1999). *Gestión del conocimiento y capital intelectual' Experiencias en España.* Comunidad de Madrid- Euroforum Escorial Madrid.

Bueno, E. (2000). *La dirección del conocimiento en el proceso estratégico de la empresa: información, complejidad e imaginación en la espiral del conocimiento.* Euroforum Escorial.

Bueno, E. (2002a). Enfoques principales en Dirección del Conocimiento (Knowledge Management) y tendencias. In R. Hernández (Ed.), *Gestión del Conocimiento: Desarrollos teóricos y aplicaciones.* Ediciones La Coria, Fundación Xavier de Salas.

Bueno, E. (2002b). La sociedad del conocimiento: un nuevo espacio de aprendizaje de las personas y organizaciones. In *La Sociedad del Conocimiento.* Monografía de la Revista Valenciana de Estudios Autonómicos.

Buttery, L.M. (1978, Jan.). New survey of U.S. Maintenance Costs. Hydrocarbon Processing.

Cárcel Carrasco, F. J. (2014b). La gestión del conocimiento en la ingeniería del mantenimiento industrial: Investigación sobre la incidencia en sus actividades estratégicas. Omnia Science. doi:10.3926/oms.197

Cárcel Carrasco, F. J., & Grau, J. (2012c). Aspectos estratégicos del mantenimiento industrial relativos a la eficiencia energética. *Revista Mantenimiento, 257*, 25–30.

Cárcel-Carrasco, F. J., Pascual-Guillamón, M., Salas-Vicente, F., & Donderis-Quiles, V. (2019a). Influence of Heat Treatment in the Microstructure of a Joint of Nodular Graphite Cast Iron when Using the Tungsten Inert Gas Welding Process with Perlitic Grey Cast Iron Rods as Filler Material. *Metals*, *9*(1), 48. doi:10.3390/met9010048

Cárcel-Carrasco, F. J., Pascual-Guillamón, M., Solano García, L., Salas Vicente, F., & Pérez-Puig, M. A. (2019b). Pitting Corrosion in AISI 304 Rolled Stainless Steel Welding at Different Deformation Levels. Applied Sciences (Basel, Switzerland), 9(16), 3265. doi:10.3390/app9163265

Carcel-Carrasco, F. J., & Rodriguez-Mendez, M. (2013a). *Industrial maintenance and tacit knowledge: An introduction about its incidence*. DYNA Management., doi:10.6036/MN5857

Cárcel Carrasco, J. (2014d). Planteamiento de un modelo de mantenimiento industrial basado en técnicas de gestión del conocimiento. Omnia Science, 2014. Advance online publication. doi:10.3926/oms.198

Cárcel-Carrasco, J., Pascual-Guillamón, M., & Pérez-Puig, M. (2016b). Effects of X-rays radiation on AISI 304 stainless steel weldings with AISI 316L filler material: A study of resistance and pitting corrosion behavior. *Metals*, *6*(5), 102. doi:10.3390/met6050102

Cárcel Carrasco, J., Peñalvo López, E., & de Murga, G. (2018). Oficinas auto-sostenibles para las agencias de ayuda internacional en zonas geográficas remotas. *Dyna (Bilbao)*, *93*(3), 272–277. doi:10.6036/8507

Cárcel-Carrasco, J., Pérez-Puig, M. A., Pascual-Guillamón, M., & Pascual-Martínez, R. (2016a). An Analysis of the Weldability of Ductile Cast Iron Using Inconel 625 for the Root Weld and Electrodes Coated in 97.6% Nickel for the Filler Welds. *Metals*, *6*(11), 283. doi:10.3390/met6110283

Cárcel Carrasco, J., Rodríguez Méndez, M., & Carnero Moya, M. C. (2016c). Impact of the knowledge management in maintenance engineering. Effects on industrial production. In *Optimum Decision Making in Asset Management*. IGI Global.

Cegarra, J., & Moya, B. (2003). Orientadores del capital relacional. *Cuadernos Americanos*, *16*(26), 79–97.

Charmaz, K. (2006). Constructing grounded theory. A practical guide through qualitative analysis. *Sage (Atlanta, Ga.)*.

Cheong, K. F. R., & Tsui, E. (2010a). The roles and values of personal knowledge management: An exploratory study. *VINE Journal of Information and Knowledge Management Systems*, *40*(2), 204–227.

Cheong, K. F. R., & Tsui, E. (2010b). Exploring the synergy between business process management and personal knowledge management. *Cutter IT Journal*, *23*(5), 28–33.

Christer, T. (1981). Ways of assessing and improving maintenance performance. PEMEC'81.

Chua, A. Y. K., & Goh, D. H. (2008). Untying the knot of knowledge management measurement: A study of six public service agencies in Singapore. *Journal of Information Science*, *34*(3), 259–274. doi:10.1177/0165551507084139

Colino, J., Martinez, J., & Martínez-Carrasco, F. (2010). La gestión de la innovación en la industria. El caso de la región de Murcia. *Economía Industrial*, *377*, 150–158.

Colino, J., & Riquelme, P. (2000). Estructura industrial y desarrollo tecnológico en la región de Murcia. *Economía Industrial*, *335-336*, 171–183.

Conde, J. (1999). El Mantenimiento efectivo: principios y métodos. Working paper, GIO-0500-UCLM, Ciudad Real.

Crespo, A. (2004). *Ingeniería de mantenimiento. Técnicas y métodos de aplicación a la fase operativa de los equipos*. AENOR.

Darnell, H., & Bert, A. (1978). *The role of maintenance management in achieving industrial efficiency, Maintenance'78*. BMCA.

Dataware Technologies. (1998). Seven Steps to Implementing Knowledge Management in Your Organization. Manual de trabajo.

Del Moral, A. (2007). *Gestión del Conocimiento*. Thompson Editores.

Dorfman, P. (2001). The accidental knowledge manager. *Knowledge Management*, *4*(2), 36–41.

Douglas, D. (2004). Grounded theory and the 'And' in entrepreneurship research. *Electronic Journal of Business Research Methods*, *2*(2).

Edvinsson, L., & Malone, M. S. (1997). *Intellectual Capital: Realizing Your Company's True Value by Finding its Hidden Brainpower*. Harper Business.

Eich, D. (2008). A Grounded Theory of High-Quality Leadership Programs: Perspectives From Student Leadership Development Programs in Higher Education. *Journal of Leadership & Organizational Studies*, *15*(2), 176–187. doi:10.1177/1548051808324099

Eisenhardt, K. (1989). Building theories from case studies research. *Academy of Management Review*, *14*(4), 532–550. doi:10.5465/amr.1989.4308385

Ferrada, X., & Serpell, A. (2009). La Gestión del Conocimiento y la Industria de la Construcción. *Revista de la Construcción*, *8*(1), 46–58.

Glaser, B. G., & Strauss, A. L. (1967). *The discovery of grounded theory*. Aldine deGruyter.

Gómez, F., Valcuende, M., Matzarakis, A., & Cárcel Carrasco, J. (2018). Design of natural elements in open spaces of cities with a Mediterranean climate, conditions for comfort and urban ecology. *Environmental Science and Pollution Research International*, *25*(26), 26643–26652. doi:10.1007/s11356-018-2736-1 PubMed

González, R., & García, A. (2010). Innovación abierta: Un modelo preliminar desde la gestión del conocimiento. *Intangible Capital*, *7*(1), 82–115. doi:10.3926/ic.2011.v7n1.p82-115

Greer, R. W. (1960). *Records Make the Difference* (Vol. 65). Coal Age.

Hanks, H. (1961). Program for Maintenance of Mobile Equipment. *Min. Congress Joum.*, *47*(9), 35–38.

Hansen, M. T., Nohria, N., & Tierney, T. (1999, Mar.). What's your strategy for managing knowledge? Harvard Business Review, 106–116. PubMed

Hardy, M., & Bryman, A. (2004). *Handbook of data analysis*. Sage Publications., doi:10.4135/9781848608184.

Hedlud, O. (1994). A model of knowledge management and the N-form corporation. *Strategic Management Journal, 15*(S2), 73–90. doi:10.1002/smj.4250151006

Hiatt, B. (1999). *Best Practices Maintenance - A 13 Step Program in Establishing a World Class Maintenance Organization - USA*. Academic Press.

Hooff, B., & Huysman, M. (2009). Managing knowledge sharing: Emergent and engineering approaches. *Information & Management, 46*(1), 1–8. doi:10.1016/j.im.2008.09.002

Idhammar, C. (1997, July). Maintenance management: moving from reactive to results-oriented. Journal Review Pima's Papermaker.

Jonsson, P. (1997). The status of maintenance management in Swedish manufacturing firms. *Journal of Quality in Maintenance Engineering, 3-4*(4), 233–258. doi:10.1108/13552519710176863

Kans, M. (2009). The advancement of maintenance information technology A literature review. *Journal of Quality in Maintenance Engineering, 15*(1), 5–16. doi:10.1108/13552510910943859

Kogut, B., & Zander, U. (1992). Knowledge of the firm, combinative capacities, and the replication of technology. *Organization Science, 7*(3), 502–517.

Kulkarni, U., Ravidran, S., & Freez, E. R. (2007). A knowledge management success model: Theoretical development and empirical validation. *Journal of Management Information Systems, 23*(3), 309–347. doi:10.2753/MIS0742-1222230311

Lazakis, I., Turan, O., & Aksu, S. (2010). Increasing ship operational reliability through the implementation of a holistic maintenance management strategy. *Journal of Ships and Offshore Structures, 5*(4), 337–357. doi:10.1080/17445302.2010.480899

Lin, H. F. (2007). A stage model of knowledge management: An empirical investigation of process and effectiveness. *Journal of Information Science, 33*(6), 643–659. doi:10.1177/0165551506076395

Lopez, P., & Centeno, G. (2006). Integrated system to maximize efficiency in transit maintenance departments. *International Journal of Productivity and Performance Management, 55*(8), 638–654. doi:10.1108/17410400610710189

López Campos, M., Fumagalli, L., Gómez Fernández, J., Crespo Márquez, A., & Macchi, M. (2010). UML model for integration between RCM and CBM in an e-maintenance architecture. Proceedings of the First IFAC Workshop on Advanced Maintenance Engineering Services and Technology (A-MEST), 133-138. doi:10.3182/20100701-2-PT-4012.00020

Martin, J. (2008). Personal knowledge management: the basis of corporate and institutional knowledge management. In J. Martin & K. Wright (Eds.), *Managing Knowledge: Case Studies in Innovation*. Spotted Cow.

Martinez, I., & Ruiz, J. (2002). Los procesos de creación del conocimiento: El aprendizaje y la espiral de creación del conocimiento. *XVI Congreso Nacional de AEDEM*. Alicante 2002.

Maslow, A. (1943). A Theory of Human Motivation. Originally Published in Psychological Review, 50(4), 370–396. doi:10.1037/h0054346

Matteson, D. (1985). Airline experience with reliability-centered maintenance. *Nuclear Engineering and Design*, *89*(2), 385–390. doi:10.1016/0029-5493(85)90076-7

Miles, M., & Huberman, A. (1994). *Qualitative Data Analysis: A sourcebook of new methods*. Sage Publications.

Moubray, J. (1991). *Reliability-Centered Maintenance*. Butterworth-Heinemann.

Muller, A., Crespo Márquez, A., & Iung, B. (2008). On the concept of e-maintenance: Review and current research. *Reliability Engineering & System Safety*, *93*(8), 1165–1187. doi:10.1016/j.ress.2007.08.006

Muñoz, J. (1999). Sobre gestión del conocimiento, un intangible clave en la globalización. Economía Industrial, 330.

Nakajima, S. (1988). *Introduction to TPM*. Productivity Press.

Nonaka, I. (1991, November). The knowledge-creating company. *Harvard Business Review*, •••, 96–104.

Nonaka, I. (1994). A dynamic theory of organizational knowledge creation. Organizatioll Saence, 5(1), 14–37. doi:10.1287/orsc.5.1.14

Nonaka, I., & Konno, N. (1998). The concept of "Ba": Building a foundation for knowledge creation. *California Management Review*, *40*(3), 40–54. doi:10.2307/41165942

Nonaka, I., & Takeuchi, H. (1995). *The knowledge-Creating Company: How Japanese Companies Create the Dynamics of Innovation*. Oxford University Press.

Nonaka, I., & Takeuchi, H. (1999). *La Organización Creadora de Conocimiento*. Academic Press.

Norton, D., & Kaplan, R. (1996). *The Balanced Scorecard: Translating Strategy into Action*. Harvard Business School Press.

O'Dell, C., & Grayson, C. J. (1998). If only we knew what we know: Identification and transfer of internal best practices. *California Management Review*, *40*(3), 154–170. doi:10.2307/41165948

OCDE. (1996). *The Knowledge-Based Economy*. OCDE.

OCDE. (2004). *Medición de la gestión de conocimientos en las empresas: primeros resultados*. Ed. OCDE.

Oiltech Analysis S.L. (1995). Mantenimiento Proactivo de sistemas mecánicos lubricados. Fluidos oleohidráulica neumática y automación, 24(208-209).

Ordoñez de Pablos, P. (1999). *Gestión del conocimiento y medición del capital intelectual en la empresa internacional*. Proyecto de Doctorado, Universidad de Oviedo.

Ordoñez de Pablos, P. (2001). *Capital intelectual, gestión del conocimiento y sistemas de gestión de recursos humanos: Influencia sobre los resultados organizativos (Tesis Doctoral)*. Universidad de Oviedo.

Ordoñez de Pablos, P., & Parreños, J. (2003). Aprendizaje organizativo en el contexto internacional: Implicaciones para la gestión del conocimiento. *Investigaciones Europeas de Dirección y Economía de la Empresa*, *9*(2), 205–216.

Ordoñez de Pablos, P., & Parreños, J. (2005). Aprendizaje organizativo y gestión del conocimiento: Un análisis dinámico del conocimiento en la empresa. *Investigaciones Europeas de Dirección y Economía de la Empresa*, *11*(1), 165–177.

Pace, S. (2004). A grounded theory of the flow experiences of Web users. *International Journal of Human-Computer Studies*, *60*(3), 327–363. doi:10.1016/j.ijhcs.2003.08.005

Palacios, D., & Garrigos, F. (2006). Propuesta de una escala de medida en la gestión del conocimiento en las industrias de biotecnología y telecomunicaciones. *Investigaciones Europeas de Dirección y Economía de la Empresa*, *12*(1), 207–224.

Partington, D. (2000). Building grounded theories of management action. *British Journal of Management*, *11*(2), 91–102. doi:10.1111/1467-8551.00153

Pauleen, D. (2009). Personal knowledge management: Putting the "person" back into the knowledge equation. *Online Information Review*, *33*(2), 221–224. doi:10.1108/14684520910951177

Pavez, A. S. (2000). *Modelo de implantación de Gestión del Conocimiento y Tecnologías de Información para la Generación de Ventajas Competitivas*. Universidad Técnica Federico Santa María.

Peluffo, M., & Catalán, E. E. (2002). *Introducción a la gestión del conocimiento y su aplicación al sector publico*. Ed. Instituto Latinoamericano y del Caribe de Planificación.

Peña, P. (2001). *To know or no to be. Conocimiento, el oro gris de las organizaciones*. DINTEL.

Peñalvo-López, E., Cárcel-Carrasco, F., Devece, C., & Morcillo, A. (2017). A methodology for analysing sustainability in energy scenarios. *Sustainability*, *9*(9), 1590. doi:10.3390/su9091590

Peñalvo-López, E., & Cárcel-Carrasco, F. J. (2019a). An Analysis about Learning to Increase Women's Participation and Employment in Europe's Energy Transition: Evidence from the European Project MEnS. *Sustainability*, *11*(16), 4345. doi:10.3390/su11164345

Peñalvo López, E., Pérez-Navarro, A., Hurtado Pérez, E., & Cárcel Carrasco, J. (2019b). Comprehensive Methodology for Sustainable Power Supply in Emerging Countries. *Sustainability*, *11*, 1–22. doi:10.3390u11195398

Peris, F., Loria, B., & Méndez, M. (2002). Creación de conocimiento y diseño de organizaciones: Equidad, confianza y objetivos compartidos como reto de la gestión del conocimiento. Cuadernos de CC.EE. y EE., 43, 41-56.

Petrides, L. A., & Nodine, T. R. (2003). *Knowledge Management in Education: Defining the Landscape*. March Institute for the Study of Knowledge Management in Education.

Polanyi, M. (1958). *Personal Knowledge: Towards a Post-Critical Philosophy*. University of Chicago Press.

Polanyi, M. (1966). *The tacit dimension*. Routledge & Kegan Paul.

Reiner, J., Koch, J., Krebs, I., Schnabel, S., & Siech, T. (2005). Knowledge management issues for maintenance of automated production systems. Proceedings of the IFIP International Conference on Human Aspects in Production Management, 160, 229—237. doi:10.1007/0-387-23078-5_18

Rivas, L., & Flores, B. (2007). La gestión de conocimiento en la industria automovilística. Estudios Gerenciales, 23(102), 83–100. doi:10.1016/S0123-5923(07)70003-7

Rodríguez, D. (2006). Modelos para la creación y gestión del conocimiento: Una aproximación teórica. *Review of Education*, *37*, 25–39.

Ruggles, R. (1998). The state of the notion: Knowledge management in practice. *California Management Review*, *40*(3), 80–89. doi:10.2307/41165944

Sánchez, M., Chaminade, C., & Escobar, C. (1999). En busca de una teoría sobre medición y gestión de los intangibles en la empresa: Una aproximación metodológica. *Ekonomiaz*, *45*, 188–213.

Serratella, C. M., Wang, G., & Conachey, R. (2007). Risk-based strategies for the next generation of maintenance and inspection programs. *International Symposium on Maritime, Safety, Security and Environmental Protection (SSE)*, 20–21.

Sols, A. (2000). Fiabilidad, Mantenibilidad, Efectividad, un enfoque sistémico. Comillas.

Tari, J., & Garcia, M. (2009). Dimensiones de la gestión del conocimiento y la gestión de la calidad: Una revisión de la literatura. *Investigaciones Europeas de Dirección y Economía de la Empresa*, *15*(3), 135–148. doi:10.1016/S1135-2523(12)60105-1

Tiemessen, L., Lane, H. W., Crossan, M. M., & Yinkpen, A. C. (1997). Knowledge management in international joint ventures. In *North American perspectives. The Cooperative Strategies Series*. The New Lexington Press.

Tirpak, T. M. (2005). Five steps to effective knowledge management. *Research Technology Management*, *48*(3), 15–16. doi:10.1080/08956308.2005.11657310

Turan, O., Lazakis, I., Judahb, S., & Incecika, A. (2011). Investigating the Reliability and Criticality of the Maintenance Characteristics of a Diving Support Vessel. *Quality and Reliability Engineering International*, *27*(7), 931–946. doi:10.1002/qre.1182

Ventura, J. (1996). *Análisis dinámico de la estrategia empresarial: Un ensayo interdisciplinar. Servicio de publicaciones*. Universidad de Oviedo.

Ventura, J., & Ordóñez de Pablos, P. (2007). Análisis de estrategias de conocimiento en la industria manufacturera española: Evidencia empíricas. *Tribuna de Economía*, *836*, 141–161.

Volkel, M., & Abecker, A. (2008). Cost-benefit analysis for the design of personal knowledge management systems. ICEIS 2008—International Conference on Enterprise Information Systems, 95–105.

Volkel, M., & Haller, H. (2009). Conceptual data structures for personal knowledge management. *Online Information Review*, *33*(2), 298–315. doi:10.1108/14684520910951221

Wah, L. (1999). Behind the Buzz. *Management Review*, *88*(4), 16–19.

Wernerfelt, B. (1984). A resource based view of the firm. *Strategic Management Journal*, *5*(2), 171–180. doi:10.1002/smj.4250050207

Xiomara, P. (2009). La gestión del conocimiento y las Tics en el siglo XXI. Revista Universitaria de Investigación y Diálogo Académico, 5(1).

Yang, J. (2006). La estrategia de gestión del conocimiento y su efecto en el crecimiento corporativo. *Economía Industrial*, *362*, 123–133.

Yin, R. K. (1995). *Case study research: design and methods*. Sage Publications.

Zaphiropoulos, E. P., & Dialynas, E. N. (2007). Methodology for the optimal component selection of electronic devices under reliability and cost constraints. *Quality and Reliability Engineering International*, *23*(8), 885–897. doi:10.1002/qre.850

## KEY TERMS AND DEFINITIONS

**Audits of Maintenance:** Maintenance assessment applied to an organization to identify which aspects are in need of improvement. There is a distinction between technical audits, which assess the condition of the facilities, and management audits, which identify the level of excellence of the maintenance department and its management.

**Experience Curve:** Assessment of improvement in efficiency or productivity through learning as an activity is repeated.

**Explicit Knowledge:** Knowledge that can be processed or transmitted easily through formal language.

**Industrial Maintenance:** Production support system whose aim is to guarantee optimum working of facilities and productive systems at minimum cost.

**Know-How:** Pre-existing knowledge in an organization, in the form of techniques, procedures, information, etc. which has in many cases been obtained by experience or repetition in carrying out activities, processes or decision making.

**Knowledge:** Stored information which has been acquired by experience, training, practice or deduction (introspection) and which can produce more knowledge or solve a problem.

**Tacit Knowledge:** Knowledge acquired through habit, experience or culture, including subjective elements such as personal beliefs, intuition, etc. this kind of knowledge is difficult to process or transmit systematically.

# Chapter 3
# Decision-Making Support on Energy Management in District Heating

**Vira Shendryk**
https://orcid.org/0000-0001-8325-3115
*Sumy State University, Ukraine*

**Yuliia Parfenenko**
https://orcid.org/0000-0003-4377-5132
*Sumy State University, Ukraine*

**Olga Aleksenko**
https://orcid.org/0000-0003-2355-6362
*Sumy State University, Ukraine*

**Sergii Shendryk**
https://orcid.org/0000-0002-2034-7920
*Sumy State University, Ukraine*

**Iryna Baranova**
https://orcid.org/0000-0002-3767-8099
*Sumy State University, Ukraine*

## ABSTRACT

*The chapter focuses on implementation of Industry 4.0 in energy management of a district heating system. The present objectives for implementation of digital technologies in district heating are analyzed. The barriers for sustainable and flexible district heating systems and challenges to overcome them are discussed. The objectives of energy management in district heating are presented. The realization of energy management stages in district heating tackling the digitalization challenge are presented. The*

DOI: 10.4018/978-1-7998-3246-1.ch003

*chapter explores approaches and techniques to information support of decision making in energy management under conditions of uncertainty. The chapter describes the decision support system "HeatCAM" as a tool of energy management in district heating. This system will provide real-time monitoring and making decisions on regulation of heat consumption modes of the building on the consumer side in the condition of uncertainty.*

## INTRODUCTION

The heating sector which makes up more than 50% of the whole energy demand is currently under transformation because of Industry 4.0 implementation (Recommendations for implementing the strategic initiative INDUSTRIE 4.0, 2013; Reducing your ICT energy costs, 2015). The main objective of the transition to District energy 4.0 is energy performance increasing. Digital technologies are aimed to make the whole energy system and, in particular, district heating system smarter, more efficient, and reliable and to boost the integration of more renewables into the system.

The impact of usage IoT devices in District Heating 4.0 such as intelligent sensors and smart meters lays in acquisition the large data amounts of monitored heat consumption and changes in external factors affecting the needs in heat energy that may be used for decision support on heat energy demand regulation.

Making decisions on the management of the heating system is carried out in the condition of uncertainty. In relation to demand side energy management the timely regulation of the heat energy demand for heating buildings eliminates the excess of the heat consumption due to the fluctuations in temperature of the environment and individual heat energy needs for each building. The implementation of decision support system (DSS) based on machine learning approaches on data processing in the management of district heating system can improve the efficiency of such decisions.

The Chapter will focus on determination the approaches of increasing district heating energy performance. The decision support models and tools for smart energy management in district heating will be analyzed. The DSS for making decision support in regulation of district heating on the consumer side will be presented. This system will provide the real-time monitoring and making decisions on regulation of heat consumption modes of the building on the consumer side in the condition of uncertainty.

## DISTRICT ENERGY 4.0 CONCEPT AND ITS IMPLEMENTATION

District heating (DH) plays a significant role in the supply of low-carbon heating and hot tap water all over the world. DH technology is most common in the countries with cold climate where the heat demand is large. It is particularly widespread in North, Central and Eastern Europe. In such countries as Iceland, Finland and Sweden the high amount of residential consumers are connected to DH system. In cities like Copenhagen, Helsinki, Warsaw, Vilnius, Riga as much as 90% of residential heat demands are satisfied by DH.

The DH system is a complex technological and socio-economic system, with heat generated in a central source and supplied for commercial and residential consumers to support comfort indoor air temperature in heated rooms. The consumers of the heat energy from the district heating systems in cities are usually houses, public sector buildings (organizations of education, science, culture and health) and some industrial enterprises.

The DH system generally consists of three main elements: the heat sources, the distribution system and the consumer. The heat carrier (hot water) circulates in pipes of the DH system and transmits the heat energy from the heat source to the consumer in heating systems. DH network as a rule is two-pipes, where there is one supply and one return pipeline.

The benefits of DH by comparison with other heating options are listed as follows (Euroheat & Power Task Force Customer Installations, 2011; the Guidance for district heating companies, 2012; Skagestad & Mildenstein, n.d.):

- **Equipment and maintenance** – the usage of DH at the customer end requires the installation equipment with compact size which is simple to run and maintain;
- **High reliability and efficiency** – DH systems are built with stand-by heating capacity to ensure that heat is always available at the central plant. The extensive range of primary fuels that can be utilized by DH makes heating supply very reliable. Renewable fuels can also be utilized.
- **Cost** – DH entails moderate investment costs and very low maintenance costs for the end consumer;
- **Enviromental issues** - using renewable and other low carbon fuels in heat generation, combined heat and power generation from a single fuel source allows to reduce primary energy use greenhouse gas emissions;
- **Comfort for consumers** - no need to purchase and handle fuels, detect problems in the heating system and eliminate them.

Despite the prevalence of DH systems, there are some challenges that need to be solved for the prosperous usage of these systems in the future. The first challenge is to decrease heat losses when its transfer from generation unit to end consumer. Heat losses in DH distribution depend on a number of factors such as the length of the system in relation to the heat load, type of insulation and temperature level. Normally, heat losses are in the range of 5-20% annually (Skagestad & Mildenstein, 2019). Due to heat losses it is not economically viable to transfer heat over long distances. Therefore, a DH system of every country is in practice composed of a large number of discrete DH networks. The second deals with decreased heat demand on the consumer side (District Heating System Management Guide, n.d.). Another challenge associated with necessity to reduce or completely avoid fossil fuel combustion in energy generating systems. The global aim for near future is to increase share of renewable and waste energy.

The DH system efficiency may be achieved through decreasing temperature levels, network geometry optimization and using local renewable energy. Implementation of discrete energy saving measures does not bring the desired result to ensure DH systems development.

Digital technology has been increasingly adopted by all major industry sectors over the last several years and the energy industry is no exception. The age of Industry 4.0 is starting to change the way goods are produced and services are provided. Industry 4.0 incorporates automation and data that is used for optimization of production, enhanced flexibility and efficiency within a smart factory environment. Implementation of the latest digital technology advances to energy sector has a great potential of sustainable DH development and energy saving. It relates to reinvention of the ways that energy and utilities companies operate and interact with their consumers.

The digital revolution in energy sector is called Energy 4.0 (From Industry 4.0 to Energy 4.0: Future Business Models and Legal Relations, 2016). It involves emerging intelligent advances and innovations to build smart grids, manage renewable energy and distributed generation. Simultaneously, innovative business-driven solutions are implemented with emphasis on internal reliability of DH system and ecological safety.

Nowadays district heating is in transformation to 4.0 generation. The District Heating 4.0 (4GDH) system in the frame of overall Energy 4.0 approach is defined as a coherent technological and organizational concept, which by means of smart thermal grids assists the appropriate development of sustainable energy systems. It may be considered as an institutional framework have been being implemented based on intelligent and digital innovations of Industry 4.0 with the aim of achieving an optimal balance between investments in savings versus heat production and an optimal integration of fluctuating renewable energy in the overall energy system. 4GDH systems development aims to provide the heat supply with low heat network losses in a way in which the use of heat sources is integrated with the operation of smart energy systems.

Within the scope of Energy 4.0 framework, various innovative concepts for the provision of heat may be implemented both technically and managerially on all stages of district heating process - production, distribution and consumption. The generalized scheme of the District Heating 4.0 is presented in Figure 1.

*Figure 1. The scheme of the District Heating 4.0*

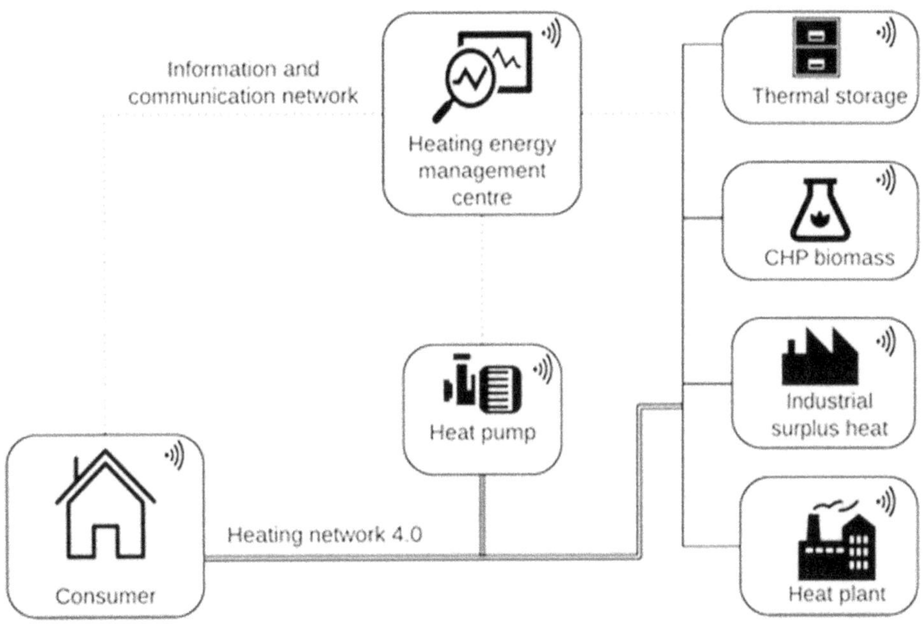


Primarily, digitalization of district heating means the large-scale rollout of smart devices and sensors, and the use of big data collection and analysis. Modern 4GDH system is inherently flexible in terms of source, size and load (Lyons, 2019).

Digitalization of district heating provides a great opportunity to efficiently manage of heat generation from different sources such as thermal storages, biomass fuel, industrial surplus heat in combination with traditional way of district heat energy production.

Heat generation units equipped with automation devices that enable data management and control relating to temperatures, flows, pressure and leak detection. Heat pumps with thermal and battery storage as a component of heating system can provide additional heating for several hours or days. They should be equipped with sensors of that can understand the necessity in heat to achieve consumer comfort and smart controls to optimize the heat demand. On the consumer side controllers would help manage heat demand, enabling more dynamic control of system temperatures, which would reduce heat loss and therefore primary energy demand. Also, consumers should be equipped with smart meters.

All installed devices should be incorporated in single network based on IoT (Internet of Things) approach. It allows their real-time remote monitoring and control by energy managers or households via smartphones or tablets. The key trends of digitalization and their impact to district heating is presented in Table 1 (Make a game-changing impact – before the others do. The impact of digitalization in district heating, 2016).

*Table 1. The Digitalization in District Heating*

| Process | District heating production | District heating distribution | District heating consumption |
|---|---|---|---|
| Actors | District suppliers, logistic partners, facility designers | Network managers, designers, housing companies | Customers, building energy managers |
| Key elements of digitalization | Intelligent sensors, augment reality, analytics and big data, robotics | IoT, blockchain technology | Smart meters, mobile devices |
| Impact | Predictive heat production, service and maintaining, more efficient investments and energy trading, the development of new revenue models, products and services | Remote control and monitoring of the heat network, predictive service and maintaining, the optimal management of network load including local heating suppliers | More complicated data management, increased transparency and comparability of prices |

To fulfil its role in future sustainable heat energy systems, the 4GDH concept implementation aims to solve the following tasks (Lund et al., 2014):

- to supply low-temperature district heating and domestic hot water to end consumers;
- to ensure energy renovation of existing buildings;
- to design new buildings as low-energy;
- to distribute heat in networks with low heat distribution losses;
- to recycle heat from low-temperature sources and integrate renewable heat sources such as solar and geothermal heat;
- to be an integrated part of smart energy systems including being an integrated part of 4th Generation District Cooling systems;
- to ensure suitable energy management and planning related to the transformation into future sustainable energy systems.

Widespread application of IoT technologies brings with it an increase the amount of data collected. For example, smart meter data are sensitive and fall under data protection laws.

The challenge for 4GDH system connected with information processing is to properly collect and handle data in compliance with data protection and digital laws. Another challenge is to protect digital components of 4GDH system from cybersecurity threats, detect potential attacks, and respond and recover from any incidents.

Digitalization of heating has many advantages and nowadays is the progressive way to reduce energy demand. However, there are some barriers to its implementation. Firstly, it requires financial investments and associated with risk of lock-in to additional costs to maintain and repair of installed equipment or shallow renovation. Secondly, it is necessary to eliminate digital divide between households to increase the number of consumers who can react to digitalization by opting it for greater convenience and comfort.

## THE CURRENT STATE OF ENERGY MANAGEMENT IN DISTRICT HEATING

There are a variety of EM definitions. It is the judicious and effective use of energy to maximize profits (minimize costs) and enhance competitive positions. Another definition for EM is the strategy of adjusting and optimizing energy, using systems and procedures so as to reduce energy requirements per unit of output while holding constant or reducing total costs of producing the output from these systems.

To carry out the successful EM within the organisation it is necessary to use the EM standards providing an organisational framework to integrate energy efficiency into the management practices.

The most general standard in the field of EM which provides a systematic approach to dealing with energy saving and specifies the requirements of an EM system of any organization is ISO 50001:2018. The purpose of this standard is to enable organizations to establish the systems and processes necessary to improve energy efficiency, use and consumption (ISO 50001:2018(en) Energy management systems – Requirements with guidance for use, 2018).

The ISO 50001:2018 EM system model is based on a concept of continuous improvement processes which is called the Plan - Do - Check - Act (PDCA) cycle (Sokovic, Pavletic, Kern Pipan, 2010). The EM actions in accordance with stages of PDCA cycle are presented in Figure 2.

The decisions on corrective/preventive action to manage the energy are made on the basis of the continuous energy monitoring, measurement and analysis.

In order to automate the processing information in the implementation of the energy management system model the Energy Management System (EnMS) should be used. The EnMS is the system of computer-aided tools which allows achieving heat energy consumption more effective. Implementation and operation of the EnMS require from persons related to realization the EM tasks, knowledge, qualifications, and capacity to perform their duties therefore it is important to conduct special trainings for staff. Also the organization should establish internal communication in relation to its energy efficiency and the EnMS implementation. According to the ISO 50001, organizations are required to document information that describes the core elements of their EnMS.

To gauge the effectiveness of the EnMS and monitor the actual energy performance, the organization is required to perform regular checking the data of measurement and analysis. The management review assists the organization to achieve continual improvement and to assess the suitability, adequacy and effectiveness of the EnMS. When conducting management review, the organization should also take into account the concept of continual improvement, which is achieved by evaluating the energy performance

*Figure 2. The Plan-Do- Check-Act (PDCA) continual improvement framework within Standard ISO 50001:2018*

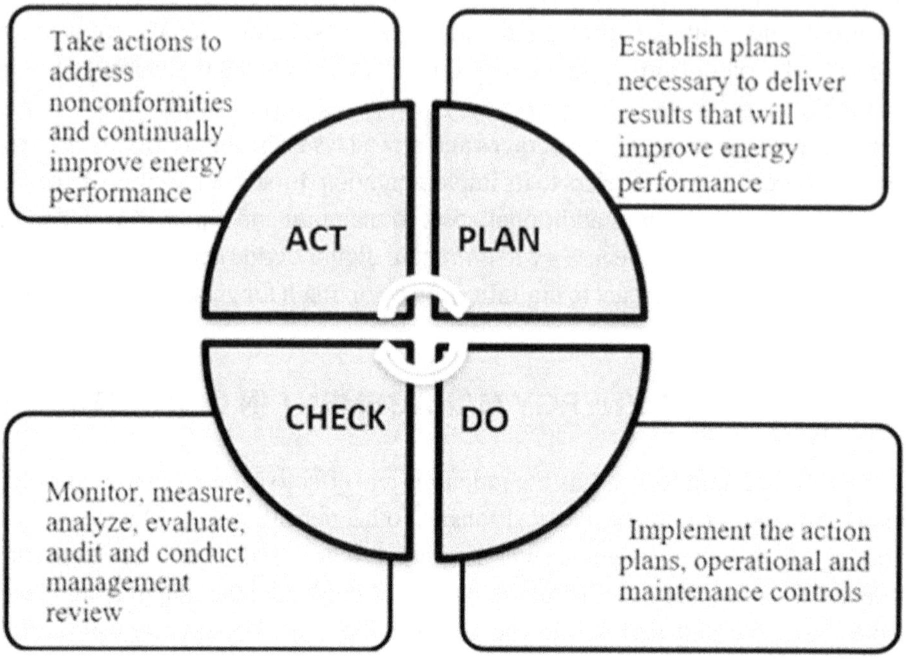


against the policies, objectives, energy targets and action plans related to its energy efficiency, energy use, and energy consumption (ISO 50001:2018(en) Energy management systems — Requirements with guidance for use, 2018).

The successful implemetation of EM in district heating is an important element for every organization which allow to reduce its operational costs. Since the costs for consumed heat energy comprise the bulk of the operational costs, the problem of improving EM of DH system is urgent.

The efficient operation of DH system requires the implementation the EM measures. The management of DH systems is a complex task which requires optimising all the stages of the DH process.

The EM of DH system is carried out on the side of DH company and on the consumer side. The DH Company management concentrates on the basic activity concerning operation, modernisation and development of the DH systems. It allows to optimize the heat sources and DH networks, reducing excessive energy losses.

Improving heat generation efficiency can substantially reduce fuel use and heat supply costs. Another way of improving heat supply efficiency is reduction of heat carrier losses from DH pipelines during its transmission.

The EM tasks on the stage of heat transmission are mainly connected with development, modernisation and operation of the DH network. Installation of pre-insulated pipes and optimisation of the heat carrier temperature reduces heat losses and water losses in DH networks.

The are two modes of DH system operation which determine the ability of management of DH operaion on the consumer side: the generation-driven mode and the demand-driven mode. In generation-driven DH Systems, which have constant water flow, the heat supply is usually regulated through changes of water temperature supplied from the heat source. The generation-driven mode of DH system operation is not able to provide rational energy use on the consumer side. This causes the impossibility to ensure comfort in heated rooms.

The demand-driven mode, when customers are able to regulate heat use according to individual needs, is the most usable today. It allows the quality and quantity regulation of water supplied from heat source to network, depending on weather conditions and schedule of organization, and automatic control of flow rate. The implementation of the demand-driven mode involves install heat meters in buildings and devices of heating mode regulation (District Heating System Management Guide, 2015).

The trend of intelligent application of information technologies to manage DH system is rapidly growing. Intelligent DH is the combination of traditional district heating engineering and modern information and communication technologies (Johansson, 2010). This enables to integrate the functions of planning, monitoring and control of the heat supply in a single system.

Now there is a tendency of Smart Cities development and implementation. Smart City is the integrated approach which is based on using intelligent technologies to make cities more efficient, healthier, more sustainable, more livable and cleaner. It is designed to achieve the highest overall efficiency of the energy system. In the context of the Smart Cities design it is necessary to develop models and tools to support of making decisions including the decisions on EM of DH systems to guarantee a proper city development. The development and implementation of EM measures of DH systems must take into account their specific features, which are discussed below.

Information technology (IT) allows managing and sharing information between people within the EnMS system. Advanced IT can help to integrate all EM functions in one tool for proper EM implementation. IT are widely used in Smart Cities development.

Using IT allows to create an organizational framework and network decision-making processes in the cities in order to establish smart technologies. It adds intelligence to the EM process. This intelligence uses for the analysis of visible and actionable information on energy use and performance and brings new opportunities of data processing. Intelligent EM can use the information provided by new technologies in decision-making. It includes energy information management, fault detection and diagnostics (FDD) and measurement and verification of energy savings (M&V). These three key components compose the Intelligent Energy Management Triangle which is shown in Figure 3 (Granderson, Piette, Ghatikar, Price, 2009).

*Figure 3. The Intelligent Energy Management Triangle*

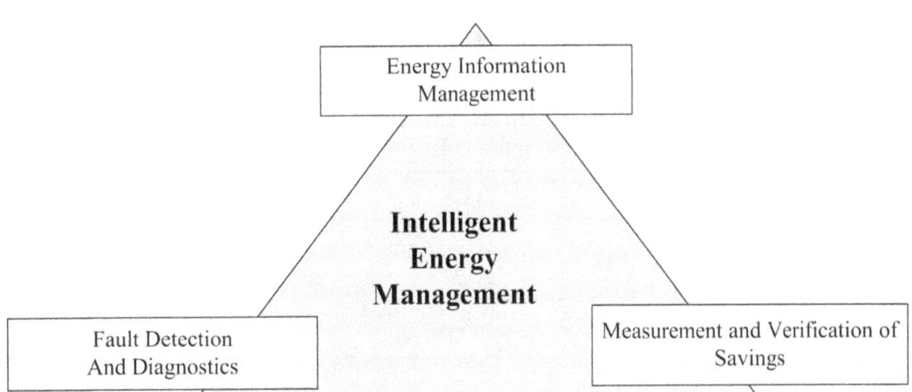

Energy information management is the presentation of data related to a building's energy consumption that allows to monitor and track energy usage for more comprehensive energy management. FDD allows more efficient operation of DH systems and equipment by identifying inefficiencies or operational problems quickly. This all provides tools to attain the high-performance of DH system functioning. FDD may also display equipment failures or inefficiencies in financial terms in addition to signaling technical issues. M&V allows determining savings from an energy-efficiency retrofit project. It is important in helping to justify or fund energy improvement or equipment purchase decisions. Combining M&V with FDD enhances the ability to detect equipment problems that affect a building's energy usage (Nix, 2011).

In general case the EM tool based on IT is energy information system (EIS). It is hardware and software complex which consist of monitor software, data acquisition hardware and communication systems used to monitor, store, analyze and display of energy data. Time-series data from meters, sensors, collected during the monitoring, and other data streams are used to energy consumption tracking and performing analysis for detection of abnormal energy consumption.

The EIS provides the follows functions (Granderson, Piette, Ghatikar, Price, 2009):

- Web accessible hourly whole building data of energy consumption;
- Data collection, transmission, storage and security;
- Graphical/visualization capabilities;
- Weatherand demand response information;
- Energy, advanced and financial analysis;
- Remote control and management.

*Figure 4. The EIS Actors and Stages of Data Transformation*

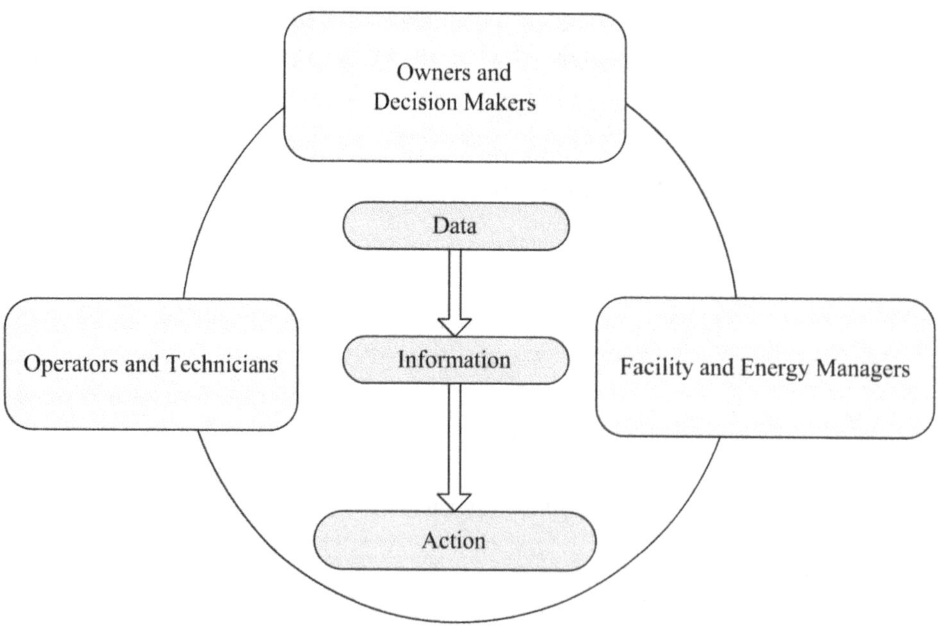

The EIS serves as the information link between persons who are involved in EM process in organization. The actors of EIS are facility and energy managers, owners and decision makers, operations and technicians (Figure 4).

The EIS operates with monitored data of DH system functioning and data from weather information services.

The monitoring in district heating performs checking the quality of DH system operation to detect the deflections from normal heat load mode, the corrosion of DH pipes and to determine the failures of installed equipment in DH plants and substations, etc. In the context on implication of Industry 4.0 on EM the DH system monitoring should be carried out real-time using IoT concept.

Now most common approaches for monitoring are web-based or agent-based. These techniques allow to provide remote control of DH system and access to monitoring data online. In web-based monitoring it is appropriate to use GPS/GPRS wireless technologies for data acquisition. At present GPS is the only satellite system with sufficient availability and accuracy for remote monitoring which allows data transferring for long distances (Peulic, Dragicevic, Jovanovic, 2013).

The agent-based monitoring technique realizes the decentralized approach to the DH management problem. The multi-agent system manages the operational decisions, e.g., reacts on shortages and performs resource redistribution. This approach to monitoring and control in DH system allows achieving coordination of production and distribution of heat energy (Wernstedt, 2003).

All monitoring data are obtained in a few steps – collection data of energy performance indicators and influencing factors, evaluation the data accuracy and reliability, estimation the data shortage (Polikarpova & Rosa, 2017). These data are the input data to solve different tasks such as simulation of DH network, forecasting of heat energy demand, optimization and decision making support on energy management of DH system The EIS users are then able to make decision and to take action based on the information provided by DSS. As follows the EIS transformes input data to actionable information.

## DECISION MAKING UNDER UNCERTAINTY IN HEAT ENERGY SYSTEMS

The organization of decision-making process in the management of DH systems aims at realization of corrective and preventive actions which are intended to reduce energy consumption and $CO_2$ emission. Moreover the DH consumers must receive the heating services of high quality. Hence the decision making criteria on DH management are:

**Social** – to satisfy heat energy demand;
**Economic** – to reduce costs on heat energy production and distribution to consumer;
**Enviromental** – to obtain carbon reduction.

The models of making decision may be single- or multi-criteria. In order to consider the criteria in the complex the models of multi-criteria decision making were developed (Bush & Bale, 2014).

The decision making process can be represented by the following stages:

- Definition of goal;
- Definition of actions;
- Establishment the evaluation criteria;

- Identification and analysis of solutions;
- Selection of the best solution.

The realization of these stages with high effectiveness requires using intelligent information technologies to create DSS. DSS is a class of information systems intended to help decision makers use the necessary data, documents, knowledge and/or models to complete decision process tasks. Decisions on regulation of DH operation are made under conditions of uncertainty.

Regardless to energy system the uncertainties and their examples are listed as the follows:

- **Market** –the volatility of the fuel price;
- **Weather** – outdoor temperature,wind speed, humidity;
- **Policies** – changes in taxation;
- **Behaviour** – energy demand;
- **Technical**– connected with system operation.

The relevant uncertainties should be considered accordingly to stages of DH process.

The most common tasks of DH system management requiring accounting uncertainties are the follows (Soroudi A. & Amraee, 2013):

- Distributed energy generation impact assessment;
- Renewable energy operation and planning;
- Heat generation and distribution planning, operation and control;
- Maintenance scheduling;
- Risk analysis.

The solution of these tasks is based on models which provides information support of making decision on EM of DH system. These models are described below.

## THE SIMULATION MODELING OF DISTRICT HEATING NETWORK

The simulation model is intended for research the dynamic properties of DH networks. It involves a full topology modeling and physical modeling of the DH network, taking into account individual pipes, dimensions, material properties etc. Such full models are computationally intensive when considering large DH systems. The DH network can be considered as a collection of nodes connected by directed edges (pipes). The topology model of DH network is complete description of the DH network structure by the graph theory. A graph contains information about pipes, nodes and the relation between them.

The DH network graph in a matrix data structure contains the incidence matrix and the loop matrix (Valdimarsson, 2014).

The dependence of the flow rate and the pressure is nonlinear. The main difference between methods of solving this problem lies in the selection of independent variables. The finite element method is widely used for fluid flow and heat transfer analysis (Gabrielaitiene, Kacianauskas, Sunden, 2003).

The simulation model for hydraulic and thermal calculation may be presented as a system of equations which defines dependence between parameters of flow in DH network (Baranova et al., 2011).

As output there are calculated:

- **For each branch**: Hydraulic resistance, mass flow, pressure and heat losses
- **For each point of junction:** Pressure and temperature.

The simulation models may determine the calculated value of water flows, temperatures and heat losses in DH networks. The uncertainty analysis of such models may determine analytically the uncertainty in the linear water flow system and the uncertainties of the estimated temperatures and heat losses (Fang & Lahdelm, 2013).

## THE HEAT DEMAND FORECASTING MODELS

To analyse a large amount of the monitoring data the forecasting techniques are used. Primarily the forecasting techniques in district heating were served to determine the required volume of heat energy supplied from the heating source to the consumers. Now the consumer-oriented forecasting models of heat consumption are increasingly being used (Grzenda, 2012). It allows the regulation of heat energy demand on the consumer side, which reduces the consumption of the heat energy.

The heat demand forecasting is typically divided into short-term, mid-term and long-term according to which time horizons are considered. In recent years a large amount of the short-term techniques for solving the problems of forecasting in heating system has appeared (Buhari, Adamu, 2012).

The heat demand forecasting approaches can be divided into two general groups – model-based and data-driven (Medojevic et al., 2018). Model-based prognostics approaches in heat demand forecasting are mainly based on analytical models to describe behavior of DH system. It is difficult ot achieve the precision of such models applicable to DH because of DH system has a complex structure, described by nonlinear models and operates in conditions of uncertainty.

In data-driven forecasting models the information from historical treated data, which can be collected by applying variety of existing IoT solutions, is used. The data-driven techniquies of heat demand forecasting are regression analyses, bayesian algorithms, neural networks.

The regression analysis is applied to establish relationships between the amount of the heat energy demand, temperature, flow rate and such parameters as the type of the building, the heated area and the time of day, the day of the week (working day or weekend), and period of the heating season (Chramcov, Baláte, Princ, 2010; Komass, Sniders, Laizans, 2012).

Now the forecasting models based on the artificial neural networks are widely used in district heating. The forecasting allows DH customers to plan their own heat consumption and possible local production in smart DH systems. The benefits of using forecasting models for DH producers are better production planning and ability of organization of customer profiles for different customer types (Kontu, 2013).

## THE OPTIMIZATION IN DISTRICT HEATING

Optimization tasks are solved to find solutions which are optimal or near-optimal with respect to some goals. In relation to district heating these goals may be to minimize cost on heat production, to minimize greenhouse gas emission or to maximize profit of heat generating company. Depending on how the optimization goals can be achieved there are the structural and the operational optimization of DH system.In general structural optimization is focused on determination of optimal DH network configuration. The basic factors for determination of the structure of DH system are economy and functionality of the tree pipe networks. In structural optimization the classical problem is how to select the optimal diameter of pipes in which a fluid is transported. In order to ðnd the optimal solution the evolutionary algorithm may be applied (Weber, Marechal, Favrat, 2006).

The definition of optimal diameter of the pipe is based on the fact that the costs of pipeline installation and heat losses increase strongly with diameter, while the pumping power drops rapidly. The dynamic simulation models of DH networks today are used for this optimization task. To solve this task the models for energy system analysis such as EnergyPlan and RETScreen International models are also applicable (Hlebnikov, Dementjeva, Siirde, 2009).

The optimal path of heat carrier with minimal transport costs may be defined by two methods of linear programming, the transport method and the simplex method. The mathematical model for defining the optimal tree path consists of the objective function of capitalized costs that is minimized (Jamsek, Dobersek, Goricanec, Krope, 2010).

The operational optimization as a rule is intended to the determination of such DH network parameters as the supply temperature and the pump pressure. The temperature optimization models may be based on the SCADA data from the DH network. This constitutes the best baseline for the DH system optimization and encompasses operational changes and any fluctuations in DH system operation during weekends and vacation periods.

To asses uncertainties in solving optimization tasks the stochastic and robust optimization techniques can be used. In stochastic optimization uncertain parameters take values in a discrete probability space represented one possible realization of the uncertainty. The set of the possible realizations are called scenarios. Stochastic optimization is applied to determine optimal heat production and to make optimal strategy of integration renewable energy sources in DH system under uncertainties of fluel prices and energy demand.

Robust optimization is used when the statistical information of input parameters is insufficient and there is no possibility of extracting probability density function of uncertain parameters. Robust optimization ensures the decision maker that even if there are errors in the prediction of uncertain parameters, the objective function value will remain optimized. Robust optimization is used for long-term energy planning including planning of renewable energy integration. It may be considered as an alternative approach to stochastic programming for a lower computational burden and a less data dependence (Mohammadi et al., 2017).

## FUZZY LOGIC BASED ENERGY SYSTEMS MODELING

Application of fuzzy logic for making decision support provides a convenient framework that can deal with ambiguity or partial process knowledge (Musee, Lorenzen, Aldrich, 2006). Fuzzy logic refers to possibilistic approah of uuncertainty modeling.

The fuzzy sets approach is an efficient method for dealing with uncertainties and conflicting objectives modeling the real behaviors. DH system directly interacts with natural processes and human behaviors and so modeling this behavior with certain and determined values is not effective because usually human behavior and natural processes are not predictability deterministic and controllable. Using fuzzy logic for expressing the uncertainties helps to achieve more accurate results in control and decision-making (Mohammadi et al., 2017).

Fuzzy logic controllers are designed on the basis of the expert knowledge of the DH system behavior. The DH system with fuzzy logic controllers provides the accurate automated regulation such a way that heater will be turned between several positions depended on room temperature and weather conditions to provide required temperature in heated room (Kobersi et al., 2013). Fuzzy controllers are used also for the demand-driven energy regulation. Smart home heating controllers are able to control the heating system to preheat the home before home users getting home. By applying such system, home users can save energy without sacriðce the quality of life (Ain et al., 2018). Besides control areas, fuzzy logic in DH systems is applied in decision-making for DH system expansion planning, in determination the load demand and when diagnosis of equipment malfunctions.

## TOOLS FOR SMART ENERGY MANAGEMENT IN DISTRICT HEATING

The District Heating 4.0 systems are becoming increasingly complex with multiple heat production sources including RES and consumers. Also such systems tend to be coupled with other energy sectors, like electricity or district cooling. Digitalization in energy sector prerequisite to achieve the objectives of the EM of District Heating 4.0 systems. It involves application various hardware and software tools on every stage of DH process.

On the generation stage the heat sources like CHPs, boilers, heat pumps should be equipped with smart controllers to provide intelligent operation of heat production. Such controllers must satisfy communication requirements and be interconnected in IoT network. Intelligent control algorithms should be developed preferably based on data driven models rather than physical models of the network (Digital Roadmap for District Heating & Cooling, 2019). In the heat distribution stage digitalisation can enable cost efficient, robust and scalable real-time data monitoring tools which will collect data for application machine learning and data mining technologies. Another tools can provide in-depth analysis of the operational characteristics of each node of DH network or leakage detection. Digital solutions at the consumer side are used for demand-driven energy management. They allow an active response to weather changes and other uncertainty factors of influence of heat demand and regulate heat consumption for more efficient use of generation capacities.

The software tools are widely used for information support the EM of District Heating 4.0 systems. Depending on the specific software they provide complex solution or aimed at solving a specific problem. The comparison review of the top EM software is presented in Table 2. In the table are used such reductions, which denote separate functions of software: 1 – monitoring, 2 – control, 3 – data management and analysis (include forecasting), 4 – reporting.

Most modern EM software have user web interface which enables the managers of the DH services to choose their preferred energy profile online. Thus, a decision on regulation of DH operation may be suggested at once by request of energy manager.

The vast majority of EM software is and only provides demo for free. More over, it can work only with specialized heat meters, sensors etc. It causes additional costs using such EM tools. This makes their usage impossible for low-profit and non-profit organizations which include public sector institutions.

Moreover, the heat energy demand of public sector buildings depends on their work schedule. The EM software taking this into account allows reducing heat consumption on weekends and provides ad-

*Table 2. The Comparison Review of Software*

| Software | Heat Supply | | | | Heat Consumption | | | |
|---|---|---|---|---|---|---|---|---|
| | 1 | 2 | 3 | 4 | 1 | 2 | 3 | 4 |
| TERMIS - a simulation tool that uses real-time data of DH network operation to analyze and track the current situation for making decisions on optimizarion heat production (Termis District Energy Management Software, n.d.) | + | + | + | + | | | | |
| Wattis - cloud-based energy analytics platform lets you manage energy use online and analyze data all in one place. (The Official Website of Wattics Company, n.d.). | | | | | + | + | + | + |
| The ePortal software enables for monitoring, analyzis energy consumption and invoice control (The Official Website of ePortal: Energy Insight Software, n.d.) | | | | | + | + | + | + |
| The ENMAT software that enables the monitoring of multiple heat meters and generation charts of heat consumption (The Heat Monitoring Solution of ENMAT Software, n.d.) | | | | | + | | | + |
| The software of the ENORO company - solutions for managing metered consumption data, forecasts, schedules and pricing (ENORO Meter & Energy Data Management, n.d.) | | | | | + | | + | |
| The Landis+Gyr company software provides smart grid applications, demand-management technologies, data analytics and renewables integration (The Official Website of Landis+Gyr Company, n.d.) | + | + | + | + | + | + | + | + |

ditional heat savings. The software for demand-side management that can be applied in public sector is destribed below.

## THE DECISION SUPPORT SYSTEM «HEATCAM» FOR ENERGY MANAGEMENT OF DISTRICT HEATING SYSTEM ON DEMAND SIDE

The DSS «HeatCAM» is a tool of the decision support for the EM of DH system on the consumer side. It is implemented through a three-tier client-server architecture of the web-based applications.

The data tier comprises of the database and DBMS MySQL for managing read and write access to a database. The application tier of the DSS «HeatCAM» consists of web-server Apache and several subsystems written in a PHP programming language, functionalities of which are listed in Table 3. The presentation tier is deployed to a users PC through a web-interface of the DSS «HeatCAM» which can be accessed through a the client's browser. Thus the functionality of the DSS «HeatCAM» is implemented as a set of software applications that are united into the monitoring, modeling, forecasting, mode determination and data visualization subsystems.

The business processes model of the DSS «HeatCAM» is presented in Figure 5.

The DSS «HeatCAM» is used for EM of the DH system of the Sumy State University in the last three heating seasons. It provides the necessary data to make decisions on regulation of heating modes in dependence of changing weather conditions and working schedule of this university.

*Table 3. The brief description of the DSS «HeatCAM» subsystems*

| Subsystem | A brief description | Input data of the subsystem | Output data of the subsystem |
|---|---|---|---|
| Monitoring | Data collection from heat meters, sensors and the weather forecast site, their validation and data store in the database. Calculations of the daily monitoring indicators that are based on the data, collected hourly. | Monitoring data collected from sensors, meters and the site of weather forecast. | Monitoring data of heating and the data of the forecast of the meteorological conditions collected daily and hourly |
| Forecasting | Definition of the forecasted data of heat demand | Data of the forecast of the meteorological conditions and data of the organization schedule for tomorrow, historical data of the daily heat demand | Forecasted value of the volume of the heat energy |
| Modeling | Carrying out of the numerical calculation of the flow in the network to determine the heat loss and the correction of the monitoring and forecasting data including heat losses | Data of monitoring and forecasting of the heat demand | Refined data of monitoring and forecasting of the heat demand |
| Decision making Support | Determining of the current mode of heating and providing recommendations for management of the heating modes | Data about temperature of the environment, the amount of the heat energy consumed for the current day, the forecasted amount of heat energy | The current heating mode and data of heating modes management |
| Visualization | Displaying data in the web-interface of the DSS | Requested data from the database | Tables and graphs on the web-page |

The DSS «HeatCAM» provides monitoring of the the current state of heating mode, short-term forecasting of heat energy demand and also an information support of the decision making on heat energy management of the social and public sector buildings.

The group of people, that manages the heating of the organization of social and public sector online has an authorized access to the DSS «HeatCAM». Each user has an access to the needed components of the system.

The key functions of the DSS «HeatCAM» users are presented in Figure 6 as a UML use case diagram.

The administrator has full rights for the web-resource management. He carries out the distribution of the access rights to the web-system between the users, has the right to make and edit all monitoring data and to correct all the output data.

The operator is responsible for data collection from the installed sensors, meters of heat consumption and of the information resources about the weather conditions, and also for entering the collected data to a single database. In the DSS «HeatCAM» the role of the operator is performed by the software modules of the monitoring subsystem, so the monitoring is automated in real time. Data collection of the heat energy monitoring is carried out by a specialized terminal (Parfenenko, Okopnyi, Nenja, 2012). The data collection of monitoring of the meteorological conditions of the environmental is done by us-

*Figure 5. The business processes model of the DSS «HeatCAM»*

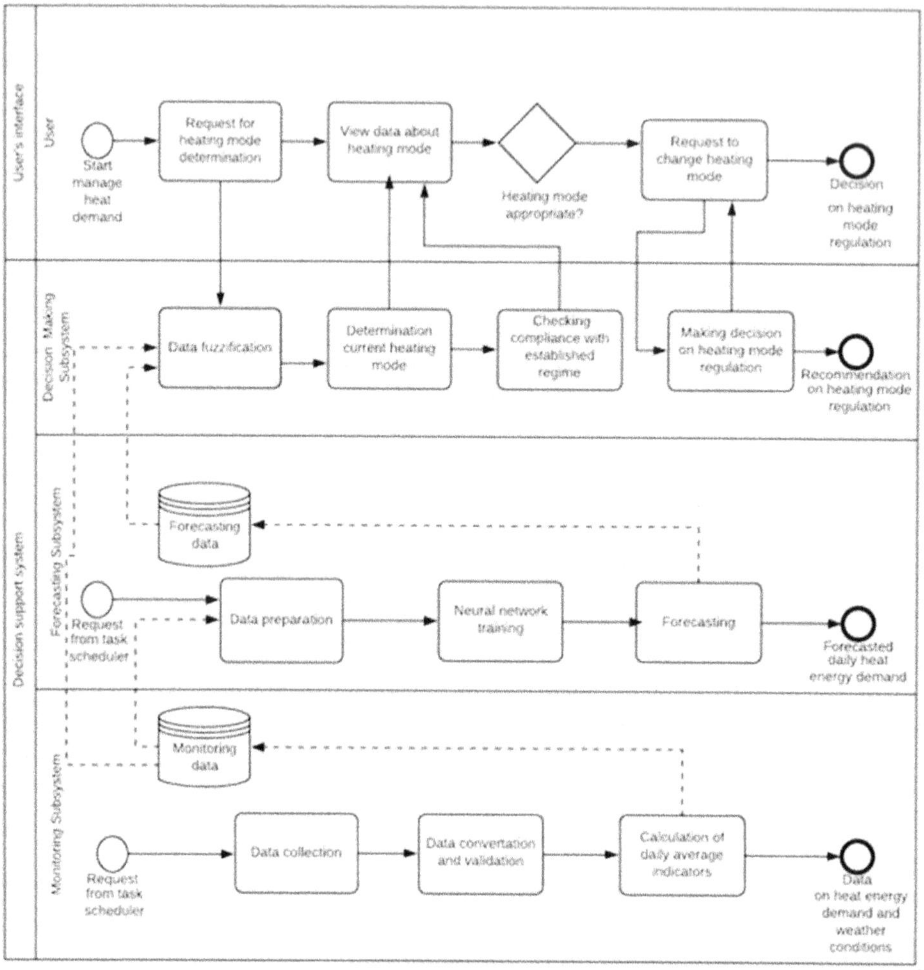


ing specialized software – parser that provides the data extract from the weather forecasting web-site (Shendryk & Vashchenko, 2011).

The role of the person, who makes decisions, usually fulfills the specialist, responsible for energy audits and energy efficiency. This person analyzes an actual level of heat demand and other indicators of functioning of the heating system and does the comparative analysis of the actual and the forecasted heat demand.

The expert on energy saving forms the base of knowledge and rules through which the determination of the current heating mode is carried out and the recommendations for setting the needed mode are given.

The person, who makes decisions, tracks the current heating mode and decides on the correction of the heating mode on the basis of the recommendations made by the DSS.

The head of the organization has the ability to view the reports on the DH system functioning and on the state of payment for the consumed heat.

The process of the DSS «HeatCAM» functionality can be represented as follows:

*Figure 6. The use case diagram of DSS «HeatCAM»*

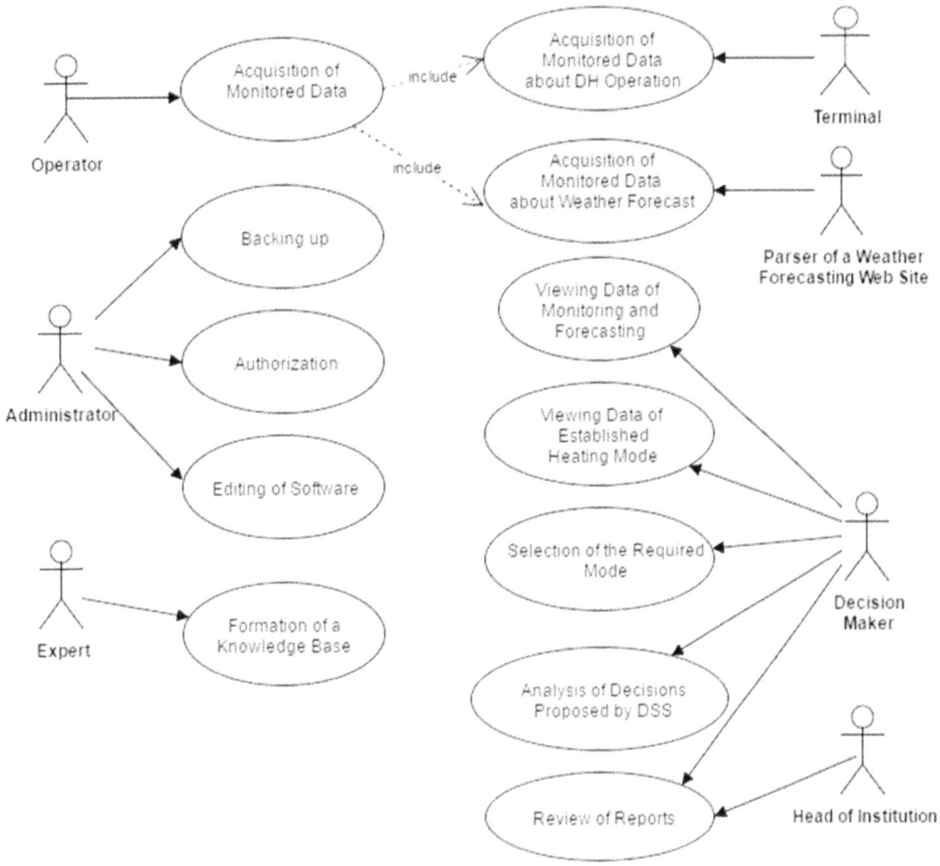


$$\text{Proc} \rightarrow O \times H_c \times H_m \times S_c \times C_{ch} \times C_s \times D_s \times Z\ ,$$

where $H_c$ – is a set of hardware maintenance, $H_m$ – is a set of meters, from which the indicators are taken, $S_c$ – is a set of software, $C_{ch}$ – is a set of communication channels, $C_s$ – is a set of communications signals, $D_s$ – is a set of data on which the decisions are made, where $O$ – is a set of people who make decisions ; $Z$ – is a set of tasks that must be executed by the DSS «HeatCAM», which is represented in the form:

$$Z = Z_m \cup Z_s \cup Z_p \cup Z_f$$

where $Z_m$ – is a set of monitoring tasks, $Z_s$ – is a set of tasks of thermal network modeling, $Z_p$ – is a set of heat consumption forecasting tasks, $Z_f$ – is a set of tasks of decision formation.

These tasks are performed by the DSS «HeatCAM» subsystems that perform tasks of monitoring, forecasting and making decision support.

## THE REALIZATION OF MONITORING TASKS IN THE DSS «HEATCAM»

The information for the analysis of DH system of the organization in real time is given by the monitoring subsystem of the DSS «HeatCAM», which performs the following functions: data collection, presented in a specialized form from the accounting devices, their converting into a format suitable for the transmission over the communication channel; archiving of the information in the database; visualization of the results of the analysis of data in the form of charts, graphs, standardized reports for the decision support; providing the possibility of taking the monitoring results from the remote terminals via web-interface, regardless of the software platform.

The formal mathematical description of the monitoring of the DH system of the social and public sector in the terms of the theory of sets is presented below. This model has the level of detail required for the making decision support.

The set of the monitoring parameters $M_p = M_{po} \cup M_{pi}$ consists of a subset $M_{po}$ that contains parameters that are taken online and the subset $M_{pi}$ containing daily monitoring parameters which are calculated based on the collected data online. The set $M_{po} = M_{ho} \cup M_{wo} \cup M_{vno}$ of the monitoring parameters collected online, consists of the subsets of parameters of the heating $M_{ho}$, the weather forecasting parameters $M_{wo}$ and the set of temperature data of the environment $M_{wno}$.

The set of the heating parameters $M_{ho}$ consist of the number of heat meter, the quantity of the consumed heat energy, Gcal; s the temperature of the coolant in the direct pipeline, ° C; the coolant temperature in the return pipeline, ° C; the temperature of the coolant in the pipeline after mixing, the pressure of the coolant in the direct pipeline kgf / cm$^2$; the pressure of the coolant in the return pipeline kgf / cm$^2$; the thermal capacity of the coolant in the direct pipeline Gcal / h; the thermal capacity in the return pipeline, Gcal / h; the volumetric flow rate of the coolant in the direct pipeline, m3 / h; the volumetric flow rate of the coolant in the return pipeline, m3 / h.

The set of monitoring of the weather forecast, carried out online, $M_{wo}$ consists of the collected data of the meteorological conditions such as the air temperature, ° C; the wind speed, m / s; the atmospheric pressure, mm. Hg. c. and the air humidity, %.

The set of the monitoring parameters $M_{pi}$ consists of a set of elements similar to the above described elements of the set $M_{po}$ and contains daily data, which are calculated based on hourly data.

The data, collected during the monitoring, after their validation are stored in the database of the DSS «HeatCAM». The monitoring data are displayed on the web-page after the user's request in the form of tables and graphs and they are also used during the report formation for the consumed heat. These data are also used to forecast the heat demand of the buildings of social and public sector.

## THE REALIZATION OF FORECASTING TASKS IN THE DSS «HEATCAM»

The forecasting subsystem of the DSS «HeatCAM» performs the determination of the daily heat energy demand based on archived monitoring data.

The data sets for forecasting include the following data for the previous heating seasons:

- The data of the daily building heat demand in accordance with the date of their monitoring and the type of the day of the week (the working day or the day off);
- The data of meteorological conditions (the ambient temperature, wind speed, the atmospheric pressure, and the air humidity) according to the date of their monitoring.

To forecast the heat energy demand in DSS «HeatCAM» the nonlinear autoregressive network with exogenous inputs (NARX) architecture of the neural network was chosen.

The NARX architecture mathematically is described as the follows:

$$\hat{y}(t) = F(y(t-1), y(t-2), ..., y(t-n), x(t-1), x(t-2), ..., x(t-n)).$$

The input vector of neural network models $x(t)$ consists of components:

- $x_1$ – the ambient temperature, C;
- $x_2$ – the wind speed, m/s;
- $x_3$ – the atmospheric pressure, mm. Hg. c.;
- $x_4$ – the air humidity, %;
- $x_5$ – the type of the day of the week (working day or weekend).

The output variable of the proposed neural network models is the daily heat energy demand $\hat{y}(t)$, Hcal. Thus the proposed models represent the dependence between the data about meteorological conditions, data of the organization schedule and the daily building heat demand.

At the stage of the parametric synthesis the training of the neural networks is performed. The procedure of training is the reflection of the monitoring data on the weights of the neural network to obtain the optimal forecasting heat demand data as the output signal of neural network.

The training of the neural network is implemented using Levenberg-Marquardt algorithm for changing the weights coefficients, until the forecasting error will be minimized (Hao Yu, Wilamowski, 2011; Roweis, n.d.).

The evaluation of forecasting accuracy of the proposed model with NIO and NAR neural network models has been carried out. The mean absolute percent error MAPE, the mean absolute percent error MAE and the root-mean-square error RMSE were calculated.

To evaluate the forecasting accuracy the data of the heat energy demand of one campus of the university which were collected during the heating season 2014-2015 have been used. The results of forecasting errors calculation showed that the NARX model has the highest accuracy among the NIO, NAR and NARX models (Parfenenko et al., 2014). The forecasting accuracy of the developed forecasting model based on NARX architecture is approximately 95%. Thus, obtained results proved the high accuracy of proposed forecasting model.

## THE REALIZATION OF MAKING DECISION SUPPORT IN THE DSS «HEATCAM»

The subsystem of mode determination of the DSS «HeatCAM» is designed to establish the current heating mode and to propose the decision on the regulation of the heating in order to establish the needed mode.

In this subsystem the decision support models are used, that are built with the usage of the methods of the theory of fuzzy sets and fuzzy logic, which allows formalizing the expert knowledge on the heat energy management.

The model of the setting of the current mode of heating is presented as expression:

$$R_c = \begin{cases} \text{f(ZY1),if Zw is Zw1} \\ \text{f(ZY2),if Zw is Zw2} \\ \text{f(ZY3),if Zw } is \text{ Zw3} \\ \text{f(ZY4),if Zw } is \text{ Zw4} \end{cases}$$

where $R_c$ - is the heating mode; ZY1, ZY2, ZY3, ZY4 – are linguistic variables that determine the amount of the consumed heat energy at different temperature conditions, Zw – is a linguistic variable that characterizes the state of temperature conditions of the environment; Zw1, Zw2, Zw3, Zw4 – are terms of linguistic variable Zw.

According to this model, initially we determine the state of the temperature conditions of the environment as a function of the average daily air temperature Zw = F (Xw), where the output variable Zw sets the affiliation of the temperature of the environment to a particular class, Xw – air temperature of the environment, °C.

Depending on the resulting variable Zw, the value of the heating mode *Rc* is determined with the help of the linguistic variables ZY1, ZY2, ZY3, ZY4. The realization of the models of fuzzy inference decision to determine the state of temperature conditions and the current heating mode is carried out according to the Sugeno algorithm.

The result of the work of the model of the setting of the current mode is the determination of the current heating mode among the possible values “Below necessary”, “Acceptable”, “Optimal”, and “Higher than optimal”. The determination of the current heating mode allows you to monitor the compliance with the appropriate temperature conditions in the heated rooms and promptly detect deviations from the planned mode.

The model of the formation decision on heating mode regulation is represented as the function Z = F (ZY, BY), where Z – is the control parameter data, ZY – is the variable, characterizing the current heating mode, BY- is the variable, characterizing the required heating mode.

For fuzzy inference decision to control the current heating mode the Mamdani algorithm is selected.

The result of the model of fuzzy inference decision is a recommendation to regulate the current mode to the necessary one among the possible values “Acceptable” and “Optimum”. The example of the rule for fuzzy inference decisions on the heating mode regulation is given in Table 4.

*Table 4. The example of the formal fuzzy inference rule for the making decision on control of heating mode*

| Prerequisite | ZY is ZYM1, BY is BYM1 |
|---|---|
| The formulation of the rule | If ZY is ZYM1 and BY is BYM1 then Z = NC; |
| Interpretation | If currently the mode “Acceptable mode1” is set and tomorrow this “Acceptable mode1” will be needed, then the change of the regulatory option is “Leave unchanged” |

The energy manager decides to regulate the heating mode based on the proposed by the DSS «HeatCAM» recommendations. The use of the DSS «HeatCAM» allows the energy manager to react promptly to the changing of the influencing factors, according to the need of the organization of social and public sector in the heat energy, which ensures the compliance of the comfortable temperature conditions in the heated rooms and avoids excess heat consumption.

Thus the DSS «HeatCAM» provides decision support for managing centralized heating of the organizations of social and public sector on the basis of monitoring data, forecasting data of the heat demand of the building and also using expert knowledge to regulate the heating mode.

## CONCLUSION

This chapter studies an urgent problem of improving energy efficiency of EM process which may be achieved through the application of IT technologies.

The district heating 4.0 concept as a result of DH systems digitalization and the current state of EM in DH were explored. The analysis of the existing models and tools which can be used for information support of EM process was carried out.

The DSS «HeatCAM» for making decision support on regulation of district heating on the consumer side was presented as a tool which is intended for the information support of the energy management stages in accordance with ISO 50001:2018. The main functions of DSS «HeatCAM» are monitoring of DH operation, short-term forecasting of heat energy demand, DH simulation and decision support on regulation of heating modes. The application of this system allows determining the required amount of heat energy for concrete building. So, it is possible to achieve energy conservation through the elimination excess heat energy consumption. Hence the DSS helps decision-makers to achieve the primary goal of EM on the consumer side which is to reduce energy consumption. Also it enables to evaluate the effectiveness of taken decisions. The obtained research results will be used to development of the EM in DH systems at the Smart Cities implementation.

Future research direction is related to development the decision support model for planning centralized heating in combination with using renewable energy sources.

## REFERENCES

Ain, Q., Iqbal, S., Khan, S., Malik, A., Ahmad, I., & Javaid, N. (2018). IoT Operating System Based Fuzzy Inference System for Home Energy Management System in Smart Buildings. *Sensors (Basel), 18*(9), 2802. doi:10.339018092802 PMID:30149631

Baranova, I. V., Parfenenko, Yu. V., & Nenja, V. G. (2011). Complex mathematical model for computer calculation of delivery and heat distribution in pipeline system. *IEEE 6th International Conference on Intelligent Data Acquisition and Advanced Computing Systems: Technology and Applications, IDAACS 2011, 2*, 556 – 559. 10.1109/IDAACS.2011.6072828

Buhari, M., & Adamu, S. (2012). Short-Term Load Forecasting Using Artificial Neural Network. *Proceedings of the International MultiConference of Engineers and Computer Scientist.* Retrieved from http://www.iaeng.org/publication/IMECS2012/IMECS2012_pp83-88.pdf

Bush, R. E., & Bale, C. S. E. (2014). Realising the Social Benefits of District Heating Throught Strategic Planning. *Proceeding of the 14th International Symposium on District Heating and Cooling*. Retrieved from http://eprints.whiterose.ac.uk/81735/

Chramcov, B., Baláte, Ja., & Princ, M. (2010). Heat Demand Forecasting for Concrete District Heating System. *International Journal of Mathematical Models and Methods in Applied Sciences*, *4*(4), 231–239.

Digital Roadmap for District Heating & Cooling. (2019). DHC+ Technology Platform. Retrieved from https://www.euroheat.org/wp-content/uploads/2018/05/Digital-Roadmap_final.pdf

District Heating System Management Guide. (n.d.). Retrieved September 24, 2019, from https://projects.bre.co.uk/DHCAN/pdf/Management.pdf

ENORO Meter & Energy Data Management. (n.d.). Retrieved September 24, 2019, from https://www.enoro.fi/mdmandedm

Euroheat & Power Task Force Customer Installations. (2011). District Heating in Buildings. Retrieved from https://www.buildup.eu/sites/default/files/content/District%20Heating%20in%20buildings_final.pdf

Fang, T., & Lahdelm, R. (2013). Uncertainty Analysis of District Heating State Estimation Model. *Proceedings of the IASTED International Conference Modelling, Identification and Control (MIC 2013)*, 380 – 387. 10.2316/P.2013.794-077

From Industry 4.0 to Energy 4.0: Future Business Models and Legal Relations. (2016). Retrieved from http://www.germanenergyblog.de/wp-content/uploads/2016/04/160317_Lang_IBE_Bochum_From_Industry_4_0_to_Energy_4_0_fin.pdf

Gabrielaitiene, I., Kacianauskas, R., & Sunden, B. (2003). Thermo-hydraulic Finite Element Modelling of District Heating Network by the Uncoupled Approach. *Journal of Civil Engineering and Management*, *IX*(3), 153–162. doi:10.1080/13923730.2003.10531321

Granderson, J., Piette, M. A., Ghatikar, G., & Price, P. (2009). *Building Energy Information Systems: State of the Technology and User Case Studies*. Retrieved from https://eta-publications.lbl.gov/sites/default/files/lbnl-2899e.pdf

Grzenda, M. (2012). Consumer-oriented heat consumption prediction. *Control and Cybernetics*, *41*(1), 213–240.

Hlebnikov, A., Dementjeva, N., & Siirde, A. (2009). Optimization of Narva District Heating Network and Analysis of Competitiveness of Oil Shale CHP Building in Narva. *Oil Shale*, *26*(3), 269–282. doi:10.3176/oil.2009.3S.09

ISO 50001:2018 (en) Energy management systems – Requirements with guidance for use. (2018). Retrieved from https://www.iso.org/obp/ui/es/#iso:std:iso:50001:ed-2:v1:en

Jamsek, M., Dobersek, D., Goricanec, D., & Krope, J. (2010). Determination of Optimal District Heating Pipe Network Configuration. *WSEAS Transactions on Fluid Mechanics*, *3*(5), 167-174.

Johansson, C. (2010). *Towards Intelligent District Heating*. Retrieved from https://www.diva-portal.org/smash/get/diva2:833722/FULLTEXT01.pdf

Kobersi, I. S., Finaev, V. I., Almasani, S. A., & Ahmed Abdo, K. W. (2013). Control of the Heating System with Fuzzy Logic. *World Applied Sciences Journal*, *23*(11), 1441–1447. doi:10.5829/idosi.wasj.2013.23.11.13156

Komass, T., Sniders, A., & Laizans, A. (2012). Statistical Analysis of Municipal Heat Supply. *Proceedings of 11th International Scientific Conference*, *11*, 1 – 8.

Kontu, K. (2013). *Forecasting district heating consumption based on customer measurements*. Report on the 36th Euroheat & Power Congress. Retrieved from http://www.lsta.lt/files/events/2013-05-27-29%20EHP%20kongresas%20Vienoje/09_130527_1600_1730_Pl_6_Kaisa+Kontu.pdf

Lund, H., Werner, S., Wiltshire, R., Svendsen, S., Thorsen, J. E., Hvelplund, F., & Mathiesen, B. V. (2014). 4th Generation District Heating (4GDH). *Energy*, *68*, 1–11. doi:10.1016/j.energy.2014.02.089

Lyons, L. (2019). *Digitalisation: Opportunities for heating and cooling*. doi:10.2760/00116

Make a game-changing impact – before the others do. The impact of digitalization in district heating. (2016). Retrieved from https://energia.fi/files/1345/2016_Impact_of_digitalization_in_district_heating_summary_ENG.pdf

Medojevic, M., Villar, P. D., Cosic, I., Rikalovic, A., Sremcev, N., & Lazarevic, M. (2018). Energy Management in Industry 4.0 Ecosystem: a Review on Possibilities and Concerns. *Proceedings of the 29th DAAAM International Symposium*, 674 – 680. 10.2507/29th.daaam.proceedings.097

Mohammadi, M., Noorollahi, Y., Mohammadi-Ivatloo, B., Yousefi, H., & Jalilinasabady, S. (2017). Optimal Scheduling of Energy Hubs in the Presence of Uncertainty-A Review. *Journal of Energy Management and Technology*, *1*, 1–17. doi:10.22109/JEMT.2017.49432

Musee, N., Lorenzen, L., & Aldrich, C. (2006). Decision support for evaluating energy demand in vinification processes using fuzzy sets theory. *Journal of Energy in Southern Africa*, *17*(4), 4–18. doi:10.17159/2413-3051/2006/v17i4a3190

Nix, O. (2011). *Building Information Technology and Management*. Retrieved from https://www.researchgate.net/publication/265555488_Building_Information_Modelling_Adoption_An_Analysis_of_the_Barriers_to_Implementation

Parfenenko, Yu., Okopnyi, R., & Nenja, V. (2012). Creation of Control Instruments for Buildings' Heating. *The Journal of the National Technical University NTU KPI. Series Information Systems and Networks*, *34*, 93–97.

Parfenenko, Yu., Shendryk, V., Nenja, V., & Vashchenko, S. (2014). Information System for Monitoring and Forecast of Building Heat Consumption. *Communications in Computer and Information Science*, *465*, 1–11. doi:10.1007/978-3-319-11958-8_1

Peulic, A., Dragicevic, S., Jovanovic, Z., & Krneta, R. (2013). Flexible GPS/GPRS Based System for Parameters Monitoring in the District Heating System. *International Journal of Computers, Communications & Control*, *8*(1), 105–110. doi:10.15837/ijccc.2013.1.174

Polikarpova, I., & Rosa, M. (2017). Energy reduction potential of the district heating company introducing energy management systems. *Energy Procedia*, *128*, 66–71. doi:10.1016/j.egypro.2017.09.016

Recommendations for implementing the strategic initiative INDUSTRIE 4.0. (2013). Final report of the Industrie 4.0 Working Group. Retrieved from https://en.acatech.de/publication/recommendations-for-implementing-the-strategic-initiative-industrie-4-0-final-report-of-the-industrie-4-0-working-group/

Reducing your ICT energy costs. (2015). Retrieved from https://www.jisc.ac.uk/guides/reducing-your-ict-energy-costs

Roweis, S. (n.d.). *Levenberg-Marquardt Optimization*. Retrieved September 24, 2019, from https://www.cs.nyu.edu/~roweis/notes/lm.pdf

Shendryk, V. V., & Vashchenko, S. M. (2011). The system of Data Collection, Allocation and Analysis. *Visnyk of the National University. Lviv Polytechnic*, *715*, 343–353.

Skagestad, B., & Mildenstein, P. (n.d.). *District Heating and Cooling Connection Handbook*. Retrieved September 24, 2019, from https://www.districtheatingscotland.com/wp-content/uploads/2015/12/DistrictHeatingAndCoolingConnectionHandbook.pdf

Sokovic, M., Pavletic, D., & Kern Pipan, K. (2010). Quality Improvement Methodologies – PDCA Cycle, RADAR Matrix, DMAIC and DFSS. *Journal of Achievements in Materials and Manufacturing Engineering*, *43* (1), 476-483.

Soroudi, A., & Amraee, T. (2013). Decision making under uncertainty in energy systems: State of the art. *Renewable & Sustainable Energy Reviews*, *28*, 376–384. doi:10.1016/j.rser.2013.08.039

Termis District Energy Management Software. (n.d.). Retrieved September 24, 2019, from http://7t.dk/products/termis/Download/product-brochures.aspx

The Guidance for district heating companies. (n.d.). *The environmental benefits of district heating: using the new Ecoheat4cities label Guidance for district heating companies*. Retrieved September 24, 2019, from http://www.bre.co.uk/filelibrary/rpts/ecoheat4cities/Ecoheat4Cities_WP4_Guidance_for_companies.pdf

The Heat Monitoring Solution of ENMAT Software. (n.d.). Retrieved September 24, 2019, from http://www.en-mat.com/solutions/heat-monitoring

The Official Website of ePortal: Energy Insight Software. (n.d.). Retrieved September 24, 2019, from https://eportal.eu/en/

The Official Website of Landis+Gyr Company. (n.d.). Retrieved September 24, 2019, from https://www.landisgyr.eu/

The Official Website of Wattics Company. (n.d.). Retrieved September 24, 2019, from https://www.wattics.com/

Valdimarsson, P. (2014). *District Heat Distribution Networks*. Retrieved from http://www.os.is/gogn/unu-gtp-sc/UNU-GTP-SC-18-27.pdf

Weber, C., Marechal, F., & Favrat, D. (2006). Design and Optimization of District Energy Systems. *Proceedings of the 10th International Symposium on District Heating and Cooling*. Retrieved from http://www.lsta.lt/files/events/16_weber.pdf

Wernstedt, F. (2003). *Multi-Agent Systems for District Heating Management*. Retrieved from https://www.diva-portal.org/smash/get/diva2:837359/FULLTEXT01.pdf

Yu, H., Yu, H., & Wilamowski, B. M. (2011). Levenberg–Marquardt Training. In The Industrial Electronic Handbook: Intelligent Systems. Taylor & Francis Group, CRC Press. doi:10.1201/b10604-15

## KEY TERMS AND DEFINITIONS

**Decision Support System:** An interactive system, that can produce data and information, in some cases, contribute to the understanding relating to this subject area in order to provide useful assistance in solving complex and ill-defined problems.

**District Heating:** An economic activity related to supplying heat produced centrally in one or several locations to a non-restricted number of customers.

**Energy Information System:** The software, data acquisition hardware, and communication systems used to store, analyze, and display building energy data.

**Energy Management:** Energy management is the proactive, organized, and systematic coordination of procurement, conversion, distribution and use of energy to meet the requirements, taking into account environmental and economic objectives.

**Energy Management System:** A system of computer-aided tools used by operators of electric utility grids to monitor, control, and optimize the performance of the generation and/or transmission system.

**Forecasting:** The use of historic data to determine the direction of future trends.

**Information System:** An integrated set of components for collecting, storing, and processing data and for delivering information, knowledge, and digital products.

**Monitoring:** An act of observing something to regulate and control processes in observed system.

# Chapter 4
# Fatigue Fracture, CTOD, and CVN Testing in Heat-Affected Zone of High Strength Low Alloy Steels

**Manuel Viseras Pico**
*University of Cadiz, Spain*

**Rafael González-Palma**
*University of Cádiz, Spain*

**María Carmen Carnero**
*University of Castilla-La Mancha, Spain*

**Carlos López-Escobar**
*Arconic, Spain*

**David Almorza**
*University of Cadiz, Spain*

## ABSTRACT

*High strength low alloy steels (HSLA) are currently used in large quantities for constructions in the North Sea and elsewhere. For success operation under such rigorous conditions, it is essential to ensure adequate weld metal toughness at temperatures of -10ºC or even lower is used. To comply with this high requirement, the heat input control is needed to avoid forming martensitic microstructure in high strength low alloy steels (HAZ) near to the fusion line during weldment cooling. This work carried out CTOD tests to ensure that, under welding conditions, the fatigue crack, toughness, hardness, and resilience in the HAZ are acceptable with the standards. Crack size in accordance with number cycles will be represented as the crack growth rate da/dN vs. stress intensity factor amplitude.*

DOI: 10.4018/978-1-7998-3246-1.ch004

## INTRODUCTION

One problem that design of these steels of high resilience have shown is their low toughness in the HAZ Schuller, R., *et al.* (2015), when they are welded with a high heat input (HI). In this work we have studied nine specimens that have been welded under three different process:

SAW (submerged arc welding), SMAW(shield metal arc welding), GSFC(gas shield flux core arc welding) controlling the welding parameters and checking in the HAZ of such specimens, critical tensions ($\sigma_{cr}$)at the ends of the cracks, the critical cracks lengths ($a_{cr}$) and stress intensity factors ($K_{IC}$).It is intended to check that the parameters that indicate the values of fracture mechanics: $(\sigma_{cr}, a_{cr}, K_{IC})$, da/dN=A $\Delta K^m$, etc.,) in the HAZ, after heat cycle to which the steel has undergone, under a process with a maximum "heat input" of 2.327kJ /mm (see Table 5 below), are still valid, with the welding parameters applied. It is checked a correlation between the theoretical values and those obtained experimentally. Experimental work, here conducted, we intend to ensure that these parameters regulating the mechanism of fracture, remain valid under the rules compatible with the design of the base material and that as far as possible be established a correlation between welding parameters and those obtained in fracture tests, so that if the results of the tests are not satisfactory, we could apply the appropriate solution in welding for the parameters governing fracture tests were acceptable. We will check as the HAZ region behave as ferritic-Pearlitic steels complying with the equation:

$$\frac{da}{dN} = 5.01 \times 10^{-12} (\Delta K)^{3,1} m / cycle$$

$$206.9 N / mm^2 < \sigma_{ys} < 344.8 N / mm^2$$

The tests of CTOD, resilience and hardness were carried out on the panel numbered as CTOD HAZ. Said panel, 60 mm thick, 980 mm weld length and 750 mm wide, was tested under "As welded" conditions (without post-weld heat treatment); and it was used to test the HAZ. The welding was done in K and V, in three different process Shield Metal Arc Welding (SMAW) in the panel 3 AW where from taken the samples 01-1(1), 01-2(2) and 01-3(3). The second process was Gas Shield Flux Core (GSFC) in the panel 7 AW where from were taken the specimens 02-01(4), 02-02(5) and 02-03(6) and the third in the process Submerged Arc Welding (SAW) the panel 01 AW where from the following panel were taken 01-10(7), 01-11(8) and 01-12(9). The base metal was BS 7191(1979) 450 EMZ TYPE 2.

The thermal cycle to which these test pieces were subjected, for the effects of the transformation products that are produced in the cooling inside the HAZ with a contributed maximum energy of 2.327 KJ / mm, a peak fusion temperature is obtained 1,527 °C, a cooling time $\Delta t_{8/5}$, practically constant for the entire HAZ, of 10 seconds, as shown by Rosenthal's theoretical cycle for this thick plate process, González-Palma *et al* (2006). The physical characteristics of the 450 EMZ micro alloy steel are given in the casting certificate.

The life of the components of a structure containing premature cracks, can be governed by the degree of subcritical crack propagation. To this end, the many destructive and non-destructive testing can collaborate to establish any cracks before commissioning. Thus, knowledge of crack propagation through the concepts of fracture mechanics calculation to determine the fatigue of the structure is necessary.

However, after the thermal cycle in a process, material suffers, the mechanism of the microstructure of the base metal can change, being able to modify the parameters such and to the extent that if not properly controlled, could lead to catastrophic failure of the structure.

## BACKGROUND

Great deal investigations of crack propagation by fatigue led to show that no crack propagation begins until an increase of the threshold of stress intensity factor, $\Delta K_{th}$, Bäcklund, J., Blom, A.F., Beevers, C.J. (1982), depends on several known factors. Initial investigations that were conducted by Frost (1960) showed a low degree of slowdown in growth due to fatigue at low stress, the data provided offered the presumption that below certain values, the crack does not spread. The existence of this threshold was predicted by Clintock (1963) works under elastic–plastic analysis. Later Lindner (1965), Carman C.M. et al (1966) and Paris (1970) showed that the propagation of fatigue crack can be analyzed in the context of linear elastic fracture mechanics, and the threshold $\Delta K_{th}$, can be determined. In the 60s, Elber (1971) demonstrated that fatigue cracks interfere each other via a locking mechanism. It was Schmidt (1973) & Wilson (1973) who later it showed that the locking mechanism has a significant effect on the threshold behavior. The information provided led to indicate that there are four locking cracks. Factors that may influence the threshold crack propagation by fatigue are yield strength, grain size and other microstructural elements as residual stresses, CTOD mode, Young's modulus, and environmental temperature. The effects of most of these factors have on $\Delta K_{th}$, they can be explained by its relationship with the locking mechanisms of the crack. Researchers as Kitagawa and Takahashi (1976), El Haddad, Dowling, Topper and Smith (1980), found a constant value of $\Delta K_{th}$ for crack lengths greater than 0.5 mm. For our steel EMZ 450, belonging to ferrite-pearlite according to Barsom (1977):

$$\Delta K_{th} = 7(1 - 0.85R)\ MN/m^{3/2} = 6.405 MN/m^{3/2} = 202.16\ N/mm^{3/2} \quad (1)$$

Life of a structural element fatigue can be considered from three different states:

1. Initiation of fatigue crack.
2. Crack Propagation.
3. Fracture.

Most testing the growth of fatigue cracks are performed on samples subjected to load fluctuations at constant amplitude. The lengths of the cracks to the cycles that are undergoing represented graphically as measured in Figure 16.

Thus, by varying the cyclic load fluctuation and initial crack size, different curves can be drawn $a =$ f (N). (Crack length according to the cycles), Tanaka K. et al. (2002) & Roiko, A. (2012). However, the curves can be summarized in a single, when expressed in terms of speed of crack growth, depending on the amplitude of the stress intensity factor which it has no geometrical dependence and summarized in a single parameter the ($\Delta\sigma$) and crack length ($a$); being $\Delta K = Y(\Delta\sigma)\sqrt{\pi a}$ normal curve representing the speed of crack growth, is determined by the expression of Paris (1970):

$$da/dN = A\left(\Delta K\right)^{m}, \text{ Region 2, Barsom 1971} \tag{2}$$

where: $a$ = crack length, $N$ = number of cycles, $\Delta K$ = range or amplitude of the stress intensity factor. "A" and "m" are constants depending on the material Quan, G., Zou, C., Hong, Y. (2001). This curve is represented in Figure 1, it consists of three different parts:)

*Figure 1. Sckematic Representation of Fatigue Crack in Steel (by John M. Barsom, Paris)*

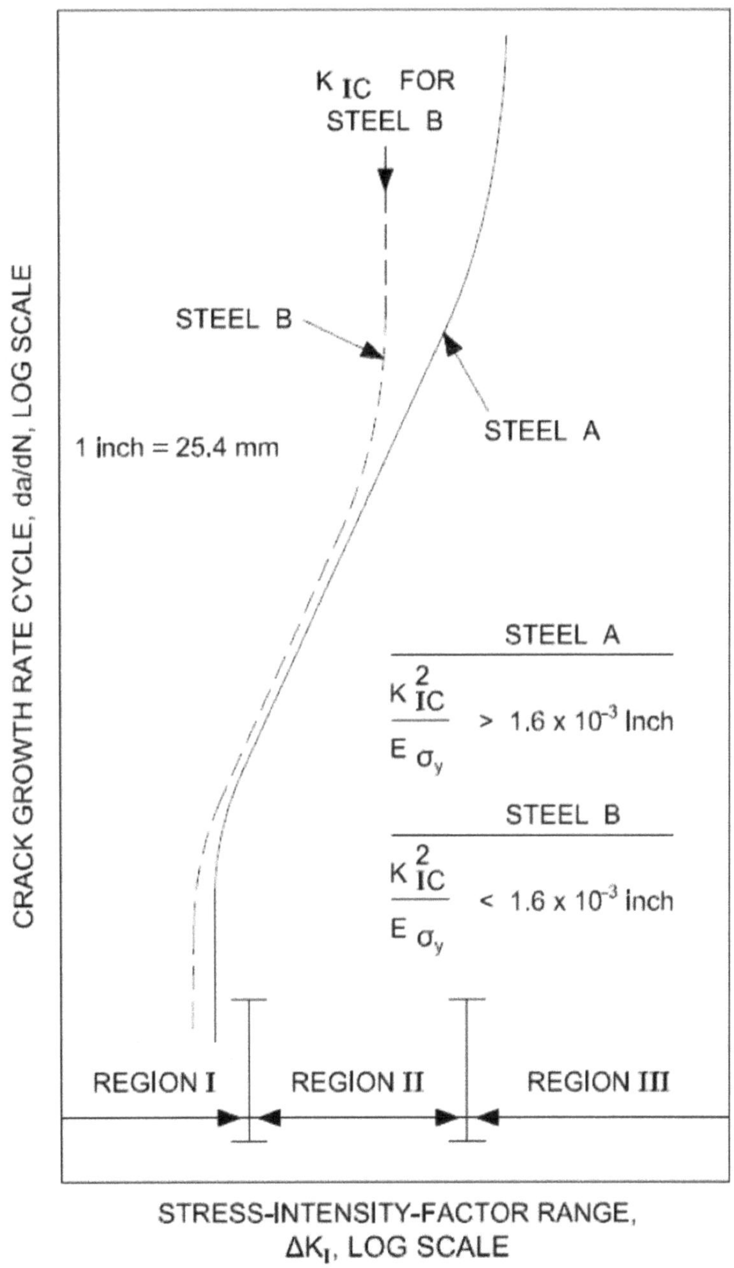

- **Region I**: Displays the threshold of fatigue fracture in which the crack does not spread under cyclic stress fluctuation (Sree, P.C. et al.,Frost, 1960; Paris, P. C., Bucci, R. J., Wessel, E. T., Clark, W. G., & Mager, T. R. (1972)).
- **Region II**: Sample propagating fatigue crack above the fluctuation threshold stress intensity factor $\Delta K_{th}$, in this region.
- **Region III**: Region where the growth cycle fatigue is greater than in the region II above. The data which the authors Barsom, (1974), Harrison (1970) and Silverio et al. (2003) presents in this region show that the growth of a fatigue crack is increased from zero stress to the load, namely $\Delta K = K$max. This increase occurs at a constant value of displacement of the end of the crack $\delta_T$ and a value corresponding to the stress intensity factor, $K_T$, which can be determined by the following expression, Barsom (1977) gives for high tension steels a constant value of crack tip displacement:

$$\delta_T = \frac{K_T^2}{E\sigma_{ys}} = 1.6 \times 10^{-3}\, inches\, (0.04064\, mm) \tag{3}$$

Where $K_T$=Point of transition from the region II to region III (Figure 1) for those materials which have high fracture toughness as is our steel 450 EMZ, in this way for to determine to the constant value $\delta_T$ in our case specimen (01-10)(7), as an example, is as follow:

$$\partial_T = K_T^{\ 2} / \sigma \mathrm{E} = 2090^2 / 205000 \times 453 = 0.047\mathrm{mm}.$$

Being $K_T$=$K_f$=2090N/mm$^{3/2}$;$\sigma$=453 N/mm$^2$; E (Young´s Module) =205000N/mm$^2$.

Where $K_f$ has being taken the maximum stress Intensity factor from experimental value of specimen 01-10(7)

Acceleration the growth rate of fatigue crack, it is determined by the transition between the region II and the region III. This seems to be caused by a combination of a ductile tearing mechanism, on a cyclic mechanism subcritical crack extension, which leaves a fatigue striation on the fracture surface. Crack tearing occurs when the crack reaches a critical value. Thus, the fatigue rate of the transition from region II to region III depends $K_{max}$ and stress ratio R.

Being:

- E = Young's Modulus
- $K_T$ = The range of the stress intensity factor, corresponding to the beginning of accelerated growth of the fatigue crack.
- $\sigma_{ys}$ = Yield strength that is closer to the yield point, and that can be taken as the average of the yield point and yield stress.

## PURPOSE OF TEST

450 EMZ steel is known and proven by tests supplied by the mill, which at the beginning of the crack and during propagation obeys the design parameters ($\sigma_{cr}$, $a_{cr}$, $K_{IC}$, $da/dN = A\ (\Delta K)^{m}$, etc.).

However, after the thermal cycle in a welding process, material suffers and the mechanism of the microstructure of the base metal can change, being able to modify the aforementioned parameters in such a way and to such an extent that if not properly controlled, could lead to a catastrophic failure of the structure.

In experimental work carried out here, we intend to ensure that these parameters governing fracture mechanism, remain valid under the rules compatible with the design of the base material and that as far as possible a correlation is established between welding parameters and those obtained in fracture tests, so that if the results of the tests are not satisfactory, we could apply the appropriate solution for welding in order to that parameters governing the fracture tests are acceptable. Also, emphasis will be placed on the behavior of fatigue fracture after a welding cycle to verify the validity of its parameters.

In HAZ we have tested the hardness at different points to the fusion line (FL) in order to verify these are found within the hardness limit and verify the composition of Martensitic structure near to FL, being this zone the more fragile of HAZ, even though sometimes the crack is initiated in other region of the HAZ as we will have the occasion to see forward.

Robert Newton method (experimental equation) will allow us to stablish a correlation between the resilience (Charpy test CVN) in the HAZ and the fracture toughness. Grain size as well as hardness at different points of the HAZ were measured. Other function that has been determined is the correlation of Number of cycles and the size of the crack verifying the proximity between the theoretical and experimental value. Sakai el al. (2009).

## PREPARATION OF THE SAMPLES PRIOR TO TESTING

From a submerged arc welded (SAW) joint panel of sixty mm thickness were prepared 3 specimens from a panel 01 AW, specimens to test CTOD fracture toughness. Testing was carried out at -10ºC being fully in accordance with BS 5762(1979) and with EEMUA 158 (2014) publication nº158 (construction Specification for Fixed Offshore Structure in the North Sea). Other 3 specimens 60 mm thickness from a panel 7 AW which were welding under a Gas Shield Flux Core (GSFC) process were prepared to be tested with a CTOD test with the same conditions as before. Finally, to complete the testing under other welding process, shield metal arc welding (SMAW) 3 specimens from a 60 mm panel referenced as 3 AW were obtained. CTOD testing at -10ºC in aforementioned standards were performed. In total 9 specimens from panels welded in three different process. All of them in 450 EMZ micro alloyed steel.

Even though the study is performed over 9 specimens it should be very repetitive to represent all research on the 9 specimens therefore we are going to describe a complete develop only on the three specimen which have been welded with a higher heat input as is the SMAW process that correspond the specimens 01-10(7), 01-11(8) and 01-12(9). Specimens preparation and CTOD fracture toughness testing were like those applied for a previous testing set of the nine specimens.

## EXPERIMENTAL WORK IN IN A SAW PROCESS

### Specimen Preparation

From the welded test panel, the specimen blanks were extracted and prepared transverse to the weld joint the specimens. They were machined to the preferred (Bx2B) specimen geometry according to BS 5762(1979). The nominal dimensions were thickness B=60 mm. with 2B=W=120 mm and length 630 mm. The weld cross section on either side of the specimens were polished and etched with 5% of nital to allow the location of the notch on both sides and parallel zone to the fusion line, i.e., coarse grains area in the HAZ.

This allowed to indicate accurately, at either side of the specimen, the weld metal centerline. In case the specimen was not perfectly cut perpendicular to the weld axis, resulting in two notch positions located in two different planes perpendicular to the longitudinal specimen axis, the location of the machined notch was taken as the average position between both indications.

The mechanical crack started notch consisted of a 2mm wide slot, sawn to an appropriate depth of 50 mm and finished with a 45º V-shape to facilitate fatigue crack initiation.

Since the test panel was delivered in the as-welded condition, a mechanical stress relief treatment of the ligament, i.e. a lateral precompression of 0.5% of the specimen thickness B, has been implemented prior to fatigue precracking. This precompression was applied to assure a regular fatigue crack tip complying with BS 5762(1979).

### Pretesting Metallography

The grain size evaluation was carried out by the method of Heyn in accordance with the ASTM 112-63 standard. Grain sizes were classified in the heat affected zone (HAZ) in the following categories, at weld start and weld finish represented in Table 13. (Leiva, Donoso, Muehlich, & Labbe, 2004).:

- Coarsed area, called GCHAZ and according to ASTM are those grains having size less than 5, i.e., over 65 μm.
- GCHAZ as above but with grain size of ASTM 5-7, that is, between 64 and 32 μm.
- Fine-grained zone (FGHAZ) and whose grain size upper of ASTM 7, i.e., less than 32 μm.
- Intercritical zone it is so called as the peak temperature is between AC1 <Tp <AC3 and grain size between 9 ASTM (15μm) and 11 ASTM (7 μm).
- Subcritical area where Tp <AC3 and size of grain above 11 ASTM and below 7 ASTM.
- Base metal 12 ASTM grain size, i.e., 12 (5μm).

The different zones at different points thickness were marked and measured according to ASTM 112 and percentages of fine grain size measured in the entire thickness and in 2/3 of the thickness (central 2/3 = 100%), this last condition is in accordance with the EEMUA 158(2014) standard requirements. Macrograph coupon initially obtained (welded panel), and its grain size corresponds to Figure 2.

*Figure 2. Macrograph of the Specimen Section.*

Of said coupon, nine samples were extracted from three different panels in the preferred geometry B x 2B as the standard BS 5762(1979). which were welded in following process:

S.M.A.W (3AW) Specimens 01-01(1), 01-02(2), 01-03(3)

G.S.F.C. (7AW) Specimens 02-01(4), 02-02(5), 02-03(6)
S.A.W. (01AW) Specimens 01-10(7), 01-11(8), 01-12(9)

In Figure 7 measured percentages of the various ASTM grain sizes for the specimen 01-10(7) are shown depicting Mapping of fatigue fracture of the specimen 01-10 (7) in the HAZ conditions AW

Given the difficulty of drawing a straight line completely along the area where the successive welding beads have not refined the grain, as GCHAZ (coarse region), this line where is the notch includes Intercritical reheated zone (IRGCHAZ) and reheated subcritical zone (SC).

## Precracking by Fatigue

One problem that design of these steels of high resilience have shown is their low toughness in the HAZ, when they are welded with a high HI (heat input).

In this work we have studied nine specimens that have been welded under three different process a mentioned before, being the specimen 01-10(7) welded in a SAW (submerged arc welding) process controlling the welding parameters and checking in the HAZ of such specimens, critical tensions ($\sigma_{cr}$)at the ends of the cracks, the critical cracks lengths ($a_{cr}$) and stress intensity factors ($K_{IC}$).

It is intended to check that the parameters that indicate the values of fracture mechanics: $(\sigma_{cr}, a_{cr}, K_{IC})$, $da/dN=A(\Delta K)^{m}$, *etc*.,) in the HAZ, after heat cycle to which the steel has undergone, under a process with a maximum HI of 2.327 kJ / mm, they are still valid, with the welding parameters applied. (González-Palma, R., Suárez-Bermejo, J. C., & Vicario, F. J., & Aurelio Muñoz. (2006)).

It is checked a correlation between the theoretical values and those obtained experimentally.

Of the nine specimens, in this section, we have represented those which have been welded in SAW, so they are the higher H. I. which have been referenced as (01-10(7), 01-11(8) and 01-12(9) that we explain as follow:

Fatigue precracking was performed in three-point bending at room temperature on a 200 kN SCHENCK resonance pulsator. Throughout, the fatigue force ratio was R=0.1 and the loading span was equal to nominally four times the specimen width W. Crack extension was followed visually at both specimens faces and was stopped once the fatigue crack had reached the appropriate depth, i.e. half of the specimen width W. The final fatigue crack depth versus the specimen width ($a_m$/w), the maximum stress intensity factor ($K_{fmax}$) and the number of cycles (N) applied during fatigue precracking are reported in Table 1.

## CTOD Testing Procedure

Displacement controlled CTOD fracture toughness testing was performed in three-point bending on a universal servo-hydraulic 1000 KN ESH testing machine. The actuator displacement rate was kept constant at $19\times10^{-3}$ mm/s and loading span S (Figure 11) was qual to four times the nominal specimen width W i.e. 480 mm. The applied load P was measured from a calibrated load cell having the same capacity as the actuator. The crack mouth opening $V_g$ was measured by means of a clip-on gauge clamped between two knife edges, firmly glued onto the specimen surface. This knife edges had a height of 3mm (z=3mm) were positioned over the crack mouth at a mutual distance of 6 mm (Figure 3). The test temperature of the CTOD testing was carried out at a temperature of -10ºC.

During testing, a diagram of the applied load (P) was continuously recorded as a function of the crack mouth opening displacement ($V_g$). Testing was interrupted as the specimen could not sustained any further increase in load. Figures (3,4 &5)

## Procedure for Validation

We hereby comment the specimens 01-10(7), 01-11(8) and01-12(9) as an example of the nine specimens here studied, which results have been represented in Table 1 to Table 4 both included.

Following CTOD fracture toughness testing, McDonald (1991), the specimens have been subjected to a detailed optical metallographic examination to determine the fatigue crack position relative to heat affected zone profile along the straight fusion boundary, i.e. to verify whether the crack tip effectively sampled significant portions of the grain coarse grain heat affected zone (GCHAZ) as intended. Kujawski et al. (2011).

A slice of nominally 15 mm was therefore removed each specimen containing the weld metal side (WMS) of the fracture surface. This piece was sectioned perpendicular to both the place surfaces and the fracture surface a small distance back from the fatigue crack front. These sections were prepared for microscopic examination to record the microstructures and ASTM E112 grain size members sampled by the fatigue crack front.

Since no significant portions of heat affected zone microstructures with an ASTM grain size number smaller than 5 could be detected in the test specimens, the observed microstructures were grouped into the following categories. Table 3:

- Weld Metal
- Grain Coarsened heat affected zone (GCHAZ) with ASTM grain size number smaller than 7. Temperature reached 1450ºC<$T_p$<1527ºC
- Grain Coarsened heat affected zone (GCHAZ) with ASTM grain size number between 7 and 9. Temperature reached 1100ºC<$T_p$<1450ºC
- Fine Grain heat affected zone (FGHAZ), ASTM grain size between 7 & 9. Temperature reached 900ºC<Tp<1100ºC
- Intercritical heat affected zone (ICHAZ). ASTM grain size between 9 & 11, Temperature reached AC1<$T_p$<AC3, (720ºC<$T_p$<900ºC)
- Subcritical heat affected zone (SCHAZ). ASTM grain size above 11, Temperature reached 600ºC<$T_p$<AC1, .
- Base Metal ASTM> 12, Temperature reached Tp<500ºC

Further, a photomicrograph of a typical grain coarsened heat affected zone region was taken from each cross section close to the fatigue crack front in the Figure 8 it is represented for specimen 01-10(7). Finally, the location of the fracture initiation point and the corresponding microstructure have been recorded as in all case a distinct crack initiation point could be identify on the fracture surfaces in the Figure 6 we can see the one for specimen 01-10(7)

Upon completion of the micrographic examinations, the weld metal side (WMS) sections have been repolished and macro etched in 5% nital to visualize the position of the fatigue crack front relative to the "straight" fusion boundary (Figure 2).

## Test Results

The CTOD fracture toughness values have been calculated though the guidelines of BS 5762 (1979), using the average crack depth, including the stable crack growth portion ($\delta_a$) (column 7, Table 16). The yield strength of the HAZ to be used for the calculating the elastic portion of the CTOD fracture toughness, was taken as the average of the specified base metal yield strength ($\sigma_y$= 429 N/mm$^2$) and the yield strength of the weld metal deposit ($\sigma_y$= 483 N/mm$^2$) of the test panel 01-AW-HAZ. The latter have been determined previously at room temperature on an all weld metal tensile test piece ($\phi$=10 mm) for earlier weld metal CTOD fracture toughness tests on the same test coupon.

The fatigue precracking data and the detail numerical results are summarized in Table 1 and Table 2. This also includes test results obtained from the original set of heat affected zone test specimens but extracted from panel 01-AW-HAZ. The CTOD value listed in this table are followed by a symbol "u", indicating the fracture mode. This symbol "u" is used when there is evidence to suggest that a pop-in (short arrested brittle fracture) or unstable fracture was preceded by ductile tearing. This symbolization conforms to BS5762(1979).

A summary of the post-testing micrographic examinations, aimed at "metallurgical validation "of the CTOD fracture toughness test specimens, is presented in Table 3. This Table lists the estimated amounts of the different microstructures intersected with the fatigue crack front both through the full specimen thickness and through the middle two thirds of the specimen thickness (excluding the outer B/6). The purpose of relative portions of the microstructure sampled across the full thickness is to demonstrate that at least in one case (specimen 01-10 (7) most of the heat affected zone target area was sampled in the outer B/6 areas, which are excluded by EEMUA 158 Publication n°158(2014).

We summarize in Table 3 both the numerical data and the main findings of the micrograph investigation, i.e. the total amount of the grain coarsened heat affected zone structure inserted with the fatigue crack front in the middle two thirds of the specimen thickness and the microstructure at point of crack initiation. Specimens 01-10(7), 01-11(8) and 01-12(9)

In the Table 4 is represented the CTOD values with the mode of fracture, the relative Amount of GCHAZ sample by the Fatigue Crack and Microstructure at Fracture Initiation Point. As we can see in the Table the specimen 01-12 (9) that shows higher quantity of GCHAZ (GC<7, 7.8%) has the lower value of CTOD=0.22mm. Nevertheless, of specimens obtained values over the standard requires 0.2 mm proving enough fracture toughness.

Fully detail information on CTOD fracture toughness testing and on subsequent metallurgical validation are given in the Figures 5-8.

- The diagram of the applied load (P) versus crack opening displacement ($V_g$), as recorded during testing at -10°C. Figure 5
- The photograph of the one fracture face (base metal side). The photograph illustrate that all test pieces could be considered as valid to BS 5762(1979) as regards both the regularity of the fatigue crack front and the fatigue crack depth (which should be between 45% and 55% of the specimen width W); Figure.6.
- The visualization of distribution of the different microstructures sampled by the fatigue crack front. The estimated amounts and positions of these macrostructures and ASTM grain size numbers are presented in bar chart mentioning the relative amounts of different microstructures sampled, excluding the B/6 (B is the specimen thickness) from each plate surface. Figure 7 (specimen 01-10(7))
- A photomicrograph of a typical grain coarsened heat affected zone area present close to the fatigue crack front. Figure.8 (specimen 01-10 (7)).

## CTOD ELASTICS AND PLASTICS COMPONENTS

The CTOD elastics and plastics components are determined as follow (we represent aa an example the specimen 01-10 (7), Elastic component:

*Table 1. Stress Intensity factor and Cycles in Specimens.*

| Specimen Number | Fatigue Crack Depth $a_m$(mm) | Ratio of Fatigue Crack Depth to Specimen Width $a_m$/W | Maximum Stress Intensity Factor $K_{fmax}$(N/mm$^{3/2}$) | Number of Cycles Applied N |
|---|---|---|---|---|
| 01-10 (7) | 59.8 | 0.491 | 2090 | 67000 |
| 01-11(8) | 60.5 | 0.503 | 2180 | 63000 |
| 01-12(9) | 59.5 | 0.487 | 2070 | 62000 |

Material: Grade 450 EMZ, Welding Identification:01-Aw-Haz, Welding Process: Submerged Arc Welding
Aimed Notch Location: Heat Affected Zone (Haz)
Notch Orientation: Through Thickness
Specimen Geometry: B×2b (Preferred Specimen Bs 5762,1979)
Frecuency:35 Hz
Fatigue Force Ratio:0.1
Temperature:20°C
Ctod Fracture Toughness Test Data Sheet

$$\delta_e = \frac{K^2\left(1-v^2\right)}{2\sigma_{ys}E} \tag{4a}$$

Being: $K$=fracture toughness (N/mm$^{3/2}$) =5683.53 N/mm$^{3/2}$

$\sigma_{ys}$= yield strength(N/mm$^2$) = 456 MPa =456N/mm$^2$
$v$= Poisson Module =0.3
E= Young Module (N/mm$^2$) =205000N/mm$^2$.

Where

$$K = \frac{YP}{B\sqrt{W}}$$

where $Y$=f(a/W) that according to BS5762(1979); $Y$(0.491)=10.31 and P=load= 376000 (Figure 3),on specimen, B= specimen thickness = 61.5 mm and W= 2B≅121.7mm according to Table 2. So

$$K = \frac{10.31 \times 376000}{61.5 \times \sqrt{121.7}} = 5714N / mm^{3/2}$$

$$\delta_e = \frac{5714^2 \times \left(1-0.3^2\right)}{2 \times 456 \times 205000} = 0.16\text{mm}$$

For CTOD plastic the following expression is to be used: Figures 3-5. Plastic component

$$\delta_p = \frac{0.4(W-a)V_p}{0.4W+0.6a+Z} \quad (4b)$$

Being: $a_m$(mm)=a = the crack length =59.8 mm.
$V_p$(mm)= the plastic opening of the crack (see Table 2 and Figure 3) = 2.76 mm
Z= high of the knife= 3mm, being $V_g$ the total opening (Figure 3 & 4)

*Table 2. Experimental Values in CTOD Testing*

| Specimen Number | B (mm) | W (mm) | $a_m$ (mm) | $a$ (mm) | P (kN) | $V_P$ (mm) | $\sigma_y$ (MPa) | $\delta_e$ (mm) (*) | $\delta_p$ (mm) (*) | $\delta$ (mm) (*) | Fract. Mode (**) | Valid (BS 5762) |
|---|---|---|---|---|---|---|---|---|---|---|---|---|
| 01-10(7) | 61.5 | 121.7 | 59.8 | 60.0 | 376 | 2.76 | 456 | 0.16 | 0.78 | 0.94 | u | Yes |
| 01-11(8) | 61.7 | 120.2 | 60.5 | 61.2 | 378 | 2,88 | 456 | 0.18 | 0.77 | 0.96 | u | Yes |
| 01-12(9) | 61.2 | 122.2 | 59.5 | 59.6 | 304 | 0.40 | 456 | 0.10 | 0.11 | 0.22 | u | Yes |

(*): $\delta_e$ and $\delta_p$ are the elastic and plastic component of the CTOD fracture toughness respectively, whilst $\delta_t$ is CTOD fracture toughness.
(**): The symbol "u" is used when there is evidence to suggest that the occurrence of a pop-in an unstable fracture was preceded by ductile tearing.
Material: Grade 450 EMZ
Weld Identification: 01-Aw-Haz.
Condition: As Welded
Welding Process: Submerged Arc Welding.
Specimen Geometry: B×2b Preferred Specimen Bs 5762(1979).
Aimed Notch Location: Heat Affected Zone (Haz).
Notch Orientation: Through Thickness.
Test Temperature: -10°C.
Displacement Rate: $19\times10^{-3}$ mm/S.

So,

$$\delta p = \frac{0.4\times(121.7-59.8)\times2.76}{0.4\times121.7+0.6\times59.8+3} = 0.78mm$$

CTOD total:

$$\delta = \delta_e + \delta_p = 0.16 + 0.78 = 0.94mm.$$

In the Figure 3, we can see the representation of the notch section during the CTOD testing where the value of the $V_p$, and r(W-a) have been represented as well as others as: W, a, Z. In the Figure 4 of below the diagram of load indicating mechanic and plastic area, the plastic displacement $V_p$ and total displacement $V_g$ are represented. The elastic displacement will come given by the difference $V_e= V_g-V_p$. WM: Weld Metal; GC<7: Grain Coarsened Heat Affected Zone with ASTM Grain Size Number Smaller than 7; GC7-9: Grain Coarsened Heat Affected Zone with ASTM Grain Size Number between 7

*Figure 3. Flection in Three Points CTOD Testing*

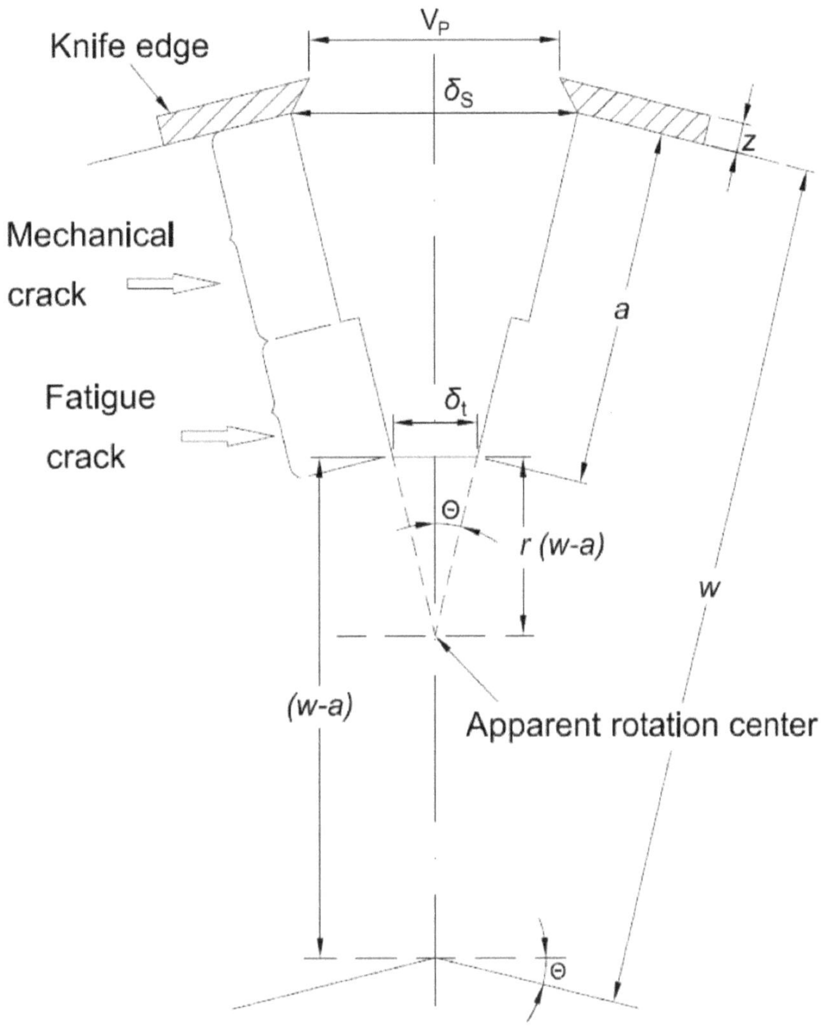


and 9; FG: Fine Grain Heat Affected Zone; IC: Intercritical Heat Affected Zone; SC: Subcritical Heat affected Zone; BM: Base Metal.

## Summary Of The Micrographic Examination

(Mapping) Of The Ctod Fracture Toughness Test Specimens.
Material: Grade 450 EMZ
Welding Identification: 01-Aw-Haz
Welding Process: Submerged Arc Welding (Saw)
Condition: As Welded

*Figure 4 Load in CTOD Testing vs. Displacement.*

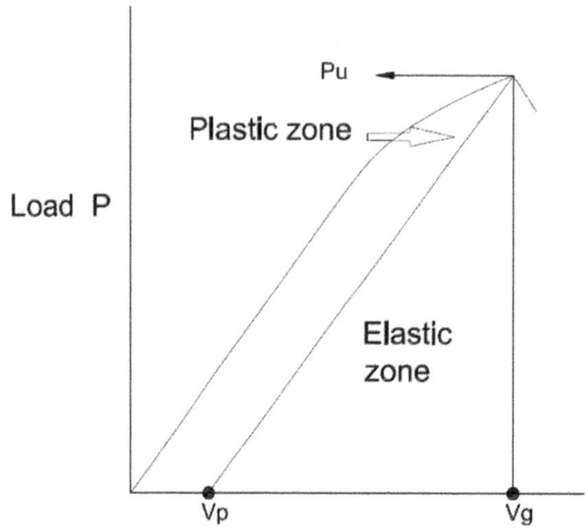


*Table 3. Amount of Microstructure Sampled by the Crack.*

| Specimen Number | CTOD Value at -10°C | Total Amounts of Sampled Microstructures Relative to the Full Specimen Thickness B (B=100%) | | | | | | Total Amounts of Sampled Microstructures Relative to the Middle Two Third (2B/3=100%) | | | | | |
|---|---|---|---|---|---|---|---|---|---|---|---|---|---|
| | (mm) | WM (%) | GC<7 (%) | GC7-9 (%) | FG (%) | IC (%) | SC+BM (%) | WM (%) | GC<7 (%) | GC7-9 (%) | FG (%) | IC (%) | SC+BM (%) |
| 01-10(7) | 0.94u | 12.7 | 4 | 2.7 | 43.7 | 20.5 | 16.4 | 0.0 | 0.0 | 4.0 | 58.1 | 21,6 | 16.3 |
| 01-11(8) | 0.96u | 44.0 | 6.2 | 1.2 | 46.0 | 2.7 | 0.0 | 46.3 | 5.0 | 1.3 | 43.4 | 4.0 | 0.0 |
| 01-12(9) | 022u | 37.0 | 7.8 | 4.0 | 32.3 | 19.0 | 0.0 | 40.0 | 7.8 | 4.7 | 30.5 | 17.0 | 0.0 |

*Table 4. CTOD and Microstructure at Fracture Initiation*

| Specimen Number | CTOD | | Mode (BS5762) | Relative Amount of GCHAZ Sample by the Fatigue Crack (2/3) B=100% | | Microstructure at Fracture Initiation Point |
|---|---|---|---|---|---|---|
| | Individual (mm) | Mean (mm) | | GC<7 (%) | GC7-9 (%) | |
| 01-10(7) | 0.94 | | (u) Fracture | 0.0 | 4.0 | FGHAZ at 30.0 mm from lower surface |
| 01-11(8) | 0.96 | 0.71 | (u) Fracture | 5.0 | 1.3 | FGHAZ at 24.5 mm from lower surface |
| 01-12(9) | 0.22 | | (u) Fracture | 7.8 | 4.0 | FGHAZ (close to GCHAZ) at 23.5 mm from lower surface |

*Figure 5. Diagram of Load Versus Clip Gauge Displacement Recorded During Testing Speciment 1-10(7) Opening $V_g$ - Load P*

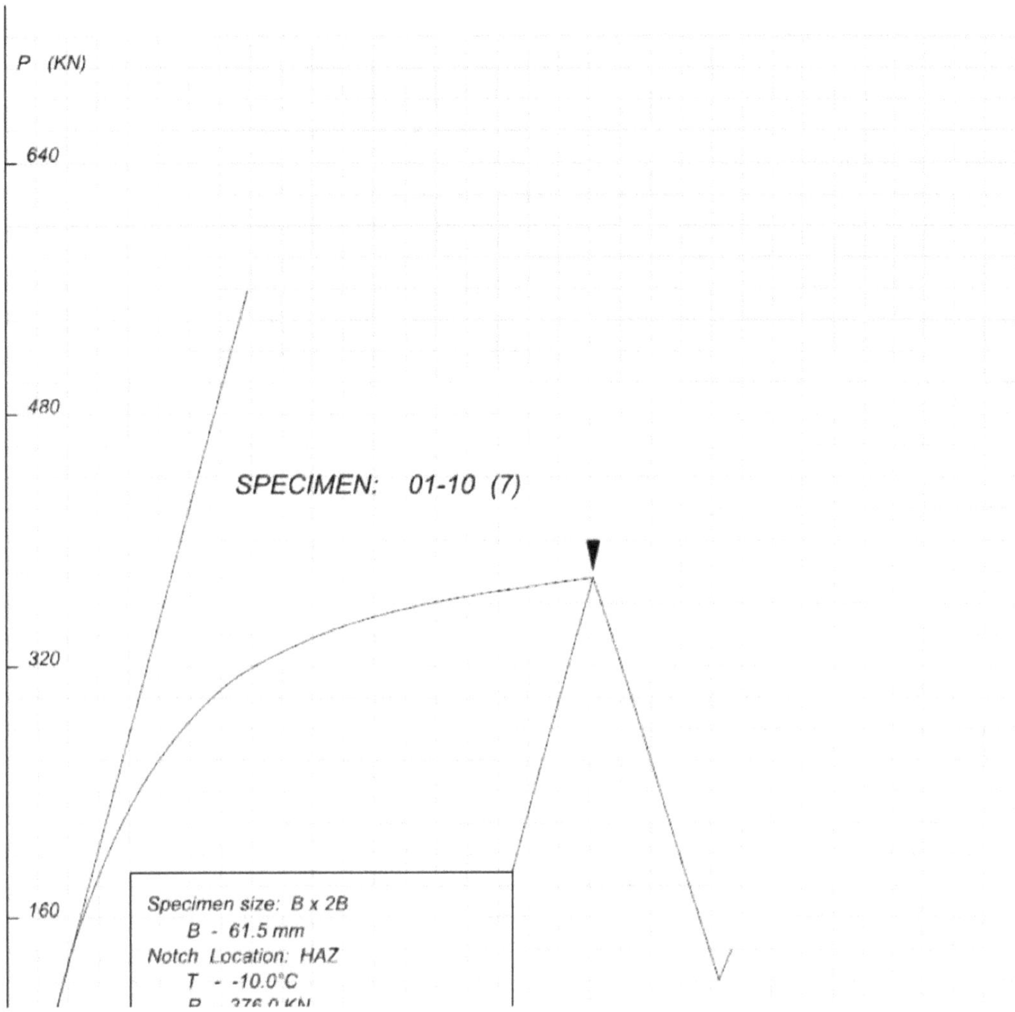


WM: Weld Metal; GC<7: Grain Coarsened Heat Affected Zone with ASTM Grain Size Number Smaller than 7; GC7-9: Grain Coarsened Heat Affected Zone with ASTM Grain Size Number between 7 and 9; FG: Fine Grain Heat Affected Zone; IC: Intercritical Heat Affected Zone; SC: Subcritical Heat affected Zone; BM: Base Metal.

Material: Grade 450 EMZ

Weld Identification: 01-Aw-Haz

Condition: As Welded

Welding Process: Submerged Arc Welding.

Specimen Geometry: B×2b Preferred Specimen Bs 5762.

Aimed Notch Location: Heat Affected Zone (Haz).

Notch Orientation: Through Thickness.

Test Temperature: -10°C.

*Figure 6. Surface of Specimen 01-10(7)*

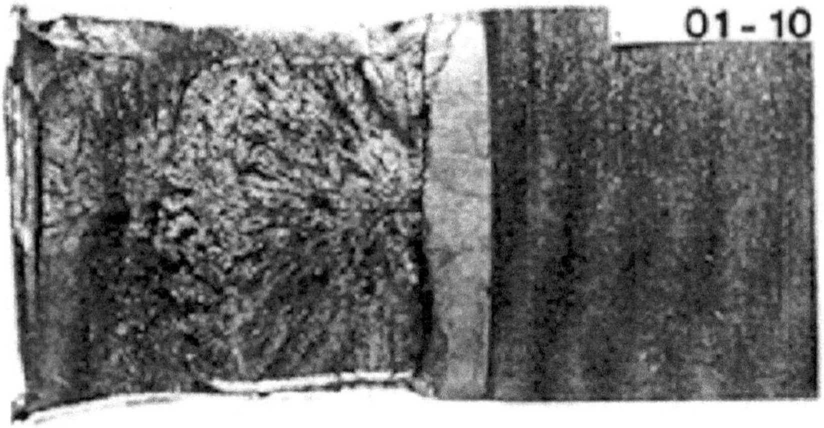


Displacement Rate: $19\times10^{-3}$ mm/S.

## CONCLUSION

In order to quantifying the resistance against brittle fracture initiation of the heat affected zone of an as-welded submerged arc welded joint in a 60 mm thick Grade 450 EMZ steel plate were carried out as additional series of welding procedure qualification CTOD fracture toughness tests.

The CTOD fracture toughness tests have demonstrated that the microstructure intersected by the fatigue crack tip possess an adequate average but a moderate minimum resistance to brittle fracture initiation. The CTOD fracture toughness values recorded at -10ºC ranged from 0.22 to 0.96 mm, been the mean value 0.71 mm (Table 4) all cases shown a ductile tearing (stable crack extension) was visible ahead of the fatigue crack tip. Which is indicative of a good resistance against brittle fracture initiation. The post testing optical metallographic examinations carry out to verify the fatigue crack tip placement relative to the grain coarsened heat affected zone (GCHAZ) microstructure, i.e. to validate the CTOD tests, have shown that the fatigue crack again did not always interest the GCHAZ region as intended.

In case of specimen 01-10 (7) the fatigue crack front was predominantly in the fine-grained heat affected zone at more than 0.6 mm from the straight fusion line, according to Table 4. No grain GCHAZ regions were intercepted in the middle two thirds of the specimen thickness 01-10(7) (Table 3). Due to the fatigue crack was not positioned within 0.5 mm from the fusion boundary and did not sample the coarsest grained heat affected zone microstructure available, this test should be considered invalid at this point according to EEMUA 158(2014).

Better results in this respect were obtained for specimen 01-11 (8) and 01-12(9), so the percentage of the target heat affected zone microstructure (i.e. the GCHAZ region at less than 0.5 mm from the fusion boundary) intersected with the fatigue crack front in the central two thirds of the specimen thickness were 5% and 7.8% respectively (Table 4.) It is acknowledged, however, that the maximum amount of coarsest grained heat affected zone microstructures available within the middle 2B/3 was not sampled with the crack tip. The most accurate crack tip placement into the GCHAZ regions was found for specimen

*Figure 7. Schematic Representation of the Amount and Distribution of Heat Affected Zone Microstructure Sampled by the Fatigue Crack in Specimen 01-10(7). Grain Zone Distribution*

| | ZONE | B % | 2/3 B % |
|---|---|---|---|
| | WM | 12.7 | .0 |
| | CG | 4.0 | .0 |
| | CG << | 2.7 | 4.0 |
| | FG | 43.7 | 58.1 |
| | IC | 20.5 | 21.6 |
| | SC + BM | 16.4 | 16.3 |

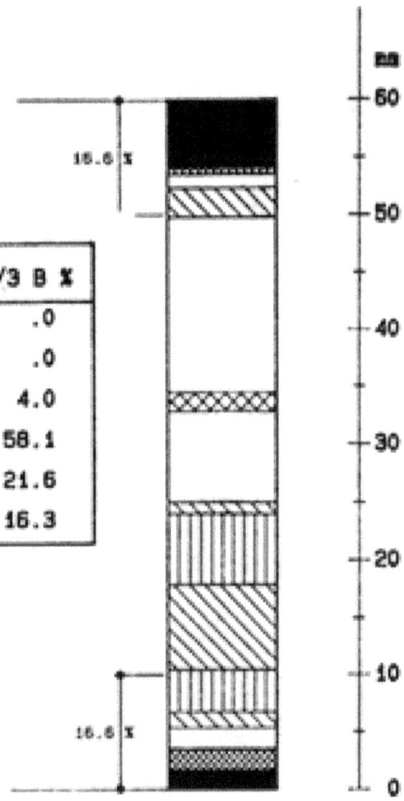


Specimen 01-10(7)

01-12(9), which yielded the lowest CTOD value ($\delta_u$=0.22 mm) and for which crack initiation occurred in a FGHAZ region, very close to a GCHAZ region (table 4).

Although it seems that the CTOD fracture toughness is not inversely proportional to the amount of GCHAZ sampled by the fatigue notch (Table 4) the lowest results in both series of specimens effectively were obtained in the specimens which contained the maximum amount of the GCHAZ areas. On the other hand, it is believed that higher sampling portions of the target area can hardly be obtained, the present series of tests would only lead to more specimens with a same low rate of sampling success.

The low rate of success of placing the fatigue crack tip in the target heat affected zone areas was due to some extent to the fact that the fusion boundary is not very straight and, moreover, not perfectly normal to the plate surfaces. In fact, the heat affected zone profile was such that it was extremely difficult, if at all feasible, to place the fatigue crack in the coarsest grained heat affected zone over a significant length. This stresses the need that, in order to obtain valid CTOD tests, special precautions should be taken to avoid irregularities of the fusion boundary.

*Figure 8. Micrograph of the Grain Coarsened Close to the Crack. Specimen 01-10(7)*

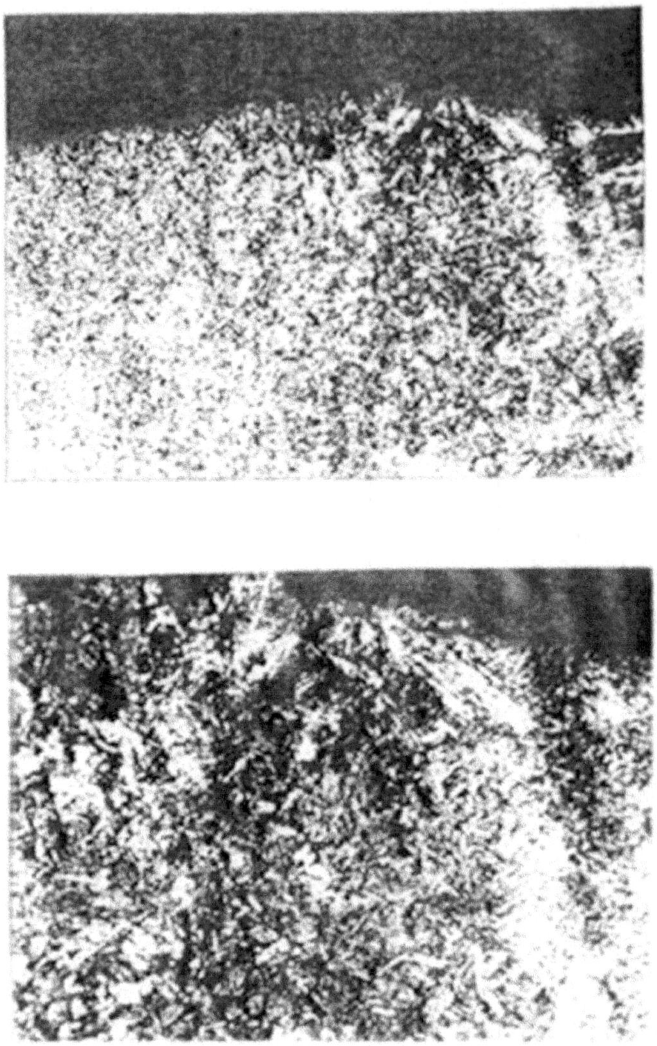

(91/1428)
Magnification: x 100

(91/1424)
Magnification: x 200

**PHOTOMICROGRAPHS OF THE GRAIN COARSENED HEAT AFFECTED ZONE AREA CLOSE TO THE FATIGUE CRACK IN SPECIMEN 01-10 (7)**

- **ASTM Grain Size Number: 5-8**
- **Distance from Lower Surface: 3 mm (see mapping chart)**

## EXPERIMENTAL WORK IN THE NINE SPECIMENS

After the machined notch 50 mm deep, it was proceeded to a lateral precompression, continuing the fatigue testing of specimens to ambient laboratory temperature in resonance pulsator 200 KN Schenck, with a ratio $R = \frac{P_{min}}{P_{max}}$ =0.1, being the span between supports four times the width of the test piece W. The crack propagation (a) was followed visually stopping when. $a_m/W$»0.5

The experimental values of the stress intensity factor based on the number of cycles, for each of the nine samples are shown in Table 5, where "*W*" is the width of the specimen and "*a*" the length of the crack.

Is represented by "*a*" the total crack length ($a_m+\delta_a$), i.e., when to maintain the value of the constant "P" load the crack extends to breakage and "$a_{cr}$" is the critical crack length corresponding to the first part of the crack extension, i.e., crack growth fatigue. The difference between values "*a*,, and "$a_m$" determines the stable growth "$\delta_a$". In the last column N x1000 are represented number of cycles required to reach the crack length "$a_m$". By "$K_{fmax}$" is represented the maximum stress intensity factor, during the vibration test corresponding to the crack formation "$a_{cr}$", all values represented in Table 5 have been taken from experimental CTOD tests. These values were taken in the HAZ, and we will compare them with the base metal values, such as steel 450 EMZ should have under the point of view of design. If we start from the specimen with three-point bending and supporting on Elices-Calafat (1992) curves $\left(\frac{a}{W}, \frac{K_1}{K_0}\right)$ Figure 9, the following empirical function is obtained:

$$\frac{K_I}{K_0} = 1.09 - 1.73\left(\frac{a}{W}\right) + 8.2\left(\frac{a}{W}\right)^2 - 14.2\left(\frac{a}{W}\right)^3 + 14.6\left(\frac{a}{W}\right)^4$$

Being, $a_m=a_0+a_{cr}$; $a=a_m+\delta a$, where $a_0$= mechanical notch in Figure 11 is represented by M, $a_{cr}$ represents the length of the crack by fatigue and it corresponds to first part of the testing which stops when $a_m/W \cong 0.5$, $\delta_a=a-a_m$, is the stable crack in the second part of the testing and it is carry out at a P constant. the testing stops when the sample is broken, pop-in or test stop.

In the above Table 5 are represented the HI with the specimens were welded in the three-welding process. The lowest cycles it is corresponded with Higher HI and the higher cycles corresponds to the lowers HI. It is obvious that the lowers welding energy is corresponded with the more fatigue cycles.

In Table 6 we can see the HI of the nine specimens and its relationship with stress intensity factor and CTOD as well as the number of cycles. We can observe that those specimens with bigger CTOD which were welded in process with lower HI as G.S.F.C. and S.M.A.W had higher of fatigue fracture cycles. This means that a lower HI makes toughness the HAZ in this specimen. No other conclusions should be taken out from this Table, since results as CTOD depend in great from metallography that shows the microstructure that crack goes through.

However, since the stress intensity factor ($K_{fmax}$) depends of the tension at crack end ($\sigma$) and of the crack length ($a_{cr}$), the conclusions are to be taken from values given on Table 7.

where $K_0 = \sigma\sqrt{\pi a}$ and $\sigma = \frac{MY}{I} = M\frac{\frac{W}{2}}{\frac{1 \cdot W^3}{12}} = \frac{6M}{W^2}$

Since the beam section is W×1, these values are substituted in:

$$K_0 = \sigma\sqrt{\pi a} = \frac{6M}{W^2}\sqrt{\pi a}\ ;\ \text{Where } M = \frac{P \times 2W}{2} = P \times W\ ;\ Y=W/2.$$

*Figure 9. Flection on Three Points. Relation a/b- $K_f/K_0$*

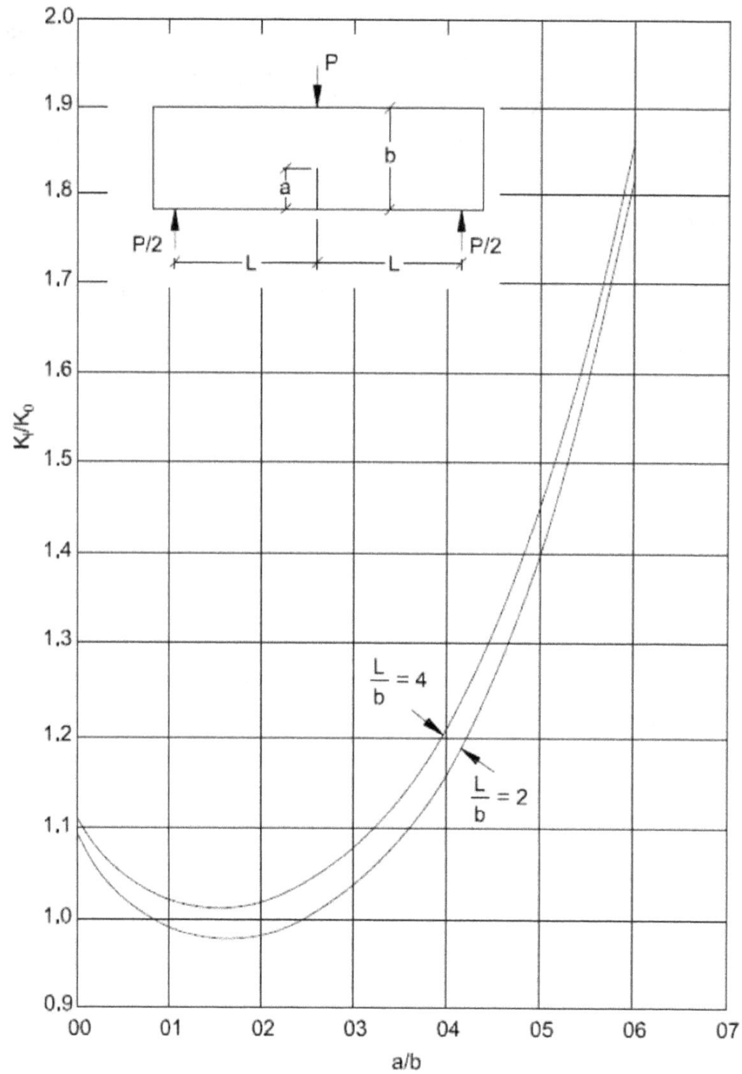


From Figure 9 it follows that the ratio a / b = 0.5W / W is approximately 0.5, this value entered in the abscissa's axis and for the curve L / b = 2W / W = 2.

$K_I / K_0$ takes a value of 1.4, then: $K_I = 1.4K_0 = 1.4\sqrt{a}$

If we start from the stress intensity factor; and we represent in Table 7 the values of $\sigma$, by this expression, the following values are obtained for "$a_{cr}$":

$$\sigma = \frac{Kf_{max}}{1.4\sqrt{\pi a_{cr}}} = \frac{Kf_{max}}{2.481\sqrt{a_{cr}}} => a_{cr} = \left(\frac{Kf_{max}}{2.481\sigma}\right)^2 \tag{5}$$

*Table 5. Fracture Propagation at" N" Cycles*

| Specimen | Heat Input HI | Crack Depth | | Relation Crack Specimen Width | Stress Intensity Factor $K_{fmax}$ | Cycles |
|---|---|---|---|---|---|---|
| | **(kJ/mm)** | **$a_m$ (mm)** | **$a$ (mm)** | $a_m/W$ | **(N/mm $^{3/}$)$^2$** | **Nx1000** |
| 01-1(1) | 1.560 | 61.70 | 62.30 | 0,513 | 1209 | 126 |
| 01-2(2) | 1.560 | 60.49 | 61.11 | 0,503 | 1155 | 118 |
| 01-3(3) | 1.560 | 61.62 | 63.19 | 0,513 | **1032** | 205 |
| 02-01(4) | **1.348** | **62.34** | 63.50 | 0.518 | 1222 | **337** |
| 02-02(5) | 1.348 | 60.51 | 61.52 | 0.502 | 1165 | 95 |
| 02-03(6) | 1.348 | 60.66 | 61.81 | 0.506 | 1173 | 128 |
| 01-10(7) | 2.327 | 59.80 | 60.00 | 0.491 | 2090 | 67 |
| 01-11(8) | 2.327 | 60.50 | 61.20 | 0.503 | **2180** | 63 |
| 01-12(9) | **2.327** | **59.50** | 59.60 | 0.487 | 2070 | **62** |

*Table 6. CTOD Experimental Values for the Nine Specimens in the Three-Welding Process.*

| Specimen Process | Heat Input | Crack depth | | Relation Crack Specimen Width | Stress Intensity Factor $K_{fmax}$ | CTOD | Cycles |
|---|---|---|---|---|---|---|---|
| **S.M.A.W.** | *(kJ/mm)* | $a_m$ (mm) | $a$ (mm) | $a_m/W$ | (N/mm $^{3/2}$) | (mm) | *Nx1000* |
| 01-1(1) | 1.560 | 61.70 | 62.30 | 0.513 | 1209 | 0.349(u) | 126 |
| 01-2(2) | 1,560 | 60.49 | 61,10 | 0.503 | 1155 | 0.794(u) | 118 |
| 01-3(3) | 1.560 | 61.62 | 63.19 | 0,512 | **1032** | 1.137(m) | 205 |
| **G.S.F.C.** | kJ/*mm* | $a_m$ (mm) | $a$ (mm) | $a_m/W$ | N/mm $^{3/2}$ | Mm | *Nx1000* |
| 02-01(4) | 1.348 | 62.34 | 63.50 | 0.518 | 1222 | 1.364(t.s.) | **337** |
| 02-02(5) | 1.348 | 60.51 | 61.53 | 0.502 | 1165 | 1.333(t.s.) | 95 |
| 02-03(6) | 1.348 | 60.66 | 61.80 | 0.506 | 1173 | 1.523(t.s) | 128 |
| **S.A.W.** | kJ/*mm* | $a_m$ (mm) | $a$ (mm) | $a_m/W$ | N/mm $^{3/2}$ | Mm | *Nx1000* |
| 01-10(7) | 2.327 | 59.80 | 60 | 0.491 | 2090 | 0.940(u) | 67 |
| 01-11(8) | 2.327 | 60.50 | 61.2 | 0.503 | **2180** | 0.960(u) | 63 |
| 01-12(9) | 2.327 | 59.50 | 59.6 | 0.487 | 2070 | 0.220(u) | **62** |

In Table 7 are represented the design stress values versus critical crack. Values obtained in this Table shows that the design stress in the heat affected zone (HAZ) based on the values from the crack obtained in the fatigue tests are acceptable values because $\sigma/\sigma_{ys}$ they are between 30.7% and 63.21% of the stress corresponding to yield strength from $\sigma_{ys}$= 557N/mm$^2$ to $\sigma_{ys}$=429 N/mm$^2$.

From these values, we can obtain the function for each specimen $\sigma = f(a)$, with these results obtained and the equation:

*Table 7. Cracking Stress/Yield Strength.*

| Specimen | $\sigma_{ys}$ (N/mm$^2$) | $K_{fmax}$ (N/mm$^{3/2}$) | $a_{cr}$ (mm) | $a_0$(machined notch) (mm) M | $\sigma$ (N/mm$^2$) | $\sigma/\sigma_{ys}$ (%) |
|---|---|---|---|---|---|---|
| 01-1(1) | 500 | 1209 | 7.30 | 54.40 | 180.33 | 36.07 |
| 01-2(2) | 500 | 1155 | 6.39 | 54.10 | 184.13 | 36.83 |
| 01-3(3) | 500 | 1032 | 7.34 | 54.28 | **153.51** | **30.70** |
| 02-01(4) | 557 | 1222 | 8.08 | 54.26 | 173.24 | 31.10 |
| 02-02(5) | 557 | 1165 | 6.37 | 54.13 | 186.02 | 33.40 |
| 02-03(6) | 557 | 1173 | 6.84 | 53.83 | 180.75 | 32.45 |
| 01-10(7) | 429 | 2090 | 9.8 | 50 | 269.12 | 62.70 |
| 01-11(8) | 429 | 2180 | 10.5 | 50 | **271.17** | **63.21** |
| 01-12(9) | 429 | 2070 | 9.5 | 50 | 270.70 | 63.10 |

Note: The machined notch $a_0$ is represented in Figure 11 as M

$$\sigma = \frac{Kf_{max}}{1.4\sqrt{\pi}} a_{cr}^{-1/2} \tag{6}$$

are derived for each of them the quantities shown in Table 8. From these values, those corresponding to specimen 01-10 (7) have been represented in the Figure 10. With these functions, for easier calculation, graphics have been obtained from which only those corresponding to specimen 01-10 (7) have been represented. Sree, P. C. R. & Kujawski, D. (2014)

This relation allows us to establish a correspondence between the crack size and nominal tension. That for the panel CTOD 01 AW that we are analysing in this work and from results obtained, are acceptable. Tension values are fluctuating between 153.51N / mm$^2$ and 271.17N / mm$^2$, as shown in Table 7. This means that these materials have acceptable fracture toughness in the HAZ.

## Crack Growth

The experimental work was carried out on a coupon of 450 EMZ HSLA steel 60 mm thick quenched and tempered steel welding in the following process and referenced as 01AW(SAW)(which is being studied in this paper), 03AW (SMAW) and 07AW(GSFC) in conditions "as welded" (without post-weld heat treatment). On the HAZ fracture mechanics test, CTOD were performed at a temperature of -10 °C. The tests were conducted conform to BS 5762(1979) standard and the publication of the EEMUA 158(2014).

Samples were prepared and tested in order to obtain at least three valid data from them in accordance with said rule. The tests were conducted on preferred geometry test pieces (B × 2B) in which was made a notch through the entire thickness length (M)=$a_0$ < W/2≅ 50 mm in this specimen, Figure 11.

It was tried to carry out the notch in the GCHAZ (coarse area, next to the fusion line) (FL). This experimental work is based on the results of tests CTOD 450 EMZ steel, on which this chapter consists, the results of nine samples were obtained according EEMUA 158 (2014) in the HAZ CTOD in trials of the panels 01 AW, 03AW and 07AW which are listed in Table 6.

*Table 8. Relation Between Tension and Crack Size*

| Specimen | $\sigma$=f($a$) (N/mm²) |
|---|---|
| 01-1(1) | 487.22×$a_{cr}^{-1/2}$ |
| 01-2(2) | 465.46× $a_{cr}^{-1/2}$ |
| 01-3(3) | 415.89× $a_{cr}^{-1/2}$ |
| 02-01(4) | 492.46× $a_{cr}^{-1/2}$ |
| 02-02(5) | 469.49× $a_{cr}^{-1/2}$ |
| 02-03(6) | 472.71× $a_{cr}^{-1/2}$ |
| **01-10(7)** | **842.25× $a_{cr}^{-1/2}$** |
| 01-11(8) | 878,52 × $a_{cr}^{-1/2}$ |
| 01-12(9) | 834,19× $a_{cr}^{-1/2}$ |

*Figure 10. Critical Crack Size $a_{cr}$ – Critical Crack Stress at Crack Tip.*
*Specimen 01-10(7)*

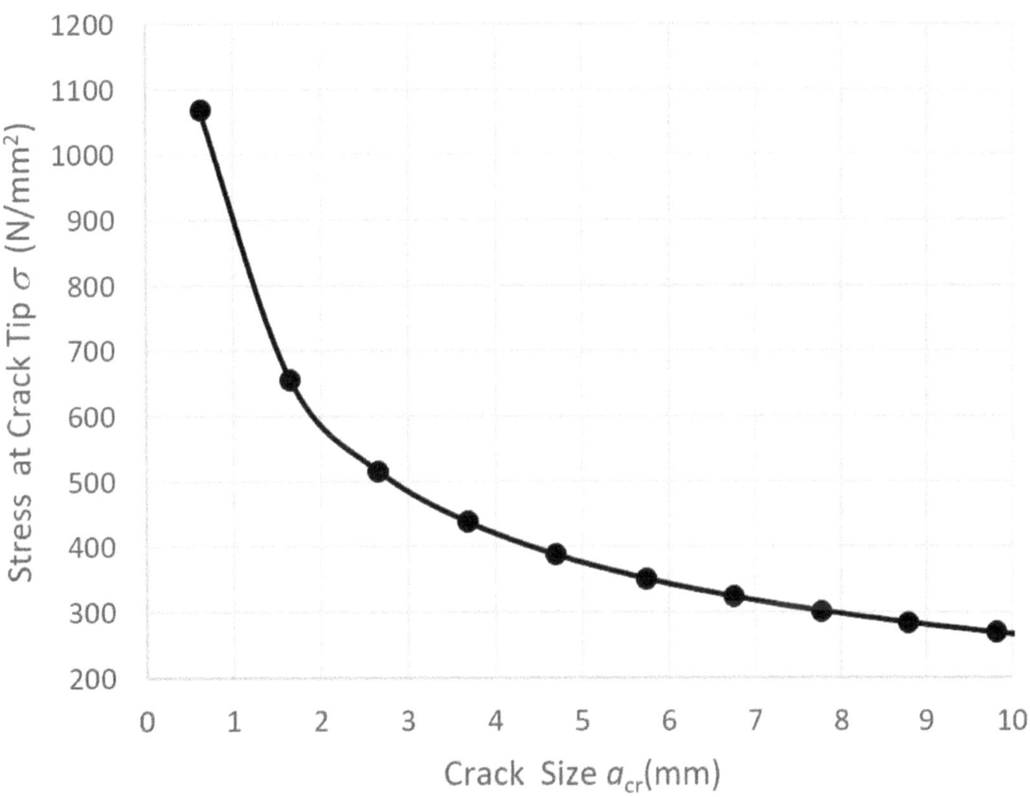

*Table 9. Table of figure 10*

| $a_{cr}$ (mm) | $\sigma$ (N/mm$^2$) |
|---|---|
| 0.62091 | 1068.88 |
| 1.64091 | 657.505 |
| 2.66091 | 516.328 |
| 3.68091 | 438.999 |
| 4.70091 | 388.463 |
| 5.72091 | 352.134 |
| 6.74091 | 324.401 |
| 7.76091 | 302.332 |
| 8.78091 | 284.231 |
| **9.80091** | **269.034** |

*Figure 11. Sample Dimensions*

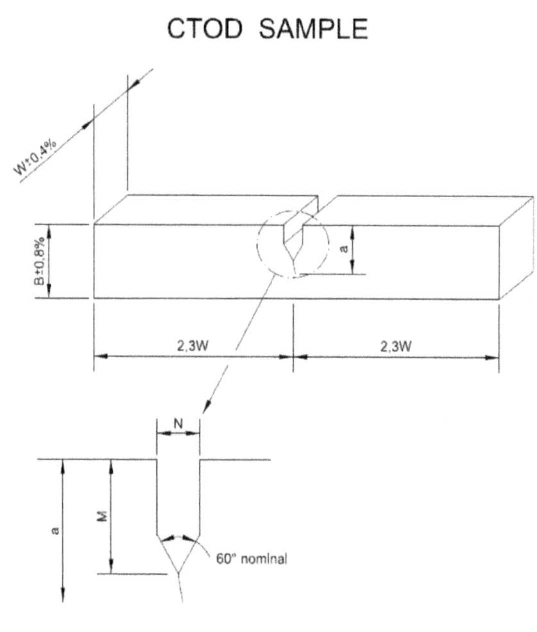


As shown in the values specified in the Table 5, the crack length ($a_m$) is between 59.5 mm and 62.34mm. A tension intensity factor between $K_{f\,max} = 1032N / mm^{3/2}$ and $K_{fmax} = 2180\ N / mm^{3/2}$ and cycles between $N$= 337000 and $N$ = 62000.

**We will take as example the specimen 01-10 (7)** which has the higher N value among the SAW group and one of the minimum values of the nine specimens. Considering that the notches made in the

*Figure 12. Number of Cycles N- Crack Length $a_{cr}$, Liu YB. et al. (2010).*
*Specimen 01-10(7)*

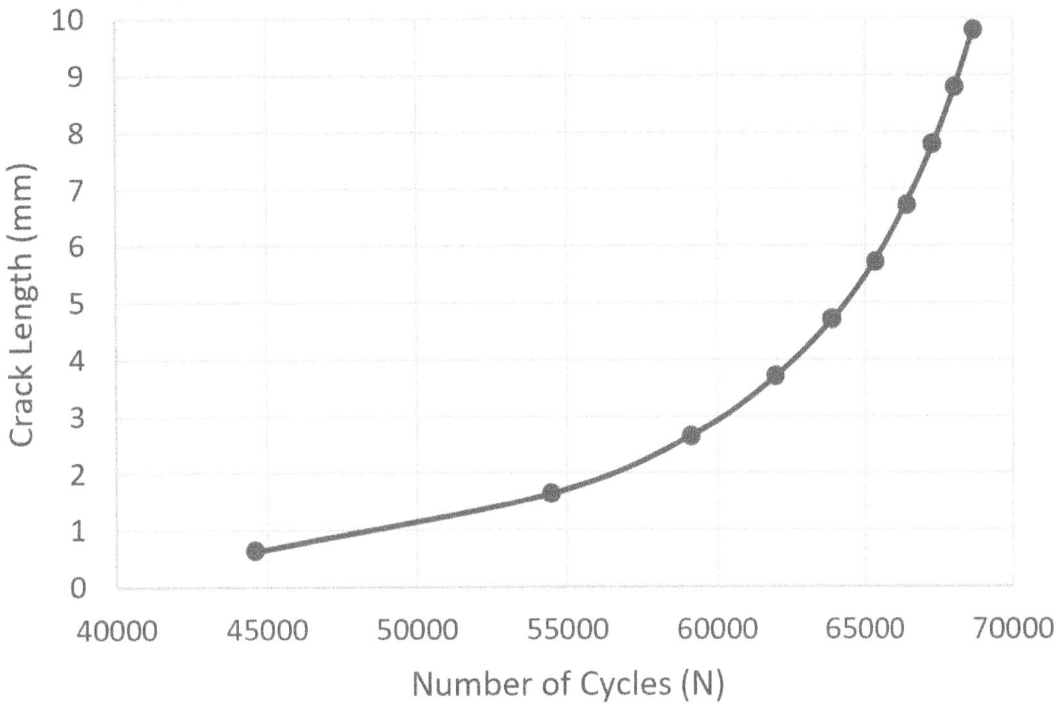


*Table 10. Table of Figure 12*

| Σ *N* (ciclos) | $a_{cr}$ (mm) |
|---|---|
| 44571 | 0.62091 |
| 54454 | 1.64091 |
| 59125 | 2.66091 |
| 61950 | 3.68091 |
| 63884 | 4.70091 |
| 65310 | 5.72091 |
| 66416 | 6.74091 |
| 67305 | 7.76091 |
| 68040 | 8.78091 |
| **68659** | **9.80091** |

specimens, have the design shown in Figure 11, we begin by determining the threshold values of crack propagation.

Researchers like Harrison (1970), Paris(1970) and Cooper(2004) among others, showed that for martensitic steels, bainitic and ferritic-pearlitic, the latter belonging to the 450 hardened and tempered

EMZ for the relationship: $R = \frac{\sigma_{min}}{\sigma_{max}} > 0.1$ meet the following function that gives the range of the stress intensity factor amplitude $\Delta K_{th}$ (threshold) below which the crack does not propagate:

$$\Delta K_{th} = 7(1 - 0.85R)MN \,/\, m^{3/2} \tag{7}$$

*Figure 13.Crack Size $a_{cr}$ - Crack Growth da/dN Specimen 01-10(7)*

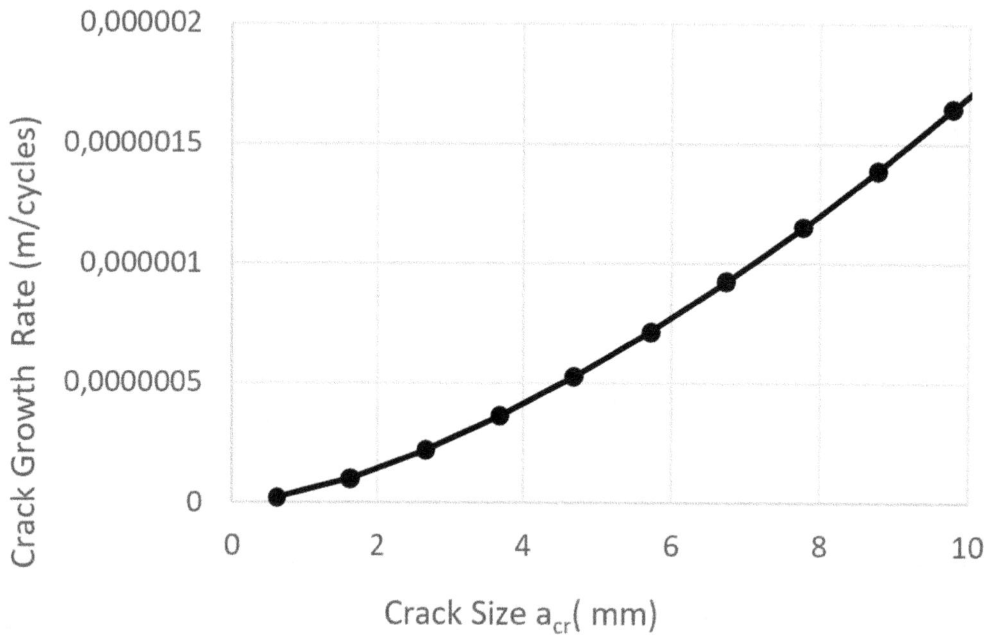


*Table 11. Table of Figure 13*

| $a_{cr}$ (mm) | da/dN (m/cycle) |
|---|---|
| 0.62091 | 2.28845E-08 |
| 1.64091 | 1.03212E-07 |
| 2.66091 | 2.18346E-07 |
| 3.680914 | 3.6106E-07 |
| 4.70091 | 5.2751E-07 |
| 5.72091 | 7.15187E-07 |
| 6.74091 | 9.22279E-07 |
| 7.76091 | 1.1474E-06 |
| 8.78091 | 1.38942E-06 |
| **9.80091** | **1.64745E-06** |

When R ≤ 0.1 our tests meet, $\Delta K_{th}$ have a constant value (Equation 1), so, below these values, the crack does not spread.

With the equation (18) and Table 22 we obtain the following graphic (Figure 12), where the $\Delta N$ have been rounded:

As we can see there are some difference between the number of cycles theoretical *N*= 68659 and the number of cycles *N*= 67000 obtained in the test for a crack size of 9.8mm, this difference can be justified for several reasons, Sun C. et al.(2012). One due to the coefficient of 1.4 that we have adopted for obtaining the *K* (stress intensity factor) other the small difference in the base material among others. Nevertheless, this difference between both value of *N* which is of 2.4% it can be acceptable. In the Figure 12 we see as the crack size varies with the number of cycles in Table 22 are all the values for the specimen 01-10 (7). The maximum $a_{cr}$ from the crack is extended takes place at 68658 Cycles. Figure 13 represents crack size $a_{cr}$ vs. crack growth rate d*a*/d*N,* as the crack size is getting bigger the crack rate increases, to obtain the graphic the equation (8) has being used.

From the results obtained in Table 8 curves of critical tension vs. critical crack are represented, (which we have plotted the values of the specimen 01-10 (7) in Figure 10, although the crack goes through the same material, different results are got, because the crack went through different zones of HAZ with different toughness values as if it the steel would have being made of different materials. The most toughness, as deduced from the values shown in Table 7 corresponds to the (SAW) process where there is a greater margin of safety design stress with a percentage of the critical yield stress.

For specimen 01-10 (7) the average value corresponds to a stress ratio of 62.7%, and $K_{fmax}$ = 2090Nmm$^{3/2}$. The expression of Paris gives the following values for ferrite – pearlite steel:

$$\frac{da}{dN} = 5.01 \times 10^{-12} (\Delta K)^{3,1} m \,/\, cycle; \quad 206.9\, N \,/\, mm^2 < \sigma_{ys} < 344.8\, N \,/\, mm^2 \qquad \text{(Ewald's, Chapter 9, section 9.5)} \qquad (8)$$

As we will see later, our results are close to the values Ferrite-Pearlite (structural steel).

The differences found are not significant and they are due to changes in thermomechanical processes.

In any case, these different values depend on the different changes in the HAZ microstructure, however we have taken an average value of the nine test pieces if we would have taken values for each would have gotten slightly different results as obtained for the critical stress vs. critical crack, Figure 10.

The same author mentioned above, consider using a single formula for ferrite- pearlite, martensitic and austenitic stainless steels, this is since the values obtained in the behavior of crack propagation are approximate, Krupp U. et al. (2014).

The study of the speed of crack growth (d*a*/ d*N*) as a function of the amplitude of the stress intensity factor equation (8), conducted by W.G. Clark (1970) under submerged arc process on Steel ASTM A 533 Grade B Class 1, it shows that the growth rates of the crack in the weld and in the heat affected zone are similar to the base metal at room temperature. Figure 14.

## Metallographic Test

The results of the metallographic test are shown in Table 13 and Table 14. These Tables links intercepted structures in a section of the sample taken by the end of the crack through the thickness and in the 2/3

*Figure 14. Intensity Factor Amplitude- Speed Crack Growth.*

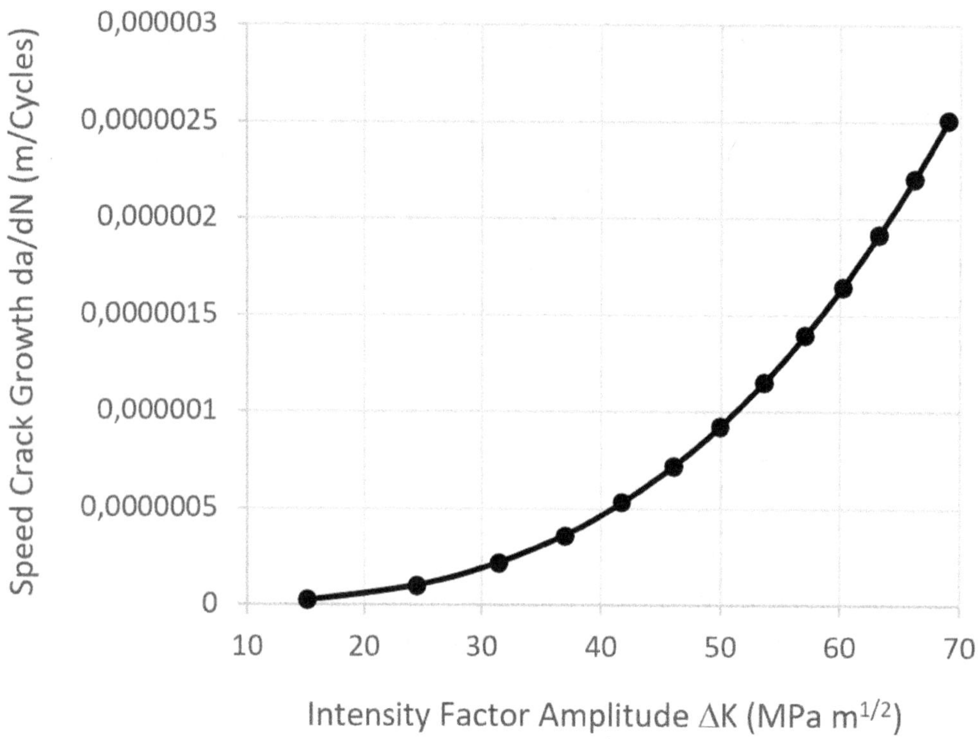


*Table 12. Table of Figure 14*

| **Δ*K* (MPam1/2)** | **d*a*/d*N* (m/ciclos)** |
|---|---|
| 15.155 | 2.28845E-08 |
| 24.636 | 1.03212E-07 |
| 31.373 | 2.18346E-07 |
| 36.899 | 3.6106E-07 |
| 41.699 | 5.2751E-07 |
| 46.001 | 7.15187E-07 |
| 49.934 | 9.22279E-07 |
| 53.579 | 1.1474E-06 |
| 56.991 | 1.38942E-06 |
| **60.210** | **1.64745E-06** |
| 63.266 | 1.92069E-06 |
| 66.180 | 2.20849E-06 |
| 68.972 | 2.51027E-06 |

thereof, as the standard relates mark and it is shown that in some cases the affected area is within the value of $B$ / 6 that the standard EEMUA 158(2014) Construction Specification for Fixed Off-Shore Structures in the North Sea, does not consider.

In relation with specimens 01-1(1), 01-2(2) and 01-3 (3) the results found of the structure sampled by the crack are given in the mentioned Table 13 and 14 being as follow: It can be seen from these results that, as expected from the pretest metallography, all three specimens sampled a high percentage of refined HAZ material (ASTM grain size >7). Consideration must be given to the fact that there was very little ASTM<5 microstructure available to aim for when notching the specimens (Table 13). The CTOD prove to be over the minimum required by the standard with values 0.35 mm, 0.79 mm and 1.14mm. In the specimen 01-3(3), the maximum force was reached, and the test stopped without broken the pieces.

For specimens 01-1(1) and 01-2(2) which exhibited brittle fracture behavior, a second section was taken through the initiation site. An optical macrograph of the 01-1(1) was taken. In this specimen fracture initiate in refined HAZ material at 1.5 mm from the fusion boundary. This specimen exhibited an extremely small pop-in, which was difficult to identify precisely. For specimen 01-2(2) an optical macrograph depicted that fracture initiation started in weld metal close to the fusion boundary.

Measurements were made along the fatigue precrack within central 2/3 specimen thickness to determine the amount of precrack that was actually within 0.5 mm of the fusion boundary (excluding sampled weld metal).The precrack specimen 01-1(1)was found to be within 0.5 mm of the fusion boundary over 63% of its length (within the actual 2/3 thickness region of interest), similarly 01-2(2) and 01-3(3) had figures of 38% and 30% respectively. The assessment of the validity of the specimen, in accordance with section A2.4.4 of the specification EEMUA 158 (2014) revealed that, all three tests were technically invalid as100% of the fusion boundary was not within 0.5 mm from the fatigue precrack.

Although the pretest metallography Table 13 found that between 15.7% (at weld start) and 9.8% (at weld finish) of ASTM<5 microstructure existed adjacent to the fusion boundary, in practice it would be impossible to sample the whole amount with a straight notch because the pretest survey followed the cups and undulation of the fusion boundary effectively taking a longer path.

As summary we can say that all test was valid accordance with to BS 5762(1979). All CTOD value $\delta_m$=1.14 mm, $\delta_u$=0.79 mm and $\delta_u$(pop-in) =0.35 mm all exceeded the specified minimum CTOD required for as welded joints 0.20 mm.

Nevertheless, none of these three specimens were fully satisfy with the EEMUA158(2014) Appendix 2 requirement that the precrack must be within 0.5mm of the fusion boundary throughout the central 2/3 thickness region. Percentages of grain coarsened material sampled and the amount of precrack within 0.5 mm of the fusion boundary were noticed.

The specimens 02-01(4), 02-02(5) and 02-03 (6) shown a high quantity of fine grained HAZ something that justify the CTOD high value with test stop (TS) without the specimen were broken.

We note that the sampling areas that cross the crack are those specimens that obtained a higher percentage of WM, GC and I.C. which CTOD values obtained in the test were the lowest, demonstrating the fragility of these three regions. Furthermore, the specimens 02-01(4), 02-02(5) and 02-03(6) with higher CTOD values have shown the more percentage of fine grained at the star and weld finish with ASTM grain size >7.The heating input(HI) of these three specimen was the lowest of the nine samples testing with a value of 1.348 kJ/mm welded in process GSFC as the Table 16 shows. The precrack in the specimen W02-01(4) was found to be within 0.5 mm of fusion boundary over thirty percent (30%) of its length, within 2/3thighness region of interest, similarly 02-02(5) and 02-03(6) had Figures of 46% and 63% respectively.

Of The values given in Table 13, one can conclude that in the samples 01-11 (8) and 01-12(9) the front of a fatigue crack was found in the weld metal (WM)as shown in the mentioned Table. As we can see on Table 14 the specimen 01-10(7),01-11(8) and 01-12(9) initiated the crack at the fine grained (FGHAZ)

The specimens 01-11(8) and 01-12(9) shown the greater weld metal intercepted by the crack in both (2/3)B=100% (Table 13). The biggest percentage intercepted by the crack corresponded to fine grained heat affected zone (FGHAZ), being the specimen 01-12(9) the lower CTOD ($\delta$) (Table 14 and 15), value that shown bigger grain coarsened heat affected zone (GC<7) and lower fine-grained heat affected zone. The three specimens broken as ultimate value $(CTOD)_u$. were specimens 01-10(7), 01-11(8) and 01-12(9)) that shown lower CTOD value are those specimens that the crack intercepted more percentage of grain coarse or WM. The welding energy (HI) of these three specimens are the higher of the nine specimens testing being welded in a SAW process (Table 16).

The $(CTOD)_u$ shown values over the minimum required by the standard with figures of 0.94 mm, 0.96 mm and 0.22mm higher than 0.2mm as the standard ask for. All were unstable fracture after an elongation, therefor the $(CTOD)_u$ was represented with the subscript (u). On spite of the sample 01-12(9) shows a great percentage of coarsened grain 40% the $(CTOD)_u$=0.22mm>0.20mm as the standard requires. So, all CTOD were within the BS rules.

We must highlight the extreme difficulty of getting the fusion line is completely straight and even perfectly normal to the surface of the plate, this helps the percentage found be of grain coarse where it crosses the crack are few, even zero. Regarding the previously noted, it is possible that the values obtained in the CTOD not represent the lower limit of the fracture toughness of the coarse-grained microstructure GCHAZ present in the testing.

Although the values deduced from Table 13, do not establish an inverse proportionality between CTOD and coarse amount found in the test can be observed that the minimum values obtained in the CTOD test correspond to samples in which they found higher percentage of grain coarse intercepted by the crack.

In the Table 14 has been represented the CTOD ($\delta$) individual each sample and average of the group, of this group the CTOD lower correspond to those specimens that were welded in process with higher HI (SAW). As we see in this Table all specimens show a significant percentage that the crack went through the weld metal due to the notch was not completely straight. The specimens welded with lower HI which in GSFC process (Table 16), gave a TS testing failure, and higher CTOD ($\delta$), furthermore they have the lower percentage of (GCHAZ) and higher percentage of fine grain (FGHAZ). Observation by optical microscope the group welded in SMAW (specimens 01-10(7), 01-11(8) and 01-12(9) initiated the crack in FG.

In the Table 15, are described all details related with the opening of the crack as well as the elastic ($\delta_e$) and plastic ($\delta_p$) CTOD that have been work out by the following formula:

$\partial = \partial_e + \partial_p$ . See equation (4a) and (4b). The crack size $a_m$ being ($a_m$=$a_{0+}a_{cr}$) is represented in column 4 and crack total ($a$) is in the column 5, the stable crack extension is: $\delta a = a - a_m$. The $\sigma$ is the yield strength of the steel. The load $P$ it is used in the second part of the CTOD testing to determine the fracture toughness $K$ given in the column 9 of the Table 16 using equation (10) where $K_f$ is $K$ and $P_f$ is $P$.

In the Table 16 are represented the nine specimens with the welding process and the HI they were welded, other mechanical values have been represented which have already been described. In the column 5 are the Interpass temperature. The crack total length($a$) is the sum of mechanical notch(M)=$a_0$ plus critical fatigue crack ($a_{cr}$) plus crack stable growth ($\delta_a$) (Figure 11). The fatigue crack ($a_{cr}$)correspond to the first part of the CTOD testing where the load is variable and applies by vibration till the $a_m/W$=0.5

*Table 13. Amount of Grain Intercepted by the Crack*

| Review Extract Micrograph (mapping) of Fracture Toughness of the Specimens Tested in Ctod | | | | | | | | | | | | | |
|---|---|---|---|---|---|---|---|---|---|---|---|---|---|
| Specimen | CTOD Value at -10 °C (mm) | Weld Start ASTM Grain Size (%) | | | Weld Finish ASTM Grain Size (%) | | | Total Number of Microstructures Tested with Two Thirds of the Thickness (2/3 *B* = 100%). | | | | | |
| | | <5 | 5-7 | >7 | <5 | 5-7 | >7 | WM (%) | < 5 (%) | 5-7 (%) | >7 (%) | | |
| 01-1(1) | 0.35 u | 15.7 | 28.6 | 53.7 | 9.8 | 17.5 | 72.7 | 27.4 | 6.2 | 9.8 | 56.6 | - | - |
| 01-2(2) | 0.79 u | 15.7 | 28.6 | 53.7 | 9.8 | 17.5 | 72.7 | 3.9 | 7.5 | 17.3 | 71.3 | - | - |
| 01-3(3) | 1.14 m | 15.7 | 28.6 | 53.7 | 9.8 | 17.5 | 72.7 | 46.4 | 3.4 | 3.6 | 46.6 | - | - |
| 02-01(4) | 1.36 TS | 29.6 | 22.1 | 48.3 | 21.8 | 30.2 | 48.0 | 11.8 | 1.3 | 13.8 | 73.1 | - | - |
| 02-02(5) | 1.33 TS | 29.6 | 22.1 | 48.3 | 21.8 | 30.2 | 48.0 | 14.5 | 6.7 | 16.0 | 62.8 | - | - |
| 02-03(6) | 1.52 TS | 29.6 | 22.1 | 48.3 | 21.8 | 30.2 | 48.0 | 19.7 | 8.8 | 32.7 | 38.8 | - | - |
| | Total number of tested microstructures throughout the thickness, *B*, of the specimen (*B* = 100%) | | | | | | | WM (%) | G.C.<7 (%) | G.C7-9 (%) | F.G (%) | I.C (%) | S.C.+Base (%) |
| 01-10(7) | 0.94 u | 12.7 | 4.0 | 2.7 | 43.7 | 20.5 | 16.4 | 0.0 | 0.0 | 4.0 | 58.1 | 21.6 | 16.3 |
| 01-11(8) | 0.96 u | 44.0 | 6.2 | 1.2 | 46.0 | 2.7 | 0.0 | 46.3 | 5.0 | 1.3 | 43.4 | 4.0 | 0.0 |
| 01-12(9) | 0.22 u | 37.0 | 7.8 | 4.0 | 32.3 | 19.0 | 0.0 | 40.0 | 7.8 | 4.7 | 30.5 | 17.0 | 0.0 |

WM = Weld Metal; GC < 7 = The grain size ASTM in HAZ lower 7
BM =Base Metal; GC 7-9 = Grain Coarse size ASTM between 7 to 9 in HAZ.
FG = HAZ Fine Grain; IC = HAZ Intercritical. SC = HAZ subcritical
MATERIAL: STEEL 450 EMZ.
WELDING PROCESS: Specimens 01-1(1),01-2(2) and 01-3(3) SMAW.Specimens 02-01(4), 02-02(5) and 02-03(6) GSFC. Specimens 01-10(7)01-11(8) and 01-12(9)9 Submerged Arc. (SAW)
IDENTIFICATION OF THE WELD: Heat Affected Zone Test Panel

giving a fracture toughness $K_f$ at a load $P_f$. From this point we follow with the second part of the testing at a load *P* constant given in column 6 Table 15 to obtain the fracture toughness *K* by the equation (10) the crack in this second part of the testing is the slow stable crack ($\delta_a$).

## FATIGUE CRACK EXPERIMENTAL STUDY AT SPECIMEN 01-10(7) HAZ.

The study of fatigue crack for testing CTOD, Methods for Crack Opening Displacement C.O.D. (Testing), fracture toughness was conducted in a push resonance SCHENCK 200 KN. The frequency of the electric motor driving the pushbutton, is adjusted by rotating an eccentric weight in order to obtain maximum load $P_{max}$ which has been predetermined beforehand. This frequency, which is unknown but controlled depends inter alia on the resistance of the test piece. During fatigue process which the specimen is subjected, and due to the crack growth rate, the frequency should be adjusted to maintain the

*Table 14. CTOD Results*

| Extract of Test Results of Fracture Toughness on CTOD | | | | | | | | | | |
|---|---|---|---|---|---|---|---|---|---|---|
| **Specimen** | **CTOD δ** | | **Type (BS 5762)** | | | **% ASTM Grain Size Tested Fatigue Failure** | | | | **Point of The Microstructure of Fracture Initiation** |
| | **Individual (mm)** | **Average (mm)** | | **WM %** | **GC<7** | **GC<5 %** | **BETWEEN 5-7%** | **>7 (%)** | | |
| 01-1(1) | 0.349 | | (u) Fracture | 27.4 | | 6.2 | 9.8 | 56.6 | | |
| 01-2(2) | 0.79 | 0.76 | (u) Fracture | 3.9 | | 7.5 | 17.3 | 71.3 | | |
| 01-3(3) | 1.14 | | (m) Fracture | 46.4 | | 3.4 | 3.6 | 46.6 | | |
| 02-01(4) | 1.36 | | (T.S) | 11.8 | | 1.3 | 13.8 | 73.1 | | |
| 02-02(5) | 1.33 | 1.40 | (T.S) | 14.5 | | 6.7 | 16.0 | 62.8 | | |
| 02-03(6) | 1.52 | | (T.S) | 19.7 | | 8.8 | 32.7 | 38.8 | | |
| | | | | W.M. % | G.C.<7 % | G.C.7-9 % | F.G % | I.C. % | S.C+BASE % | |
| 01-10(7) | 0.94 | | (u) Fracture | 0.0 | 0.0 | 4.0 | 58.1 | 21.6 | 16.3 | FGHAZ a 30.0 mm from LOWER surface |
| 01-11(8) | 0.96 | 0.71 | (u) Fracture | 46.3 | 5.0 | 1.3 | 43.4 | 4.0 | 0.0 | FGHAZ a 24.5 mm from LOWER surface |
| 01-12(9) | 0.22 | | (u) Fracture | 40.0 | 7.8 | 4.7 | 30.5 | 17.0 | 0.0 | FGHAZ (next to GCHAZ) a 23.5 mm from LOWER surface |

MATERIAL: STEEL 450 EMZ; PROPORTION OF DIVERSION: $19 \times 10^{-3}$ mm / s; WELDING PROCESS: Specimens 01-1(1),01-2(2) and 01-3(3) SMAW.Specimens 02-01(4), 02-02(5) and 02-03(6) GSFC. Specimens 01-10(7), 01-11(8) and 01-12(9)9 Submerged Arc. (SAW)

CONDITION: Butt weld; Specimen geometry: B × 2B Specimen BS 5762: LOCATION NOTCH: HAZ; Test temperature: -10 °C; ORIENTATION NOTCH: SIDE BY SIDE THICKNESS

$P_{max}$ and $\Delta K$ (amplitude of the stress intensity factor (Clarck W.G., How J.R. (1974)) Murakami, Y., & Yamashita, Y(2014)

Keep in mind that the stress intensity factor for fatigue fracture, according to BS 5762(1979) standard is adjusted by the following equation **(for specimen 01-10(7)):**

$$K_f \leq 0.63 \times \sigma_y \times B^{1/2} = 0.63 \times 456 \times 61.5^{1/2} = 2252.9\text{N/mm}^{3/2} > 2090\text{N/mm}^{3/2} \quad (9)$$

Where:

*Table 15. CTOD Fracture Toughness Results.*

| CTOD Data Table Fracture Toughness. | | | | | | | | | | | | |
|---|---|---|---|---|---|---|---|---|---|---|---|---|
| **Specimen** | *B* (mm) | *W* (mm) | $a_m$ (mm) | *a* (mm) | *P* (KN) | $V_p$ (mm) | σ (MPa) | $\delta_e$ (mm) (*) | $\delta_p$ (mm) (*) | δ (mm) (***) | **Fracture Kind (**)** | **Valid (BS 5762)** |
| 01-1(1) | 58.66 | 120.16 | 61.70 | 62.30 | 270.82 | 0.955 | 500 | 0.0960 | 0.2530 | 0.349 | U | Yes |
| 01-2(2) | 59.16 | 120.34 | 60.49 | 61.11 | 316.58 | 2.487 | 500 | 0.1130 | 0.6770 | 0.790 | U | Yes |
| 01-3(3) | 58.84 | 120.13 | 61.62 | 63.19 | 307.93 | 3.822 | 500 | 0.1238 | 1.0162 | 1.140 | M | Yes |
| 02-01(4) | 58.78 | 120.34 | 62.34 | 63.5 | 335.30 | 4.798 | 557 | 0.1060 | 1.2540 | 1.360 | test stop | Yes |
| 02-02(5) | 58.31 | 120.59 | 60.51 | 61.52 | 331.66 | 4.508 | 557 | 0.1100 | 1.2200 | 1.330 | test stop | Yes |
| 02-03(6) | 59.18 | 120.59 | 60.66 | 61,81 | 349.79 | 5.007 | 557 | 0.1500 | 1.3700 | 1.520 | test stop | Yes |
| 01-10(7) | 61.50 | 121.70 | 59.80 | 60.00 | 376.00 | 2.760 | 456 | 0.1600 | 0.7800 | 0.940 | u | Yes |
| 01-11(8) | 61.70 | 120.20 | 60.50 | 61.20 | 378.00 | 2.880 | 456 | 0.1900 | 0.7700 | 0.960 | u | Yes |
| 01-12(9) | 61.20 | 122.20 | 59.50 | 59.60 | 304.00 | 0.400 | 456 | 0.1055 | 0.1145 | 0.220 | u | Yes |

Are the elastic and plastic components of CTOD fracture toughness respectively, while (***) is the CTOD fracture toughness.

(**): The symbol "u" is used when there is suggestive evidence that a pop-in or an unstable fracture was preceded by a ductile elongation, however "m" indicates the maximum force was reached. Test stop without broken (TS)

*Table 16. H.I. and Experimental CTOD Values.*

| Ctod, data table fracture toughness. | | | | | | | | | | | | | | |
|---|---|---|---|---|---|---|---|---|---|---|---|---|---|---|
| **Specimen** | *B* (mm) | *H.I.* (kJ/mm) | **Welding Process** | **Temp. Between Beads °C** | **Crack Total Length mm $a_m+\delta_a$** | **Crack Stable Growth $\delta_a$ mm** | **CTOD δ mm** | **Fracture Toughness K (N/mm$^{3/2}$)** | **Fatigue Fracture Toughness $K_{fmáx.}$ (N/mm$^{3/2}$)** | **Crack Initiation From CAP (mm)** | **WM %** | **Grain Size <5 ASTM %** | **Grain Size 5-7 ASTM %** | **Grain Size >7 ASTM %** |
| 01-1(1) | 58.66 | 1.560 | S.M.A.W. | 102.70 | 62.30 | 0.60 | 0.349 | 4754 | 1209 | 26(V.G.) | 27.4 | 6.2 | 9.8 | 56.6 |
| 01-2(2) | 59.16 | 1.560 | S.M.A.W. | 102.70 | 61.11 | 0.62 | 0.790 | 5311 | 1155 | 16(V.G.) | 3.9 | 7.5 | 17.3 | 71.3 |
| 01-3(3) | 58.84 | 1.560 | S.M.A.W. | 102.70 | 63.19 | 1.57 | 1.140 | 5295 | 1032 | | 46.4 | 3.4 | 3.6 | 46.6 |
| 02-01(4) | 58.78 | 1.348 | G.S.F.C. | 124.70 | 63.50 | 1.16 | 1.360 | 6042 | 1222 | | 11.8 | 1.3 | 13.8 | 73.1 |
| 02-02(5) | 58.31 | 1.348 | G.S.F.C. | 124.70 | 61.52 | 1.02 | 1.330 | 5671 | 1165 | | 14.5 | 6.7 | 16.0 | 62.8 |
| 03-03(6) | 59.18 | 1.348 | G.S.F.C. | 124.70 | 61.81 | 1.14 | 1.520 | 6046 | 1173 | | 19.7 | 8.8 | 32.7 | 38.8 |
| 01-10(7) | 61.50 | 2.327 | S.A.W. | 118.91 | 60.00 | 0.20 | 0.940 | 5714 | 2090 | 30(F.R.) | 0.0 | 0.0 | 4.0 | 96.0 |
| 01-11(8) | 61.70 | 2.327 | S.A.W. | 118.91 | 61.20 | 0.70 | 0.960 | 6108 | 2180 | 24.5(F.R.) | 46.3 | 5.0 | 1.3 | 47.4 |
| 01-12(9) | 61.20 | 2.327 | S.A.W. | 118.91 | 59.60 | 0.10 | 0.220 | 4588 | 2070 | 23.5(F.R.) | 40.0 | 7.8 | 4.7 | 47.5 |

Lower surface = Bottom side in the respective mapping diagrams showing the distribution of the microstructures of the HAZ tested for fatigue failure. MATERIAL: STEEL 450 EMZ; PROPORTION OF DIVERSION: 19 × $10^{-3}$ mm / sec.; WELDING PROCESS: Specimens 01-1(1),01-2(2) and 01-3(3) SMAW.Specimens 02-01(4), 02-02(5) and 02-03(6) GSFC. Specimens 01-10(7), 01-11(8) and 01-12(9)9 Submerged Arc. (SAW)

Butt Weld; Specimen geometry: B × 2B Specimen BS 5762: 1079.

LOCATION NOTCH: HAZ; TEST TEMPERATURE: -10 °C;

ORIENTATION NOTCH: SIDE BY SIDE THICKNESS.

$\sigma_y = 456$ N/mm$^2$, yield strength of the HAZ which is the average of base metal $\sigma_{y=}$429N/mm$^2$ and the yield strength of weld deposit $\sigma_y$=483 N/mm$^2$. B = thickness of the specimen.

From this stress intensity factor, and by the expression $K_f = \frac{YP_f}{B\sqrt{W}}$ (10)

for specimen 01-10(7).

Y = f($a/W$)=10.31According to BS 5762(1979); $P_f$=137.533KN according to that rule, where $P_f$ is the values of the maximum load to which the machine must be set for the fatigue test (**N**)
B = thickness of the specimen (mm)=61.5 mm.
W = width of the specimen (mm)=121.7 mm.
Gives a $K_f$=2090 N/mm$^{3/2}$

Y is a factor that depends on $a_m$ /W, that is, the ratio of the length of the crack to specimen width. These values are expressed in Table 1 of BS 5762(1979), and they are obtained through experimental polynomial formula, which (Broek´s,1982) gives in his book of fracture mechanics by the following expression:

$$Y' = 2.9(a_m/W)^{1/2} - 4.6(a_m/W)^{3/2} + 21.8(a_m/W)^{5/2} - 37.6(a_m/W)^{7/2} + 38.7(a_m/W)^{9/2} \quad (11)$$

From which, the increase of the stress intensity factor by expression is obtained:

$$\Delta K = \frac{\Delta P 4WY'}{BW^{3/2}} = \frac{\Delta P 4Y'}{B\sqrt{W}} \quad (12)$$

where: S = span between supports = 4W (as standard). So that:
substituting $4Y' = Y$, we obtain the value of the BS 5762(1979) standard

$$\Delta K = \frac{\Delta PY}{B\sqrt{W}}$$

In our case, and following the results obtained for one of nine similar specimens comprising 01 AW CTOD test which is corresponding to the authenticated as the specimen 01-10(7), and the results are around a mean value of the nine composing this test. We obtain a value of $Y'$=2.58946776; where $a_m$/W = 0.491, and according to experimental results, this value $4Y'$= $Y$=2.58946776×4=10.358, agrees approximately with $Y$ = 10.31, which is obtained from Table 1 of the BS, to $a_m$/W = 0.491.

However, throughout the test, no control of the point values of $\Delta P$ were taken, as well as neither of $\Delta K$, as it would be impossible to detect the continued growth of the crack during it.

So, we only know the initial value of the depth of the crack, $a_0$, which corresponds to the length of the crack mechanically obtained by saw and the depth value, $a_{cr}$, of the final crack obtained by fatigue.

The fatigue fracture test, stops when $a_m$ /W, is near of following values: 0.45 <$a_m$ / W <0.55. In the case of the specimen 01-10(7), $a_0$=50 mm

From $K_f = 2090\ N / mm^{3/2}$ (obtained above), the $P_{fmax}$ is giving by the expression:

$$K_t = 2090 = \frac{YP_{f\max}}{B\sqrt{W}} \tag{13}$$

substituting the known values:

$Y$=10.31; $K_f$ =2090N/mm$^{3/2}$; B=61.5mm, W=121.7mm

it is got to a $P_{fmax}$ = 137533 N, (predetermined beforehand) this is the load when the first part of the testing stop the vibration (fatigue crack) is finished. It means when $a_m$/W=0.491. where $\Delta P$ is obtained by the following expression:

$\Delta P= P_{fmax}\ 0.9 = 123780$ N.

Putting the above function as their increments, we obtain:

$$\Delta K = \frac{\Delta PY}{B\sqrt{W}}$$

Substituting the values listed, $\Delta K$ =2090×0.9= 1881 N / mm$^{3/2}$ is obtained. In this paper, we have presented the expression:

$$\Delta K = 1.4\Delta\sigma\sqrt{\pi a_{cr}}$$

in which:

$a_{cr}$ = mean value of the crack and $\Delta\sigma$= increasing tension applied

The following relationships must be considered:

$$\frac{P_{\min}}{P_{\max}} = 0.1; \frac{\sigma_{\min}}{\sigma_{\max}} = 0.1; \Delta\sigma = \sigma_{\max} - \sigma_{\min} = 0.9\sigma_{\max} \tag{14}$$

From the above expressions we get:

$$\Delta K = 1.4\Delta\sigma\sqrt{\pi a_{cr}} \tag{15}$$

resulting the following:$\Delta K$=1881 N/mm$^{3/2}$

$a_{cr}$ = 9.8 mm which corresponds to the depth of the fatigue crack, since $a_0$ = 50 mm and $a_m$= 59.8 mm for the specimen 01-10 (7) which are considered in this calculation.

$a_{cr}= a_m$-$a_0$=9.8 mm; $\Delta\sigma \cong 242$ N/mm$^2$; $\Delta\sigma=0.9\sigma=0.9\times269.12=242.21$ N/mm$^2$ in which $\sigma$ represents tension of design for HAZ obtained from fatigue tests at the crack tip (see Table 7).

Therefore, being $a_{cr}$ = 9.8 mm, is obtained: Being $a_{cr}$ in meters

$$\Delta K = 1.4\Delta\sigma\sqrt{\pi a_{cr}} = 1.4\times 242.21\sqrt{\pi a_{cr}} = 601.028\sqrt{a_{cr}}MN / m^{3/2} \tag{16}$$

$MN/m^{3/2} = MPa\ m^{1/2}$

Where:
$da/dN$ = crack rate per cycle of charge (m/cycle).
$\Delta K$= range or amplitude of stress intensity factor (MPa $m^{1/2}$)

$$\frac{da}{dN} = 5.01\times10^{-12}(\Delta K)^{3.1} m / cycle \tag{17}$$

for ferrite-pearlite steels 286.85N/mm$^2$ $\leq \sigma_{ys} \leq$ 551.6 N/mm$^2$; BS 4360 (1990) Structural Steel (H.L.Ewalds1984).

Writing the above formulas in the form of increase, we can calculate integrating the curve of cycle - crack growth, by the expression:

$$\Delta N = \frac{\Delta a}{5.01\times10^{-12}(\Delta K)^{3.1}} \tag{18}$$

Taking account that for ferritic-pearlitic steels with R = 0.1, the threshold value $K_{th}$ researchers Harrison et al. (1970), Paris et al. (1972), Schweizer et al. (2011) and Kujawski (2013), E.K. Priddle (1982) can be estimated at:

$\Delta K_{th}$=7(1-0.85R) MN/m$^{3/2}$

=7(1-0.85×0.1) =6.405MN/m$^{3/2}$

*Figure 15. Stress Ratio (R) - Threshold of the Amplitude of the Stress Intensity Factor(.deltaKth) in the Specimen 01-10(7)*

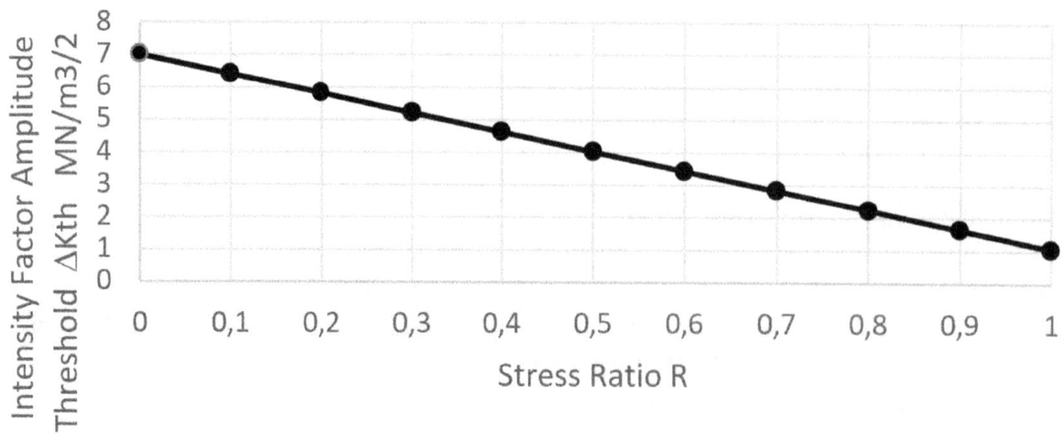

*Table 17. Table of Figure 15*

| **R** | **$\Delta K_{th}$ ($MN/m^{3/2}$)** |
|---|---|
| 0 | 7 |
| **0.1** | **6.41** |
| 0.2 | 5.81 |
| 0.3 | 5.22 |
| 0,4 | 4.62 |
| 0.5 | 4.03 |
| 0.6 | 3.43 |
| 0.7 | 2.84 |
| 0.8 | 2.24 |
| 0.9 | 1.65 |
| 1 | 1.05 |

for R = 0.1 we obtain: Figure 15

$\Delta K_{th}$=7(1-0.85R) =6.405 $MN/m^{3/2}$=200.16 N mm $^{3/2}$ <202.54 $N/mm^{3/2}$

The threshold value of the crack will be:

$$a_{01} = \left(\frac{\Delta K_{th}}{1.4\Delta\sigma\sqrt{\pi}}\right)^2 = \left(\frac{200.16}{1.4\times 242.21\times\sqrt{\pi}}\right)^2 \cong 0.110908 \quad (19)$$

Where it has been considered a $\Delta\sigma$=242.21$N/mm^2$

Correspond to the size of the crack from which it begins its propagation. Based on these thresholds, we obtain the frequency curves - crack length to the specimen 01-10 (7) and to a convenient value $\Delta a$ = 1.02 mm is obtained:

$$\Delta N = \frac{1.02\times 10^{-3}}{5.01\times 10^{-12}\Delta K^{3.1}} \quad (20)$$

Starting from the experimental values given in the Table 6 referencing the specimen 01-10(7) and expressions derived above, we can determine the values listed in Table 22 which have been calculated for the values of $a_{cr}$ to $\Delta N$ in cycles, summing them, we obtain the total cycles for the crack extends i.e. for 68658.7 5 cycles, corresponding to a crack size of 9.8 mm (Table 22), the difference with the 67000 cycles, experimental value, can be due to the value of 1.4 obtained in the graphic or the Figure 9 for a $K_I/K_0$ or for the value sampled of $\Delta$a=1.02 mm. Nevertheless, can almost coincide both for the specimen 01-10 (7).

The speed of crack growth d*a*/d*N* given in m/cycles theoretically determining the stretch of the formula corresponding to Paris (1970). The amplitude of the stress intensity factor and its logarithm allow us to check the nature of the ferritic- pearlite steel of our specimens. All these values are plotted for the corresponding conclusions.

The first column of Table 22 was obtained starting from the crack threshold value $a_0 = 0.11091$ mm which was obtained by equation (19) and $a_{cr} = a_0 + \Delta a / 2$, where $a_f = a_{cr} + \Delta a / 2$ being $\Delta a = 1.02$ mm, which was mentioned above and is taken in an amount with a quantity acceptable enough for the thirteen values shown in Table 22, reach a total amount of $N$ and $a_{cr}$ this corresponds to the experimental results. The amplitude of the stress intensity factor $\Delta K$ was obtained from equation (16) in MPa $m^{1/2}$. The quantity represented by $\Delta N$ is given by equation (18) where $\Delta a$ is expressed in meters. $\Sigma N$ cycles total are obtained by adding to the column $\Delta N_1$ successive values $\Delta N$i. The rate of crack growth da/d$N$ in m/cycles is obtained from the equation (17) in which $\Delta K$ values are expressed in MPa $m^{1/2}$ based in column 4 of Table 22 (equation 16).

Concerning the fatigue crack growth, there are three aspects to be determined, Sander et al. (2013).

1.-The initial crack size "$a_d$". Determined as specified below.
2.-The maximum permissible or critical crack size "$a_{cr}$"
3.-The period of crack growth between "$a_d$" and "$a$cr"

Even though exists a value "$a_0$" given by a threshold Intensity factor when we apply the equation:

$$\int_0^N dN = \int_{a_d}^{a_{cri}} \frac{da}{5.01 \times 10^{-9} (\Delta K)^{3.1}} \tag{21}$$

(Do not mistake this $a_0$ with the mechanical notch given in Figure 11 where $a_0$=M)

For a "$a_{cr}$" we can work out the minimum size $a_d$ which can be detected by non-destructive testing (NDT) techniques. The initial flaw size given above, in the case of the specimen 01-10 (7) is $a_0 = 0.11091$ mm (Table 22), being $a_d$=0.276 mm (Bäcklund et al., Fatigue-Thresholds-Fundamentals and Engineering, 1982), which is the first detectable threshold of the crack and $a_{cr}$= 9.8 mm is the size at the crack is extended, as it is shown in Table 22. Because we can consider an intensity factor rate $\Delta K$ constant a correlation between this and d$a$/d$N$ are to be used. Practically the difference between $a_0$ and $a_d$ are to certain discontinuity in the structure as: voids, flaws, damages and inhomogeneities. Murakami, Y. (2012).

In Figure 16 has been depicted the number of cycles - crack size ($a_0$(mm), $a_{cr}$(mm), $a_f$(mm)).

Corresponding to the region II of the Paris equation graph is drawn representing $\Delta K$ = f (d$a$/d$N$) Table 22, to a $\Delta K$ value of 60.21 MPa$m^{1/2}$ where propagation occurs at 68658.66 cycles. This region has a higher incidence microstructure, medium tension and a small environmental impact surrounding the specimen. We see that the profile of the graph of Figure 18 corresponds to the values of the pearlitic-ferritic steels Barsom (1971) vs. specimen values, on being parable very laid with a regression line given by Barsom (1971) Figure 22 at log. Scale. Thus, it is found that with the obtained experimental values of the BS 4360 (1990) Structural Steel (H.L. Ewalds1984). By equation (17) but in (mm)

$$\frac{da}{dN} = 5.01 \times 10^{-9} (\Delta K)^{3.1} mm / cycle$$

stress intensity factor applied to the function; the graph corresponds to the pearlitic-ferritic steels for the values obtained in the HAZ. Function is shown in Figure 18.

*Figure 16. Number of Cycles - Crack sizes. Specimen 01-10(7)*

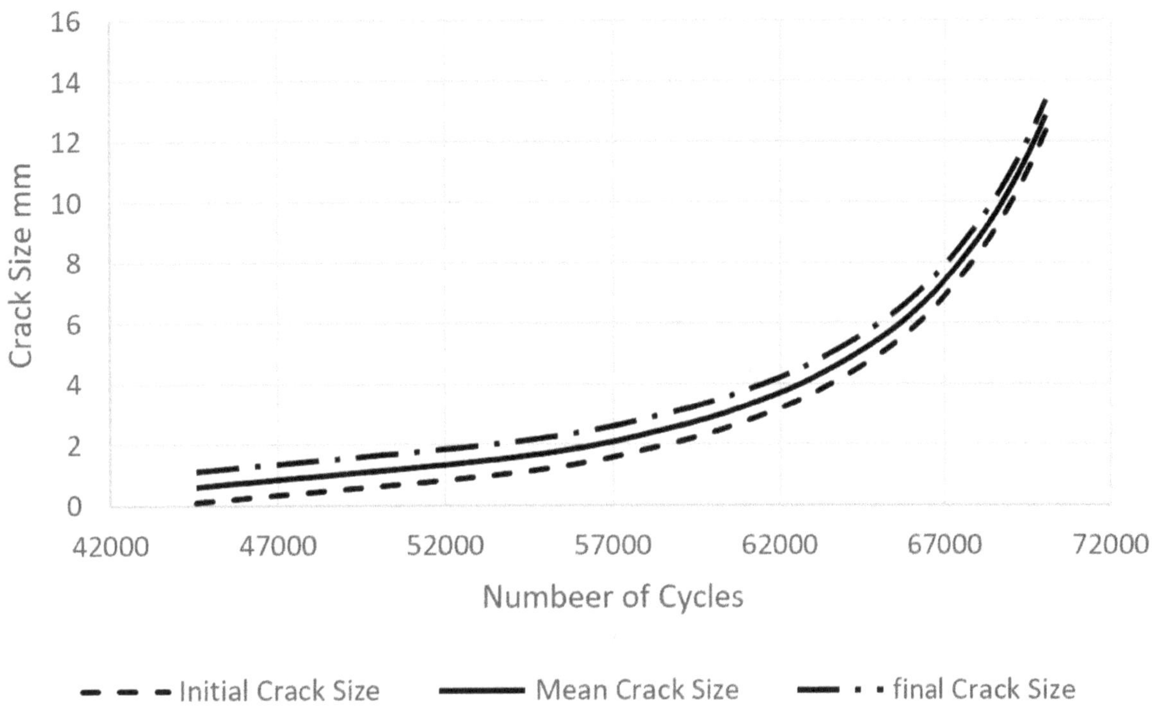


*Table 18. Table of Figure 16*

| Σ *N*(ciclos) | $a_0$(mm) | $a_{er}$(mm) | $a_f$(mm) |
|---|---|---|---|
| 44571.58 | 0.110909 | 0.620909 | 1.130909 |
| 54454.16 | 1.130909 | 1.640909 | 2.150909 |
| 59125.65 | 2.150909 | 2.660909 | 3.170909 |
| 61950.66 | 3.170909 | 3.680909 | 4.190909 |
| 63884.27 | 4.190909 | 4.700909 | 5.210909 |
| 65310.47 | 5.210909 | 5.720909 | 6.230909 |
| 66416.43 | 6.230909 | 6.740909 | 7.250909 |
| 67305.40 | 7.250909 | 7.760909 | 8.270909 |
| 68039.52 | 8.270909 | 8.780909 | 9.290909 |
| **68658.66** | **9.290909** | **9.800909** | **10.310909** |
| 69189.71 | 10.310909 | 10.820909 | 11.330909 |
| 69651.57 | 11.330909 | 11.840909 | 12.350909 |
| 70057.90 | 12.350909 | 12.860909 | 13.370909 |

*Figure 17. Stress σ at Tip Crack - Intensity factor K. Specimen 01-10(7). K=7.768×σ*

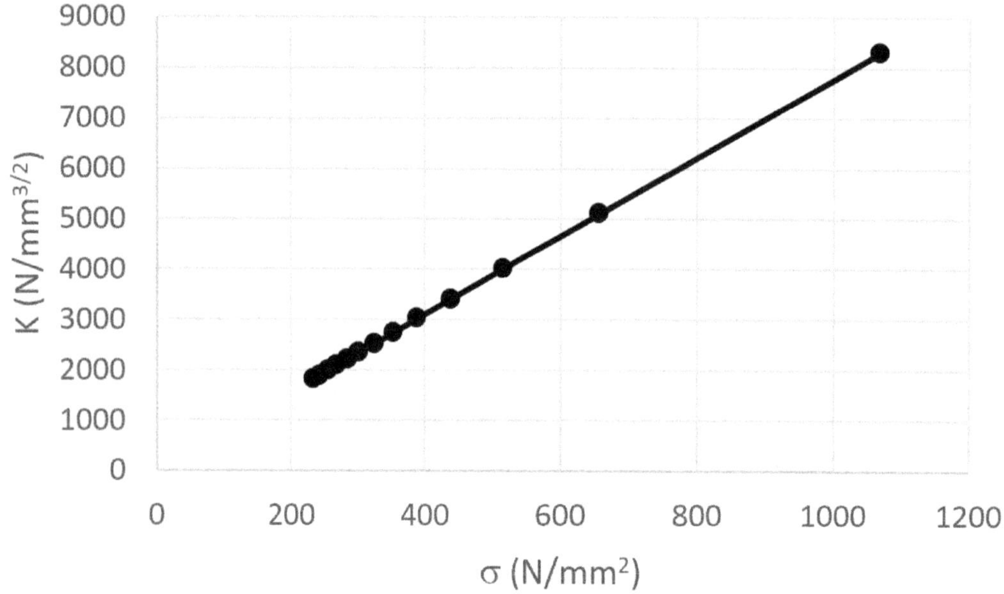


*Table 18. Table of Figure 17*

| $\sigma$ (N/mm²) | $K$ (N/mm$^{3/2}$) |
|---|---|
| 1068.88 | 8303.02 |
| 657.50 | 5107.50 |
| 516.33 | 4010.84 |
| 439.00 | 3410.14 |
| 388.46 | 3017.58 |
| 352.13 | 2735.38 |
| 324.40 | 2519.94 |
| 302.33 | 2348.52 |
| 284.23 | 2207.91 |
| **269.03** | **2089.86** |
| 256.04 | 1988.92 |
| 244.76 | 1901.33 |
| 234.86 | 1824.38 |

The stress intensity factor ($K$) in function of stress at crack tip come represented by equation:

$$K= (\Delta K/\Delta\sigma)\sigma =(1.4\times\pi^{1/2}\times\Delta\sigma\times a_{cr}^{\ 1/2})/242.21)\times\sigma = (1.4\times\pi^{1/2}\times 9.801^{1/2})\times\sigma=7.769\sigma \quad (22)$$

*Figure 18. Speed Crack Growth da/dN – Stress Intensity factor Amplitude ΔK, sspecimen 010-10(7)*

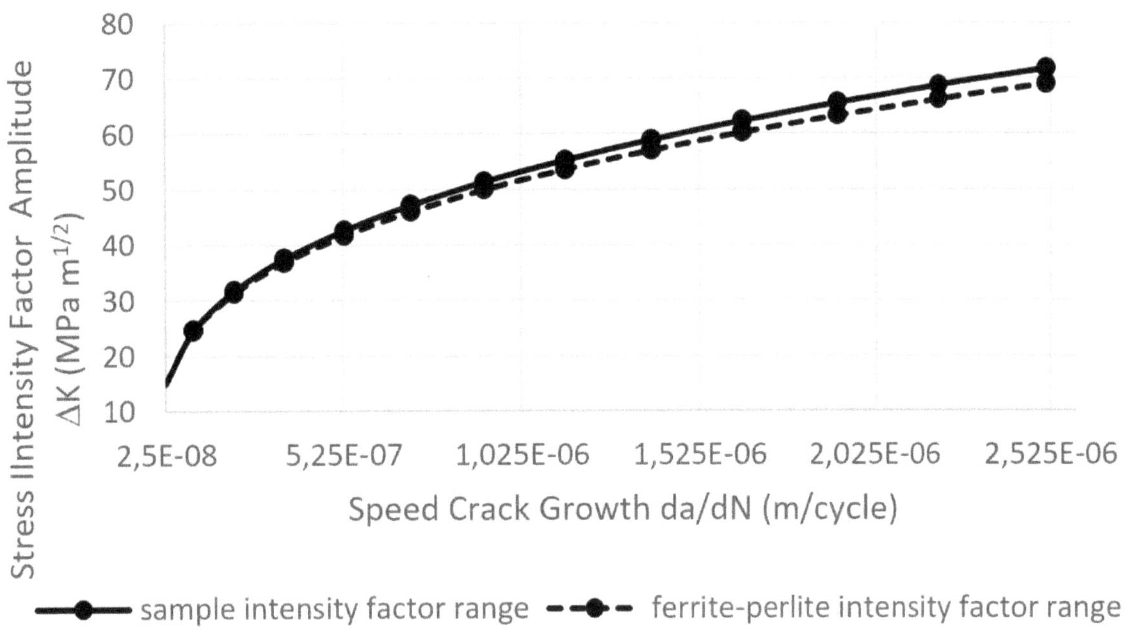


*Table 19. Table of Figure 18*

| *da/dN*(m/cycle) | Δ*K*(MPam$^{1/2}$) Barsom | Δ*K*(MPam$^{1/2}$) Specimen (01-10(7)) |
|---|---|---|
| 2.288E-08 | 14.981 | 15.155 |
| 1.032E-07 | 24.751 | 24.636 |
| 2.183E-07 | 31.774 | 31.373 |
| 3.611E-07 | 37.573 | 36.899 |
| 5.275E-07 | 42.635 | 41.699 |
| 7.152E-07 | 47.188 | 46.001 |
| 9.223E-07 | 51.362 | 49.934 |
| 1.147E-06 | 55.241 | 53.579 |
| 1.389E-06 | 58.880 | 56.991 |
| **1.647E-06** | **62.320** | **60.210** |
| 1.921E-06 | 65.590 | 63.266 |
| 2.208E-06 | 68.715 | 66.180 |
| 2.510E-06 | 71.712 | 68.972 |

The above function (22) come represented in Figure 17 for the fatigue stress of the material, Shiozawa, T. et al. (2009), at the end of the crack as is indicated in Table 7 is 269N/mm$^2$ corresponds to an acceptable value of 62.7% stress specimen 01-10(7) of this Table 7, being a *K* value of 2089.8N/mm$^{3/2}$as calculated before. This equation is represented in Figure 17

*Table 20. Table of Figure 19*

| **Σ *N* (cycles)** | ***σ* (N/mm²)** |
|---|---|
| 44572 | 1068.88 |
| 54454 | 657.50 |
| 59126 | 516.33 |
| 61951 | 439.00 |
| 63884 | 388.46 |
| 65310 | 352.13 |
| 66416 | 324.40 |
| 67305 | 302.33 |
| 68040 | 284.23 |
| **68659** | **269.03** |
| 69190 | 256.04 |
| 69652 | 244.76 |
| 70058 | 234.86 |

So, check the experimental values obtained stress intensity factor, applied to the function $da/dN = 5.01\text{x}10^{-12} (\Delta K)^{3,1}$ m/cycle, and drawn the curve of the crack propagation cycles obtained experimentally, that the behavior in the HAZ are satisfied from a viewpoint of fracture mechanics corresponds to that of the aforementioned steels.

In the Figure 18 it checks that the values obtained from [(d*a*/d*N*), Δ*K*], corresponds to the ferritic-pearlitic steels according Barsom (1971) where values have been converted in MPa $m^{1/2}$compare with the results of our specimen 01-10(7). This claim becomes both results those given by (1971) and the given for the specimen 01-10 (7) of our study, almost coincide when the (d*a*/d*N*,ΔK) are expressed in (m/cycle, $MPam^{1/2}$).

Where **Barsom** (1971) gives the following fatigue crack propagation rate:

**$da/dN=3.6\times10^{-10}(\Delta K)^{3.0}$ (inch/cycle),** where *a*=inches;$\Delta K$=ksi$(in)^{1/2}$, Barsom (1971) for ferrite perlite Steels

and **Ewalds** (1984) for structural steels BS 4360 (1990):

**$da/dN=5.01\times10^{-12}(\Delta K)^{3.1}$m/cycle,** where *a*=meter; $\Delta K=MN/m^{3/2}=MPa(m)^{1/2}$.

Interestingly is to observe, the results of the amplitude factor stress intensity when expressed in International units as H.L Ewalds for structural steels, second equation of above, for experimental value of our sample, coincides approximately with the results of Barsom expressed in International units for ferrite-pearlite steels, first equation of above. So, we conclude the nature ferrite-pearlite of our steels based on the experimental results obtained in HAZ. One can observe the parallelism of the results depicted in the graph of Figure 18. Both represent a straight equation virtually identical.

Representing the number of cycles in the first part of the CTOD test one can see in the Figure 19 the stress at the tip of the crack in accordance with the number of cycles Lu, L.T., Zhnag, J.W., & Shiozawua, K. (2009). When the crack is extended at 68659 cycles the stress at the tip of the crack is 269.03 $N/mm^2$. In the Figure 10 was depicted the same graphic but in function of the crack length in this the stress at the

*Figure 19. Number of Cycles N - Stress at the Crack Tip σ.*
*Specimen 01-10(7)*

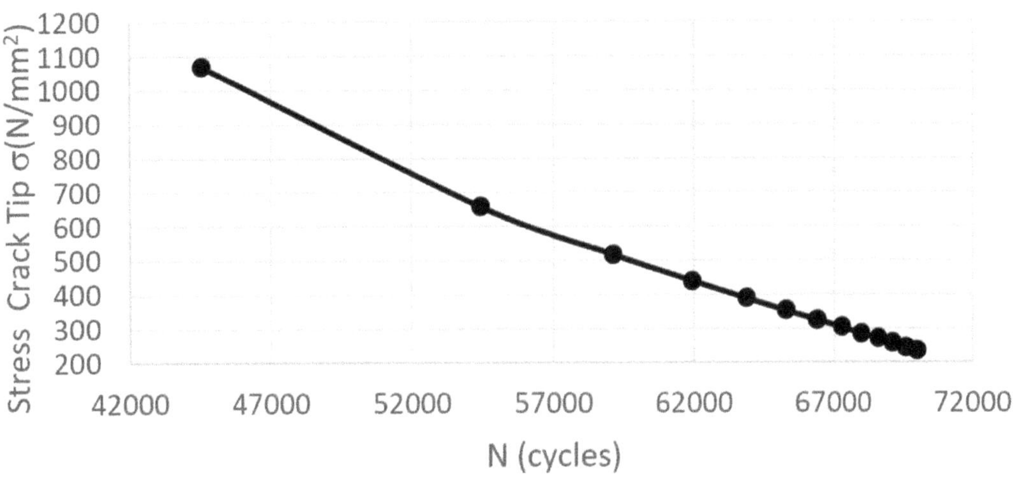


*Figure 20. Number of Cycles- Stress Crack Tip Size-Fatigue Crack Tip Size. Specimen 01-10(7)*

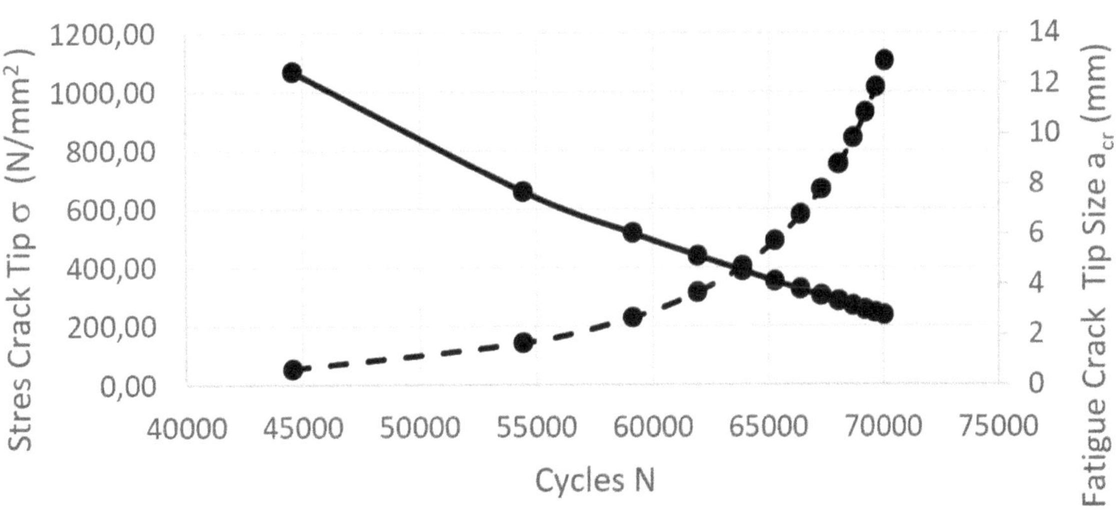

*Table 21. Table of Figure 20*

| Σ N(ciclos) | σ (N/mm²) | $a_{cr}$ (mm) |
|---|---|---|
| 44572 | 1068.88 | 0.621 |
| 54454 | 657.50 | 1.641 |
| 59126 | 516.33 | 2.661 |
| 61951 | 439.00 | 3.681 |
| **63884** | **388.46** | **4.701** |
| 65310 | 352.13 | 5.721 |
| 66416 | 324.40 | 6.741 |
| 67305 | 302.33 | 7.761 |
| 68040 | 284.23 | 8.781 |
| ***68659*** | ***269.03*** | ***9.801*** |
| 69190 | 256.04 | 10.821 |
| 69652 | 244.76 | 11.841 |
| 70058 | 234.86 | 12.861 |

crack tip has a value of $a_{cr}$=9.8 mm. At the beginning, when the crack is been initiated, at cycles 44572 the stress at the tip is high, but as the size of the crack getting bigger the stress goes down. It is due the energy at the tip of the crack is being discharged. To describe the function given in the graph the mean crack size of the tip is "$a_{cr}$" is related with the number of cycles thru the Table 22 where $N$=f($a_{cr}$), Lei Z. et al. (2014). This result corresponding to the specimen 010-10 (7).

At the Figure 20, we have represented for specimen 01-10 (7) the stress crack tip and fatigue crack tip size at a number of cycles. The stress at tip crack is decreasing as the number of cycles are increasing so the stress is releasing tension. On the contrary the size of crack is getting bigger as the cycles increase. There is a point where both curves cross which happens at 63884 cycles with a crack size of 4.701 mm and a tension of 388.46 N/mm$^2$at the crack tip.

In this Table are represented all parameters previously determined for the specimen 01-10(7). The columns 2 by the equation (19) the

Column 3 adding to the second column $\Delta a/2$ and the column 4 adding to the third column $\Delta a/2$, being $\Delta a$=1.02mm, column 5 ($\Delta K$) by equation

(16), column 7 determined by equation (20), column 8 adding the values column 7, column 9 by equation (17) and column 10 by equation (6)

At the Figure 21 are in logarithm scale represented the speed crack growth in (m/cycle) and the amplitude stress intensity factor in (MPam$^{1/2}$) these correspond to the values given in the two first columns of Table 23 (Specimen 01-10 (7). The trend is the typical in ferrite-perlite steels.

$da/dN=5.01\times10^{-12}(\Delta K)^{3.1}$m/cycle (Structural steels, BS 4360, H.L. Ewalds,1984)

Specimen 01-10 (7). In this Figure 22 in log. Scale are represented the following crack growth rate:

**Log Scale.** $da\,.\mathrm{dN} = 3.610^{-10}\left(\Delta K_I\right)^{3.0}$ **.Barsom (1971) ferrite pearlite steels**

Log Scale. $da/dN=5.01\times10^{-12}(\Delta K)^{3.1}$m/cycle H.L. Eawalds (1984) structural steels

*Table 22. Crack Size-Amplitude Stress Intensity Factor-Cycles. Specimen 01-10(7.)*

| | $a_0$ (mm) | $a_{cr}$ (mm) | $a_f$ (mm) | $\Delta K$ (MPam$^{1/2}$) | Ln $\Delta K$ | $\Delta N$ (cycles) | $\Sigma N$ (cycles) | $da/dN$ (m/cycle) | $\sigma$ (N/mm$^2$) |
|---|---|---|---|---|---|---|---|---|---|
| 1 | 0.11090894 | 0.62090894 | 1.13090894 | 15.1547859 | 2.718316386 | 44571.5817 | 44571.5817 | 2.28845E-08 | 1068.88 |
| 2 | 1.13090894 | 1.64090894 | 2.15090894 | 24.6364478 | 3.204226965 | 9882.58068 | 54454.1624 | 1.03212E-07 | 657.50 |
| 3 | 2.15090894 | 2.66090894 | 3.17090894 | 31.3726248 | 3.44593569 | 4671.48324 | 59125.6456 | 2.18346E-07 | 516.33 |
| 4 | 3.17090894 | 3.68090894 | 4.19090894 | 36.8988971 | 3.608181662 | 2825.01625 | 61950.6619 | 3.6106E-07 | 439.00 |
| 5 | 4.19090894 | 4.70090894 | 5.21090894 | 41.6991083 | 3.730479744 | 1933.61259 | 63884.2745 | 5.2751E-07 | 388.46 |
| 6 | 5.21090894 | 5.72090894 | 6.23090894 | 46.0011158 | 3.828665653 | 1426.20026 | 65310.4747 | 7.15187E-07 | 352.13 |
| 7 | 6.23090894 | 6.74090894 | 7.25090894 | 49.933853 | 3.91069919 | 1105.95563 | 66416.4304 | 9.22279E-07 | 324.40 |
| 8 | 7.25090894 | 7.76090894 | 8.27090894 | 53.5786963 | 3.981151533 | 888.969385 | 67305.3998 | 1.1474E-06 | 302.33 |
| 9 | 8.27090894 | 8.78090894 | 9.29090894 | 56.9909092 | 4.042891767 | 734.116748 | 68039.5165 | 1.38942E-06 | 284.23 |
| **10** | **9.29090894** | **9.80090894** | **10.3109089** | **60.2100552** | **4.097839369** | **619.139435** | **68658.6559** | **1.64745E-06** | **269.03** |
| 11 | 10.3109089 | 10.8209089 | 11.3309089 | 63.2656129 | 4.147341941 | 531.058684 | 69189.7146 | 1.92069E-06 | 256.04 |
| 12 | 11.3309089 | 11.8409089 | 12.3509089 | 66.1802448 | 4.192382001 | 461.853779 | 69651.5684 | 2.20849E-06 | 244.76 |
| 13 | 12.3509089 | 12.8609089 | 13.3709089 | 68.971819 | 4.233698001 | 406.331263 | 70057.8997 | 2.51027E-06 | 234.86 |

*Table 23. Speed Crack Growth - Stress Intensity Factor Amplitude in Specimen and According Barsom. Specimen 01-10(7)*

| $\Delta K$(MPam$^{1/2}$) Specimen | $da/dN$(m/ciclos) | $\Delta K$ (ksi in$^{1/2}$) Specimen | $dN/da$(inch/cycle) | $\Delta K$ (ksi in$^{1/2}$) Barsom |
|---|---|---|---|---|
| 15.155 | 2.288E-08 | 13.734 | 9.010E-07 | 13.577 |
| 24.636 | 1.032E-07 | 22.327 | 4.063E-06 | 22.432 |
| 31.373 | 2.183E-07 | 28.431 | 8.596E-06 | 28.796 |
| 36.899 | 3.611E-07 | 33.440 | 1.421E-05 | 34.052 |
| 41.699 | 5.275E-07 | 37.790 | 2.077E-05 | 38.639 |
| 46.001 | 7.152E-07 | 41.689 | 2.816E-05 | 42.766 |
| 49.934 | 9.223E-07 | 45.253 | 3.631E-05 | 46.549 |
| 53.579 | 1.147E-06 | 48.556 | 4.517E-05 | 50.064 |
| 56.991 | 1.389E-06 | 51.648 | 5.470E-05 | 53.362 |
| **60.210** | **1.647E-06** | **54.565** | **6.486E-05** | **56.480** |
| 63.266 | 1.921E-06 | 57.334 | 7.562E-05 | 59.444 |
| 66.180 | 2.208E-06 | 59.976 | 8.695E-05 | 62.276 |
| 68.972 | 2.510E-06 | 62.506 | 9.883E-05 | 64.992 |

*Figure 21. Speed Crack Growth-Stress intensity factor Log. Scale. Specimen 01-10(7).*

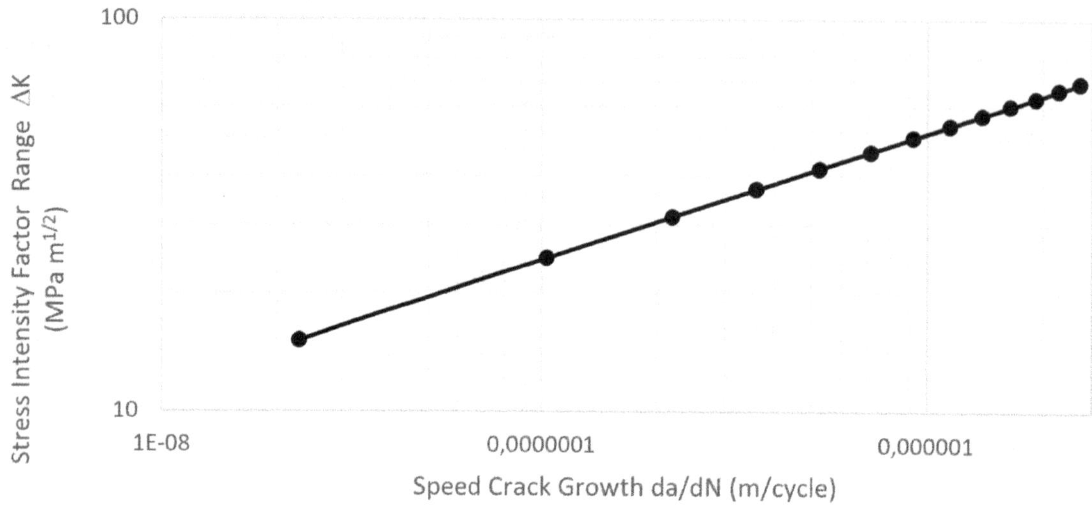


In the Figure 22 it is represented the speed crack growth (inch /cycle) and the amplitude of stress intensity factor in the specimen and the values given by Barsom (1971) for ferrite-pearlite steels. It is obvious that both values are very similar, so it means that our steel behaves as a ferrite-perlite at the HAZ. In the graphic both lines are very near almost together only when the rate increases the values are slightly different.

Specimen 01-10 (7) Barsom (1971) for Ferrite-Pearlite Steels (Table 25):

$$\frac{da}{dN} = 3.610^{-10}\left(\Delta K_I\right)^{3.0}. \tag{23}$$

In the Figure 23 has been represented in Log. Scale the crack rate - the stress intensity factor range which data are in the column 4 and 5 respectively of the Table 25, when the amplitude of the stress intensity factor of the sample (column 3 Table 25) is compared with the value given by Barsom (column 5 Table 25) are very similar. This probes the character ferritic-pearlitic of the sample in HAZ.

## CHARPY TEST

From the panel 3 AW which has being welded in the process shield metal arc welding (SMAW) where from the specimen 01-1(1), 01-2(2) and 01-3(3) were taken specimens at different distances of the fusion line with dimensions of 10mm× 10 mm×55mm in FL, WM, +2FL and +5FLwithout heat treatment has been sample at Charpy Impact Test at -40°C (Bs EN 10045-1,1990) taken measure of at the top bead and at the root bead. The results have been represented in a graph (Figure 24) and the Charpy values are in Joules Bs EN 10045-1. (1990) . In the abscissa the Charpy values are given in the ordered the distances from fusion lines are given, where"-1" means the value in the weld metal in this zone as well as in the fusion line the resilience down due to martensitic structure is quite fragile as the distance is getting

*Figure 22. Crack Growth Rate - Stress Intensity Factor Amplitude ΔK in Specimen 01-10(7) and Ferrite-Perlite ΔK by Barsom. Log. Scale*

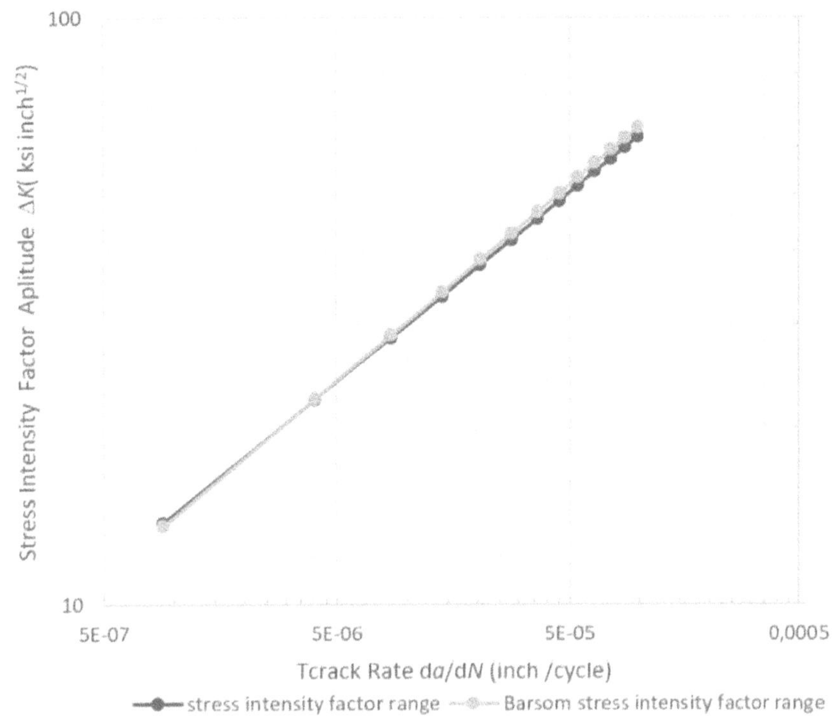


*Table 24. Table of Figure 22*

| $\Delta K$ (Ksi in$^{1/2}$) Specimen | da/dN (inch/Cycle) Specimen | $\Delta K$ (Ksi in$^{1/2}$) Barsom | da/dN (inch/Cycle) Barsom |
|---|---|---|---|
| 13.734 | 9.33E-07 | 13.577 | 9.01E-07 |
| 22.327 | 4.01E-06 | 22.432 | 4.06E-06 |
| 28.431 | 8.27E-06 | 28.796 | 8.59E-06 |
| 33.440 | 1.35E-05 | 34.052 | 1.42E-05 |
| 37.790 | 1.94E-05 | 38.639 | 2.08E-05 |
| 41.689 | 2.61E-05 | 42.766 | 2.82E-05 |
| 45.253 | 3.34E-05 | 46.549 | 3.63E-05 |
| 48.556 | 4.12E-05 | 50.064 | 4.52E-05 |
| 51.648 | 4.96E-05 | 53.362 | 5.47E-05 |
| **54.565** | **5.85E-05** | **56.480** | 6.49E-05 |
| 57.335 | 6.79E-05 | 59.444 | 7.56E-05 |
| 59.976 | 7.77E-05 | 62.276 | 8.70E-05 |
| 62.506 | 8.79E-05 | 64.992 | 9.88E-05 |

*Figure 23. Rate of the Crack Growing - Stress Intensity Factor Range for Ferrite-Perlite steels According to Barsom (1971) Specimen 01-10(7)*

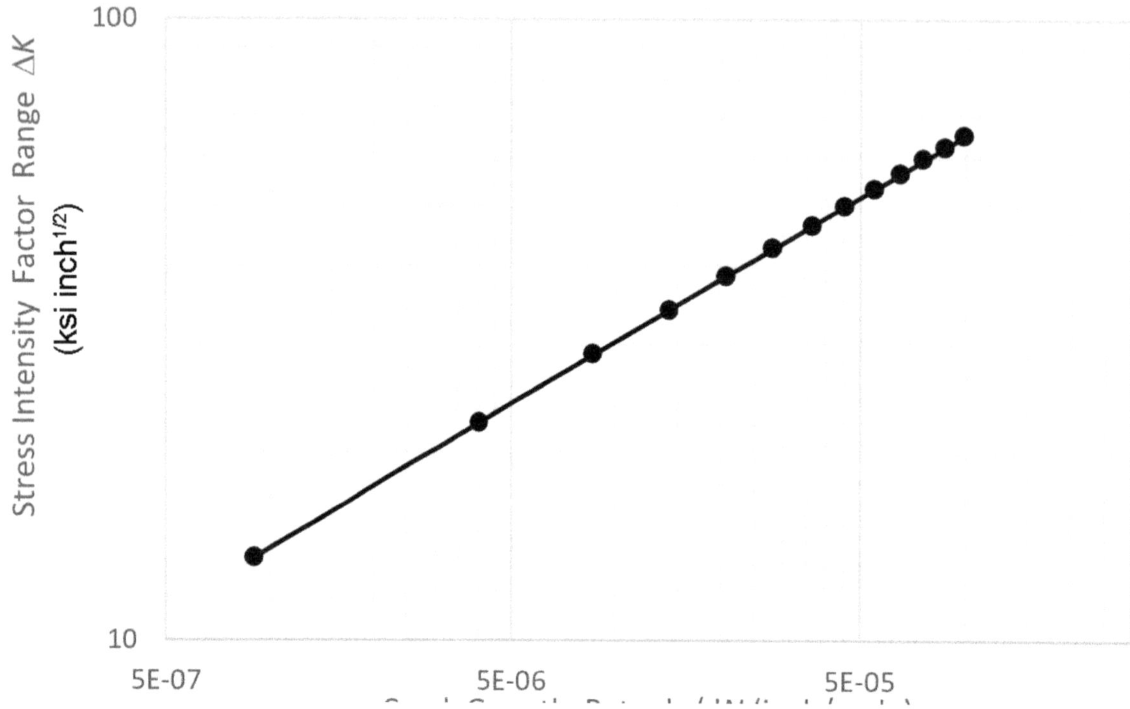


*Table 25. Crack Growing Rate- Stress Intensity Factor Amplitude Barsom. Specimen 01-10(7)*

| $\Delta K$(MPam$^{1/2}$) | d$a$/d$N$(m/cycle) | $\Delta K$ (ksi in$^{1/2}$) | d$a$/d$N$(inch/cycle) | $\Delta K$ (ksi in$^{1/2}$) Barsom |
|---|---|---|---|---|
| 15.155 | 2.288E-08 | 13.734 | 9.010E-07 | 13.577 |
| 24.636 | 1.032E-07 | 22.327 | 4.063E-06 | 22.432 |
| 31.373 | 2.183E-07 | 28.431 | 8.596E-06 | 28.796 |
| 36.899 | 3.611E-07 | 33.440 | 1.421E-05 | 34.052 |
| 41.699 | 5.275E-07 | 37.790 | 2.077E-05 | 38.639 |
| 46.001 | 7.152E-07 | 41.689 | 2.816E-05 | 42.766 |
| 49.934 | 9.223E-07 | 45.253 | 3.631E-05 | 46.549 |
| 53.579 | 1.147E-06 | 48.556 | 4.517E-05 | 50.064 |
| 56.991 | 1.389E-06 | 51.648 | 5.470E-05 | 53.362 |
| **60.210** | **1.647E-06** | **54.565** | **6.486E-05** | **56.480** |
| 63.266 | 1.921E-06 | 57.335 | 7.562E-05 | 59.444 |
| 66.180 | 2.208E-06 | 59.976 | 8.695E-05 | 62.276 |
| 68.972 | 2.510E-06 | 62.506 | 9.883E-05 | 64.992 |

*Figure 24. Charpy Value at Distances from Fusion Line. Panel 3AW (SMAW)*

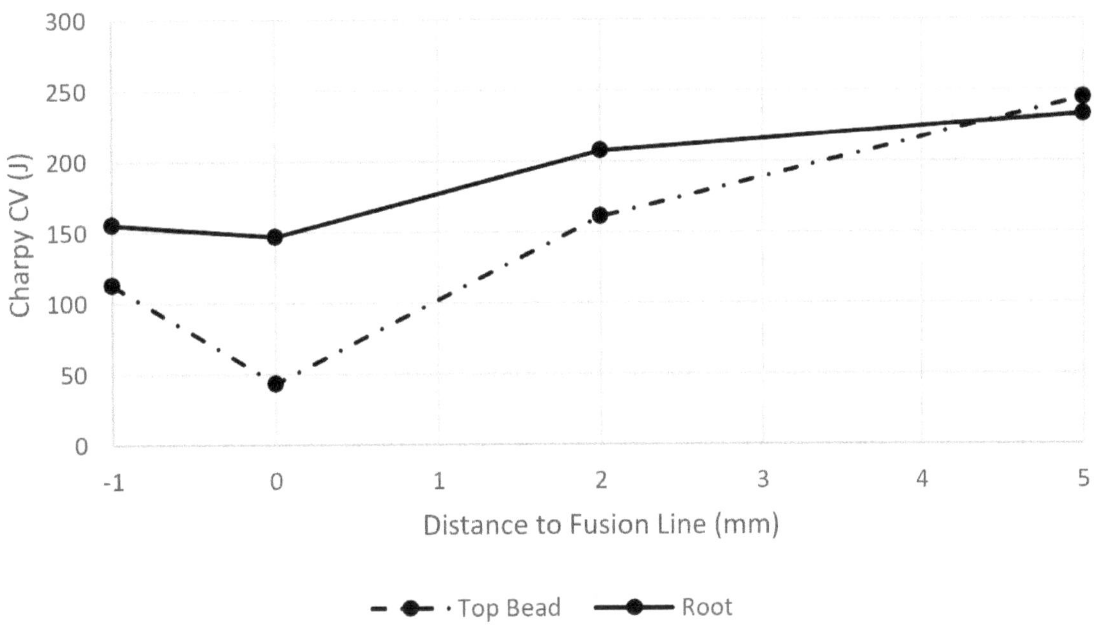


*Table 26. Table of Figure 24*

| Specimen 3AW (SMAW) | | Charpy test |
|---|---|---|
| **Distance to FL(mm)** | **Charpy CVN (J) Top Bead** | **Charpy CVN(J) Root** |
| -1 (WM) | 112.5 | 155.2 |
| 0 (FL) | 43.2 | 147.2 |
| 2 (HAZ) | 160.83 | 207.3 |
| 5 (HAZ) | 244.6 | 233.2 |

bigger the resilience is growing up so the structure is less fragile. Two curves have been represented, one corresponds to the root bead with higher resilience and other the beads at the top of the weld with lower Charpy values, the root has higher resilience due to the different weld beads that improve the weld metal quality in this zone. The multilayer welds in the root increases the HAZ toughness as the reheating provided by the subsequent passes in accordance with the distance to fusion line grows up the resilience in both lines trend to be equal.

The Charpy (impact test) is performed with the samples at a temperature of -40°C (BS 131,part 2) and if test fail giving a poor resilience, no CTOD test is required so the panel is directly rejected so is considered too fragile in the HAZ. The aging effect get fragile the HAZ caused by free Nitrogen(N). In the microalloy steels the nitrogen combines with microalloy elements origin AlN, TiN, VN, among other, however these are dissolved in the zones near to the fusion line being the nitrogen free causing a bigger embrittlement being the Hydrogen from welding consumables that enters in HAZ cause of embrittlement too, nevertheless owing to multipass this is considerably reduce at the root, this effect is reduced outside of HAZ.

## SLOW STABLE CRACK EXTENSION-CTOD

In the graph of Figure 25, we can see the Slow stable crack extension ($\delta a$) with the CTOD ($\delta$) in the nine specimens which represent the three-welding process sampled. It is not observed a higher CTOD value in accordance with the welding process used, this test comes to represent how the crack is extended in the HAZ and in accordance with the Standard (BS5762: 1.979) a minimum value of CTOD "$\delta$=0.20 mm" it is stablished. As we can see in the case of the nine samples represented in the three-welding process comply with this requirement. Since this test represents the toughness in the HAZ, these results prove the HAZ has toughness enough. The CTOD test is carry out at a temperature of -10°C. The stable crack extension is the length of the crack when the sample is tested at a fix load before the sample be broken, this is the second part of the CTOD testing and in accordance with the graph (Figure 25) it is can see almost the proportionality with the CTOD values. The bigger CTOD corresponding to those the sample was stop without broken (TS). The "m" represents the maximum force was reached. So, the Test Stop (TS) and the maximum force "m" are those the maximum toughness (CTOD) reached. Anyway, the slope of the trend is ascending.

## HARD VICKERS 10 PANEL 3AW SPECIMEN 01-2(2)

As we can see in the graph Figure 26 the harder points are those closer to fusion line (FL) so the metallurgy transformation in this area corresponding to martensite, so this is the most critical area which correspond to Grain Coarsened Heat Affected Zone (GCHAZ), with a grain size of 7 ASTM (31.8μm)

*Figure 25. Slow Stable Crack Extension- CTOD*

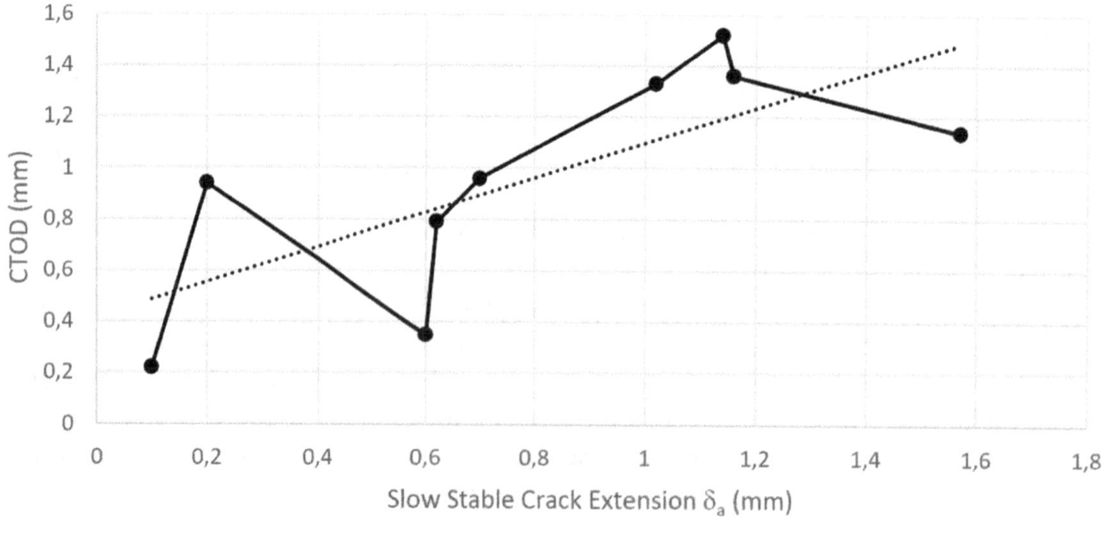

*Table 27. Table of Figure 25*

| Process | Specimen | $\delta_a$ (mm) | CTOD $\delta$ (mm) | Sample stop |
|---|---|---|---|---|
| S.A.W. | 01-12 (9) | 0.1 | 0.22 | u |
| S.A.W. | 01-10(7) | 0.2 | 0.94 | u |
| S.M.A.W. | 01-1(1) | 0.6 | 0.349 | u |
| S.M.A.W. | 01-2(2) | 0.62 | 0.79 | u |
| S.A.W. | 01-11(8) | 0.7 | 0.96 | u |
| G.S.F.C. | 02-02(5) | 1.02 | 1.33 | T.S. |
| G.S.F.C. | 02-03(6) | 1.14 | 1.52 | T.S. |
| G.S.F.C. | 02-01(4) | 1.16 | 1.36 | T.S. |
| S.M.A.W. | 01-3(3) | 1.57 | 1.14 | M |

where the crack can be initiated and propagated easier. In the graph has been represented three situations: Top beads where the maximum hardness is produced at the (FL) with an acceptable value of 201 HV10, the Central Root with a maximum value of 233 HV10 and in the lower beads with a maximum of 199 HV10. The three maximum were at the FL so this is the more fragile point of the HAZ, there is just a point at the top beads with 218 HV 10 at the Fine Grain of HAZ with a ASTM grain size of 13 (4µm), in spite of fine grain point in this HAZ. The grain size suffers a reduction in HAZ which correspond to the refine grain area (ICHAZ) 13.55 ASTM increasing the grain size to 10.55 ASTM at Subcritical HAZ (5 mm from FL), at the spheroid carbide area. The panel where these specimens come from was welded in SMAW process with a heat input of 1.56 KJ/mm, as it can be seen the grain size decrease as the distance to the FL grows up. The maximum hardness is 233 HV10 in the FL and the minimum hardness is 171 HV10 in the ICHAZ, All the HV 10 results were acceptable with the standard BS 4360.

At Figure 27 it is represented the beads distribution of the weld sample at the top, central and lower root. At below Figure 28 we have represented the distance to FL with the Vickers 10 hardness vs. ASTM grain size. As it is logic the maximum hardness it is produced at those points which are closer to the weld, so in these points are the martensitic grain transformation. It can be observed, as well, the points with the biggest ASTM (grain size smaller) are corresponded with lowers Vickers 10 values, been the grain size is getting smaller as we go away from the fusion line where the metallurgic transformation gives values nearer to Ferrite-Perlite grain. The grain size is being reduced up to the non-affected zone (BM)where the hardness has lower values that in points close to FL. As we can see at the graphic and in the Figure 28 the beads hardness are corresponded to those beads which are at the central root, when they are close to FL, but as the distance to the weld increases, the hardness at the central bead improve, given lower values, so the number of beads improve the quality of the grain by annealing process. At 4 mm from FL, are registered the finest grain (ASTM13.55) and lower hardness. All Vickers 10 results are inside the standard that requires a maximum value of 325.

*Figure 26. Distance to Fusion Line -Hardness Vickers 10. Panel 3AW. Specimen 01-2(2)*

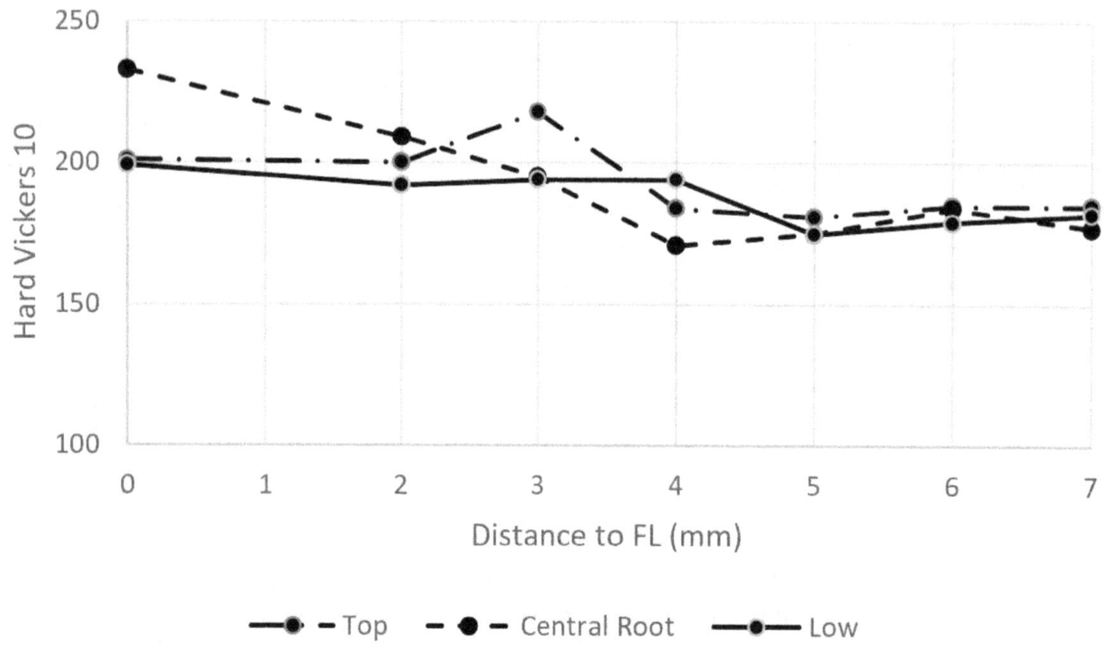


*Table 28. Table of Figure 26*

| Hardness Vickers10 | | | | | | | |
|---|---|---|---|---|---|---|---|
| Distance to FL(mm) | **0** | **2** | **3** | **4** | **5** | **6** | **7** |
| Top beads | 201 | 200 | 218 | 184 | 181 | 185 | 185 |
| Central root beads | 233 | 209 | 195 | 171 | 175 | 184 | 177 |
| Lower beads | 199 | 192 | 194 | 194 | 175 | 179 | 182 |
| ASTM grain size | 7 | 12.52 | 13 | 13.55 | 10.55 | 11.03 | 11.03 |
| Grain size (μm) | 31.80 | 4.70 | 4 | 3.30 | 9.40 | 7.90 | 7.90 |
| HAZ (AREA) | Grain Coarsened GCHAZ | Growing grain & Fine grain FGHAZ | Fine Grain FGHAZ | Intercritical ICHAZ | Subcritical SCHAZ | Subcritical SCHAZ | Base Metal BM |

## CHARPY TEST- ROBERT NEWTON CORRELATIONS

A study of correlation between the results of Charpy test (CVN) and fracture toughness through Roberts. R, Newton C. (1984), it is an empirical equation $K_{IC}$ (Ksi in$^{1/2}$) = 9.35 (CVN)$^{0.63}$(ft-lb), where the $K_{IC}$ have been expressed in MPam$^{1/2}$, in the graph of the Figure 29, the results have been plotted. In this

*Figure 27. Cross Section of the Weld of the Sample*

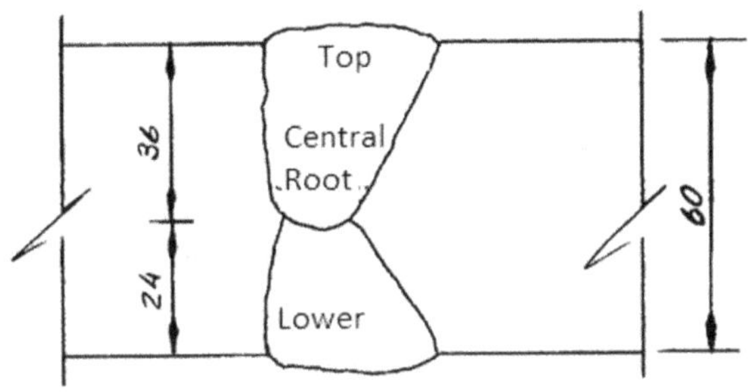


*Figure 28. Distance to FL.- HV10-ASTM Grain Size. Panel 3AW. Specimen 01-2(2)*

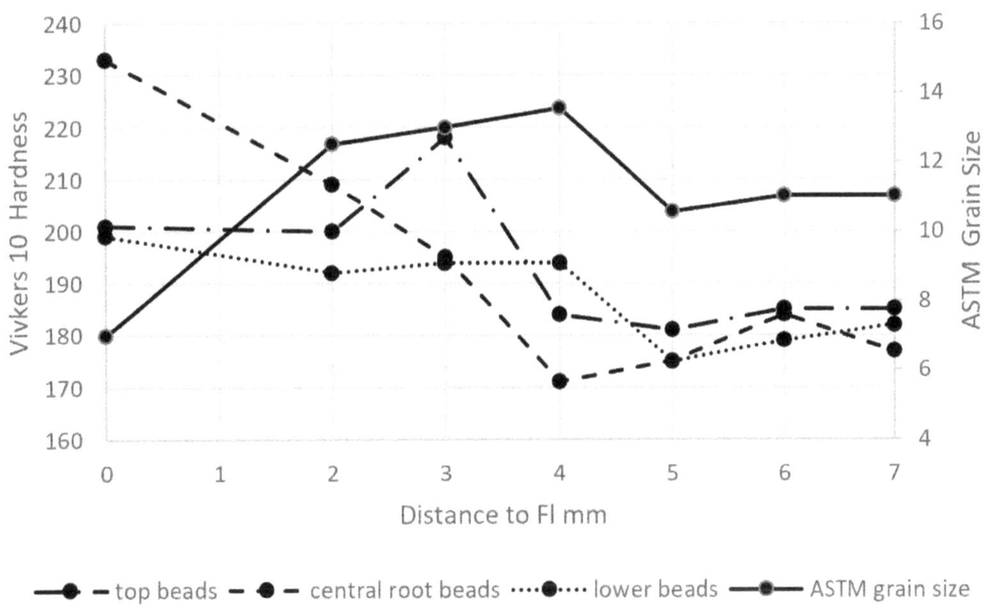


*Table 29. Table of Figure 28*

| Distance to FL (mm) | Top beads. HV10 | Central root beads HV10 | Lower beads HV10 | ASTM Grain Size |
|---|---|---|---|---|
| 0 | 201 | 233 | 199 | 7 |
| 2 | 200 | 209 | 192 | 12.5 |
| 3 | 218 | 195 | 194 | 13 |
| 4 | 184 | 171 | 194 | 13.55 |
| 5 | 181 | 175 | 175 | 10.55 |
| 6 | 185 | 184 | 179 | 11.03 |
| 7 | 185 | 177 | 182 | 11.03 |

graph has been represented the Charpy values that were obtained in the Charpy test at -40°C. This testing is perfumed before CTOD testing and only just in case the Charpy give acceptable values, CTOD will be carried out, otherwise the panel will be rejected-The Charpy specimen were taken at FL, at 2mm from FL and 5mm from FL, as represented in in Figure 29 and Table 32. The test will be in accordance with BS 131 Part 2.For the samples obtained from the panel 3 AW at HAZ in the Tables 30 values at International system and English system, obtained directly from the formula are represented. In the Table 30, one can observe the low value (43.16 J) of the resilience at 0 (FL) is in the limit according to the standard BS 4360 & BS 7191(1989) that ask for a minimum of 45 J average. In weld root both the resilience and the stress intensity factor are higher, this is due the weld passes are refining the grain in this area. As the distance to FL gets further both values improve.

In the Table 30 are represented the Charpy CVN value at the top and root beads stablishing a correlation by Robert Newton between Charpy and stress intensity factor. As Robert Newton considers the $K_{IC}$-CVN do not account for strain-rate-effects, these correlations would be expected to be quite conservative and not applicable to low strength steels. Having in mind that the above correlation should have account the transition temperature for determining the $K_{IC}$ from the CVN impact values, being this temperature shift:

*Table 30. Charpy CV and Stress Intensity Factor and Root and Top Beads. Panel 3 AW at HAZ*

| distancia To FL(mm) | Charpy Top Beads CVN (J) | Charpy Root Beads CVN(J) | $K_{IC}$ Top (MPam$^{1/2}$) | $K_{IC}$ Root (MPam$^{1/2}$) |
|---|---|---|---|---|
| -1(WM) | 112.49 | 155.21 | 166.83 | 204.33 |
| 0(FL) | 43.16 | 147.15 | 91.23 | 197.58 |
| 2mm from FL | 160.88 | 207.32 | 209.02 | 245.21 |
| 5mm from FL WM | 244.60 | 233.15 | 272.14 | 264.04 |
| **Distance To FL(mm)** | **Charpy CVN (ft lbf) Top Beads** | **Charpy CVN (ft lbf) Root Beads** | **$K_{IC}$ Top (Ksi in$^{1/2}$)** | **$K_{IC}$ Root (Ksi in$^{1/2}$)** |
| -1(WM) | 82.91 | 114,39 | 151.19 | 185.18 |
| 0(FL) | 31.81 | 108.45 | 82.68 | 179.06 |
| 2mm from FL | 118.57 | 152.79 | 189.42 | 222.22 |
| 5mm from FL | 180.27 | 171.83 | 246.63 | 239.29 |

$T_{shift}$=215-1.5$\sigma_{ys}$ for steels 36ksi<$\sigma_{ys}$<140ksi as it can be considered HSLA we are working on lower energy in all beads are in in FL. CVN impact test values as stress intensity factor ($K_{Ic}$) growths, as above equation the shift temperature would be about 117ºF (48°C) being for this temperature a CVN impact value about 60 ft-lb (81.4 J).

In the figure 29 we see a correspondence between Charpy and fracture toughness where there is a drop in the FL for both Charpy and Toughness maintaining a proportionality in both, being the lower values in the line of top beads.

In the Figure 30 and Table 32 have been represented the hardness Vickers 10 with the resilience Charpy CVN test in order to compare both results. As we can see in the graphic the minimum resilience is in FL being lower at the top beads up to 4.5 mm from the FL where both values, top and root resilience, are approximately the same. The higher resilience at the root it is due that number of beads

*Figure 29. Distance to FL- Charpy CVN (Resilience)- Fracture Toughness Robert Newton Correlation in HAZ in Panel 3AW.*

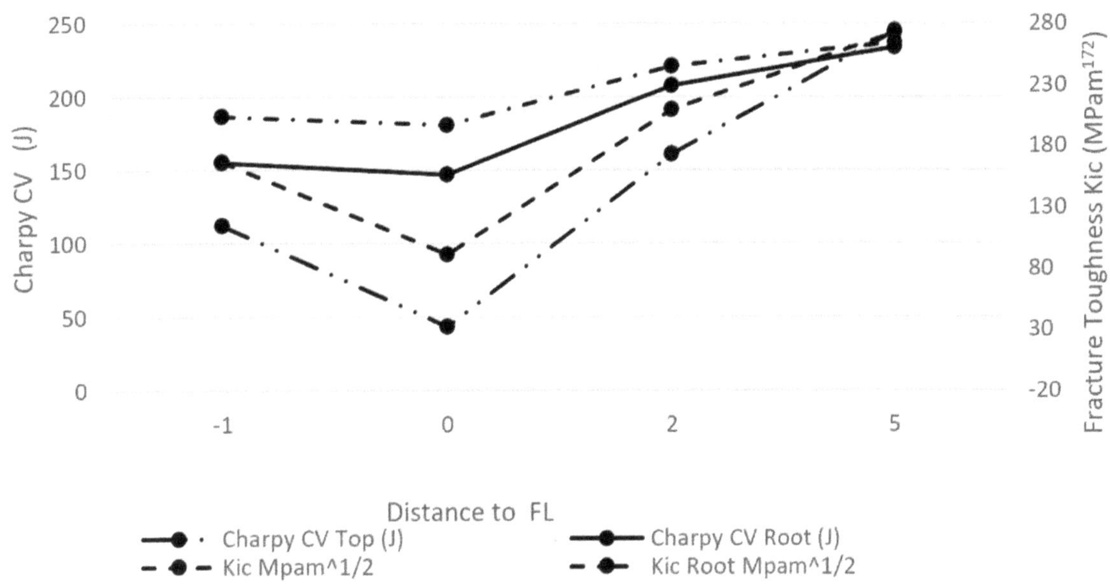


*Table 31. Table of Figure 29*

| FL Distance | -1(mm)(WM) | 0(FL) | 2(mm) | 5(mm) |
|---|---|---|---|---|
| Charpy at Top beads CVN (J) | 112.49 | 43.16 | 160.88 | 244.60 |
| Charpy at Root beads CVN(J) | 155.20 | 147.15 | 207.32 | 233.15 |
| $K_{IC}$ at Top (MPam$^{1/2}$) | 166.84 | 91.29 | 209.02 | 272.14 |
| $K_{IC}$ at Root (MPam$^{1/2}$) | 204.33 | 197.58 | 245.21 | 264.04 |

at the line of root improve the quality of the weld by annealing. All the hardness is with values within the standard BS 4360(1990).

In the closer points to FL we can see that the grain coarse has the lowest resilience corresponding with the most hardness, which is reducing as the FL distance increases being reducing the hardness, as correspond the transition from an area of biggest grain size with a martensitic microstructure to an area where the microstructure is less fragile, Chai, G. et al.(2015) up to reach the zone no affected by the heat, (base Metal, ≅7 mm from FL).

So, we can deduct from above that the most fragility is near to FL which is the region where crack normally initiates, being this area where the resilience and toughness are lowest. So, in order to make less brittle the closest structure to FL, some attention is to be taken, as to limit the HI below 2.5 kJ/mm to prevent a fast cooling and to avoid all moisture in the consumable among other cares.

The maximum HV10 is in the CR with a value of 233 HV 10, close to FL, the harder points are in central root (CR) but a bigger distance from FL are in this area where register lower value due to

*Figure 30. Charpy CVN - Vickers Hardness. Panel 3AW Specimen 01-2(2)*

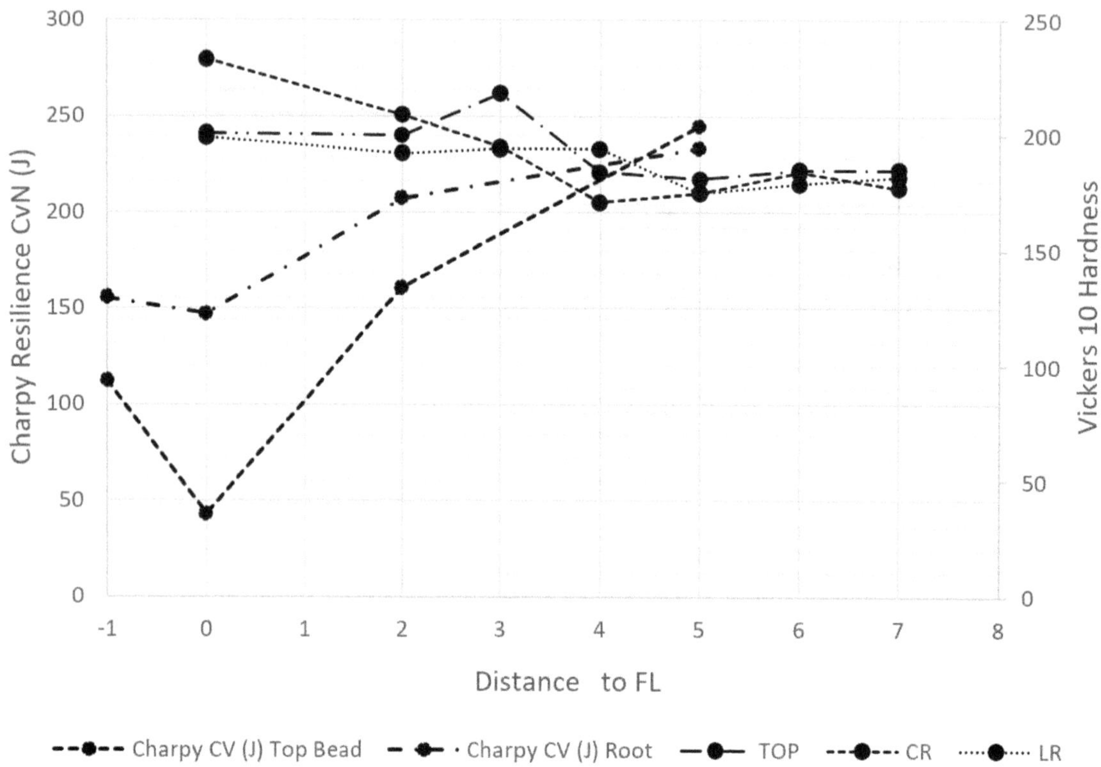


*Table 32. Panel 3AW Vickers Hardness Panel 3AW specimen 01-2(2)*

| Charpy CVN (J) Top Beads | Charpy CVN(J) Root | Distance to FL (mm) | Temperature °C | Beads TOP Hardness Vickers 10 | Beads CR Hardness Vickers 10 | Beads LR Hardness Vickers 10 |
|---|---|---|---|---|---|---|
| 112.5 | 155.2 | -1(WM) | T>1527 | | | |
| 43.2 | 147.2 | 0(FL) | 1527£T | 201 | 233 | 199 |
| | | 1(HAZ) Grain coarsened | 1450<T<1527 | | | |
| 160.8 | 207.3 | 2(HAZ)Grain growth | 1100<T<1450 | 200 | 209 | 192 |
| | | 3(HAZ)Fine grain | 900<T<1100 | 218 | 195 | 194 |
| | | 4(HAZ)Intercritical | 720<T<900 | 184 | 171 | 194 |
| 244.6 | 233.2 | 5(HAZ)Subcritical | 600<T<720 | 181 | 175 | 175 |
| | | 6 (HAZ)Subcritical | 600<T<720 | 185 | 184 | 179 |
| | | 7(Base Metal) | T£500 | 185 | 177 | 182 |

multipass effect. The maximum resilience gives a Charpy CVN value of 244.6 J top beads at a 5 mm from FL in Subcritical area. In Table 32 are depicted the different regions of the HAZ with maximum temperature reach during the welding.

*Figure 31. Crack Growth Rate- Stress Intensity Factor Amplitude. Specimen 01-10(7) Decimal Scale*

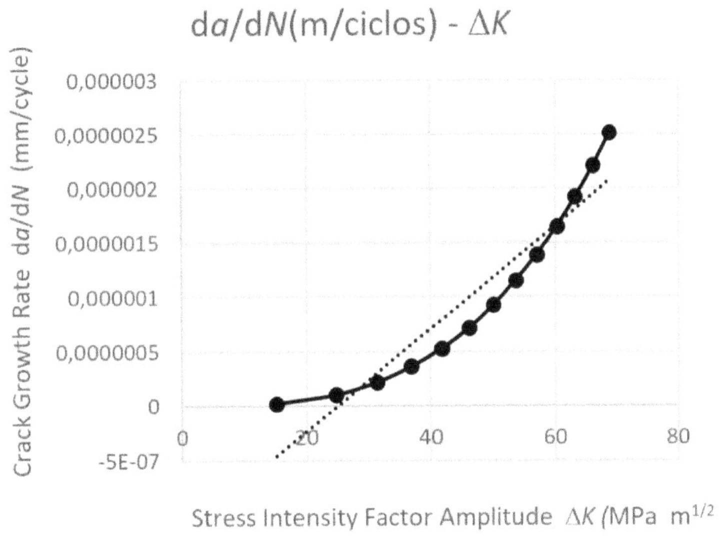


*Table 33. Table of Figure 31*

| $\Delta K$ (MPam$^{1/2}$) | da/d$N$ (m/cycle) |
|---|---|
| 15.155 | 2.288E-08 |
| 24.636 | 1.032E-07 |
| 31.373 | 2.183E-07 |
| 36.899 | 3.611E-07 |
| 41.699 | 5.275E-07 |
| 46.001 | 7.152E-07 |
| 49.934 | 9.223E-07 |
| 53.579 | 1.147E-06 |
| 56.991 | 1.389E-06 |
| **60.210** | **1.647E-06** |
| 63.266 | 1.921E-06 |
| 66.180 | 2.208E-06 |
| 68.972 | 2.510E-06 |

In the Figure 31, we have represented the crack growth rate da/d$N$ for the specimen 01-10(7) given the stress intensity factor range $\Delta K$ which correspond to region II of the sigmoidal shape for the steels Ferrite-Perlite that has been calculate by equation: H.L. Ewalds(1984)

$$\frac{da}{dN} = 5.01 \times 10^{12} (\Delta K)^{3.1} m / cycle;$$
$$206.9N / mm^2 < \sigma_{ys} < 344.8N / mm^2$$

The sigmoidal shape is divides in three regions in accordance with the curve shape which has been represented in the Figure 1. this correspond to The Paris equation region II of the crack growth curve that usually is represented in Logarithm scale (Figure 23). Attempts have been made to determine a crack growth law which are wholly, or semi empirical formulae fitted to a set of data. The crack rate $da/dN$= 1.647E-6 m/cycle corresponding to $\Delta K$=60.210 MPa m$^{1/2}$ is to $a_{cr}$=9.8mm specimen 01-10(7)

## DISCUSSION

The chapter finalizes with a conclusion about all maters dealt along its. As summary we can say the HSLA has to pass all those testing that according to the standards these steels have to pass successfully. In The first part the chapter are tested three samples which have been welded in a SAW procedure being analysing and discussing the results obtained in the HAZ. The testing carry out in these three samples are: Specimens preparation, stress intensity factor and cycles determination in fatigue fracture, theoretical number of cycles at which the specimens brake by fatigue, stable crack, CTOD testing in the HAZ of the three specimens. Micrography study of the crack giving percentage of the grain sample thrown by the crack were performed. This analysis is finished with the discussing and conclusions stating the validity of the results obtained in the samples.

The second part of the chapter nine specimens are testing, where the three dealt before are included, being adding three samples welded in GSFC and three samples in SMAW procedures. Testing for validate the nine specimens have been carried out, studding CTOD in accordance with heat input were welded a complete results data are tabled including metallography study of the nine specimens. Of these results one can conclude that, when heat input is below the recommended value (2.3kJ/mm) and having in mind a good welding performance the results are acceptable. Of these nine samples we have focused the study in specimen 01-07 to perform fatigue crack propagation comparing the theoretical cycle with those given in the testing.

Finally, in the third part of the chapter other testings as a complement to those given in the first and the second part of the chapter have studied, we have referenced this study to the specimen 01-07. These are: verify the character ferritic-pearlitic of our steel has by comparison of the stress intensity factor amplitude with the results given by authors as Barsom. To determine the stress at tip crack vs. Number of cycles and crack tip size find them acceptable values.

The hardness was determined in different points of the HAZ of the specimen 01-2 as to verify the Vickers 10 hard gives acceptable values, mainly in grain coarse zone close to fusion line, the high values would be a prove that a very hard martensitic structure is formed in this area with the consequence of a brittle zone. Charpy test, this is a resilience testing performed before CTOD so if the Charpy fails no CTOD is needed and the sample is rejected. The Charpy test results corresponding to the panel 3 AW welded in a SMAW and the results are acceptable only one result at the top beats in the fusion line resulted low. A stress intensity factor starting from Charpy values by Robert Newton method was studied having a proportional correlation in the points of the HAZ. Nevertheless, the Robert Newton values shown to be very conservative value, so they no consider the transition temperature. Deeply details can be viewed in the chapter or in the conclusions.

## CONCLUSION

1. In the Table 4 is represented the CTOD values with the mode of fracture, of the three specimens welded in SAW process, and the relative amount of GCHAZ sample by the Fatigue Crack and Microstructure at Fracture Initiation Point. As we can see in the Table the specimen 01-12 (9) that shows higher quantity of GCHAZ (2B/3) has the lower value of CTOD, being the Fracture Initiation Point at FGHAZ but close to GCHAZ. Nevertheless, of specimens obtained values over the standard requires 0.2 mm proving enough fracture toughness, so the minimum CTOD=0.22mm<0.71mm average. This proves that the heat input (2.327kJ/mm) the sample was welded was right and the microalloy worked correctly (Table5).
2. We have studied nine specimens which have been welded in three different process, SMAW, GSFC and SAW, as we can see in Table 6. The heat input (H.I) is between 1.348 kJ/mm and 2.327 kJ/mm corresponding to the minimum energy to GSFC and the maximum to SAW, nevertheless all were below 2.5 kJ/mm as recommended for this kind of steels (HSLA). All CTOD values were over the 0.2 mm as the standard requires. The specimens that found higher CTOD values were those which were welded with lower HI.

In Table 6 we can see the HI of the nine specimens and its relationship with stress intensity factor and CTOD as well as the number of cycles. We can observe that those specimens welded in process with lower HI as G.S.F.C. and S.M.A.W. had a higher fatigue cycles against the specimens welded with higher HI as the process S.A.W. So, the lower HI provides higher HAZ cycles.

3. At the Table 7 have represented fatigue stress intensity factor with the stress at the tip of the crack for the nine specimens, the values of this with yield strength($\sigma/\sigma_{ys}$) give proportion between 30.7% and 63.21% which are acceptable. The maximum $a_{cr}$ =10.5 mm correspond to maximum crack tip stress $\sigma$=271.17 N/mm$^2$.
4. Although the values deduced from Table 13, do not establish an inverse proportionality between CTOD and coarsed grain amount found in the test, can be observed that the minimum value obtained in the CTOD test correspond to samples in which they found higher percentage of grain coarsened intercepted by the crack of the specimen 01-12(9).
5. In the Figure 10 is represented the tension at the crack tip vs. the length of the crack for specimen 01-10 (7). At the beginning when the crack is small (0.62091mm) the crack tip has a high tension 1068,88 N/mm$^2$, but as the crack size is growing up the stress is being reducing up to reach a value of 269.034 N/mm$^2$ for a crack size of 9.8 mm, this is due the crack is releasing energy as the crack grows. So, this is the tension at the crack tip when the crack is extended reaching a value $a_{cr}$=9.8 mm in the specimen mentioned above.
6. In the Figure 12 has been depicted the number of cycles vs. crack size for the specimen 01-10 (7) the first crack represented is 0.62091 mm at 44571 fatigue cycles the last size when the crack is extended at 9.8 mm is 68658cycles of fatigue theoretical. The difference with the experimental value 67000 cycles is due to difference in the base material as well as the stress intensity factor

coefficient 1.4, and effects of the temperature reached in HAZ, anyway the difference between experimental cycles and the theoretical cycles is a 2.4% that is acceptable.

7. 7. Compliance with the function of Paris is checked for values of ferritic - pearlitic steels (structural steel)

$$\frac{da}{dN} = 5.01 \times 10^{-12} (\Delta K)^{3.1} m / cycle;$$
$$206.9N / mm^2 < \sigma_{ys} < 344.8N / mm^2$$

The initial flaw size given above, in the case of the specimen 01-10 (7) is $a_0 = 0.11091$ mm (Table 22), being $a_d$=0.276 mm (Backlünd Fatigue-Thresholds-Fundamentals and Engineering, 1982), the first detectable threshold of the crack. The critical crack $a_{cr}$= 9.8 mm as it is shown in the column 3 of Table 22. Because we can consider an intensity factor rate $\Delta K$ constant a correlation between this and $da/dN$ are to be used. Practically the difference between $a_0$ and $a_d$ are to certain discontinuity in the structure as: voids, flaws, damages and material inhomogeneities.

8. Nine specimens were tested in three welding process: SMAW with a HI of 1.560kJ/mm. GSFC with a HI of 1.348kJ/mm and SAW with HI of 2.327kJ/mm. The welding energy recommended, for this high strength low alloy steel (HSLA) below 2.5kJ/mm were observed and the CTOD results were all inside the standard $\delta$>0.2mm. As in the Table 6 can be seen specimens welded with lower energy, gave higher values of fracture toughness as well as bigger frequency of breaking. As depicted on Table 7 the tension at the tip crack was $153.51 N/mm^2 < \sigma < 271.1 N/mm^2$ which correspond with a range of the yield strength $30.7\% < \sigma/\sigma_y < 63.21\%$ with a wide safety margin in HAZ.
9. All specimen gave CTOD>0.2mm, value inside the Standards BS 5762 (1979) and BS 4360(1990). However not all specimens complied with EEMUA 158 (2014), that requires that crack initiate within 0.5 mm from FL. In Table 13 CTOD values and percentages of grain are represented for each of the samples. Where one can see that the most of samples with higher percentage of grain coarsened sampling by the crack gave lower CTOD value as well as unstable fracture.
10. In the Table 14 has been represented the CTOD ($\delta$) individual each sample and average of the group, of this group the CTOD lower correspond to those specimens that were welded in process with higher HI (SAW). As we see in this Table all specimens show a significant percentage that the crack went through the welled metal due to the notch was not completely straight. The specimens welded with lower HI which in GSFC process (see Table 16), gave a TS testing failure, and higher CTOD ($\delta$), furthermore they have the lower percentage of GC and higher percentage of fine grain (FG). Observation by optical microscope the group welded in SMAW (specimens 7,8 y 9) shows that the crack initiated in FG
11. At the table 23 it can see the proximity of the stress intensity factor amplitude ($\Delta K$) between the values given by Barsom(Barsom, J.M 1971) for ferrite-pearlite steels in column 5 and the obtained in our samples in column 3, according to (Eawalds, H.L 1984). This prove the character of ferritic-pearlitic of our steel in the HAZ after welding, as it has been proved in results which have been depicted in the Figures 18 and Figure 22.

12. At the Figure 28 we have represented the values of the Vickers 10 hardness vs. ASTM grain size against the distance to FL As it is logic the maximum hardness it is produced at those points which are closer to the weld, so in these points are the martensitic grain transformations. It can be observed, as well that these points are corresponded with the biggest grain size ASTM <7, it means lowers values, been the grain size smaller as we go away from the fusion line, where the metallurgic transformation gives values nearer to Ferrite-Perlite grain, going growing up to the non-affected zone distance to FL>7mm. As we can see at the graphic and the Table 29 the beads of more hardness are corresponded to those beads which are at the central root, when they are close to FL but as the distance to the weld increases, the hardness at the central bead improve, given lower values, so the number of beads improve the quality of the grain annealing by the welding process.
13. At the Figure 29 and Table 31 are represented the Charpy test results and fracture toughness vs. distance to FL, according to Robert Newton correlation, biggest resilience and stress intensity factor showed those beads at roots than the top beads, lowest resilience and fracture toughness are at FL. As correspond to this type of weldment in HAZ all parameters both Charpy CVN and stress intensity factor improve, as the distance to FL increases.
14. In the Figure 30 and Table 32 have been represented the resilience (Charpy CVN) in the line of top beads, root beads and hardness Vickers 10 for distance to FL as well as temperature reached in every zone of the HAZ. The biggest resilience was shown at the line of the root beads up to the subcritical area (5 mm to FL), from this point line of the top beads gave higher resilience than the root beads. All the values proved to be acceptable as the standard requires. About the Vickers hardness the higher values were obtained at the line of CR beads up to a distance to the FL of 3mm (Fine Grain Area). The higher Resilience (Charpy)are corresponded with the lower hardness values Vickers 10
15. It has been proved the character of Martensitic structure at the FL so it is the most brittle area having the lower resilience and biggest hardness, being this region together with the grain coarsened the most critical and the main reason why the heat input is to be limited. Other phenomena such as brittle fracture for hydrogen by diffusion when the welding must be taken into account.

## REFERENCES


Bäcklund, J., Blom, A. F., & Beevers, C. J. (1982). *Fatigue thresholds, Fundamental and Engineering Applications* (vols. 1-2). Engineering Materials Advisor Services Ltd.

Barsom, J. M. (1971). Fatigue-Crack Propagation in Steels of Various Yield Strengths, *Transactions of the ASME Journal of Engineering for Industry, Series B, 93(4)*. 10.1115/1.3428061

Barsom, J. M. (1974). *Fatigue Behaviour of Pressure-Vessel Steels. WRC Bulletin, 194*. Welding Research Council.

Barsom, J. M., & Rolfe, S. T. (1977). *Fracture and Fatigue Control in Structures. Applications of Fracture Mechanics*. Prentice-Hall, Inc.

Broek, D. (1982). *Elementary Engineering Fracture Mechanics*. Kluwer Academic Publishers., doi:10.1007/978-94-011-9055-8

Carman, C. M., & Katlin, J. M. (1966). *Low Cycle fatigue crack propagation of high- strength steels.* ASME paper N° 66-MET.

Chai, G., Forsman, T., Gustavsson, F., & Wang, C. (2015). Formation of the grained area in martensitic steel during very high cycle fatigue. *Fatigue & Fracture of Engineering Materials & Structures*, *38*(11), 1315–1323. doi:10.1111/ffe.12345

Chai, G., Zhou, N., Ciurea, S., & Andersson, M. (2012). Author. *Scripta Materialia*, *66*, 769–772. doi:10.1016/j.scriptamat.2012.02.003

*Charpy Impact Test on Metallic Materials- Test Method (V and U notches), Bs EN 10045-1.* (1990). British Standards Publication.

Clark, J. R. (1970). *Fatigue Crack Growth Characteristics of Heavy Section. ASTM A533 Grade B Class 1 Steel Weldments. ASME Paper N° 70-PVP 24.* American Society of Mechanical Engineers.

Clark, W. G., & How, J. R. (1974). *August. Fatigue Crack Initiation and Growth Properties Affect the Material Selection and Design Criteria.* Metals Engineering Quarterly.

Clintock, M. C. (1963). On the Plasticity of the Growth of Time Fatigue Cracks. In *Fracture of Solids* (p. 65). Willey-Interscience.

Construction specification for fixed offshore structures in North Sea EEMUA 158. (2014). *Engineering Equipment and Materials Users Association, 158.*

Cooper, R. J., Silva, H. F., & Trevisan, R. E. (2004). Influencia del precalentamiento en las propiedades de uniones soldadas de acero API 5L-X80 soldadas con alambre tubular autoprotegido. *Rev. Metalurgia-Madrid, 40*(4), 280–270. doi:.2004.v40.i4.275 doi:10.3989/revmetalm

El Haddad, M. H., Dowling, N. E., Topper, T. H., & Smith, K. N. (1980). J Integral Application for Short Fatigue Cracks at Notches. *International Journal of Fracture*, *16*(1), 15–30. doi:10.1007/BF00042383

Elber, W. (1971). *The Significance of Crack Closure. Damage Tolerance Aircraft Structure. ASTM STP 486.* American Society for Testing and Materials., doi:10.1520/STP26680S

Elices, M. (1992). *Mecánica De La Fractura Aplicada A Sólidos Elásticos Bidimensionales.* Departamento de Ciencias de Materiales E.T. Superior de Ingenieros de Caminos de Madrid.

Ewalds, H. L., & Wanhill, R. J. H. (1991). *Fracture Mechanics.* Edward Arnold Edition.

Frost. (1960). Notch Effects and the Critical Alternating Stress Required to Propagate a Crack in an Aluminium Alloy Subjected to Fatigue Loading. *Journal of the Mechanical Engineering Society, 2*(2), 109-119.

González-Palma, R., Suárez-Bermejo, J. C., & Vicario, F. J., & Muñoz, A. (2006). Procesos metalúrgicos de aceros microaleados de alta resistencia 450 EMZ tipo II en la zona afectada térmicamente debido al efecto multipasada en soldaduras por arco sumergido. *Rev. Metal. Madrid*, *42*, 256–269.

Harrison, J. D. (1970, March). An Analysis of Data on non-Propagating Fatigue Cracks on a Fracture Mechanics Basis. *British Welding Journal*, *2*(3).

Kitagawa, & Takahashi, S. H. (1976). Applicability of Fracture Mechanics to Very Small Cracks or the Cracks in the Early Stages. In *Proceedings of the Second International Conference on Mechanical Behaviour of Materials.* American Society for Metals.

Krupp, U., & Álvarez-Armas, I. (2014). Short Fatigue crack propagation during low-cycle, high cycle and very-high cycle fatigue of duplex steel a unified approach. *International Journal of Fatigue, 65*, 78–85. doi:10.1016/j.ijfatigue.2013.06.018

Kujawski, D. (2013). Environment and Stress Interaction in Corrosion Fatigue and Stress Corrosion Cracking. *Metallurgical and Materials Transactions. A, Physical Metallurgy and Materials Science, 44A*(3), 1340–1352. doi:10.100711661-012-1507-8

Kujawski, D., & Sadananda, K. (2011). Effect of Crack-Tip Stresses on SCC Behaviour. *Metallurgical and Materials Transactions. A, Physical Metallurgy and Materials Science, 42A*(2), 377–382. doi:10.100711661-010-0373-5

Lei, Z., Hong, Y., Xie, J., Sun, C., & Zhao, A. (2012). Effects of inclusion size and location of very-high-cycle fatigue behaviour for high strength steels. *Materials Science and Engineering A, 558*, 234–241. doi:10.1016/j.msea.2012.07.118

Leiva, R., Donoso, J. R., Muehlich, U., & Labbe, F. (2004). Un enfoque de dos parámetros para el análisis del efecto del cordón de soldadura sobre el constreñimiento. *Rev. Metal Madrid, 40*(1), 280–287. doi:10.3989/revmetalm.2004.v40.i1.237

Liu, Y. B., Yanga, Z. G., Li, Y. D., Chena, S. M., Li, S. X., Hui, W. J., & … . (2010). Prediction of the S-N curve of high- strength steels in the very high cycle fatigue regime. *International Journal of Fatigue, 32*(8), 1351–1357. doi:10.1016/j.ijfatigue.2010.02.006

Lu, L. T., Zhnag, J. W., & Shiozawua, K. (2009). Influence of inclusion size on S-N curves characteristics of high-strength steels in very high cycle fatigue regime. *Fatigue & Fracture of Engineering Materials & Structures, 32*(8), 647–655. doi:10.1111/j.1460-2695.2009.01370.x

McDonald, K.A., & Barnes. (1991). *Crack tip opening displacement fracture mechanics testing for project topsides.* Academic Press.

*Method for crack opening displacement (COD) testing Bs 5762.* (1979). British Standard Institution.

Methods for Crack Opening Displacement C.O.D. (Testing). (1979). *Methods for Crack Opening Displacement C.O.D. (Testing) DO: 19, (1.972).*

Methods for Plane Strain Fracture Toughness KIC Testing. DD3. (1971). British Standard British standard 5762.

Murakami, Y. (2012). Material defects as the basis of fatigue design. *International Journal of Fatigue, 41*, 2–10. doi:10.1016/j.ijfatigue.2011.12.001

Murakami, Y., & Yamashita, Y. (2014). Prediction of life and scatter of fatigue failure originated at non-metallic inclusions. *Procedia Engineering, 74*, 6–11. doi:10.1016/j.proeng.2014.06.214

Paris, P. C. (1970). Testing for Very Slow Growth of Fatigue Cracks. *MTS Close Loop Magazine, 2*(5).

Paris, P. C., Bucci, R. J., Wessel, E. T., Clark, W. G., & Mager, T. R. (1972). *Extensive Study of Low Fatigue Crack Growth Rates in A533 and A508 Steels. ASTM STP513*. American Society for Testing and Materials.

Priddle, E. K. (1982). *The Threshold Stress Intensity Factor for Fatigue Crack Growth in Mild Steel Plate and Weld Metal: Some Effects of Temperature and Environment. Fatigue Threshold-Fundamentals and Engineering Applications*. Engineering Materials Advisory Services Ltd.

Quan, G., Zou, C., & Hong, Y. (2001). Experimental and theoretical investigation of environmental media on very-High cycle fatigue behaviour for a structural steel. *Acta Materialia, 59*(4), 1321–1327. doi:10.1016/j.actamat.2010.10.064

Roberts, R., & Newton, C. (1984). Report on small-scale test correlations with $K_{IC}$ data. *WRC Bulletin, 299.*

Roiko, A., & Murakami, Y. (2012). A design approach for components in ultralong fatigue life with step loading. *International Journal of Fatigue, 41*, 140–149. doi:10.1016/j.ijfatigue.2011.12.021

Sakai, T. (2009). Review and prospect for current studies on very high cycle fatigue of metallic materials for machine structural use. *J Solid Mech. Materials & Design, 3*, 425–439.

Sander, M., Müler, T., & Lebahn, J. (2013). Influence of mean stress and variable amplitude loading on the fatigue behaviour of a high-strength steel in VHCF regime. *International Journal of Fatigue, 62*, 10–20. doi:10.1016/j.ijfatigue.2013.04.015

Schmidt, R. A., & Paris, P. C. (1973). Threshold for Fatigue Crack Propagation and Effect of Load Ratio and Frequency. *ASTM Special Technical Publication*, 536.

Schuller, R., Fitzka, M., Irrash, D., Tran, D., Pennings, B., & Mayer, H. (2015). VHCF properties of nitride 18 Ni of maraging steel thin sheets with different Co and Ti content. *Fatigue & Fracture of Engineering Materials & Structures, 38*(5), 518–527. doi:10.1111/ffe.12251

Schweizer, C., Seifert, T., Nieweg, B., Von Hartrott, P., & Riedel, H. (2011). Mechanisms and modelling of fatigue crack growth under combined low and high cycle fatigue loading. *International Journal of Fatigue, 33*(2), 194–202. doi:10.1016/j.ijfatigue.2010.08.008

Shiozawa, K., Hasegawa, T., Kashiwagi, Y., & Lu, L. (2009). Very high cycle fatigue properties of bearing steel under axial loading condition. *International Journal of Fatigue, 31*(5), 880–888. doi:10.1016/j.ijfatigue.2008.11.001

Silverio, R. C., & Moura, T. (2003). Evaluation of cold crack susceptibility on HSLA steel welded joints. *Rev. Metal. Madrid, 39*(2), 83–89. doi:10.3989/revmetalm.2003.v39.i2.318

*Specification for weldable structural steels BS 4360.* (1990 September). British Standard Institution.

*Specification for weldable structural steels for fixed offshore structures Bs 7191.* (1989). British Standard Institution.

Sree, P. C. R. & Kujawski, D. (2014). A Deviatoric Stress-Strain Function for Fatigue Life Analysis. *Advanced Materials Research, 891-892*, 789–802. doi: 10.4028/www.scientific.net/./AMR.891-892.797

Sun, C., Xie, J., Zhao, A., Lei, Z., & Hong, Y. (2012). Accumulation damage model for fatigue life estimation of high strength steels in high-cycle and very high-cycle fatigue regimes. *Fatigue & Fracture of Engineering Materials & Structures*, *35*(7), 638–647. doi:10.1111/j.1460-2695.2011.01658.x

Tanaka, K., & Akinawa, Y. (2002). Fatigue crack propagation behaviour derived from S-N data in very high cycle regime. *Fatigue & Fracture of Engineering Materials & Structures*, *25*(8-9), 775–784. doi:10.1046/j.1460-2695.2002.00547.x

Wilson, W. K. (1973). *Elastic-Plastic Analysis of Blunt Notched CT Specimens and Applications.* Research Report 73-1 E7-MAGRR-P1. Westinghouse Research Laboratories.

## KEY TERMS AND DEFINITIONS

**AC1:** Region of peak temperature at 720°C.

**AC3:** Region of peak temperature at 900°C.

**ASTM Standard E 112:** Stablish a correspondence with the grain size classifying them by a number that this standard represents between 0 (508 µm) and 14 (2.8µm) average diameter. Considering that base metal (non-affected zone) has a size of 12 ASTM (5µ) average diameter approximately.

**AW:** As welded conditions, metal without post-weld heat treatment.

**Charpy CVN:** It is a testing to take measure of resilience, if this testing fails, the CTOD testing not need to be and the sample is rejected.

**Critical Crack:** It is represented as $a_{cr}$ and is the crack is produced in the first part of the testing is made by fatigue the testing stop when $0.45<a_m/W<0.55$.

**CTOD:** Crack Tip Opening Displacement this testing it is to measure toughness in HAZ or weld metal, this testing require Charpy CVN has passed the testing.

**FGHAZ:** Fine grain, it is region within the heat affected zone of the metal that is being welded, where the temperature reached is between 900°C and 1100°C

**FL:** Fusion line. It is the line between the welding and heat affected zone.

**Fracture Toughness:** It is a property which describes the ability of a material containing a crack to resist fracture, and is one of the most important properties of any material for many design applications. Here it is measured by CTOD testing.

**GCHAZ:** Grain coarsened heat affected zone, it is region within the heat affected zone of the metal that is being welded, where the temperature reached is between 1100ºC and 1450ºC. It is a fragile zone.

**GSFC:** Gaz Shield Flux Core It is semiautomatic welding where the weld it is protected with gaz

**Hardness:** Hard Vickers 10 to determine hard points in HAZ or weld metal.

**HEYN:** It is a method to measure the grain size in the heat affected zone of steel after welding, observed by microscope, grain size in accordance with the ASTM (112-63) standard.

**HI:** Heat input, heat energy, it depends of electrical Intensity and voltage of the welding arc and the speed of the torch. in the HSLA steel is limited to 2.6 kJ/mm. (V×I/v).

**HSLA:** High strength low alloy steel, they are microalloyed steels with high yield point. they have low content of carbon compensated with microalloys.

**Initial Crack:** It is the mechanical notch made by saw represented as $a_0$. The length (M)=$a_0$ equal to the width of the specimen (W) see Figure11.

**IRGCHAZ:** Intercritical grain coarse zone, it is region within the heat affected zone of the metal that is being welded, where the temperature reached is between 720°C and 900°C. Reheated by passes of welding.

**Resilience:** Is the ability of a material to absorb energy when it is deformed elastically, and release that energy upon unloading.

**SAW:** Submerged Arc Welding. Automatic welding process.

**SC:** It is region within the heat affected zone of the metal that is being welded, where the temperature reached is lower than 720ºC.

**Slow Stable Crack Extension:** It is represented as $\delta_a$ is difference between $a$-$a_m$, being $a_m=a_0+a_{cr}$, where $a_{cr}$ is the critical crack.

**SMAW:** Shield Metal Arc Welding. It is manual welding using electrodes.

**Stress Intensity Factor:** It is a dimensional constant which gives the magnitude of the elastic stress field. It is represented by *K*. Dimensional analysis shows that *K* must be linearly related to stress and directly related to the square root of a characteristic length is a dimensionless parameter that depends on the geometric of the specimen and crack.

**Stress Intensity Factor Amplitude:** It is the fluctuation of the stress intensity factor $\Delta K$, between a $K_{max}$ and $K_{min}$, which can be called as stress intensity factor range. At the threshold it represented as $\Delta K_{th}$.

**Stress Intensity Factor Threshold:** It is the stress intensity factor at which the crack propagates.

**Total Crack:** It is represented as $a=\delta_a+a_m$.

**Toughness:** The elasticity and hardness of a metal object; its ability to absorb considerable energy and undergo considerable deformation before cracking.

# Chapter 5
# Internal Control and Assets Management

**María de Lourdes Eguren Martí**
*Tecnocampus. Universitat Pompeu Fabra, Spain*

## ABSTRACT

*Risk management is a core element from a strategic and operational point of view. Conditions of the environment, financial scandals like Enron and Worldcom, globalization, and new technologies make risk a priority for companies that want to keep a sustainable competitive advantage under uncertainty. In line with these, there are several frameworks, initiatives, and ways to approach risk under a structure organization and integrated approach, including the development of internal control standards like Sarbanes-Oxley (SOX). One of the elements to be considered when taking optimum decisions is a proper risk management. In this chapter, the key elements of assets risk management will be exposed, starting from main concepts, constructing the elements that are applicable to build a risk management system, and developing a practical example under SOX framework with a management and system approach.*

## INTRODUCTION

Environment of companies, are composed of challenges, risks and change. These elements are a source of uncertainty and are strongly related with high volumes of information in a globalized environment and which sometimes are not fully reliable; considering enterprises are open systems, the turbulence of the environment generates an impact from risk management perspective (Febles & Oreja, 2008) .

In this sense, to have into consideration all the variables and elements which can have incidence into the matter of study, is a key aspect for a proper decision making, as there are factors like reputation risk, information availability, process design among others which increase the intrinsic complexity given by the environment (Leisinger, y otros, 2010). In addition to this knowledge, is important to understand this needs to come together with having in place the proper procedures and methodology, otherwise, this will open the door for not taking into account, all the aspects that must have being considered as well as lack of consistency over the applied criteria's; in this sense the need of an integrated approach considering existing variables, open the door to defining new methodologies for an improved management, with a

DOI: 10.4018/978-1-7998-3246-1.ch005

more efficient focus and adapted to reality (Porter & M., The Big Idea. Creating Shared Value. How to reinvent capitalism and unleash a wave of innovation and growth., 2011)

As stated above, current turbulence of the environment and the occurrence of several financial scandals (like Enron explained in detail in a further section), have increased the firm´s interest on risk management procedures and requirements. As a consequence, academic researches on this topic are increasing however these researches are still under an early step of evolution, which give opens the door for new and original developments in this area (Pfister, 2009)

More specifically when studying decision making on asset management, one of the key aspects to have into consideration for being able to take optimum decisions, is an adequate management of the risk considering all the types of risks that can be applicable over asset management under a holistic approach (Ittner & Oyon, 2014)

However, the driver of the decision is sometimes affected by the fact of having related penalties due to regulatory constraints instead of understanding risk management as an added value. "If penalties are sufficiently large, the manager perfectly complies with Internal Control regulation. In contrast, if the penalty of Internal Control noncompliance is sufficiently small, the manager has no incentive to invest in Internal control". (Schantl & Wagenhofer, 2019)

This gap between understandings internal control proactively and as an strategic element or having a reactive approach, generates a difference on the potential added value that the overall framework will be able to create. In addition, the Tone at the Tope and the talent are also key elements to be considered. Having a management structure that do not emphasize reporting quality and integrity increase the exposure of having fraud or other type of risks materialized. (Donelson, Ege, & McInnis, 2017)

In order to assure risk assessment is done effectively, the methodology and clear understanding of the nature of each of them complement the knowledge over the assets management process. With this being said, we can consider that the following aspects need to be considered from risk perspective, in order to contribute with an optimum decision making by understanding entity objectives, business processes, organizational resources, structure roles and responsibilities(Arens, Elder, & Beasley, 2014)

- Process: A deep knowledge over the applicable processes and procedures is required as mentioned above.
- Information availability: As in any decision making process, the quality and completeness of information will be directly correlated with the level of the decisions taken, from an optimal point of view. External information for stakeholders, provided by the company will be affected as well by the quality of the internal information coming from proper risk management (Feng, Li, & McVay, 2009)
- Understand the risks: risk mapping and a proper identification of the potential risks, is the first basic step for defining an efficient risk strategy. Due to this, to have the right understanding on risk meaning and proper approach over risk coverage are key aspects to be able to capture the main exposures and be able to define the proper controls. The relationship between risk and processes is fundamental, only one risk may be really relevant affecting all processes and areas(Scandizzo, 2005)
- Internal Control: Once risks have been defined, the related measures in order to have them "under control" need to be implemented. In this sense, a proper internal control environment adapted to each of the cases (e.g. company size, legal requirements, etc.) must be defined, taking into consideration the available resources, applicable laws, financial impact of the risks, ability to adapt

in case of changes over the process or strategy among other factors that will be detailed in this chapter. Management of internal control, is usually referenced to an "internal control framework", recent studies evidence companies select and adapt their internal control frameworks according to their needs, by taking into account specific constraints, contingencies and firm's characteristics. (Jokipii, 2010)

- Change management: Processes are constantly subject to changes. In this sense, the ability to capture these changes and adapt to the new situation and potential new risks, is basic in order to generate the adequate information for decision making. Management control systems, can support the firm on adapting to changes (from a paradigm point of view or by generating new scenarios), or become a growing issue for the firm. (Kloot, 1997)
- People management: Risks and people are two elements that can't be separated. Several risks can come from aspects not directly related to the people involved in the process (as it could be the case over uncontrolled aspects coming from the environment). However, a significant amount of the risks that finally are not detected and generate a financial impact, come from the fact of having human errors not detected, issues on process design, lack of training or fraud. "to conclude that internal control is effective in any category of objectives - operations, financial reporting, or compliance, requires that a substantial number of other control elements were present, such as the ethics and competence of people and management's philosophy"(Simmons, 1997). In fact, several internal control frameworks, like COSO have a dedicated section to those controls which are common to all the company (Entity Level Controls) covering aspects like people hiring process, performance management, on boarding procedures, training plans, job description definition among other aspects.
- Information Technology (IT): "In today's technologically advanced business environments, IT has become the center of most, if not all, business activities" (Goosen & Rudman, 2013), in this sense is key to consider IT as a core element rather and a support activity, having a key role in process improvement and risk management

Each of the mentioned topics will be detailed in the body of this chapter, together with a practical application under Sarbanes- Oxley (SOX) framework. The reason of selecting this framework instead of other internal control approaches is due to the fact that a significant amount of companies are obligated by Law to comply with SOX requirements, so this could be of help to the readers of this chapter. In addition to this, SOX approach is based on a flexible and process oriented framework, which can be adapted to different companies with independence of their complexity and dimensions.

However, risk management can't be properly performed without taking into consideration that resources are limited, a margin needs to be achieved and the probability of implementing non-added value control instead of helping to minimize the risks can generate additional risks. Due to this, value management needs to be considered in junction when defining an asset management internal control environment. Recent studies evidence the relationship between internal control and enterprise value (Lin, Zheng, & Wang, 2007), moreover firm´s governance is directly linked with company culture, values and ethics, which are supported by those firm elements which assure firm transparency, as it's the case of auditing processed, reporting structures and internal control.(Wieland, 2005)

Having these aspects into consideration, the objective of this chapter is to present with a practical approach the elements and the methodology, in order to implement a proper internal control framework, sustainable and adaptable to each company or case, considering internal control over financial reporting has become an integral source for economic decision-making and assuring a proper level of reliability (Dwinanto, Siregar, & Anitawati, 2019)

Figure1, includes a summary of the structure of this chapter and main elements covered

*Figure 1. Chapter Content*
*Source: own elaboration*

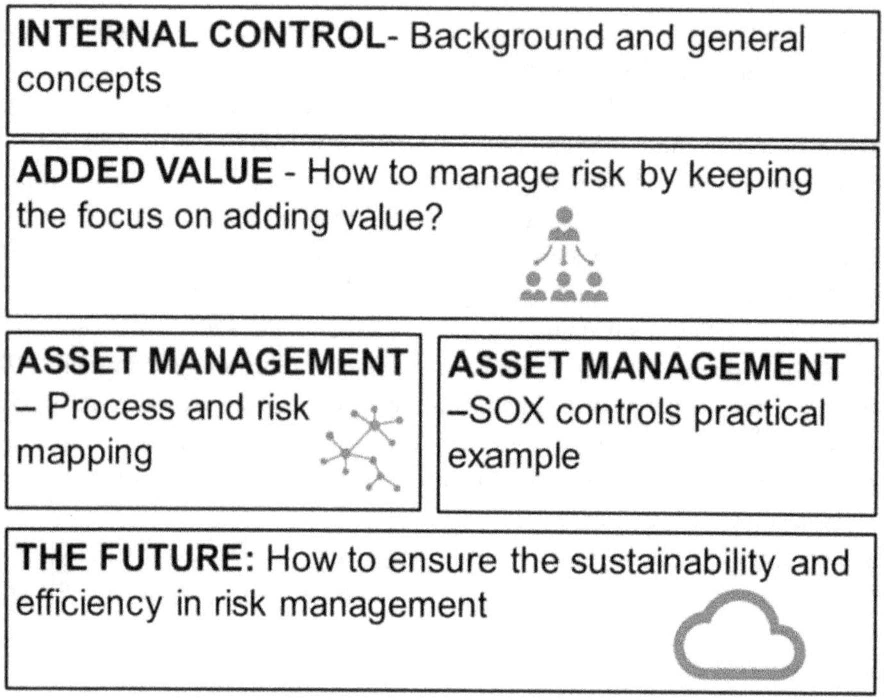


## INTERNAL CONTROL AND ASSETS MANAGEMENT

### Issues, Controversies, Problems

Risk management, is inherent to human activities. When it comes to enterprises risk management, the relation of this concept with internal control activities has a direct relation and this link has been increased in the recent years:

*"The concept of risk has become central to corporate governance and linked to the idea of internal control. In the process, the meaning of both terms has shifted. The shift is symptomatic of important changes in the way corporations govern themselves." (Spira & Page, 2001, pág. 641)*

Having this into consideration and focusing the discussion over this aspect, internal control detractors and supporters had increased their differences in their postures especially after the Sarbanes-Oxley implementation in 2002.

Main critics on internal control are related to the increase of costs, empiric evidence on risk contention and advantages for the company (Ribstein, 2002). The increased audit cost, the lack of quantitative evidence showing a relationship between internal control and risk occurrence as well of "real" advantages for the company, are the main aspects defended by this position.

By other hand, defenders support their opinion over the financial impact of scandals like Enron's, the evidence of a need of an improvement on internal control structures, as well mandatory nature of Sarbanes-Oxley Act.

Both positions have valid and fair positions regarding this topic; however the added value which can be generated by a proper internal control structure is an aspect which is generally missed on the existing literature. The same happens when we refer to the human factor that represents the implementation of an internal control model.

In line with this point of view, we will cover all those aspects, given the tools to management to perform a proper risk management implementation on assets management processes.

## Internal Control - Background and General Concepts

In order to be able to define the proper measure for managing risks over assets, in this section the main concepts over internal control will be defined and clarified, setting up the basis for the next topics in this chapter.

According to the Committee of Sponsoring Organizations of the Treadway Commission (COSO, 2013), Internal Control can be defined as:

"Internal control is a **process**, effected by an entity's board of directors, management and other personnel, designed to provide **reasonable assurance** regarding the achievement of **objectives** in the following categories:

-effectiveness and efficiency of **operations**
-reliability of **financial reporting**
-**compliance** with applicable laws and regulations"

If we read carefully each of the topics mentioned in this definition, we can identify the main aspects to be considered for defining the basis on risk and internal control management.

Internal control is a process itself, meaning that is not an isolated entity but a complete set of elements which together are defined over a common objective and under a series of attributes. According to this statement, we can consider that internal control is a system that's needs to be understood and defined according to its own characteristics and objectives.

Going to the definition, a system can be defined as a set of linked elements, with a common objective (Von Bertalanffy, 1989). If we consider this split, on the below definition we can make the following deductions:

- Internal control is process, which can be modeled as a system.
- Internal control main element is people. In this sense, every employee of the company must be an active part of the internal control process, in order to align with the "Tone of the Top" principle, a definition that we'll develop in detail afterwards. In addition to this, all other resources (material

and immaterial) which are related to the company need to be considered as will be an active part of the process, as for example fixed assets, IT applications, brands, etc.

- The objective of internal control process is to assure a level of assurance over a sub set of objectives defined. If we analyze this in detail, this is related with the concept of risk which has been introduced in the previous section. Risk can be minimized and monitored under a certain level of probability; however it is not possible to prevent risk in all scenarios as there will always be the possibility of having a risk becoming a reality. Due to this, internal control needs to minimize the exposure on risk, by aligning over the mentioned sub-objectives. In this line:
    - Risk must be controlled, in a efficient way assuring an effective risk coverage according to available resources, business specifics and real exposure.
    - Internal control process must assure that the financial reporting presented by the company, is reliable and adjusted to the real situation of the company and its processes.
    - Internal control activities must be defined in a way that they not only comply with risk coverage and internal control principles, but also with the company statements and external legislation and regulations.

In this line, internal control is not:

- A written procedure
- A law or a quality standard.
- A set of controls
- A finance topic / issue.
- A theory with no link to operations and business
- Sarbanes-Oxley, we will detail this aspect. As an introduction, Sarbanes-Oxley is a framework in order to manage internal control.

So, if we summarize the above, we can say that internal control is a set of activities defined to manage the risk that comes from the processes within the company, across all functions and involving all the personnel with independence of their job.

Having this is mind; let's go back to the risk concept and its meaning under a internal control framework. According to the Concise Oxford English Dictionary (The Concise Oxford Dictionary, 2012), risk can be defined as a "hazard, a chance of bad consequences, loss or exposure to mischance".

As can be observed, this definition only reflects the "negative or undesired" side of the event, however risk can also be considered as an opportunity and result of this "unpredicted events". In this, sense the effect of the uncertainty should be considered in junction to the added value which can be generated thru an adequate risk management. This is a key aspect to have into consideration for taking proper management decisions as well as from an efficiency and effectiveness perspective.

In order to make a proper risk management, the different typologies and factors which influence risk need to be considered according to the scope and effects of the exposure, understanding the business and risks applicable to every level of the organization (Heise, Strecker, & Frank, 2014). Some of the main risks that can be considered are the following:

- Financial
- Operational

- Reputational
- Other

Financial risk includes all the cases related to the probability of having an unpredicted, unexpected or not desired event with effects over the financial statements of the company. In this category we can include specific examples as credit risk with the associated probability of default as well as other examples like over estimation of sales, losses not accounted for the period, etc. In general terms in this typology we will include all the risks with a direct effect on finance. As we will see, Sarbanes-Oxley will be focused on these cases.

Operational cases are related to the potential occurrence of any event which affects the operations of a process. In this case, we can include a wide typology, for example IT disruptions, process breakdowns, natural disasters, among others. It is important to remark, that some operational risks can generate a financial risk due to the effect over the processes behind. This is the reason why process design and systems approach is a key factor on risk management and risk approach.

With reputational risk, we can consider all those cases were the external and internal perception of the company can generate an unexpected impact over the company. There are several well-known examples over the effect of market information over a certain subject with direct effect over the sales or shares of a company, as an illustrative examples we can comment about a bottled water company from Spain, that after facing some rumours over water contamination significantly reduced their sales and never recovered although it was shown rumours were not demonstrated. Other examples can be linked to corporate social responsibility with negative effects on shareholders perception, with direct financial effects on the company value and strategic positioning (Porter & Kramer, Strategy and Society: The link between competitive advantage and corporate social responsability, 2006).

Other risks include a wide variety of diverse situations that can affect a specific event. This variety over risks is part of the risk nature itself, unpredictable at some cases, difficult to measure and with an impact that in a lot of cases can't be measured.

Once we have defined the risk concept, we need to understand the measures on how risk can be controlled, in general terms, "controls" are the elements defined to cover this aspect.

Controls can be defined as a task or process designed to mitigate the probability of occurrence of a specific risk.

From a process perspective, applicable steps for control performance should include the following steps:

- <u>Control preparer</u>: person who executes and performs the control.
- <u>Control reviewer / approver:</u> Approver or reviewer according to process.
- <u>Control owner:</u> person responsible to assure control is working properly.

Based on this, controls should comply with a series of attributes in order to be aligned with the objectives:

- <u>Objective</u>: Clearly defined in order to minimise / mitigate the risks. SMART criteria can be used in order to be sure, proper objectives are defined considering the following criteria's:
  a. Specific
  b. Measurable
  c. Achievable

   d. Realistic
   e. Time-Bound

- Owner: Person who is responsible for control performance according to the defined parameters.
- Scope: Area and subject of application of the control.
- Description: How the control is performed. Should be clear and easy to understand.
- Evidence: Items that demonstrate that the control has been performed. It should include who has performed it, who has reviewed, who has approved and corresponding dates. Reports are a key element in order to evidence control performance.
- Frequency: How often the control is performed (e.g. monthly).

In this sense, internal control needs to have evidence on control and processes performance. If evidence is not available, from internal control perspective is the same as not having the control in place. Control performance cannot be evidences, if documentation is not available. There are different types of evidence that should be considered when designing a control:

- Event based: usually includes transaction level evidence like invoices, purchase orders, credit assessments, etc.
- System parameters check: to assure configuration is aligned with requirements, so process and embedded controls work properly. Can be done by checking system set-up or process walk-through by trying to generate an issue.
- Procedures: Written procedures, policies and authorization tables.
- Reports – key for the company: When tested, a significant part of the process is covered. Due to this, completeness of these reports is a key piece in effectiveness testing process.

Another key concept, which needs to be defined, is the roles definition over processes and controls. The following cases needs to be taking into consideration:

- Roles: Functional description of the tasks performed by a person.
- Functional roles: Based on process and job description with independence of scope
- System roles: Roles granted in system or IT applications, based on requirements. This must be aligned with functional roles. For example, in the case of a company using SAP as ERP, the assignation of the roles from the system according to functions must be the applicable criteria.

Roles clarity, allow to have clear processes, minimize the probability of having controls or processes with no owner, coordinate changes and action plan on a centralised way and with an expert in the area.

Written procedures are another of the aspects, to be considered when defining a control. Policies, standard operating procedures, workflows and other type of documentations, support on describing, regulating and controlling the company activities which they are referred to.

In this sense, the responsibility of every employee is to know each of the written procedures applicable to their activities. Management leadership is in this line a fundamental aspect in order to assure the proper training, distribution, of these procedures to their teams as well as for assuring they are periodically updated to reflect the business and process changes.

There are a few concepts that must be remarked.

- Responsible: each written procedure must have a responsible, who should be expert in the area and in charge of updating the procedures according to the changes received by business or other sources.
- Relation between procedures: all the procedures must be related, as they are not isolated elements. As long as these references are included in this procedures, more opportunities will be created to improve processes and create added value.
- Internal control references: specific references to the internal control check points, should be include improving the efficiency during control assessment. The fact of having internal controls linked to operational procedures increase the awareness, as well as the supervision and execution tasks for the related controls. If the basis is clear, operations, efficiency, and effectiveness are improved.
- Legal aspect: written procedures should be aligned with applicable laws and legislation.
- Roles: functions and roles should be defined and linked into written procedures, in a way to assure the three action "who-what-how" are related and formalized with transparency and clarity.

*Figure 2. Core elements on internal control*
*Source: own elaboration*

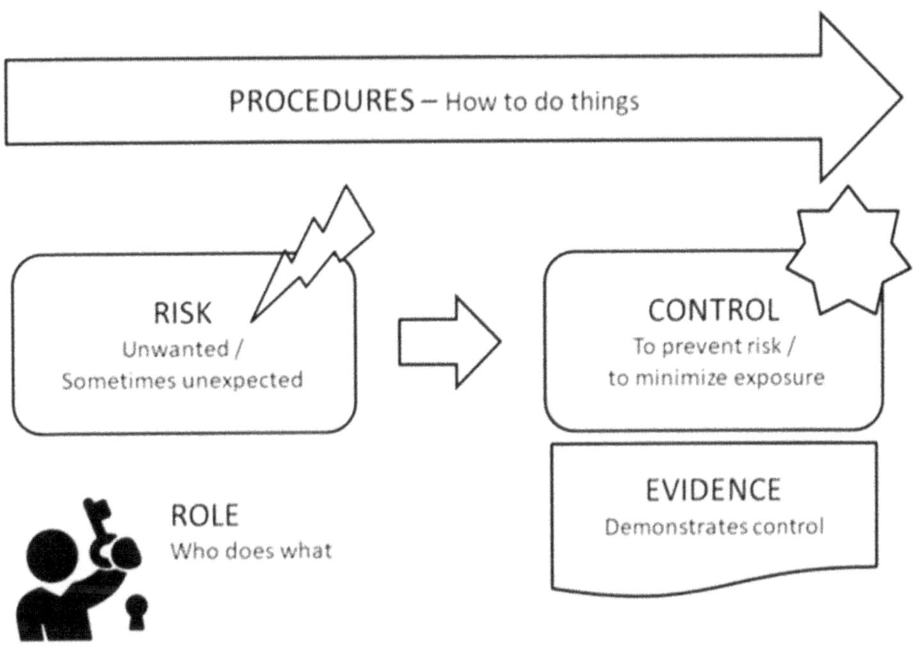


So, summarizing we may say that risk, controls, evidence, roles and written procedures are the core elements to be considered when defining and describing an internal control system as shown at Figure 2.

After defining controls, roles and evidence, control type understanding needs to be achieved as well in order to assure a proper control and risk management. In general terms, there are two types of controls, preventive and detective.

- Preventive: Focus on preventing errors and exceptions. Designed for identifying the risk in advance to its occurrence. Preventive approach is efficient and once implemented have less impact from a workload perspective. IT controls tend to be preventive in nature.
- Detective: Designed to identify issues after it has occurred. This controls are also called "monitoring controls" as according to their design support business in process monitoring, giving also additional information like performance indicators, trend, etc.. This type of controls stop issues before they start being recurrent and minimizing the impact, so mitigating the risk. Common examples include exception reports, reconciliations, and periodic audits among others.

In the Figure 3, we can find one example of the application of the criteria's below, showing how preventive controls are built as well as detective or monitoring control.

Preventive controls have the ability to avoid risk situation to materialize, so as possible they're preferred to detective controls. Preventive controls, avoid risk occurrence, are usually embedded in the process itself and are the aspects to be considered from resources perspective, as once implemented do not require additional changes unless process changes.

Preventive controls can also be automatized taking advantage of IT systems. This approach is considered as a best practice among the industry, and is a key element considering all the new trends and new technologies.

*Figure 3. Preventive vs. Detective – practical example*
*Source: own elaboration*

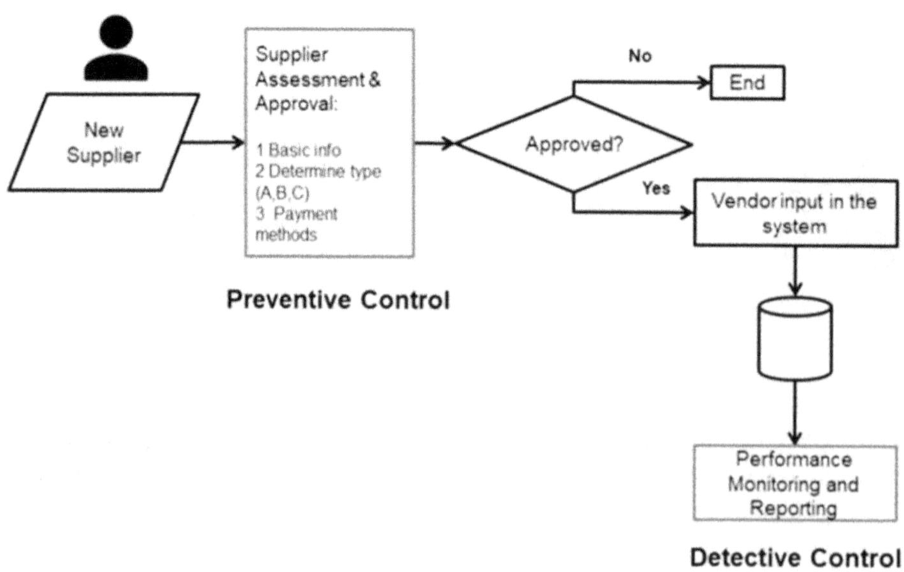

By other hands, detective control requires ad-hoc work to define, implement and execute the control on the defined frequency. This type of controls needs to have a careful design as well to assure added-value is generated as possible and "bureaucratic" controls are avoided.

This is one of the topics that generate more controversies over internal control. If the implemented actions are not aligned with business and embedded in the processes, they will be considered as an imposed requirement with no added value for the company.

## How to Manage Risk by Keeping the Focus on Adding Value?

According to the Oxford Advanced Learner's Dictionary, value can be defined as:

"1. The regard that something is held to deserve; the importance, worth, or usefulness of something

2. Principles or standards of behavior; one's judgment of what is important in life
3. The numerical amount denoted by an algebraic term; a magnitude, quantity, or number
4. Music .The relative duration of the sound signified by a note.
5. Linguistics. The meaning of a word or other linguistic unit.
6. The relative degree of lightness or darkness of a particular colour"

As we can see from the definition below, value is related to perception, relationship, understanding and calibration.

If we extend this concept to the internal control activities, it can be observed that value is related to the stakeholder's perception over internal control risk, linked to the values and mission of the company and how these values are cascaded to each process, as well as the financial materiality behind each process and risk; considering not only the exposure of the specific controls but also to the potential effect over company value.

Within the existing methodologies, one of the most popular and recognized at strategic and process level is the analysis of the value chain from Porter, which provides a useful model for analyzing the advantages held by the company, allowing fitting the sources of differentiation from the side of demand and supply (Martínez Marín & Eguren Martí, 2013). Under this vision, the company can be conceived as a series of activities, each contributing to the creation of value. In other words, the value chain of a company incorporates a number of activities, processes, resources and objectives that relate to each other provide "value" to products manufactured or services provided, represented by the amount that buyers are willing to pay for "goods and / or services" that the company provides. (Porter M., 1985)

As can be seen in the Figure 4, all the activities and processes of the company must be included when analyzing how the company is generating value. Moreover and likewise to the "Internal Control" characteristics, the value chain methodology presents the attributes to be considered as a system, so that the existing multidisciplinary tools over this field of knowledge can be applied to their analysis (Eguren Martí & Castán Farrero, 2011).

According to (Fernandez, Avella, & Fernandez, 2006) value chain is defined as the set of activities performed by the company with the aim of transforming raw materials into finished products that are then distributed to final consumers, and this activities can be classified into two main groups primary activities and support activities.

*Figure 4. Porter's Value Chain Model*
*Source: adapted from Porter, 1980*

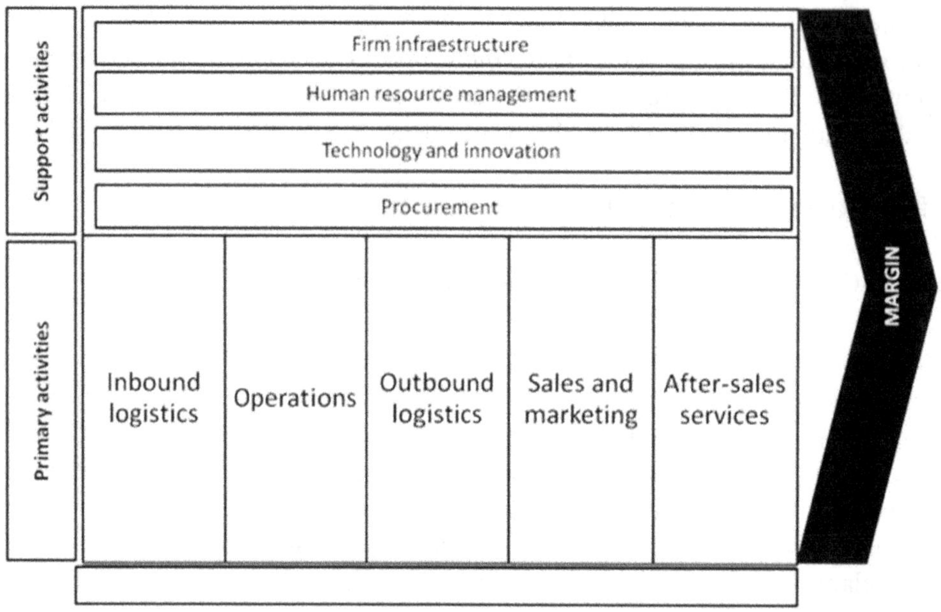


Primary activities are directly involved in the production process of the company, covering the physical transformation and creation of the product or services and the after-sales activities (Guerras & Navas, 2015). More specifically the activities included in this area are the following:

- Inbound logistics or input factors: including all the activities related with the receipt, storage, inventory control and internal distribution of raw and auxiliary materials, as well as any element received for developing the services to be provided.
- Operations / Production: covering the transformation of the raw material into products or services.
- Outbound logistics: warehousing and distribution of finished product or services to the customer.
- Marketing and Sales: activities aimed at selling the product.
- After-sales service: activities performed after the sale, in order to assure the service level to the customer and proper feedback on the company performance and perception. Quality and claims management, are among the elements to be considered in this category.

The supporting activities are those that complement the primary activities and are not applicable to a specific area but to all the company, including the following:

- Procurement: acquisition of productive and non-productive material including services and advertising)
- Technological development (IT): activities involved in the development and improvement of new products, research & development (R&D), and innovation.
- Human resource management: activities that include all the administration of the human resources of the company (recruitment, training, etc).

- Firm Infrastructure: company business activities oriented management, planning and control of all primary and support activities that sustain it.

Considering this activities, the value chain model considers the capacity of the company of creating competitive advantages thru each of its activities, using two complementary approaches: cost reduction and product differentiation.

A competitive advantage by cost reduction is achieved when the costs of the product or service offered are lower than the offered by the competence for a similar product or service and with an equivalent quality (Guerras & Navas, 2015). According to this strategy, if two competitors offer the same product at a certain price, the one with the lowest cost will have a greater range and therefore increased profitability.

By the other hand, a competitive advantage by product differentiation can be obtained when a product or service has a set of attributes that makes it unique for the customer. (Guerras & Navas, 2015).

In the past, both strategies were considered separately, however recent researches evidence the fact that additional competitive advantage can be created by joining both approaches (Hamel & Prahalad, 1989)

So, going back to the initial subject of this section, risks management will need to be managed in a way that allows to the company to create competitive advantages in comparison to the competence.

In this sense, a system's approach will allow to assure the needed flexibility as well as the completeness required in order to capture the value generators in line with the requirements for risk factor. In this line, the following aspects should be taken into consideration:

- Completeness of the scoping by mapping risks vs. the applicable activities according to the value chain model. This aspect will be detailed in the next section.
- Control level analysis in order to identify value generator elements for each process or cycle, based on the key variables shown on Table 1.
- Mission, vision and company values and its application on processes and people that will be explained with the "Tone of the Top" premise.

*Table 1. Internal control and value management strategy*

<table>
<tr><th>Cost Reduction</th><th>Product Differentiation</th></tr>
<tr><td>Lower audit costs<br>Reduced timing for controls execution and performance</td><td>Automated controls<br>Differentiated process design<br>Use of new technologies</td></tr>
<tr><td colspan="2">Innovation<br>Information with added value for the business<br>Efficiency, assuring cost reduction and differentiation.</td></tr>
</table>

Source: own elaboration

In summary, risk management should be approached from a system focus in order to be able to identify all the opportunities that may come from how are controls designed, performed, executed and monitored in order to create a competitive advantage considering risk coverage perspectives.

## Asset Management Process and Risk Mapping

As previously commented, risk mapping vs, process and activities is one of the key steps to be performed before starting a risk analysis project.

We can summarize the key steps to perform this process as follow:

- Process knowledge: a full understanding of the applicable activities must be required to the person responsible of risk management. A basic step for this is to perform personal interviews with the process owners as well as checking the available documentation like written procedures.
- Process documentation: risk manager, must document the processes in order to be sure to have a proper tool to perform the process analysis and risk identification
- Process analysis: based on process documentation
- Identification of risks: having in mind the type of risk, project objective, resources and reasonability.
- Controls design: considering all the control attributers mentioned on previous sections. Some examples will be shared on the Sarbanes-Oxley practical example section.

Once this is done, controls should be implemented and monitored to complete the cycle over internal control management.

Coming back to assets management and according to the steps mentioned above, the first step to perform risk mapping is to understand the processes around asset management. Each company will have their own particular aspects that will need to be considered and documented in particular, and which are a relevant aspect on this topic. However, there are several core processes that can be considered when analyzing assets management:

- Acquisition
- Accounting / Reporting
- Disposal
- Safeguarding
- Plus specific processes on: Leases and Intangibles management.

Acquisition management includes:

- Acquisition approval by management in order to authorize the investment from a financial, operational and technical perspective.
- Acquisition analysis from accounting point of view (asset or expense). Based on this, asset type (and related useful life) must be defined, as well as the required cost elements to manage depreciation and investment status.
- Creation of all the elements in the company IT system.
- Definition of the schedule for managing the asset, including the date for the asset to be placed in service (so changed from construction in progress to the adequate asset count account and initiate depreciation).

Accounting/Reporting management includes:

- Monthly review of current projects and investments in order to:
    - Review of construction in progress assets in order to change (when applicable) to the corresponding asset type and initiate depreciation.
    - Identify assets subject to disposal
    - Other actions related to assets monitoring (errors, wrong asset grouping, etc)
- Depreciation run on existing assets
- Year-end accounting activities for assets

Disposal management includes:

- Identification of assets subject to disposal. Two cases can be considered: scrapping or disposal of the assets or sale.
- Disposal approval by management
- Accounting of the activity
- Physical activities (scrapping or delivery to customer)

Safeguarding management includes:

- Access to the company facilities
- Assurances
- Process participants definition
- Operational monitoring over assets.

Having this high level process description, the risk mapping exercise must understand each of the processes behind and detect the key risks that may arise according to process flow. In this risk mapping exercise, we will focus over the financial risks in order to set-up the basis for the Sarbanes-Oxley case example that will be explained in the following sections.

In line with this, the following risk mappings have been provided for each of the processes:

- Acquisition, please refer to Figure 5
- Accounting / Reporting, please refer to Figure 6
- Disposal, please refer to Figure 7
- Safeguarding, please refer to Figure 8

## DEFINITION AND IMPLEMENTATION OF INTERNAL CONTROLS FOR ASSET MANAGEMENT. A PRACTICAL EXAMPLE UNDER SOX ACT.

### SOX Background and Requirements

As mentioned by the United Exchange Commission (U.S. Securities and Exchange Comission, 2013):

*"On July 30, 2002, President Bush signed into law the Sarbanes-Oxley Act of 2002, which he characterized as "the most far reaching reforms of American business practices since the time of Franklin Delano Roosevelt." The Act mandated a number of reforms to enhance corporate responsibility, enhance financial*

*Figure 5. Risk mapping of assets acquisition*
*Source: own elaboration*

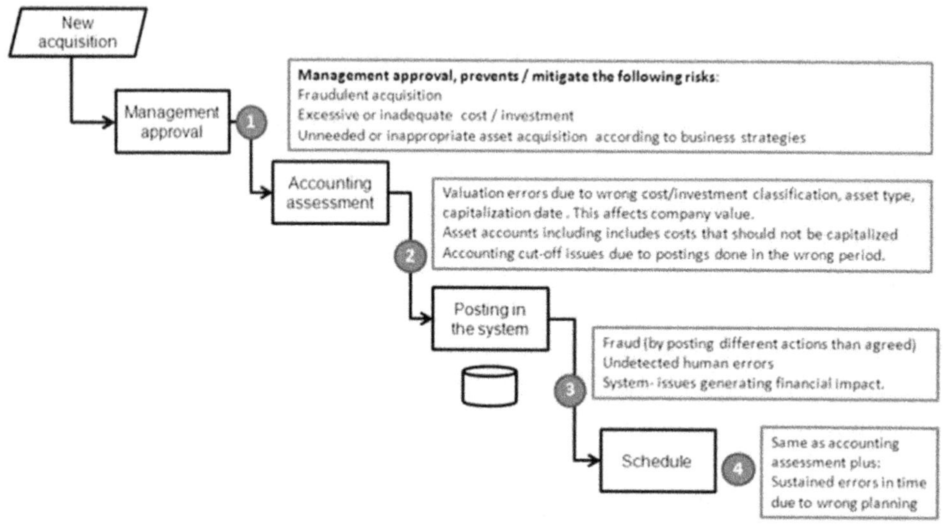


*Figure 6. Risk mapping of assets reporting and accounting*
*Source: own elaboration*

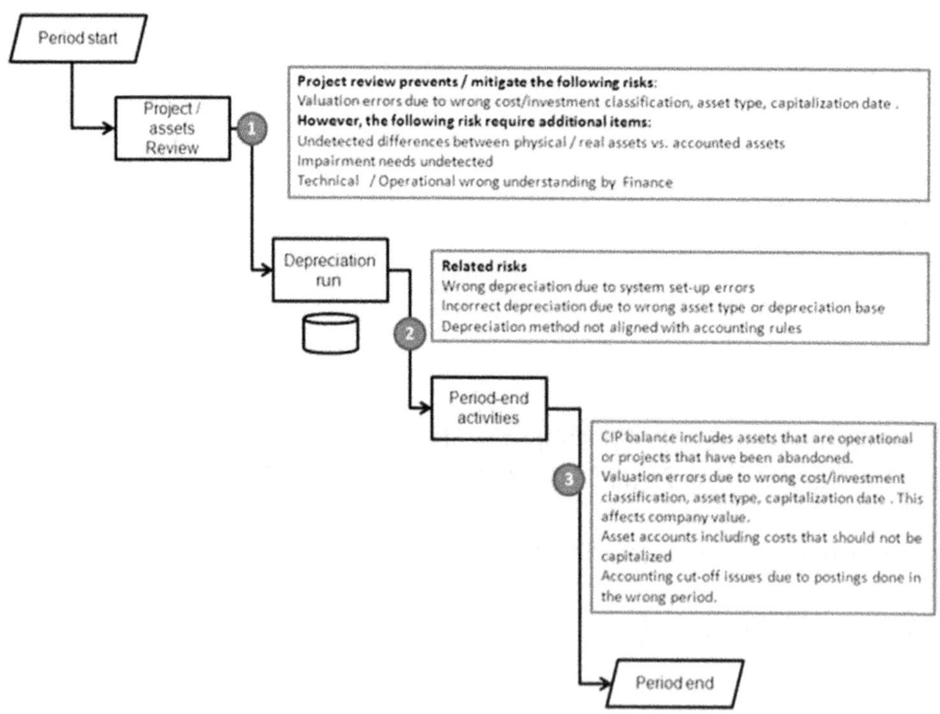

*disclosures and combat corporate and accounting fraud, and created the "Public Company Accounting Oversight Board," also known as the PCAOB, to oversee the activities of the auditing profession."*

The reason behind the signature of this law was related to the financial scandals occurred in several companies like Enron and WorldCom which had a big impact as can be seen below:

- WorldCom:

*Figure 7. Risk mapping of assets disposal*
*Source: own elaboration*

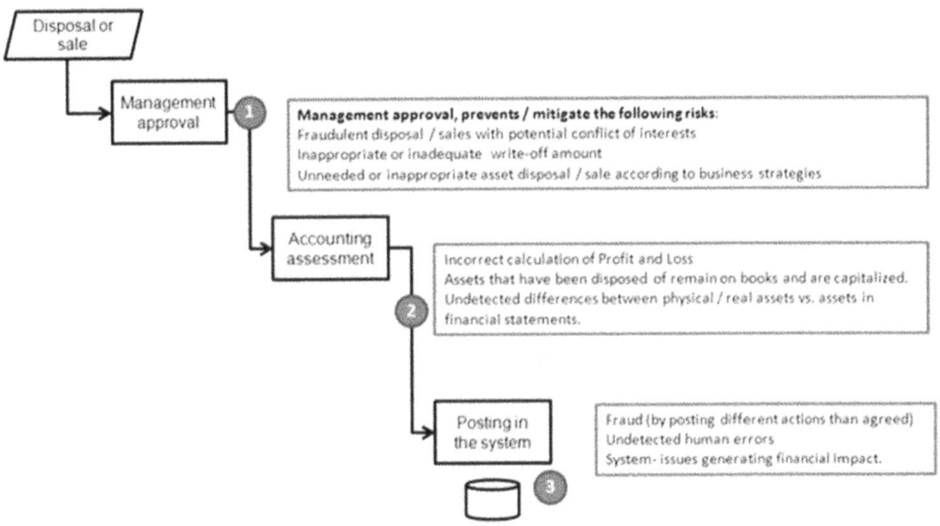


*Figure 8. Risk mapping of assets safeguarding and controls*
*Source: own elaboration*

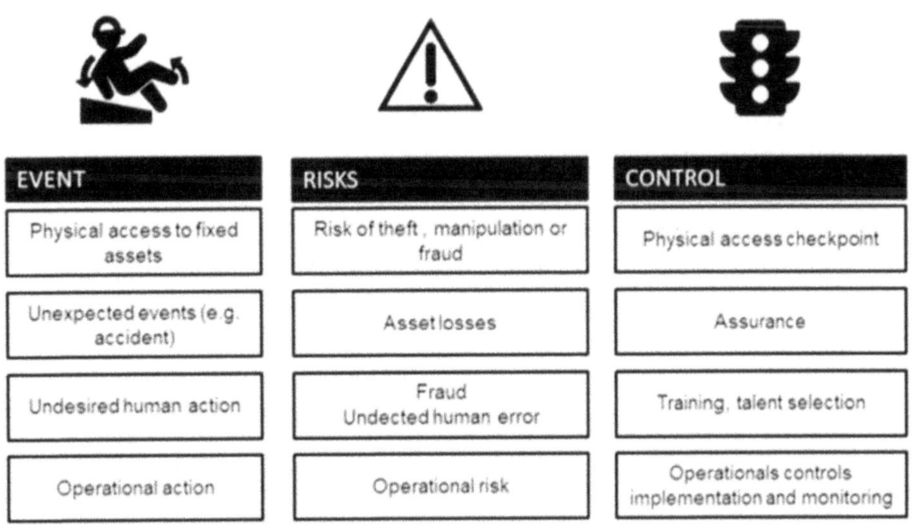

- Financial impact: $100 billion
- Incorrect capitalization of operating expenses.
- CEO sentenced to 25 years in jail. CFO sentenced to 5 years in jail.
- Controller, Accounting Director and Accounting Managers all plead guilty.
- 300.000 people lost their jobs

- Enron:
    - Financial impact: 73 billion
    - Off Balance Sheet transactions utilizing Special Purpose Vehicles to hide billions of losses.
    - CEO sentenced to 24 years in jail. CFO sentenced to 6 years in jail.
    - 15.000 people lost their jobs

Public companies are required to comply with Sarbanes-Oxley Act, which dramatically changed the approach over financial risk management, based on purely accounting and limited reviews, over a complete framework which affects the whole activities of the company and their processes design, impacting the relationship between different areas as for example IT control governance and IT control quality (Li, Lim, & Wang, 2007)

SOX Act. is composed by 11 sections. Below you will a summary of the most relevant ones:

- Section 404 of the Sarbanes-Oxley acts requires an annual internal control report in addition to the financial report.
    - Includes detailed results on the effectiveness of controls and processes
    - External audit firm should attest and report on the assessment on the effectiveness of the internal control procedures and structure
- Section 201: Services outside the scope of Practice of Auditors
    - Limit the activities that can be offered from the external auditor responsible for giving a opinion over SOX.
- Section 302: Corporate responsibility for Financial Reports
    - CEO and CFO must approve and authorize the financial reports to certify internal controls are in place to record, process, summarize and report financial data.
    - Up to $5 million in fines and 20 years in jail for CEO and CFO.
- Section 409: Disclosures
    - Timely disclosure of material changes to financial condition or operations.
- Section 806: Protection for employees who provide evidence of fraud
    - Whistleblower policies and protection

All the sections below are strongly related and depended of a key concept named "Tone of the Top". This term is used to refer to the top management's commitment towards assuring ethics, compliance, integrity, ethical behavior and cascading this message into all the levels of the company, who should retrofit this principle.

Although the execution of this framework is mandatory for a very specific set of companies, since it was defined it has also changed the mindset of internal control management, making its general principles applicable over "non SOX" companies who use or adapt this framework in order to manage their internal risks.

Having this into consideration, the specific steps required to manage internal control thru SOX will be explained in the next sections, by studying the specific case of assets management. This section, does not intend to be a case of study presentation, but a practical example of the most common SOX controls according to the author´s professional experience.

## Scoping

The first step on SOX is to make first assessment over the business areas and business unit which will be a priority for SOX purposes. Although SOX will be applicable for the whole company, internal and external audit will concentrate on those areas were risk exposure is higher due to materiality and other business reasons.

Once this is defined, a work plan must be defined and agreed with Financial Managers (usually the CFO) in order to coordinate all the internal control activities for the year assuring they are aligned with the deadlines required by SOX. In the specific case of Assets Management, the following aspects must be considered:

- Materiality
- Existence of intangible and/or leases
- Nature of the area. Is strategic for the business?
- Do you expect business changes in the year that will require new assets acquisition or disposals? For example, a plant expansion, a new product line, etc.

An internal communication must be done in order to have all the involved team aware of the next actions.

## Risk Assessment and Control Definition

The first step to be performed in this area is a risk mapping over the applicable process, as explained on the previous section.

Once this is done, the controls should be designed based on SOX standards in order to cover the identified risks. Some of the most common controls in this area cover the following:

Acquisition approval: Assure proper management level of approval according to company policies, including capitalized interest calculation, asset type class and other relevant aspects.

Asset recording: Assure new assets are accounted under the corresponding asset type, related useful life, and depreciation method. Assure the are monitoring controls in place (e.g. annual reviews), to validate the accuracy of the existing assets.

Asset identification: Assure fixed assets can be physically identified in relation to the recorded information.

Fixed Asset Count: Validate that an existing policy exist over this subject. Review fixed assets counts are done based on the agreed policies and adjustments made over identified differences

Construction in progress (CIP): Monthly review of CIP accounts, to assure all applicable projects are included and only projects are posted in this account and are transferred when asset is in place to the corresponding asset account.

Expenditures - Monitoring: Variation analysis of approved budget and actual PPE related capital expenditures.

Fixed Assets Review: Monthly controls to validate performed transactions (acquisitions, transfers, disposals and adjustments) are accurate and performed according to agreed procedures.

Depreciation method: Assure depreciation method and useful life linked to the asset classes are reviewed to assure a proper depreciation calculation. If performed thru a system, this control will require making a set-up check.

Depreciation monitoring: Perform analytical review on actual depreciation amount vs. previous months or budget in order to identify anomalies or any potential issues.

Accounts reconciliation: Perform monthly fixed and intangible accounts review vs. general ledger accounts.

Disposal approval: Assure proper management level of approval according to company policies, including profit / loss analysis, reasons for disposal and any other relevant aspects for the business.

Impairment analysis: Validate that an existing policy exist over this subject. Assure there are procedures in place in order to identity trigger events that can lead to an impairment action. Perform as a minimum an annual impairment check and assure proper postings are done after this analysis.

Segregation of duties: Check that a proper segregation of duties has been defined over the personnel involved with assets management, including a functional and system accesses approach.

In addition to the mentioned controls, it's recommended to have written procedures over the mentioned areas: acquisition, recording, CIP, disposal, physical count, impairment, accounts reconciliation and monitoring activities.

## Implementation

Once controls have been designed they should be implemented.

In order to succeed with this activity, is important to clearly define the control attributes explained at the beginning of this chapter:

- People: control preparer, reviewer, approver and owner
- Objective
- Scope
- Description
- Evidence
- Frequency
- Written procedures

A project leader should be nominated to manage control implementation and monitoring in order to assure proper implementation and internal control understanding. This function can generate into a full-time position depending on company size and business complexity.

## Monitoring

SOX methodology clearly defines the steps to be performed in order to monitor and assure that internal control environment can be reliable. This is the main steps:

- Scoping: priority areas, testing plan for the year, key events expected for the year
- Self-assessment and management testing: to be performed by the company including:
  - Walkthrough: Control description update according to current existing procedures, interview to control performers, control design test .Usually as part of this process, the person who is performing the walkthrough traces the process with one evidence of each step in order to assure control is designed according to what has been agreed and documented. This allows detecting design deficiencies before performing the full-sample operational test.
  - Testing of the applicable controls according to sample size selection method, control type, and specific applicable attributes. This testing must be done over a full sample and in line with agreed deadlines. There are two events to have in mind when planning the testing:
    - Interim testing period: Usually covers from January to September. Must be coordinated with the external auditors, to assure a efficient work
    - Year end results: Which covers the testing required covering the period between October and December, annual controls performed at year end as well as the controls which deficiencies which have been remediated and need to be re-tested by year end.
- Action plan and remediation: for those controls considered as deficient, a detailed action plan must be defined, so control will change thru the following status:
  - Failed: control with deficiencies detected on design or operational level
  - Remediated: Action plan implemented, but not tested
  - Closed: Remediation actions tested, by testing a full sample set of evidences
- External auditor assessment and testing: As commented before external auditors will make an opinion on internal control effectiveness. In practice this means, auditors will make their own walkthrough and testing over selected controls, also aligned with the two main periods mentioned before: interim and year-end.

## HOW TO ENSURE SUSTAINABILITY AND EFFICIENCY IN RISK MANAGEMENT

As discussed, risks managements implies to perform a deep analysis on processes, understand the business, align with core elements which are related with company strategy and design and implement an internal control environment.

The implementation of this control environment, is a project itself that will have several levels of impact in the company that will depend on the way this is managed.

From one side there will be an increase in costs (implementation, execution and audit) and from another side there will be an improvement on efficiency and risk contention. Both factors are fully inter-related and will depend on the ability of management of defining an internal control framework that is sustainable and which gives added value to the company.

The costs of the implementation of an internal control framework, would require a specific analysis by itself, however at high level these are the main elements included:

- Expert consultancy fees for designing internal control environment or linking to selected framework (e.g. SOX)
- Internal control structure, which includes internal headcount to monitor internal control and resources (physical assets and IT tools)

- People training on internal controls and how to execute them (this affects all the company employees at different levels)
- Execution costs, performance of internal controls requires people to make new activities and this means time and money.
- Audit costs, which depending on the selected framework can be represent a significant amount

As can be seen, those cost elements can become a relevant element from a financial perspective, and if they're not managed efficiently can became an endemic problem than create more issues than what is solved.

A classic example if the implementation of internal controls with zero added value, high level of bureaucracy, lack of understanding from involves personnel on the reason why they're doing them that at some point generate a time consuming circle that: do not address the risk, low's down motivation, and represents a cost in time and resources that do not provide anything on behalf.

Other cases of bad practices, include the definition of controls to solve a situation in the present without thinking how that would impact in the future or how the control should change or evolve.

In order to manage those elements, to have a sustainable internal control environment is the key. Sustainability is related with two main concepts:

- The fact that a process can be maintained during time by itself with no significant external intervention
- Development that meets the needs of the present without compromising the ability of future generations to meet their own needs. (Presidio Graduate School, 2006), in other words the fact of keeping a process in place with an ecological approach, not harming the environment or depleting the existing resources.

So in simple words a sustainable internal control environment, requires a long term strategy that allows to the company to keep model running with minimal changes and making sure that solving a problem in the present do not compromise the company ability in the future to keep risks under control and perform well from a financial point of view.

When analyzing sustainability, there are three areas that needs to be taken into consideration from management point of view (Presidio Graduate School, 2006), each of these areas has their own attributes in order to facilitate the analysis:

- *Social Criteria:*
  - *Socially desirable*
  - *Culturally acceptable*
  - *Psychologically nurturing*
- *Financial Criteria:*
  - *Economically sustainable*
  - *Technologically feasible*
  - *Operationally viable*
- *Environmental Criteria:*
  - *Environmentally Robust*
  - *Generationally Sensitive*

- *Capable of continuous learning*

In this sense, internal control sustainability is closely related with the ability on generating added value to the company at the same time it assures proper risk management.

Social criteria is related to the fact that employees "should believe in what they're doing". A key element to make sure this happens is the "Tone at the Top" concept mentioned above and the added value that these processes can generate to the business if are adequately implemented. Open communication, internal control training, will support on the process of having a "social" acceptance over the internal control structure and future changes.

From Financial criteria, the sustainability of internal control is related with the ability to use existing resources on defining efficient controls in line with the existing information systems, minimizing the exposure on financial risks and updating according to the requirements. Having in place a proper plan in order to manage audit costs efficiently, investment the right amount in talent that assures a results plus risk driven concept as well as investing in the related support resources (like IT) is the minimum a company should do in order to make sure their internal framework is sustainable from a financial perspective.

As internal control is a system, the capacity of learning and retrofitting is a key element which needs to be monitored and take the opportunity of improving current processes or detecting new competitive advantages. This element can be summarized in the statement that companies needs to understand internal control as a strategic element, making sure this is cascaded from mission and vision level to each of the operational levels in the company.

People management is the key, so this is probably the most relevant aspect to consider, as internal control is a system driven by people. Know-how maintenance, clear role definition, proper training plans in internal control, motivation on the team for internal control performance are the key aspects to consider.

## FUTURE RESEARCH DIRECTIONS

Risk management and especially internal control, are areas which are under a continuous change focus from an industry point of view. Professional forums on this area are continually performing sector meetings to brainstorm, improve and have a better capacity to improve the way risks are managed and have a proper contention.

However, for companies is not so easy some time to keep this flexibility once the internal control is implemented and they need to keep running from an operational point of view with the day by day work. So, this element is a topic that should require more research and creativity in order to find ways to help companies to have robust environment controls but keeping at the same time the enough malleability to be able to generate new competitive advantages.

By other hand, from an academic perspective, most part of the existing documentation and books have been written based on empirical experience rather that purely scientific approach, so this opens a big door for researchers interested in this area and the potential risk type driven topics that could be covered:

- Risk management from a quantitative point of view
- Risk management from a qualitative point of view and related elements that are linked to strategy, communication, reputation, share prices.
- Taxonomy studies on available literature

- Business cases and academic cases that cover best practices as well as nonsuccess factors from the industry
- New technologies

This last topic requires a specific mention. As long as new elements and technologies like blockchain, IoT(Internet of the things), RPA (robot process automation) are coming into the table and becoming the present of the companies, the concept of internal control needs to adapt in order to determine the right controls and actions which are required in order to manage risk properly.

## CONCLUSION

The objective of this chapter was to cover from a practical perspective the key elements that needs to be considered when managing risk and internal control, starting from a general approach and determining the specific elements or activities that needs to be in place having focus on the specific case of assets management.

Risk is an element that will be always around as is part of nature and human behavior, so having this in mind the key is to make sure that from an industry point of view we implement a proper set of controls in order to prevent the risk or minimize the exposure if risk materializes.

Internal control is a system, which is composed of several elements: processes, resources, information and people. All of them needs to be managed together in order to make sure we contain risk at the same time that value is created.

Between all these factors, people is the key. To have a "Tone at the Top" culture and the right talent is the mix a company needs in order to make sure the rest of the internal activities run as desired. So also from this point of view, this cultural need requires that top management considers internal control as strategic element and not a secondary or regulatory factor. With independence of the reference factor used to implement internal control (which also has its relevance and were SOX has constituted a reference and benchmark), is even more relevant to make sure that the internal control environment is adapted to the company needs, own risks and is sustainable among time.

Current environment, and new technologies related to artificial intelligence opens a new challenge for companies and researchers in order to find better, flexible and efficient ways on controlling old risks, new risks and risks that required different and new ways of approach.

## REFERENCES

Arens, A., Elder, R., & Beasley, M. (2014). *Auditing and assurance services-An integrated approach; includes coverage of international standards and global auditing issues, in addition to coverage of.* Aufl.

*COSO*. (2013). *Internal Control - Integrated Framework*. COSO.

De Quevedo, E., De La Fuente, J., & Delgado, J. (2005). Reputación Corporativa y Creación de Valor. Marco Teórico De Una Relación Circular. *Investigaciones Europeas de Dirección y Economía de la Empresa*, *11*(2), 81–97.

Donelson, D., Ege, M., & McInnis, J. (2017). Internal Control Weaknesses and Financial Reporting Fraud. *Auditing: A Journal of Practice & Theory, 36*(3), 45-69.

Dwinanto, I., Siregar, S., & Anitawati, A. (2019). Internal Control Over Financial Reporting, Organizational Complexity and Financial Reporting Quality. Int. *Journal of Economics and Management, 13*(2), 331–342.

Eguren Martí, M. L., & Castán Farrero, J. M. (2011). La cadena de valor bajo un enfoque sistémico y sus implicaciones en el análisis de la creación de valor en la empresa. In *XXI Congreso Nacional de la Asociación Científica de Economía y Dirección de la Empresa (ACEDE).* Barcelona: ACEDE.

Febles, J., & Oreja, J. (2008). Factores externos e internos determinantes de la orientación de la cultura estratégica de las empresas. *Investigaciones Europeas de Dirección y Economía de la Empresa, 14*(1), 13–32. doi:10.1016/S1135-2523(12)60009-4

Feng, M., Li, C., & McVay, S. (2009). Internal control and management guidance. *Journal of Accounting and Economics, 48*(2-3), 190–209. doi:10.1016/j.jacceco.2009.09.004

Fernandez, E., Avella, L., & Fernandez, M. (2006). Estrategia de Producción (2nd ed.). Madrid, España: Editorial McGraw Hill.

Goosen, R., & Rudman, R. (2013). An integrated framework to implement it governance principles at a strategic and operational level for medium-to large-sized South African businesses. *International Business and Economics Research Journal, 12*(7).

Guerras, L., & Navas, J. (2015). *La dirección estratégica de la empresa. Teoría y aplicaciones.* Thomson Reuters.

Hamel, G., & Prahalad, C. (1989). Strategic intent. *Harvard Business Review*, 63–76. PMID:10303477

Heise, D., Strecker, S., & Frank, U. (2014). ControlML: A domain-specific modeling language in support of assessing internal controls and the internal control system. *The International Journal of Accounting.*

Ittner, C., & Oyon, D. (2014). *The Internal Organization of Enterprise Risk Management.* Available at SSRN 2486588

Jokipii, A. (2010). Determinants and consequences of internal control in firms: A contingency theory based analysis. *The Journal of Management and Governance, 14*(2), 115–144. doi:10.100710997-009-9085-x

Kloot, L. (1997). Organizational learning and management control systems: Responding to environmental change. *Management Accounting Research, 8*(1), 47–73. doi:10.1006/mare.1996.0033

Leisinger, K. M., Blasco, J., Trigo, J., Petitbò, A., Álvarez, J. M., & Argandoña, A. P. (2010). *Hacia una nueva ética económica global. Innovación vs Status Quo.* Editorial Huygens.

Li, C., Lim, J., & Wang, Q. (2007). Internal and external influences on IT control governance. *International Journal of Accounting Information.*

Lin, Z., Zheng, J., & Wang, S. (2007). Study on Relation of Internal Control and Enterprise Value: An Empirical Analysis from Shanghai and Shenzhen Stock Markets. *Journal of Finance and Economics.*

Martínez Marín, J. E., & Eguren Martí, M. L. (2013). Maritime Transport: A Theoretical Analysis Under a System's Approach. *Journal of Maritime Research*, 61-68.

Pfister, J. (2009). *Managing Organizational Culture for Effective Internal Control: From Practice to Theory*. Physica-Verlag. doi:10.1007/978-3-7908-2340-0

Porter, M. (1985). *Competitive Advantage: Creating and Sustaining Superior Performance*. Free Press.

Porter, M., & M., K. (2011, Jan.). The Big Idea. Creating Shared Value. How to reinvent capitalism and unleash a wave of innovation and growth. *Harvard Business Review*.

Porter, M., & Kramer, M. (2006). Strategy and Society: The link between competitive advantage and corporate social responsability. *Harvard Business Review*. PMID:17183795

Presidio Graduate School. (2006). *The dictionary of sustainable management*. Recuperado el 18 de 10 de 2015, de https://www.sustainabilitydictionary.com/

Ribstein, L. E. (2002). Market vs. Regulatory Responses to Corporate Fraud: A Critique of the Sarbanes-Oxley Act of 2002. *The Journal of Corporation Law*, *28*(1), 1–74. doi:10.2139srn.332681

Scandizzo, S. (2005). Risk mapping and key risk indicators in operational risk management. *Economic Notes*, *34*(2), 231–256. doi:10.1111/j.0391-5026.2005.00150.x

Schantl, S., & Wagenhofer, A. (2019). *On the Economics and Optimal Design of Internal Control Regulation*. SSRN.

Simmons, M. (1997). *COSO based auditing*. Internal Auditor.

Spira, L. F., & Page, M. (2001). Risk management The reinvention of internal control and the changing role of internal audit. *Accounting, Auditing & Accountability Journal*, *16*(4), 640–661. doi:10.1108/09513570310492335

*The Concise Oxford Dictionary*. (2012). Oxford University Press.

U.S. Securities and Exchange Comission. (1 de 10 de 2013). The Laws That Govern the Securities Industry. Recuperado el 18 de 10 de 2015, de Sarbanes-Oxley Act of 2002: http://www.sec.gov/about/laws.shtml#sox2002

Von Bertalanffy, L. (1989). *Teoría General de los Sistemas* (7th ed.). George Braziller.

Wieland, J. (2005). Corporate governance, values management, and standards: A European perspective. *Business & Society*.

## KEY TERMS AND DEFINITIONS

**Control:** A task or process designed to mitigate the probability of occurrence of a specific risk.

**SOX:** Sarbanes-Oxley Act.

**Exposure:** Set of activities, which includes financial exposure that has a probability to be affected by a defined risk.

**Fraud:** Deliberated use of current knowledge on processes and activities, in order to obtain profit by misuses or misapplication of the company resources or assets.

**Internal Control:** System composed of a group of activities defined to manage the risk that comes from the processes within the company, across all functions.

**Risk:** Is the potential exposure to an unexpected event or unintended situation or outcome.

**Role:** Set of functions assigned to an individual in a specific process or activity.

**SOD:** Segregation of duties. Principle that has the objective of preventing the risk of errors or fraud. One person cannot perform two functions that if performed together can compromise the "independence" premise.

**Value Management:** Process defined with the objective of managing the value applicable under the selected scope, having into consideration that value includes those actions defined in order to control cost as well as those actions defined to create value (differentiation, innovation, perception aspects among others).

# Section 2

# Chapter 6
# Fundamentals for the Challenge of Controlling Risk Through Standardization

**Vicente González-Prida**
https://orcid.org/0000-0001-5257-8224
*University of Seville, Spain & Universidad Nacional de Educación a Distancia, Spain*

**Jesús P. Zamora**
*Universidad Nacional de Educación a Distancia, Spain*

**Pablo Viveros Gunckel**
*Universidad Técnica Federico Santa María, Chile*

## ABSTRACT

*This chapter is intended to deal with the topic of standardization as a solution for risk control and decision making, mainly in high technologized context. For that purpose, this document starts introducing the reason of controlling risk as well as the uncertainty background that lays in different sectors. After that, the risk management evolution is outlined, defining briefly the risk concept, mentioning some current management models and enumerating the more relevant norms, standards, and guidelines for risk management. From this point, an overview on the standard ISO 31000 about risk management is provided in order to finally summarize some possible future research lines and some conclusions about this groundwork on risk and uncertainty.*

## INTRODUCTION

Nowadays, digital transformation allows the reduction of risk and uncertainty associated with decision making and the choice between alternatives. However, in order to meet properly this goal, it seems to be necessary a kind of standardization or levelness on processes and management frameworks. As a consequence, controlling risk is a task that has to be consistent with organization's values and policies.

DOI: 10.4018/978-1-7998-3246-1.ch006

In order to manage risk, the organization's processes have to provide tools for the identification and assessment of risks to be avoided, eliminated or controlled (BS25999-1:2006). Since risk can change with time, condition and usage, mitigation actions should be documented and monitored for a proper control. Therefore, the use of nowadays disrupted and practical technologies related to data mining and analytics is a requirement for managing risk, aligned to the overall company management process. Of course, risk management shall be proportionate to the level of risk under consideration so, every risk should not be treated in the same way (Jakoubi & Tjoa, 2009). On the contrary, it is understandable to vary risk management effort to reflect the level of risk.

As commented, risk can vary with time, boundary condition, environment, customer expectations etc. Thus, the method or models for risk management should be consistent with the organization operating experience and capability (BS25999-2:2007). In other words, it is important to manage properly the organizational knowledge, which reflects basically how the organization manages risk. Risks have to be classified and managed appropriately i.e. avoided, eliminated or controlled. Consequently, risk management includes monitoring of control or mitigation actions in order to ensure they are implemented in an effective way and in timely manner. Of course, it is clearly better to be proactive applying preventive tasks, rather than reactive. Hence, establishing the context of the risk provides alignment with asset strategy and objectives and the organizational strategic plan. Additionally, a monitoring and review process needs to be in place to ensure that the assessment process is appropriate and to create feedback if revision is required. At each stage there needs to be appropriate consultation and communication with affected stakeholders (Tjoa et Al., 2011).

In search of that purpose, this chapter analyzes the state of the art today, observing recent trends and studies regarding risk management and uncertainty control, mainly in its application to different areas or sectors. With this, it will be reviewed the concept of risk, the evolution of its management, and some application models that can be improved or complemented by the use of emerging technologies. After that, norms and standards will be indicated that aim to facilitate and / or standardize decision making, in order to achieve a better knowledge management and, consequently, risk control. Finally, possible future research lines will be discussed, together with some conclusions about the key aspects of this document.

## Background

The application of a standard for controlling uncertainty can provide principles and guidelines for a clear system and reliable management of any sort of risk within any application and context. An integrated approach is the best solution for the problem of risk management in organizations of all kinds (Canterla, 2016). This ensures compliance with international standards and those regulatory requirements of sectors under study. Here below is some background on risk management in different sectors or areas:

### Financial Sector

The advantage of a standardized model (in comparison to other existing methods), is its amplitude, with room for continuous improvement, usability and detailed description of the successive stages. Such stages for the application of an integrated risk management system in organizations, promotes a better structure and consistency in decision-making related to risk management (Tumenbayeva & Zhaksybekova, 2016).

## Supply Chain

An example applied to the supply chain can be seen in referece (Fontoura et Al., 2016). Here, a basis is provided in order to link the risks of the supply chain with corporate risk management, ensuring all the necessary resources for its implementation. The study concludes that the use of standards such as ISO 31000: 2010, is a valid requirement to ensure compliance and ensure the security of the supply chain against the different types of risk that threaten in that sector

## Construction Sector

The assessment of a building conservation state and its durability over time is related to the components that make up the building. These components are sensitive to events that generate one or several consequences and vulnerabilities. Extrinsic actions are therefore considered effects of uncertainty or risk in buildings (Prieto et Al., 2016). That paper proposes a method where an ordered classification of architectural elements is achieved through a probabilistic estimation of the useful life in years. With this, priority actions are established according to the durability of the buildings. For an efficient management in the useful life assessment and, it is necessary to study the risk and its potential impact, providing sufficient information to the parties involved, contributing to the understanding of the risk and helping to establish priorities in order to avoid incidents in conservation preventive buildings.

## Project Management

Risk management is considered a good tool to increase the probability of achievement and success in the development of any project. As indicated by Olechowski et Al., 2016, the new ISO 31000 risk management standard was introduced with the promise of universal applicability, including eleven principles for effective risk management. The study shows that the acquisition of risk management principles is a significant factor in achieving the cost, chronological, technical and customer requirements associated with the project. That is, there is a high relationship between risk management and successful projects. Therefore, risk management linked to ISO 31000 can provide the success of the planned purpose.

## Safety

The use of risk management for safety legislation in sectors like nuclear and oil & gas industry is deeply explained in Aven & Ylönen, 2016. In particular, the analysis indicates that the regulation based on a risk management is established by pillars as robustness, resistance and precautionary principles. The adoption of risk management will allow a stronger and more valid regulation in order to increase the level of safety in sectors like nuclear and oil & gas industry.

## Manufacturing

The use of risk assessment techniques may also be relevant in the manufacturing area. In fact, these techniques take an important role in the field of decision making, as indicated by Chemweno et Al., 2015. This study presents techniques in line with ISO 31000: 2010, which has become a consolidated tool to support decision-making in this area. Other research in this regard is for example Pulido-Rojano

A., 2015, which proposes a methodological design based on the development, creation, adaptation and implementation of some analysis tools and the deployment of risk management process contained in ISO 31000. This proposal allows the design of a monitoring and quality control system for critical variables associated with operating conditions, in order to identify and mitigate potential sources of product deviation, harmonizing work philosophies with international management standards. In general terms, the content of this paper offers professionals and industrialists in the manufacturing sector a way to address and improve any type of manufacturing process in any kind of organization through standardized steps.

### Other Areas

In the case of real-time embebbed systems, risk assessment in the development phase can be critical. The risk assessment method proposed by Siru et Al., 2016, is the Risk Assessment Method for Embebbed Systems, which is a quantitative risk assessment, capable of reducing potential security failures. A publication that presents examples, tools and approaches that can be applied to marine systems is the one observed by Gibbs & Browman, 2015.

## OUTLINE ABOUT RISK MANAGEMENT EVOLUTION

### Brief Summary on the Risk Concept

The concept of risk has evolved from an origin based on mysticism and superstition related to good or bad luck, fortune or misfortune, to a term that considers factors related to science, planning and decision making (Canterla, 2016). During this evolution of the risk concept, it is possible to observe within the Modern Age the beginning of the idea of personal responsibility as a result of own decisions and, with it, the idea of risk. In fact, there is a desire to better define the foundations of society (Rosés, 2002). The scientific study of society serves to rethink and revalue the present and to plan a better future. Therefore, the evolution of risk perception and its treatment is parallel to the personal, social and business evolution in this modern age where, for instance, great events also took place in the field of exploration with their inherent threats and associated risks.

This scientific and materialist philosophy, whose purpose was to make a more perfect society, continued influencing till the second half of the nineteenth century. It is important to highlight the importance of knowing the risks associated with any activity and its proper treatment, in order to obtain expected profits (Canterla, 2016). In other words, within the evolution of risk, the view of risk became as something inherent to an organization as a whole, and not only to partiuclar activities, or individualized processes.

Currently, risk management is taking great relevance in business and industrial world, in order to have a better performance in decision-making and the successful achievement of objectives. Risk management is an effective tool that increases the competitiveness of organizations, and usually acts under the umbrella of continuous improvement (Canterla, 2016). Risk management is, consequently, a new way of doing things within the framework of continuous improvement. The following bullets describe important milestones related to risk management:

- 1928. The International Commission on Radiation Protection is established in Sweden.
- 1939. UK establishes 99.999% reliability for 1 hour of commercial flight.

- 1944. Von Neumann and Morgestern write Theory of Games and Economic Behavior.
- 1949. Callendar speculates that a 10% increase in CO2 between 1850-1940 is linked to the warming observed in Europe and North America since 1880.
- 1961. Bell Labs develops fault trees for USAF to assess safety in the launch of the Minuteman.
- 1972-1973 Articles by Kahnemann and Tversky on Subjective Probability and Psychology of Prediction.
- 1975 The US Nuclear Regulatory Commission publishes the first nuclear safety ARP
- 1981 Kaplan and Garrick characterize risk in terms of events, consequences and probabilities
- 1999 Haimes presents the role of risk analysis dealing with emerging threats on critical infrastructure, including cyber sabotage and terrorism.

## Some Risk Management Models

In this section, some relevant risk management models are enumerated, according to Rosés, 2002 and Canterla, 2016:

### Cadbury Integrated Model: British View.

The Cadbury committee was created in 1991 in Great Britain, describing a risk control system with several components. One of these components refers to risk identification, evaluation and control. It is based on the fact that the managing board identifies the risks that arise in the company's business, establishing priorities for control, as well as budgets and other indicators. All this together with the appropriate information and communication systems, will allow the board to control the effectiveness of the company's activities. The likelihood of materializing risk must be considered, not underestimating the importance of its final impact the business. In addition, priorities should be established regarding the distribution of available resources for the control, definition and communication of clear control objectives.

### COSO Integrated Model: American View.

In 1992, the Committe of sponsoring Organizations of the Treadway Commission (COSO) was created in the USA. It is made up of the American Institute of Certified Public Accounts, the American Accounting Association, and the Institute of Internal Auditors among others. The mission is to prepar a report with a control system proposal, consisted in this case of five components. The Risk Assessment component is based on the fact that every entity faces a variety of risks from external and internal sources that must be assessed. A prerequisite for assessing risk is to establish objectives. They must be set objectively and integrated with sales, production, marketing, finance, etc. Zero risk is a chimera. The risk assessment is the identification and analysis of the relevant risks to meet the company's objectives. Because economic, industrial, legislative and operational conditions continue to change, mechanisms are needed to identify and treat special risks associated with the change. The risk analysis process includes: estimation of the type of risk, evaluation of probability, frequency, and actions to be taken.

### CoCo Integrated Model: Canadian Proposal

Se crea en 1995. Establece que los riesgos significativos, internos y externos, afrontados por la organización para el logro de sus objetivos se deberían identificar y evaluar.

### The Commitee from Basilea

It was created in 1998 in reference to risks associated with the Bank, but provides principles applicable to any company and activity. It considers that risk assessment must cover all those risks that affect the organization.

### The Turnbull Report

It is created in September 1999. The objectives pursued are to reflect good management practices, taking into account that internal control must be integrated into the business processes, in order the company to fulfil its objectives. It is applicable in a changing business environment and can be adapted to the peculiarities of each company. The approach is oriented to knowledge and risk management in order to establish a good internal control system. That control system depends on a periodic assessment of risk nature and dimension to which the organization is exposed. The main purpose of internal control is to help for a correct risk management, rather than its absolute elimination. The internal control system is integrated into the company's operations, and considered part of its culture by effective procedures and effective internal communication, in order to reduce risks to acceptable levels. An integrated business risk management system would consist of the following phases; identification and evaluation of risk; identification and evaluation of control measures; optimization of risk indicators; continuous improvement and supervision.

### Basilea II

It was created in 2001. It is a system of basic indicators for operational risks. Standardization method is made for effective risk management and control. The organization must create rigorous processes for estimating parameters such as impact or probability of occurrence.

## Norms, Standards and Guidelines on Risk Management

Once observed the evolution of risk concept and some management models, the following chart summarizes the norms, standards and guidelines that can be found in recent literature (Canterla, 2016).

## Overview on the iso 31000: Risk Management

The risk is considered by the ISO 31000 as the effect of uncertainty in achieving the objectives. Uncertainty should be understood as a state of information deficiency related to the compression or knowledge of an event, its consequence or the probability of occurrence (it may never occur). Risk must be managed because it has an effect. This effect is a deviation from the established objectives. The objectives can be presented in different forms (safety, environment, health, financial) and can be applied at different levels

*Table 1. Summary of standards and guidelines*

| Norm, standard, guideline | Description | Date |
|---|---|---|
| ISO 27005:2018 | Information technology -- Security techniques -- Information security risk management | 2018-07 |
| PMBOK | A Guide to the Project Management Body of Knowledge (PMBOK GUIDE) | 2017 |
| ISO 31000:2010 | Risk management. Principles and guidelines | 2010-07 |
| ISO 31010:2011 | Risk management. Risk assessment technics | 2011-05 |
| AS/NZS 4360:2004 | Risk management | 2004-08 |
| BS 31000:2010 | Risk management. Principles and guidelines | 2010-03 |
| FERMA | Risk management standards | 2003 |
| ISO GUIDE 73:2009 | Risk management. Vocabulary | 2009-11 |

(products, processes, project organization, etc.). The risk is therefore expressed in terms of combining the consequences of an event and the associated probability of its occurrence.

*Figure 1. Scheme of risk factors*

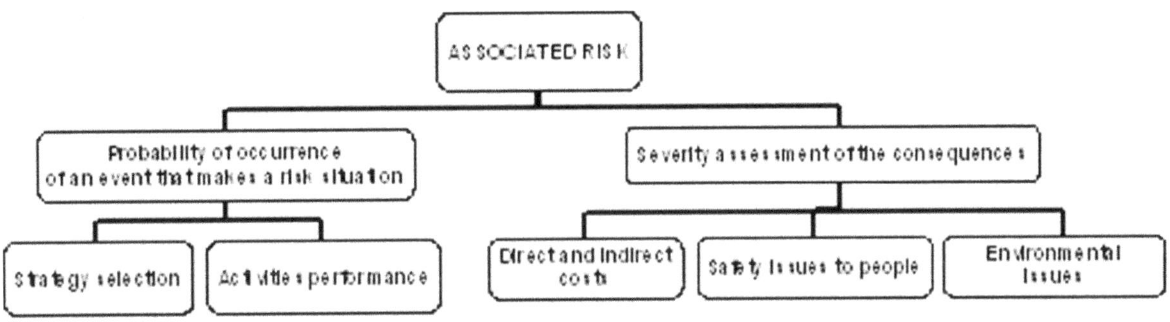


Some sorts of risk may be operational, technological, strategic, legal, compliance, environmental, safety, etc. etc. This international standard (ISO 31000) provides the generic principles and guidelines on risk management. It can be used by any public, private or social company, association, group or individual. Therefore, it is not specific to a particular industry or sector. A key feature of the standard is the establishment of the context as an activity at the beginning of the generic risk management process. This standard aims to help organizations of all types and sizes to manage their risks effectively. The ISO standard provides the principles, the framework and a process to manage any kind of risk in a transparent, systematic and credible way within any scope or context. The standard recommends that organizations develop, implement and continuously improve the risk management framework as a component of the organization's comprehensive management system. ISO 31000 is a practical document

*Figure 2. Risk management — Principles, framework and guidelines [ISO 31000]*

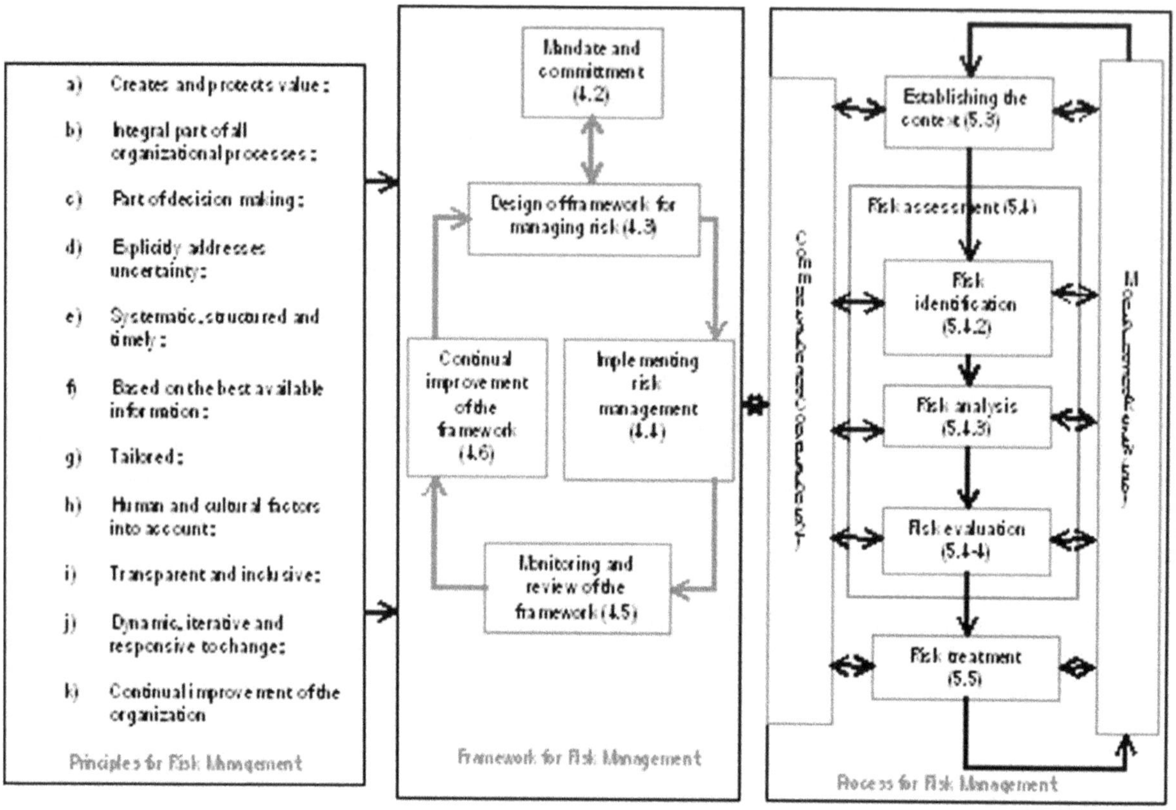

that seeks to help organizations in the development of their own strategy to manage their risks, but it is not a certifiable standard.

ISO 31000 is designed to help organizations to:

- Increase the probability of achieving its objectives, promoting proactive management
- Be aware of the needs in order to identify and treat risk throughout the organization
- Improve the identification of opportunities and threats
- Comply with legal and regulatory requirements and international standards
- Improve the financial information and governance of the organization
- Increase the trust of stakeholders
- Establish a reliable basis for decision making and planning
- Allocate and effectively use resources for risk treatment
- Improve efficiency and operational efficiency
- Increase health and safety as well as environmental protection
- Improve prevention and incident management
- Minimize losses
- Improve learning and 'resilience' of the organization

The general structure of risk management according to the ISO standard follows the following steps:

- Clause 3: Principles for Risk Management
- Clause 4: Framework for Risk Management
- Clause 5: Process for risk management.

Figure 2 shows the risk management model as identified in ISO 31000, where risk assessment is the overall process of risk identification, risk analysis and risk evaluation:

## Risk Identification

The organization should identify sources of risk, areas of impacts, events (including changes in circumstances) and their causes and their potential consequences. The aim of this step is to generate a comprehensive list of risks based on those events that might create, enhance, prevent, degrade, accelerate or delay the achievement of objectives (Suriadi et Al. 2014).

## Risk Analysis

Risk analysis involves developing an understanding of the risk. Risk analysis provides an input to risk evaluation and to decisions on whether risks need to be treated, and on the most appropriate risk treatment strategies and methods (Suriadi et Al. 2014). Risk analysis can also provide an input into making decisions where choices must be made, and the options involve different types and levels of risk.

## Risk Evaluation

The purpose of risk evaluation is to assist in making decisions, based on the outcomes of risk analysis, about which risks need treatment and the priority for treatment implementation. Risk evaluation involves comparing the level of risk found during the analysis process with risk criteria established when the context was considered. Based on this comparison, the need for treatment can be considered (Suriadi et Al. 2014).

Here below are some of the techniques that can be used for the identification, analysis and assessment of risk. The use of one instead of others will depend on the sort of risk:

- Asset Criticality Analysis
- Hazard Analysis studies (HAZAN)
- Hazard and Operability studies (HAZOP)
- Fault Tree Analysis (FTA)
- Event Tree Analysis (ETA)
- Root cause analysis (RCA)
- Failure Modes Effects and Criticality Analysis (FMECA)

As considerations to take also into account, it is important to establish correctly the context and to ensure that the interests of the parties are understood and taken into consideration, so the risks can be properly identified. In addition to this, experts from different areas should be gathered to analyze the risks, so that different points of view can be considered when defining risk criteria and when assessing them. The group of experts should ensure the approval and support for the treatment plan, encourage adequate change management during the risk management process, as well as develop an appropriate external and internal communication and consultation plan. Especially relevant in the risk management process is, as commented, their identification, analysis and assessment. Risk assessment provides a knowledge and understanding of the risks, their causes, consequences and probabilities. This provides ideas for decision making about:

- Priorities of treatment options;
- More appropriate selection of critical risk treatment strategies to reduce them to a tolerable level;
- If an activity must be carried out;
- If the risks must be treated or accepted

The purpose of risk assessment is to facilitate decision-making, based on the results of that analysis, about which risks need treatment and the priority for treatment implementation. Risk assessment involves comparing the level of risk observed during the analysis process and the risk criteria established when considering the context. Based on this comparison, the need for treatment can be considered. Decisions should be made in accordance with legal, regulatory and other requirements. The risk assessment can also result in answering questions as if the risk requires treatment, what priority should have such treatment?

## USABILITY AND APPLICABILITY OF RISK MANAGEMENT METHODS

In order to add more details about possible key elements for usability and applicability of the methods and models, it is recommended reference like Ravizza et Al. (2019) or Hartung (2018) among others. Risk management has been already defined and treated throughout this document. On the other hand, usability engineering is usually depicted as the evaluation of risks associated with use and, particularly, in operation, with the aim of developing user interfaces that reduce risks as much as possible. Its assessment usually starts with a formative phase in order to explore strengths, weaknesses, and unanticipated use errors (IEC 62366, 2015), and a second phase relative to a summative evaluation that verifies an acceptable risk-benefit profile under a usability point of view.

Reference IEC 62366 (2015) defines techniques for usability assessment like: System Description; Task Analysis; Human Error Analysis; Human Error Reduction and Mitigation. Each usability evaluation technique allows to identify risks and potentially hazardous situations. Some of the above-mentioned techniques are described in Ravizza et Al. (2019) in terms of capability of the assessment, being easily linked to a formal risk analysis as per ISO 14971 (2007), in that case, for medical solutions. Te commented references provide good practices to illustrate examples of application and usefulness. Some of the evaluation techniques are for instance:

- Advisory panel reviews
- Brainstorm use scenarios
- Cognitive walkthrough
- Expert reviews
- FMEA
- Focus groups
- Function analysis

- Heuristic analysis
- Observation
- One-on-one interviews
- Participatory design
- PCA analysis
- Simulation
- Standards reviews
- Surveys
- Task analysis
- Time-and-motion studies
- Usability tests
- Workload assessment

## FUTURE RESEARCH DIRECTIONS

Oftenly, uncertainty is associated with risk analysis. Understanding uncertainties is necessary to understand and communicate effectively the results of any risk analysis. The uncertainties linked to data, methods and models that are used to identify and analyze risks, play an important role in their application as a result of the collective variation in the parameters and assumptions used to define the results. Therefor, an area closely related to the uncertainty study is the sensitivity analysis (Canterla, 2016). The sensitivity analysis implies the variation of the size and importance of the risk level by changing individual input parameters. It is used to identify the data that must be accurate (are more sensitive), and those ones that have lower effect (less sensitive). The sources of uncertainty should be identified, if possible. A focus on the data and the applied model or method shall aim to identify the parameters for which the analysis is sensitive, as well as the degree of such sensitivity. Consequently, future lines of research may be precisely focused on seeking methodologies that reduce uncertainty and improve knowledge management, in order to improve decision making.

## CONCLUSION

Risk is an exposure to the uncertainty that anyone has to understand, accept, and to know how to handle when executing a strategy with the purpose of achieving objectives and gain value. On the contrary, in case of inappropriate processes, people or systems, risk will be the direct or indirect result of obtaining losses. Therefore, it is useful and relevant to standardize a syste to manage risk. The methods developed and the lessons learned from the studies, references and standards commented throughout this chapter can be used to guide researchers, practitioners, administrators, etc., in risk management. In fact, an overview of the current state in risk assessment applied to different areas has been provided here, seeking to encourage new reflections on how the various approaches to risk management can be applied. Risk management is obviously linked to all sectors and activities of our society and business world as seen here along. Ultimately, risk management currently allows anticipating situations, occurrences, in order to avoid unwanted losses, improving at the same time the decision making. This risk management process is optimized today thanks to the new information, communication and computer technologies, providing an agile and effective tool for the improvement of organizations.

## ACKNOWLEDGMENT

Authors want to acknowledge the support provided by the research projects: FFI2017-89639-P and RTI2018-097709-B-I00.

## REFERENCES

Aven, T., & Ylönen, M. (2016). Safety regulations: Implications of the new risk perspectives. *Reliability Engineering & System Safety*, *149*, 164–171. doi:10.1016/j.ress.2016.01.007

British Standard Institute (BSI). British Standard – BS25999-1:2006: Business Continuity Management – Part 1: Code of practice. Available at http://www.bsigroup.com

British Standard Institute (BSI). British Standard – BS25999-2:2007: Business Continuity Management – Part 2: Specification. Available at http://www.bsigroup.com

British Standard Institute (BSI), British Standard – ISO 31000:2010. (n.d.). *Risk management. Principles and guidelines*. BSI.

Canterla, J. P. (2016). Methodology for the Development of a Quality Management System, in accordance with the new ISO 9001:2015 Standard, integrated with a Risk Management System (ISO Standard 31000: 2010), in Companies depicted to Transport of Perishable Goods by Road. Degree project for the master's in industrial management (MOIGE). University of Seville, Spain.

Chemweno, P., Pintelon, L., Van Horenbeek, A., & Muchiri, P. (2015). Development of a risk assessment selection methodology for asset maintenance decision making: An analytic network process (ANP) approach. *International Journal of Production Economics*, *170*, 663–676. doi:10.1016/j.ijpe.2015.03.017

FERMA. (2003). *Risk management standards*. Federation of European Risk Management Associations.

Fontoura, M., Hoeflich, S. L., Dias, E. M., & Wee, H. M. (2016). A note on supply chain risk classification: Discussion and proposal. *International Journal of Production Research*, *54*(5), 1568–1569. doi:10.1080/00207543.2015.1067375

Gibbs, M. T., & Browman, H. I. (2015). Risk assessment and risk management: A primer for marine scientists. *ICES Journal of Marine Science*, *72*(3), 992–996. doi:10.1093/icesjms/fsu232

Hartung, P. (2018). *Link Between Usability Engineering and Risk Management*. https://www.regulatory-affairs.org/en/quality-management/news-page/link-between-usability-engineering-and-risk-management/

IEC 62366-1:2015 *Medical devices -- Part 1: Application of usability engineering to medical devices (1st ed.)*. International Electrotechnical Commission.

ISO 14971:2007 *Medical devices -- Application of risk management to medical devices (2nd ed.)*. International Organization for Standardization.

ISO 31000:2010. *Risk management. Principles and guidelines*. International Standard Organization

ISO 31010:2011. *Risk management. Risk assessment technics*. International Standard Organization ISO guide 73:2009. *Risk management. Vocabulary*. International Standard Organization

ISO/IEC 27005:2018. *Information technology - Security techniques - Information security risk management*. International Standard Organization

Jakoubi, S., & Tjoa, S. (2009) A reference model for risk-aware business process management. In Kalam, A., Y. Deswarte, & M. Mostafa (Eds.), *Risks and Security of Internet and Systems, 4th International Conference, CRiSIS'09, Toulouse, France, October 19-22, 2009, Proceedings*. IEEE Computer Society. 10.1109/CRISIS.2009.5411973

Olechowski, A., Oehmen, J., Seering, W., & Ben-Daya, M. (2016). The professionalization of risk management: What role can the ISO 31000 risk management principles play? *International Journal of Project Management, 34*(8), 1568–1578. doi:10.1016/j.ijproman.2016.08.002

Prieto, A. J., Macías, J. M., Chávez, M. J., & Alejandre, F. J. (2016). Expert system for predicting buildings service life under ISO 31000 standard. Application in architectural heritage. *Journal of Cultural Heritage, 18*, 209–218. doi:10.1016/j.culher.2015.10.006

Project Management Institute. (2017). PMBOK 6th Edition. Project Management Body of Knowledge. PMI Standards Committee.

Pulido-Rojano, A. (2015). Methodological design for the prevention of risk in production Processes. *Dyna (Bilbao), 82*(193), 16–22. doi:10.15446/dyna.v82n193.42903

Ravizza, A., Diaz Lantada, A., Ballesteros Sánchez, L. I., Sternini, F., & Bignardi, C. (2019). *Techniques for Usability Risk Assessment during Medical Device Design*. Conference BIOSTEC Biodevices., doi:10.5220/0007483102070214

Rosés F., (2002). *Risk Management. A new way to assure business success*. ACV Editions.

Siru, N., Zhuang, Y., Jingjing, G., & Ying, H. (2016). A formal model and risk assessment method for security-critical real-time embedded systems. *Computers & Security, 58*, 199–215. doi:10.1016/j.cose.2016.01.005

Standards from Australia / New Zealand. (2004). AS/NZS 4360:2004. *Risk management*.

Suriadi, S., Weiss, B., Winkelmann, A., Hofstede, A., Conforti, A. M., Fidge, R., La Rosa, C. M., Ouyang, C., Rosemann, M., Pika, A., & Wynn, M. (2014). Current research in risk-aware business process management: Overview, comparison, and gap analysis. *Communications of the Association for Information Systems, 34*(1), 933–984. doi:10.17705/1CAIS.03452

Tjoa, S., Jakoubi, S., Goluch, G., Kitzler, G., Goluch, S., & Quirchmayr, G. (2011). A formal approach enabling risk-aware business process modeling and simulation. *IEEE Transactions on Services Computing, 4*(2), 153–166. doi:10.1109/TSC.2010.17

Tumenbayeva, O. S., & Zhaksybekova, G. N. (2016). Implementation of the Integrated System of Risk Management in the Banks of Kazakhstan. *Indian Journal of Science and Technology*, *9*(5), •••. doi:10.17485/ijst/2016/v9i5/87607

## ADDITIONAL READING

BS 6079-1:2010 Project management. *Principles and guidelines for the management of projects.* © British Standards Institution.

Datta, S., & Mukherjee, S. K. (2001). Developing a Risk Management Matrix for Effective Project Planning—An Empirical Study. *Project Management Journal*, *32*(2), 45–57. doi:10.1177/875697280103200206

González-Prida, V., & Teira, D. (2014). Book review: "Handbook of risk theory. Epistemology, decision theory, ethics, and social implications of risk"(edited by Springer Verlag). *Reliability: Theory & Applications.* ISSN 1932-2321 http://www.madrimasd.org/blogs/analisis-riesgos/2011/08/31/16/

González-Prida, V., & Zamora, J. (2019). *Decision making under uncertainty and risks in the face of rapidly advancing technologies. Handbook of Research on Industrial Advancement in Scientific Knowledge.* IGI Global., doi:10.4018/978-1-5225-7152-0.ch003

González-Prida, V., & Zamora, J. (2019). A Risk Indicator in Asset Management to Optimize Maintenance Periods. *WCEAM (World Congress on Engineering Asset Management)*, Stavanger, Norway, 24 – 26 September 2018

## KEY TERMS AND DEFINITIONS

**Challenge:** Something that by its nature is a test or a difficult thing to accomplish.

**Fundamentals:** Basic principle, essential rule, or law that serves as the groundwork of a system.

**Risk:** Degree of probability of a hazard or chance of loss occurrence.

**Standard:** Something considered by an authority or by general consent as a basis of comparison or an approved model.

**Technology:** Branch of knowledge that deals with the creation and use of technical means and their interrelation with life, society, and the environment, drawing upon such subjects as industrial arts, engineering, applied science, and pure science.

**Uncertainty:** The fact of not being known precisely, nor fixed, confident, assured or clearly determined. In other words, it refers to something hesitant, unknown, unstable, or likely to change.

# Chapter 7
# Risk and Uncertainty on Technology and Science Under Bayes and Popper's Statements View

**Vicente Gonzalez-Prida**
https://orcid.org/0000-0001-5257-8224
*University of Seville, Spain & Universidad Nacional de Educación a Distancia, Spain*

**Jesus Zamora Bonilla**
*Universidad Nacional de Educación a Distancia, Spain*

**Christopher Nikulin Chandia**
*University Federico Santa María, Chile*

**Antonio Guillén**
*University of Seville, Spain*

## ABSTRACT

*The overall perspective of this study is related to the concept of risk and uncertainty in the world today. In this sense, it considers Popper's contributions together with the deductive method, contrasted Bayes' contributions with the inductive method. On one hand, induction allows to generate results considered probabilistically true. This is basically the method used by supervised predictive methods of machine learning, where a general rule is inferred from particular examples in which solutions are known, inducing consequently to possible results for new inputs. On the other hand, deduction is a process in which general hypotheses are proposed, and from them, particular statements are obtained. These particular statements can obviously generate the rejection of those initial hypotheses. Under these considerations, Bayes' and Popper's postulates should not be understood as opposed methods. With this, the specific objectives of this chapter states on an overview about technology and its relationship with science, being analyzed from the Popperian and Bayesian perspective.*

DOI: 10.4018/978-1-7998-3246-1.ch007

## INTRODUCTION

The scientific method applies a set of methodological rules. These rules or principles can be summarized as the fact of proposing assumptions in order to tests them by experience. The logic allows deducing from the hypothesis a series of results or predictions that shall be confronted with facts or evidences (Rojas et al., 2014). That means, in the scientific method, hypotheses are proposed as plausible solutions to a problem, by confronting the results deduced from the hypothesis with the true experience, the hypothesis may be rejected or not by the facts (Banegas et al., 2000). In this line, the proposed theory will remain true if it resists the attempt to reject it. The concept of contrasting raised by Popper lays therefore in the criticism or attempts of falsification, which is in other words, the review of errors within a theory, in order to reject it in case of being false. Those theories that pass the process of falsification, can be regarded as true (Renovell, 2015).

In few words, from general statements to particular facts, refers to a scientific method that uses deductive arguments (Rosales. 2002). On the contrary, induction goes in the other direction, from particular facts to general statements. . That refers when particular data are used to state hypotheses, reaching a series of overall conclusions. Hence, induction is applied to evaluate if a deductive hypothesis is false or may be true. To be fals can be demonstrated with just one case. On the other hand, the complete verification of general hypotheses or theories would require to explore all the possible situations.

In order to consider if a theory is maintained as true, casualty criteria appear as rules of the method designed to eliminate or reduce the error (random biases) to contrast a specific hypothesis (Popper, 1959). Causal inference, which is the step from particular evidence to a general theory, can be considered therefore as a process of decision-making based on the hypothesis assessment performed by logic rules (causality criteria).

With all these aspects taken into account, the objective of this chapter relies in the logical sequence of considerations addressed towards the description of those aspects provided by Popperian and Bayesian statements, in order to introduce the link between risk and technology and its connection to deductive or inductive analyses, combining when possible philosophical aspects with others much more practical. It is important to reiterate that this chapter provides a conceptual view on the topic related to risk and technology, and does not lead to tools with immediate applications in an industrial environment. In other words, this document discusses different theories associated with diagnostic or prognostic techniques and the basis of the concepts that they convey. In fact, theorists will refresh thoughout the following sections already known concepts which are connected to the concerns of many technology end-users.

## BACKGROUND

### On Deductive and Inductive Scientific Methods

In accordance to inductive thought, natural science is characterized by a method that starts from long series of observations and experiments, in order to set general behaviors or theories (Popper, 1972). That means inductive method can provide a criterion of demarcation between science and speculation, stablishing or verifying those theories as almost sure or very likely to be sure. In other words, induction is a method of trial and error from a hypothesis raised by scientists in order to explain a problem to resolve (Howson & Urbach, 1991), (Popper, 1972). If evidences obtained are in favor of the hypothesis, it will be accepted. That means, the credibility of the theory is based on a Bayesian probability of the results from the hypothesis obtained by a mass of data (Howson & Urbach, 1991).

On the contrary, Popper opted for truth approach of a hypothesis by the deductive method (Popper, 1959). That is, from a hypothesis, predictions are logically deduced; then these are confronted with experience and rigorously contrast to the hypothesis. If predictions agree with results of the study, it is decided that the hypothesis is verified and it is provisionally corroborated; but if the decision is negative, the theory from which they have been deduced is falsified (rejected) (Renovell, 2015).

In short, the difference between Popper and Bayes lies in the acceptance or rejection that can be learnt inductively from experience (Rojas et al., 2014). That is, on Bayes side, as far as data are accumulated confirming the theory, it increases the likelihood that such theory is true. On the contrary, from Popperian point of view, this aspect is rejected since there is no logic foundation from the always finite data of experience, as it is always possible to find a new observation that refute the proposition.

In any case, both methods provide provisional results. That means, in case of logic reasoning (deductive), a set of finite observations that are true (not biased), is able to falsify a universal hypothesis but not to verify it completely (it is not possible to assure that result every time, everywhere, in order to see if a counterexample appears). On the other hand, if a theory agrees with the facts, the theory can still be false and its predictions explained by other alternative hypotheses. With this, what can be expected is a provisional plausibility (Banegas et al., 2000). For that reason, scientific hypotheses, in a valid way, can be falsified but never confirmed. Those hypotheses that resist repeatedly attempts of refutation (empirical tests) can be provisionally preferred (Montañez, 2017).

However, if it is a probabilistic (non-deterministic) hypothesis, it is not verifiable (being universal) from a logic point of view, and not strictly falsifiable (since a hypothesis does not logically exclude deviations in a large sample of data) (Popper, 1959). Nevertheless, it can be falsifiable in accordance with methodological rules (e.g. large deviations or levels of statistics significance) (Rojas et al., 2014), maintaining a logical asymmetry.

The Popperian assertion about scientific hypothesis to be or not justified, nor even likely by induction, has been just answer by the Bayesian philosophical framework as can be observed in references Howson & Urbach, 1991. Similarly, contrary assertions can be found in references such as Greenland, 1988.

## Bayes, Induction and Probability

The so-called statistical inference or inductive statistics refers to the contrast of hypothesis or parameters estimation in a population from observations of samples taken randomly. In that statistical tests, it is proceeded from evidence in the sample to hypothesis in the population, i.e. from particular issues to general statement. This is the inductive direction (Montañez, 2017).

In a random experiment, error and probability values are inferred or calculated by logical reasoning from truth supposition of hypothesis zero and alternative hypothesis (Renovell, 2015), (Rojas et Al., 2014). After this, it is a question of decision-making (which is an act of will and not of deductive or inductive inference), rejecting or not the tested hypothesis (hypothesis null) with results observed under a stablished level of confidence or, in other words, after deciding if the result is statistically significant (Banegas et Al., 2000).

In a probabilistic estimation, intervals of confidence are deducted from certain statistics premises (central limit theorem, normal distribution of a random variable, large numbers, etc.). Starting from there, again, it is a matter of the decision-maker to accept as a true value the parameter located within the limits calculated from observations (Neyman, 1952). Precisely, decisions and not deductive conclusions are in fact what, in last term, decides the fate of a statistical hypothesis. It indicates that evaluation of the hypothesis is provisional and not conclusive or definitive (such as it is stated in by Popper 1959).

Although probabilistic statements and intervals of confidence are used to contrast hypothesis, what is calculated on frequentist statistics is the probability of data under certain assumptions, not the probability of a hypothesis in front of a mass of favorable data (Oakes, 1990). This is the reason why the preference for one particular hypothesis based on their Bayesian probability is not admitted in the Popperian epistemology (Howson & Urbach, 1991).

From Bayesian perspective, the theorem of Bayes applied to probability of hypothesis can be consider as the modern paradigm of inductivism. Such theorem establishes that each collection of events numerable $\{A_k\}$ $k \in K$, with $K \subset N$ exhaustive and mutually exclusive, with $P(A_k) > 0$ for all $k \in K$, and given A with $P(A) > 0$, it is verified that:

$$P(A_k|A) = \frac{P(A_k \cap A)}{P(A)} = \frac{P(A|A_k)\cdot P(A_k)}{\sum_{k\in K} P(A|A_k)\cdot P(A_k)} \; \forall k \in K .$$

If two or more hypotheses in competition are available, and there is evidence in favor of both, the theorem of Bayes does not allow to discriminate between them. In the theorem of Bayes, *posteriori* relationship between probabilities of both hypotheses (i.e., after considering the evidence) is equal to the ratio of probability *a priori*, since the probability of evidence against both hypotheses is the same. Therefore, the evidence does not support the first hypothesis more than the second one (Popper & Miller, 1987).

## Popper, Deduction and Causality

Popper defines the objective of science as the search for explanations (causes) of phenomena occurred in cosmos. That means, it looks for true laws that govern the universe (Popper, 1985). A debate appears here whether causal inference, or causality criteria, are an inductive process or not.

Causal criteria are methodological rules that help in the process of decision. In a prediction process contrasted with experimental study, it is observed that obtaining the value p (probability) and the interval of confidence is performed according to frequentist statistics (Renovell, 2015). In addition, it is necessary to decide whether the evidence (facts or data) is accepted or not as true, assuming the consequent risks of error. This is a process methodology guided by statistical and non - statistical rules (e.g., bias consideration). In other words, this is a decision and, as such, is an element taken by hefting reasons in accordance with rules (Susser, 1973).

From evidence to causal hypothesis, it seems clear that it occurs in inductive direction, from singular to general (in statistical terms, from sample to population). For those Neobayesian inductivists, it is possible to estimate the probability that a causal hypothesis is true. That means, it is expected that cases not observed will continue confirming probabilistically the hypothesis. The probability assumes that data not observed are similar to those ones observed (law of large numbers or long - term stability of frequencies) (Popper, 1959).

However, in scientific analysis, from finite favorable data is not possible to infer with certainty assumptions that go beyond that experience. In fact, it is decided on evidence and this helps to estimate or assess the hypothesis (Banegas et Al., 2000). For this, if the hypothesis is not rejected but supported by positive evidence, then, it is provisionally confirmed and accepted.

The fact that decision or judgment on one hypothesis is a methodological process does not mean that logic does not play any role. When one hypothesis is supported by evidence, only the part observed (i.e. the ratio observed in studied subjects, at a specific place and time) is deductively confirmed (Popper & Miller, 1987). However, it remains pending to be revised by the fallibility of the observations or existence of better alternatives hypothesis (Montañez, 2017). The inductive component of hypotheses (cases not observed) and, with it, the whole hypothesis, theoretically must keep on testing, especially if arise new contexts or fields of application, more accurate measurement technics, etc. Therefore, do not confuse logical (or deductive) corroboration with the empirical one, which is always unfinished. In all cases, even when a confirmed theory remains falsified to date, it never should be left unless there is a better substitutive theory, which is more informative (Popper, 1959) (Popper, 1988).

## POPPERIAN FALSIFIABILITY AND BAYESIAN INFERENCE IN REFERENCE TO RISK AND UNCERTAINTY

### Risk and Technology according to Popper

The current technological development seems to manifest itself as a global process. Technology, seen from the Popperian perspective, is one instrument to the service of values more elevated from theory and society (Rosales. 2002). From there, the technology development is not autonomous. On the contrary, it has its place in culture. Technology, as a human product, may incur in the course of its development and growth, a series of unforeseen problems and collateral effects beyond the limits originally planned (Renovell, 2015).

According to his concept of gradual engineering, Popper acknowledges the impossibility of a total effective intervention in history and its processes (Montañez, 2017). That is to say, every rational construction or planning is also an unpredictable and multidimensional area in its causes and effects (Popper, 1994).

The technology development can be seen therefore as a particular manifestation of general history. Technology is presented as a collective and continued advance according to its own 'laws of motion' (Popper, 1997). The Popperian position on technology ranges:

1. On one hand, the relationship between science and technology that accentuates purely neutral and instrumental look. Then, it is intended just to serve diligently the wishes of the theory.
2. On the other hand, a more complex perspective than the links technology with historical dynamic as a collective process for better or worse, where unforeseen results may appear with contrary consequences to those intended ones (Rosales. 2002).

Popper tends to support the second position suggesting the possibility of accepting reality and admitting it as constituted as a product of human activity (Popper, 1988). From the Popperian point of view, it is always possible to learn from tradition, culture and experience, with the view of forging criteria not only strictly technical (Rosales. 2002), (Banegas et al., 2000), but overall, moral criteria, as more democratic as possible, in order to deal successfully and effectively with processes that tend to offer an advance seemingly autonomous as it is the technology development (Popper, 1983).

## Risk and Technology according to Bayes

In general terms, Bayesian statistics is considered as a subset in statistics science. It arose when Thomas Bayes in the mid -1700 formulates his concept of inverse probability. The statistics are a powerful tool to organize, interpret and use information in order to make better decisions (Montañez, 2017). With the theory of large numbers, Bayes argues his view of probability by introducing the concept on how to change judgments of a person with occurrence of an event in the light of new additional information (evidence) (Banegas et al., 2000).

Mathematics science evolved by more empirical paths based on the implementation of measurement instruments that resulted in a frequentist interpretation of probability and the notion of likelihood (Rosales. 2002). This approach proved to be relatively easy to implement in practice and much more with the appearance of specialized software in order to carry to out quickly common computing processes.

Nowadays, with the so-called Technologies 4.0, it comes on scene Big Data Analytics, which is an excellent support for Bayesian approaches. Bayesian statistics has proved to be effective to treat matters on econometrics and social sciences where less scientific inputs as philosophy, feelings and the analysis of behavior have to be considered (Renovell, 2015). Before that, in order to a classical statistics to have validity, it was required to consider a minimum universe as more representative as possible. That kind of input is subjective in itself, since the results obtained will depend on how well done it is the sampling. Obviously, larger samples will allow better assessment and conclusions more appropriate to reality (Rojas et al., 2014).

For Bayesian statistics, a previous knowledge has to be considered formally and explicitly, incorporating subjective probabilities. That is, those that emerge from own intuition, expertise and experience of the researcher. In this way, Bayesian statistics can operate in one scenario where the reduction of required sample size does not distort the results, as it has in mind not only prior general data, but also new events that are occurring at every moment, as well as the expectations on how reality behaves in the light of statistical analysis (Renovell, 2015).

## How can Technology Represent a Risk?

In reference to technology and science, risk aspect is quite wide to deal with. The above sections try to address the link established between Popperian and Bayesian assessment methods on technological or scientific risk evaluation. Nevertheless, there are references in the literature that establish very well that matter. For instance, Acker R. (2020) considers this topic in regards to vulnerability processing and storing personal information. In fact, as far as society and industry are adopting sophisticated technologies, advanced threats are also appearing. Hence, it is relevant for organizations to assess the threat landscape in order to develop effective strategies and systems to minimize risk and reduce the gap between technology and data protection.

Other focuses connected to technology and risk appear for example in areas like device threats (where mobile devices, apps and technologies are utilized by early-adopting users); cloud security (where threats may rely on a less business control that depends on service-level agreements with external cloud service providers); the adoption of IoT devices (due or example to poorly configured devices developed by manufacturers or unsecured devices connected to vulnerable networks); the global supply chains (that are often in the target of attackers); etc. In addition to this, it is important to highlight as a sensitive area the human element. Similarly to many sectors, the use of high-tech must recognize human threats, which take many different forms. Insider threat, social engineering, process failures…, they all imply significant risks with multiple way to appear under different scenarios (Acker, 2020).

Technology introduces complexity, and complexity means risk. Equipment and information system become more and more complex. The complexity of systems and organizations, which will be new and higher. A complexity both exogenous - the organizations will evolve in environments of increasing demands and pace of change - and endogenous - the systems themselves, converted into Cyber-Physical Systems (CPS) (Lee, Bagheri, & Kao, 2015) will be much more complex, with unprecedented interaction with the rest of the systems and with the human. Increasing capabilities (intelligence, knowledge, interconnection, etc.) increases the possibilities of action. New collective behaviours may appear that are difficult to predict due to the tremendous interdependencies between entities (humans, systems and organizations) (Zio, 2018)

On the other hand, technology is introducing more chances to deal with riks control. Predictivie analytics solutions supporte by IoT technology allow arising aspects that were or were hidden remain visible to the organization (Lee, Holgado, Kao & Macchi 2014). Being able to monitor and predict behaviors, as well as record and formally translate them as explicit knowledge of the organization, constitutes the process towards transparency. As progress is made in these solutions, the organization will become increasingly transparent and, as a consequence, will be able to make a conscious management of risk, a real management. One of the principles of risk management according to ISO:31000 "*Risk management-Guidelines*" is precisely that it must be transparent to the entire organization and its stakeholders (ISO, 2018)

## DECISION MAKING IN HIGH-UNCERTAINTY CONTEXTS

Risk is usually understood as the possibility that damage may happen, linked with the danger and vulnerability of such event. According to Popper, science must take risks. Engineering measures the risk as the product between probability or frequency that an event may happen, and consequences or severity of that danger, limiting its scope to a specific environment during a determined period of time. Also from de engineering perspective ISO 31000 define risk as "*the effect of uncertainty on objectives*" (ISO, 2018), while the SRA (Society of Risk Analysis) argues "*consider a future activity (interpreted in a wide sense to also cover, for example, natural phenomen), for example the operation of a system, and define risk in relation to the consequences (effects, implications) of this activity with respect to something that humans value. The consequences are often seen in relation to some reference values (planned values, objectives, etc.), and the focus is often on negative, undesirable* consequences. *There is always at least one outcome that is considered as negative or undesirable.* (SRA, 2018)

Thus, strategies against risk can be reactive or proactive, being risk categories related for instance to: product quality, impact on the organization and/or client, effect on production process and environment, development of technology, etc. In conditions of certainty, the engineer can at least provide facts and their results (Banegas et al., 2000). However, while making decisions that can cause uncertainty and risk, there are phenomena not attributable directly to that decision-making, as for example (Rojas et al, 2014):

- Nonexistence of Historic data, directly related to the alternatives that are studied
- Biases in the calculation of data or in possible events

- Wrong analysis and interpretation of the available information, low reliability of data statistics, etc.
- Obsolescence due to rapid technological change
- Natural catastrophes
- Etc.

In order to provide an effective response to the population challenges, engineering must consider and manage uncertainties and trends of own organization culture, as well as the technology it is used. The admissible risk is no longer a technical or expert concept only, but it is a concept that also involves the opinion of the company or, more generally, organization's stakeholders objetives or espectatives (ISO, 2014). The level socially acceptable should be determined by both technology and the level of social alarm. Depending on the population awareness, the same risk is perceived in different ways by different perspective of different countries and cultures.

Engineering and technological evolution is based on the method of trial and error, supported by the scientific method. Whenever there is a new technological development, there are risks, since risk zero does not exist. When more experience history is available, the greater will be the ability of engineering to minimize risks (Montañez, 2017) When a new project is embodied, for which there is no background, the initial risk is high and, conforming the knowledge increases, the risk decreases due to obtaining of more experience, tests results in operation, accidents, etc.

Risk management implies uncertainty management. Technology introduce new risk but also new capabilities to control or even reduce uncertainty. It is interesting to note how to Roper et al. (2011), classified uncertainty methods according to determined families considered frequently used by practitioners, this classification aims to understand in which scenarios engineers should work in order to reduce uncertainty:

- Methods constituted by those ones who have the knowledge, well known as "Expert Opinion", where the opinion of the expert is obtained and analysed. The main assumption of these methods is that "some individuals know a lot about some topic than others, their forecast will be substantially better" (Roper et al., 2011). However, a limitation of the method is the identification of the appropriate expert for the forecast. Moreover, if the interaction among experts is allowed, the forecast may be affected by several social and psychological factors (Levary et al.,1995);
- Methods that rely only in prediction of future values, also known as "Trend analysis": Mathematical and statistical techniques used to extend the time series data into the future. The main assumption for this family is "past conditions and trends will continue in the future more or less unchanged". The limitations of these methods are availability and goodness of the data (Phillips et al., 2007).
- Methods focused on collecting data and process them to find relevant patterns and trends recognized as "Monitoring". This family is similar to the one mentioned above (trend analysis) but, actually and strictly speaking this case is not considered as a forecasting method; however, it is used to gather data and in some cases information. The main assumption is "there is information useful for a forecast, and it can be obtained"; the main limitation emerged from the quality of the data and their results, because several times it is performed without adequate selectivity and filtering (Coates et al., 2001);
- Methods that try to represent the reality by models, well known as "Modelling", the main assumption is "the basic structure and important aspects of part of the world can be captured by simplified

representations". However, a limitation emerges from the sophisticated techniques that may drive to faulty assumptions, giving a poor credibility to the forecasting results (Armstrong, 2001). In particular, models usually favour on quantifiable over no-quantifiable parameters, thereby neglecting potentially important factors (TFAWG, 2004);

According to the considered methods, it is possible to calculate the risk of each scenario, as the product between probability and consequences. Engineering has well defined methods and sufficient mathematical tools to calculate risk with a good approach degree. Correspondingly, accident events and scenarios in a risk assessment model have been classified according to the knowledge available at the time of the assessment (Flage R, Aven T): (i) *Unknown-unknown*: identifies those events and scenarios that were unknown to everyone, at the time of the risk assessment; (ii) *Unknown-known*: indicates those events and scenarios unknown to the risk analysts performing the assessment, but known to someone else; (iii) *Known-unknown*: identifies situations of awareness where the background knowledge is weak but there are in- dications or justified beliefs that a new, unknown type of event or scenario (new in the context of the activity posing the risk) could occur in the future; (iv) *Known-known*: indicates events and scenarios that are known to the analysts performing the risk assessment, and for which evidence exists.

The computer models to calculate the consequences have evolved significantly based on the new Technologies 4.0. The knowledge of experts, achieved after long years of study and experience, must be extracted using techniques capable of avoiding contamination' with personal opinions. The Bayes Theorem, or the so-called Bayesian transformation, is the tool generally used to combine various probabilities. Bayesian approach parts from *a priori* distribution of probabilities in which the analyst has certain degree of confidence before obtaining data targets. Then, objective measurements of reality (trials) modify the degree of initial conviction.

Popper criticized the use of Bayesian approach by considering subjective the probabilities, and proposed replacing the classic theory of probabilities by an objective theory of past probabilities, which is called theory of propensity. The propensities or tendencies are not mere possibilities, but physical realities. The current methods of inspection based on risk are some of the guidelines of the theory of propensities, combining them with historical frequencies of occurrence. In general terms, there are no questions found in engineering texts, in the sense of Popper in order to use the subjective probabilities, whose use trends to grow in the normative context. Examples of such a normative context can be found in (American Petroleum Institute) or ASME (American Society of Mechanical Engineers), as well as many other sectors.

The technical risk, as objective as it has been possible to calculate it, is not the real risk', but it is a measure of the perception of risk. As more information is obtained (tests results, observations, etc.), the perceived risk can be modified closer to real risk which is unknown. Having analyzed the risks, they should be compared with an acceptable risk value to be in conditions to make decisions. A concept very used is the As Low As Reasonably Practical (ALARP) risk, which considers that risk can be minimized only till a certain level under current technology and with a reasonable cost. Risks are "acceptable", only if it has taken all the reasonable measures and practices to reduce it (Banegas et al., 2000).

In this sense, and considering the use of predictive analytics, it is plausible to increase knowledge, expanding the sample and obtaining thus model results that approach best to real behavior. In consequence, the uncertainty and risk degree decreases, although it is possible to increase the pollution of errors due to other reasons supervening by the use of technology (Rojas et al., 2014). The advancements in knowledge, methods and techniques, the increase in information sharing, data availability and computational capabilities, and the advancements in knowledge that these can bring, offer new opportunities of development for the analysis and assessment of risks. An evolution of risk assessment is in the making, or perhaps even a "revolution" that takes the form of new approaches and methods for risk assessment (Zio, 2018)

## CONSIDERATIONS AND FUTURE RESEARCH DIRECTIONS

Sections above have observed that inductive inferences (those from particular statements) cannot establish general theories with universal validity. The justification of inductive inference is based on the fact that consideration of a hypothesis with good results in the past, infers good results in the future (Montañez, 2017).

These induction problems led the philosopher Karl Popper to propose falsification instead of induction, as the reasoning way in which science should be based. However, Popper was aware that, although a singular experimental result should not be used to assert with certainty any general statement, singular observation itself can serve to demonstrate rigorously that the statement is false. The conception of science falsificationism, in its more sophisticated version, is based on the premise (Popper, 1959) that scientific laws should be falsifiable, that is to say, there must be some possible experimental fact to be incompatible with them.

In other words, according to Popper, for a statement to be scientific there must be some hypothetical experimental result that refutes it. If after numerous and elaborate attempts, it is not possible to disprove a concrete theory, then it has to be accepted, but always provisionally. In this manner, theories are not derived from experience, but they are created by the human intellect in order to be later subjected to rigorous and relentless tests by observation and experimentation.

Theories that do not past experimental test have to be removed and replaced by others that have to be tested as well (Banegas et al., 2000). These new theories should be more falsifiable than those ones rejected. A new theory proposal is considered as worthy if it is more falsifiable than its rival and, in particular, if it predicts new type of phenomena that its rival did not mention. In this way, according to Popper, science progresses in the sense that only the most apt theories survive. Of course, it cannot be stated that a surviving theory is rigorously true, but can be stated that it is the best available.

Future researches on this topic may deepen in the matter related to Bayesian epistemology in a highly technologized post-Popperian world, as well as the analysis of standardization as a solution for risk control in high-tech contexts. In addition to this, it would be interesting to continue deploying conceptual aspects in order to show their practical contribution in the context of technological and scientific risk assessment. Additionaly, transversal discussions may be focused on the Uncertainty Principle applied to random processes, its relationship with the Second Law of Thermodynamics and the irreversibility of the processes, as well as the claim to make decision-making as a deterministic action as possible.

## CONCLUSION

In this chapter, scientific reasoning has been analyzed from both Popperian perspective of falsifiability and Bayesian conception of probability. It has been placed intentionally much emphasis on the difference between deductive and inductive methods respectively supported respectively by the Popperian and Bayesian approaches. Indeed, there can be found redundancies in the expression of the same ideas on the meaning and scope of these approaches in order to leave them clear. It happens mainly in those sections related to the deductive and inductive scientific methods, as well as considering future research directions.

In any case, it is important to underline that the research works by Kuhn, 1962, and Feyerabend, 1975, have shown that, not only scientists in the past have achieved progress ignoring falsifications, but the progress of science itself would be impossible if scientists abandon a theory before the first test experimental with a contrary result (Rojas et al., 2014). It has been observed how from Falsifiability, it is possible to reject hypotheses in case of evidence that refutes them, while with Bayesian inference is possible to accept them assigning a degree of confidence to the different hypotheses (Banegas et al., 2000). These methodologies allow distinguishing science from pseudoscience (Montañez, 2017).

Finally, whereas Popper does deal with technological risks (while this philosopher was coetaneous to the technological evolution of the 20th century), Bayes' theorem serves to observe the management of risks. Induction is intended to perform a mental process from observed evidence, to universal generality, assuming the validity of a hypothesis everywhere, everytime, and in every circumstance. Due to the lack of precision on this assumption, the combination of Bayes and new technologies offers a very powerful tool. That is due to the fact that new technologies, by the possibility o gathering and dealing big data, better limited and monitored events can be defined, reaching more precise conclusions.

## ACKNOWLEDGMENT

Authors want to acknowledge the support provided by the research projects: FFI2017-89639-P and RTI2018-097709-B-I00.

## REFERENCES

AckerR. (2020). https://www.infosecurity-magazine.com/blogs/big-data-big-risks/

Armstrong, J. S. (Ed.). (2001). *Principles of forecasting: a handbook for researchers and practitioners* (Vol. 30). Springer. doi:10.1007/978-0-306-47630-3

Banegas, J. R., & Artalejo, R. (2000). Popper and the Problem of Induction in Epidemiology. *Revista Espanola de Salud Publica*, *74*(4). Advance online publication. doi:10.1590/S1135-57272000000400003 PMID:11031841

Coates, V., Farooque, M., Klavans, R., Lapid, K., Linstone, H. A., Pistorius, C., & Porter, A. L. (2001). On the future of technological forecasting. *Technological Forecasting and Social Change*, *67*(1), 1–17. doi:10.1016/S0040-1625(00)00122-0

Feyerabend, P. (1975). *Against Method*. Verso.

Flage, R., & Aven, T. (2015). Emerging risk – conceptual definition and a relation to black swan type of events. *Reliability Engineering & System Safety*, *144*, 61–67. doi:10.1016/j.ress.2015.07.008

Greenland, S. (1988). Probability versus Popper: an elaboration of the insufficiency of current popperian approaches for epidemiologic analysis. Chesnut Hill, MA: ERI.

Howson, C., & Urbach, P. (1991). Bayesian reasoning in science. *Nature*, *350*(6317), 371–374. doi:10.1038/350371a0

ISO. (2014). International Organization for Standardization ISO 55000: Asset management - Principles and guidelines.

ISO. (2018). International Organization for Standardization ISO 31000: Risk management - Principles and guidelines.

Kuhn, T. S. (1962). *The Structure of Scientific Revolutions* (3rd ed.). Chicago: University of Chicago Press.

Rojas Lazo, O., Mavila Hinojoza, D., & Cortez Vasquez, A. (2014). Teoría de Popper y los riesgos en la ingeniería. *Industrial Data*, *16*, 58. . doi:10.15381/idata.v16i1.6390

Lee, J., Bagheri, B., & Kao, H. A. (2015). A Cyber-Physical Systems architecture for Industry 4.0-based manufacturing systems. *Manufacturing Letters*, *3*, 18–23. doi:10.1016/j.mfglet.2014.12.001

Lee, J., Holgado, M., Kao, H., & Macchi, M. (2014). New Thinking Paradigm for Maintenance Innovation design. *Proceedings of the 19th IFAC World Congress, IFAC Proceedings Volumes (IFAC-Papers Online)*, 24–29. 10.3182/20140824-6-ZA-1003.02519

Levary, R. R., & Han, D. (1995). Choosing a technological forecasting method. *Industrial Management (Des Plaines)*, *37*(1), 14–18.

Montañez Naz, S. (2017). Del Falsifiabilityo a las inferencias bayesianas. *Blog: Educación y divulgación científica en la Comunidad de Madrid*. http://divulgamadrid.blogspot.com/

Neyman, J. (1952). *Mathematical statistics and probability*. U.S. Department of Agriculture.

Oakes, M. (1990). Statistical Inference. Chesnut Hill, MA: E.R.I.

Phillips, J. G., Heidrick, T. R., & Potter, I. J. (2007). Technology futures analysis methodologies for sustainable energy technologies. *International Journal of Innovation and Technology Management*, *4*(02), 171–190. doi:10.1142/S0219877007001041

Popper, K. R. (1959). *The Logic of Scientific Discovery*. Routledge.

Popper, K. R. (1972). *Conjeturas y refutaciones. El desarrollo del conocimiento científico*. Paidós.

Popper, K. R. (1983). *Conjeturas y refutaciones. El desarrollo del conocimiento científico*. Barcelona: Paidós, edición revisada y ampliada.

Popper, K. R. (1985). *Realismo y el objetivo de la ciencia. Postscriptum a la lógica de la investigación científica* (Vol. 1). Tecnos.

Popper, K. R. (1988). *Conocimiento objetivo. Un enfoque evolucionista*. Madrid: Tecnos, tercera edición.

Popper, K. R. (1994). *En busca de un mundo mejor*. Paidós.

Popper, K. R. (1997). *El mito del marco común. En defensa de la ciencia y la racionalidad*. Paidós.

Popper, K. R., & Miller, D. W. (1987). Why probabilistic support is not inductive. *Philosophical Transactions of the Royal Society of London*, *321*, 569–591.

Renovell, J. (2015). Estadísticas bayesianas vs. frecuentistas. *Blog: Tecnología para Negocio.* https://www.paradigmadigital.com/category/techbiz/

Rodríguez, R. A. (2002). Popper y la tecnología. *Revista de Filosofía, 27*(1), 135-159.

Roper, A. T., Cunningham, S. W., Porter, A. L., Mason, T. W., Rossini, F. A., & Banks, J. (2011). *Managing the Forecasting Project* (2nd ed.). Forecasting and Management of Technology. doi:10.1002/9781118047989

Society for Risk Analysis (SRA). (2018). *Society for Risk Analysis Fundamental Principles.* Retrieved from https://www.sra.org/resources

Susser, M. (1973). *Causal thinking in the health sciences. Concepts and strategies of epidemiology.* Oxford University Press.

Technology Futures Analysis Working Group, (2004). Technology futures analysis: toward integration of the field and new methods, *Technology Forecasting and Social Change, 71*, 287-303.

Zio, E. (2018). The future of risk assessment. *Reliability Engineering & System Safety, 177*, 176–190. doi:10.1016/j.ress.2018.04.020

## ADDITIONAL READING

Nikulin, C., Gonzalez-Prida, V., Barbera, L., & Carreño, C. (2017). Enchancing the decision-making process: Integration of OTSM-TRIZ and AHP proposal. *Dyna (Bilbao)*, 18. doi:10.6036/8120

Rivadulla, A. (1991). *Probabilidad e inferencia científica.* Anthropos.

Rothman, K. J. (1988). Causal Inference. Chesnut Hill, MA: E.R.I.

## KEY TERMS AND DEFINITIONS

**Bayesian:** Pertaining to statistical methods that regard parameters of a population as random variables having known probability distributions. It proceeds from Thomas Bayes (1702–1761), English mathematician.

**Falsifiability:** Epistemological current founded by the Austrian philosopher Karl Popper (1902-1994). For Popper, contrasting a theory means trying to refute it using a counterexample. If it is not possible to refute it, this theory is corroborated, and can be provisionally accepted, but not verified; that is, no theory is absolutely true, but at most "not refuted." This concept is one of the pillars of the scientific method.

**Inference:** Process of drawing a conclusion from reasoning.

**Popperian:** Related to Karl Popper (1902-1994) an Austrian-English philosopher and professor, considered as one of the 20th century's greatest philosophers of science. Popper is known for his rejection of the classical inductivist views on the scientific method in favor of empirical falsification.

**Risk:** Degree of probability of a hazard or chance of loss occurrence.

**Technology:** Branch of knowledge that deals with the creation and use of technical means and their interrelation with life, society, and the environment, drawing upon such subjects as industrial arts, engineering, applied science, and pure science.

**Uncertainty:** The fact of not being known precisely, nor fixed, confident, assured or clearly determined. In other words, it refers to something hesitant, unknown, unstable or likely to change.

# Chapter 8
# A Model for Assessing the Widening of the Predictive Maintenance Strategy

**María Carmen Carnero**
*University of Castilla-La Mancha, Spain*

**Francisco Javier Cárcel-Carrasco**
*Universitat Politècnica de València, Spain*

## ABSTRACT

*The essential aim of Industry 4.0 is to enable industries to be more productive, efficient, and flexible. A predictive maintenance strategy can make a positive contribution to all these things, as it uses industrial IoT technologies to monitor asset health, optimise maintenance schedules, provide real-time alerts about operational risks, and maximise uptime, and can provide digital services to customers based on data from its machines. It improves productivity, improves customer satisfaction, and therefore gives the company a competitive advantage. Nevertheless, decision making in relation to a predictive maintenance strategy is not systematised, and this may lead to some inappropriate decisions, which do not achieve the goal sought. This chapter describes a multicriteria model, designed with the analytic hierarchy process, to systematise decision making with respect to a predictive maintenance strategy.*

## INTRODUCTION

Predictive Maintenance (PM) is a maintenance strategy in which parameters correlated with the condition of physical assets are monitored periodically or continuously, to identify their state, and, if the state is not good, to provide a diagnosis of the fault and the remaining lifetime of the machine. PM is sometimes called Condition-Based Maintenance; however, PM also has a prognostic phase to its application (Shafiee, 2015)

PM is thus fundamental to Industry 4.0 as it predicts asset failure by analysing data on vibration, temperature, the state of the lubricant used, ultrasound, corrosion, etc., at times combined with produc-

DOI: 10.4018/978-1-7998-3246-1.ch008

tion data, to identify patterns and predict failures before they happen or can affect production, quality and safety. Thus, PM can make industries more productive, efficient and flexible, which are the end goals of Industry 4.0. It should also be considered that the structure and behaviour of productive systems are both increasing in complexity, and so the volume of available data is also increasing, meaning that it is possible that all the available information cannot be accurately analysed by those making the diagnosis before the asset fails. It is thus useful to find alternatives to handle these quantities of data efficiently, and perform prognostic tasks automatically (Cao, Giustozzi, Zanni-Merk, de Bertrand and Reich, 2019). Smart Predictive Maintenance has recently emerged going beyond traditional PM in three ways (Durmus, 2019):

- A network of connected assets is controlled via the Internet of Things (IoT). The IoT connects PM information to the Internet. This allows all the assets to be monitored simultaneously. Also, by combining network data from operational technologies, patterns may be found between asset failures, and machine learning used to optimise the prediction algorithms over time.
- Some maintenance tasks can be automated. When a fault is detected, a maintenance work order can be emitted, a technician assigned to it, and a ticket programmed within a Computerized Maintenance Management System (CMMS). The spare parts inventory can then be checked for the necessary replacement for the defective component in an Enterprise Resource Planning (ERP) system, and the work order included within it. If there is no spare to replace the defective component, a purchase order can be generated automatically in the ERP.
- It can be integrated with other maintenance management systems. The PM platform could be introduced into other systems, such as ERP, CMMS, or Manufacturing Execution System (MES) to automate certain specific tasks. Integrating Smart PM with other management systems can automate all the maintenance processes over time.

Nevertheless, despite the significant improvements that a PM strategy can bring to the areas of safety, quality, productivity, availability, cost, company image, the environment (reducing energy consumption and greater use of assets) (Carnero, 2005) and therefore, in competitive businesses, contributions that analyse matters related to decision making in the introduction and management of a PM strategy are very limited (Carnero, 2006; Veldman et al. 2011); in fact, the costs and benefits of PM strategies are often not explicitly defined or assessed (Tiddens et al. 2017) and so in general, industries that apply it are still in the rear guard of the theoretically advertised benefits (Bianchini, Pellegrini, & Rossi, 2019). However, the technical aspects of applying a PM strategy are widely represented in the literature as new diagnostic techniques, case studies of fault diagnosis on different parts and devices, and prognosis studies (Carnero, 2016). Introduction will therefore be inefficient and will not achieve the goals sought, nor reach the full potential of the strategy. And so, this chapter describes a multicriteria model built on an Analytic Hierarchy Process (AHP), to systematise decision making in the widening of a PM strategy.

Among the contributions to different aspects of decision making in a PM strategy is Carnero (2004), which describes an indicator system to control the early detection of anomalies during the implementation stage of a PM strategy. It thus describes a predictive control of the PM strategy. The system comprises a series of indicators organised into four categories: economic evaluation, external and internal quality of the CBM, organizational structure and evolution over time. Carnero (2005) proposes a model combining AHP and factor analysis for choosing diagnostic techniques and instrumentation in PMs, integrating lubricant and vibration analyses. Carnero (2006) describes a model built by AHP and a series of decision rules, defined following the methodology of experts on rules-based systems, to assess the viability of introducing a PM strategy. The hierarchical structure company size and the productive process as setting variables. Two goals are considered: the ability of the company to set up a PM strategy and the need for this strategy in the company. The model uses 47 criteria to choose between the alternatives: setting up a PM strategy internally, developing prior requisites for the setting up of a PM strategy, not setting up a PM strategy and outsourcing the strategy. Veldman, Klingenberg and Wortmann (2011) set out eight key postulates for managing a PM strategy based on industrial practice; importantly, companies use more diagnosis than prognosis, and create autonomous organisational units to put the PM strategy into practice. Bana e Costa, Carnero and Oliveira (2012) propose a multicriteria model built with the Measuring Attractiveness by a Categorical Based Evaluation Technique (MACBETH) approach to audit the introduction of a thermography-based PM strategy in a hospital; this audit can measure the performance of the PM strategy by subject but also by criterion, identifying deficiencies and carrying out corrective measures by following a process of continuous improvement. Tiddens, Braaksma and Tinga (2015) build a framework for introducing prognostic technologies for decision making in maintenance; matters considered include identifying the corrective parameters to be controlled for, transformation of the data into useful information for optimal management decision making, and that can assist in designing the prognostic technology route. Carnero (2016) describes a model, built with fuzzy AHP, for choosing the most suitable technology among portable vibration analysers, by considering the different alternatives available on the market. Tiddens, Tinga, Braaksma and Brouwer (2017) develop a hybrid business case to evaluate and justify the introduction of a PM strategy; they use a combination of strategic multicriteria analysis related to non-financial subjects and Monte Carlo simulations for financial elements. Rastegari (2017) develops frameworks and guidelines to support the introduction of a PM strategy in manufacturing companies, especially in car manufacturing plants; the paper assesses the factors used in evaluating the cost effectiveness of a PM strategy, and also the taking of decisions about the choices of components to be monitored, and the techniques to be applied, and how to introduce them. Tiddens, Braaksma and Tinga (2018) describe a three-stage method to choose the ideal assets to apply a PM strategy to. The first step is a criticality classification, the second is to identify showstoppers, and finally a focused feasibility study. Banks, Reichard, Hines and Brought (2008) use the 10 most important cost drivers, with the same goal. The literature review therefore shows that there are no models analysing the widening the introduction of a PM strategy, like that described here.

The rest of this chapter is structured as follows. First there is a description of the main conflicts in introducing a PM strategy; next there is a description of the life cycle of a PM strategy, followed by an introduction to AHP; then the multicriteria model for the evaluation of a PM strategy is presented, including the scene variable, decision criteria, eigenvalues, decision rules and sensitivity analysis. The paper ends with future lines of research, conclusions, acknowledgement and references.

## CONFLICTS IN THE INTRODUCTION OF PREDICTIVE MAINTENANCE STRATEGY

During introduction of PM's, many conflicts may emerge with a negative effect on the final result. Among these problems, the following are noteworthy:

- The inefficiency of a PM strategy at the design and planning and adjustment stages, when there are no reliable measurements for the condition of industrial equipment, contributes to a decline in confidence in the strategy, with a consequent decrease is resources apportioned to PM, or its withdrawal. A period of adjustment between the models and theoretical procedures of PM should be established, in its technical and management aspects, to adapt them to the specifics of the industrial premises. This stage should be characterised by unconditional support from the whole company.
- The process of acquisition of predictive data may be influenced by the productive process. Areas to be examined include: the lack of information, or information obtained under different process conditions, the influence of traditional policies on the predictive control parameters, etc. The company may attempt to keep to the policies and structures previously adhered to, at the same time as the new ones. However, the perception that should be transmitted by those introducing the PM strategy is the suppression of the traditional maintenance model, and the start of a dynamic system of maintenance policies which depend directly on the predictive policy.
- When a variety of predictive techniques is applied, it is unusual to integrate the information to achieve better results in the PM (Mobley, 2001). This lack of integration may be motivated by the operation of each predictive technique by a different department of the company, and so with different personnel, who do not exchange information with each other. It may also be due to difficulties of availability of computer programmes to integrate different predictive techniques, and which do not work exclusively as databases.
- Sizing of the project as a function of the human and technical resources available to the company. This generally leads to the restriction of PM to one type of machinery or sector of the plant. The overall benefits of PM can be seen by extending it to the whole plant.
- Simple, and especially, complex mathematical models require the gathering and analysis of historical data over the period in which PM is introduced. The industrial plant can act well in advance of a catastrophic failure via a programme of Preventive Maintenance. However, PM requires historical data on the breaking points that bound evolution towards a new condition of the equipment, that is, unsatisfactory and unacceptably serious states.
- PM may lack the computing tools or efficient automatic means for gathering and handling predictive information. The potential of a PM strategy can be increased if computing tools at the technological level of the relative predictive technique are used. However, it is common to use lower-quality computing tools due to lack of confidence in the Programme, or in order to take advantage of those tools used to monitor the traditional maintenance policies.
- PM is considered to be a way of eliminating flaws within the company or caused by other maintenance policies (Carnero et al. 2000a). The traditional flaws of Maintenance Departments may arise due to poverty of information, handling, document paths, procedures, etc., and so the likelihood that these defects will be passed on to the PM programme is significant.
- Deficiencies in the passing on of information to the CMMS and ERP (Carnero et al., 2000b). In some cases, these deficiencies are due to problems in the choice of the CMMS and ERP, which do not have suitable modules for handling this kind of information.
- Problems in quantifying the benefits of a PM strategy. The limitations described by Shafer et al. (2000), to do with evaluating investment in information technologies, can be extended to technological maintenance programmes: the impossibility of quantifying inputs and outputs, differences between the costs covered and the benefits achieved, the redistribution and diffusion of benefits,

and problems with handling. Among the benefits of complex quantification in a PM strategy are: an increase in plant safety, improvements in quality, the ability to programme maintenance activities and optimisation of other maintenance policies. Therefore, these aspects are not taken into account when the assessing the return to investment.

- One of the benefits of applying PM is that it reduces maintenance costs. However, when the activity normally carried out under other maintenance policies is not altered, and a PM strategy is used, the costs of the Maintenance department increase.
- Inefficiency in the use of the predictive diagnosis technique. The simplicity of these predictive diagnosis methods makes it easier to apply PM, reduces the complexity of the training and data required, and the time given over to its application. A sample of this concept is the general use of trend analysis as opposed to spectral analysis. The use of complex analytic methods leads to a much greater number of benefits than simple analyses. On the other hand, the material and human resources required are higher in the former case.
- Accessibility of the procedures applied by those who set up the PM to those who must maintain the strategy. The success of a predictive strategy contributes to the agreement of Management, which recognises the work of the person in charge of set-up. If the appropriate document path is not used, and Programme management has been defined, the benefits of PM can determine whether the person who introduces it is retained. Application of a PM is not an individual act and it is not limited in time; maximum benefits occur when documents and staff in the organisation are updated.
- Evaluation of the cost of changing the traditional system. A breakdown was carried out of the costs of including the innovative technological project, which involves using a multiparametric PM strategy, for investment, introduction mand maintenance costs of the Programme. Applying dynamic maintenance models means higher maintenance costs than static models, with a similar level of investment and introduction costs. As described by Conde et al. (1999), applying a PM strategy leads to a decrease in maintenance costs, although inexperience at the introductory stage may lead to increased failure of risk. This may be compensated for by an improvement programme.
- PM optimisation can be achieved if combined with a proactive or TPM philosophy, which includes training for continuous improvement. Among the things to be considered are the following, which are related to standard ISO 9000, on introduction of a system of quality assurance.
    - Incorporating a process of continuous improvement via a PM control system, audits, and a system of corrective actions.
    - Looking at the data as essential, not only the staff involved in the PM strategy, but also the direct customers: Production, Quality and Safety Departments, and Management.
    - Producing a manual for the PM strategy that acts as a document backup, bringing together the what, who, how, when, how much and where of all the activities of the PM strategy.

## THE LIFE CYCLE OF A PREDICTIVE MAINTENANCE STRATEGY

Carnero et al. (1998) defines the following stages in the introduction of a PM strategy:

- *Design and planning stage.* This is the stage from identification of the need to change the maintenance policy until the human and technical resources are available for its introduction. This stage is considered to last 8 months.
- *Adjustment stage.* This lasts until reliable predictive data can be obtained. This stage may last 9 months.
- *Widening and globalising stage.* Increasing the machinery and goals of the PMP, once the period of return to investment is over. This stage can last 22 months. The goals established for the PM strategy can thus be achieved some three years from the start of introduction.
- *Integration stage*, which determines the connection of the PM strategy to the management systems of the plant.

Although a timespan has been given for some of the above stages, the real timespan depends on the predictive technique to be applied, its technological level, the human and technical resources assigned to the project, the characteristics of the company, the support of consultants, etc.

Different technological levels can be identified within the specific characteristics of the introduction, with regard to cost and diagnostic capacity (Carnero, 2005):

- *Level 0.* Introduction performed by control of sensitive variables. The cost is practically nil, and the diagnostic capacity is very low.
- *Level 1.* This means the use of basic instruments, such as vibrometers or devices for performing the crackle test.
- *Level 2.* More sophisticated instruments are available, such as vibration analysers, data handling software, or viscometer
- *Level 3.* The cost is high, since the analytic equipment is sophisticated; the diagnostic capacity is excellent.

As many widening stages may be applied as are required by the industrial plant, and the integration stage may be entered at any one of them, or a single integration stage may be applied after all the necessary widening stages for the PM strategy.

In addition to the prior introduction stages, there may be a decline stage which marks the end of the life cycle of the PM programme.

The design and planning stage is characterised by investment aimed at research into predictive techniques, diagnostic tools and making improvements in the plant to assist with introduction of the programme. The adjustment stage is when all possible predictive data is collected and the PM programme is prepared for the globalisation, when the results of the strategy should have stabilised, and the set goals been achieved, and the PM strategy is widened to the whole plant. The integration stage aims to quantify the benefits of the PM strategy in each department of the company. This last stage is the mature stage. Nevertheless, the obsolescence stage characterises the technological systems involved as a feedback process, such that, instead of eliminating the PM programme, there should be a re-engineering of the company's conditions and the suitability of the predictive techniques and the technological level to the conditions existing at any time.

Thus, the PM programme should be periodically re-engineered, assessing the need to maintain or close the programme; this latter case would mean ending its life cycle. On the other hand, introducing the

PM strategy means analysing whether changes have taken place in the plant which require the predictive techniques and their technological level to be modified.

If there is a positive result for a programme similar to that from the original application, the PM programme can be considered not to enter a decline phase, in a similar way to products that earn a position as basic goods in the market. The new predictive techniques, the technological level of these techniques, and the diagnostic tools necessary to update the strategy should be identified.

The new PM programme will have a design and planning stage during which is exists alongside the old PMP. Some of the benefits of the first PM programme may be used in introducing the new one. In this way, a continuous re-engineering and updating process of the PM strategy can be defined (see Figure 1).

Given that new predictive techniques and diagnostic tools are constantly under development, the only requirement for applying the models is updating the extra predictive techniques and diagnostic tools.

*Figure 1. Behaviour of a PM strategy based on lubricant analysis in updating the technological level from 2 to 3.*

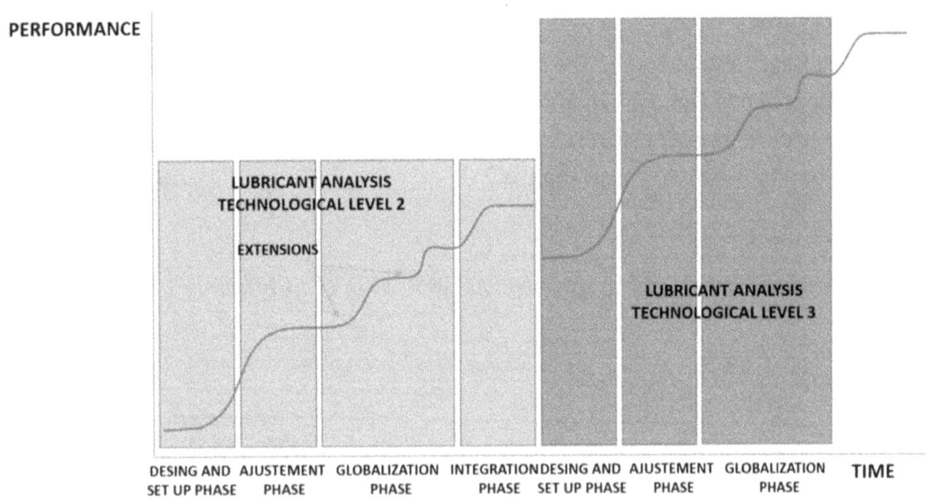


If the decision is taken not to introduce a PM strategy, and the company considers that the predictive technology applied is obsolete (decline stage), this may have arisen for the following reasons:

- A change in the needs of the plant with respect to availability, quality, safety, maintenance costs, etc.
- Withdrawal of strategic support PM strategy.
- Changes in the maintenance policies applied by the plant, with an increase in corrective as against preventive maintenance.
- Worsening of conditions in the company with respect to level of information technology, document management, human and technical resources of the Maintenance Department, etc.

Chase et al. (1978) set out the factors whose modification interacts with changes in the technology of the production process. These factors are: the environment, the social system and the exploitation and control systems. Foster (1986) give guidelines for assessing how near a technology is to maturity, which should be added to these factors:

- Improvements are infrequent.
- Changes among the people who manage the technology do not affect its efficiency.
- The productivity of the system is questioned by those who introduced it; this lowers confidence in the system.
- I+D costs rise and benefits fall.
- Innovation is decreased.
- Other competitors gain more benefits by applying unobvious and radical concepts.

In any of these situations, the efficiency of the PM strategy will decrease until the programme is cancelled.

Increasingly rapid development of new technologies also brings new products to market at a higher rate. Each new management technology brings with it a most appropriate Predictive Maintenance technology. It is thus necessary to determine the best moment for changing the technology of the Predictive Maintenance technique in use since, as mentioned in Ivancevich et al. (1997), as technology reaches the top of the curve S, it takes a greater effort to achieve improvements in a product or process. This concept can also be applied to a PM strategy (see Figure 2).

*Figure 2. Evolution of the search for improvement with respect to efficiency when one technology is replaced by another.*

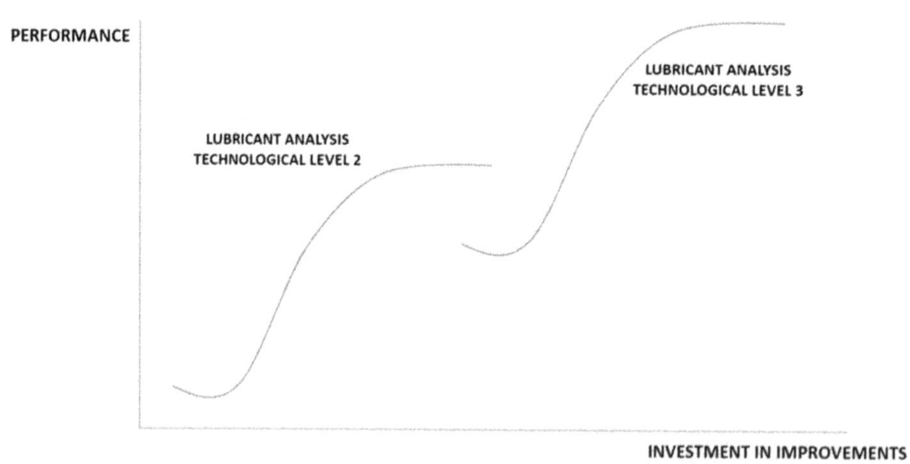


Since a PM strategy is a dynamic project, and that it must evolve, after a certain period of time the need for it should be reassessed, along with the usefulness to the company of widening the original programme. The following model is recommended for this purpose.

## ANALYTIC HIERARCHY PROCESS

AHP has been used successfully in a large number of real decision problems. The American Society for Testing and Materials (ASTM), for example, has adopted it as standard in building investment analysis, while Xerox uses AHP in its decisions on technology implementation, and British Columbia Ferries uses it in its selection processes for products, suppliers and consultants.

AHP solves decision problems handled by traditional mathematical programming, which gives a series of feasible solutions, such as those that satisfy the problem constraints. These solutions are ordered in accordance with given criteria, which include the preferences of the decision group. Mathematical programming techniques are then applied to this objective function, to give the optimal solutions. However, in real decision-making processes, the decision groups assess the possible alternatives according to a number of criteria.

AHP is based on three basic principles: pairwise comparison, decomposition and composition hierarchy or priority synthesis. Pairwise comparisons are applied to the alternatives $A_i$ with respect to the decision criteria $C_j$, and between the different criteria. These weightings or priorities come from the judgements, based on experience, data and knowledge, of the decision centre or group.

AHP facilitates complex decision making via a hierarchical structure which represents the relations between the objective, criteria, events, actors and alternatives. A variety of scenarios, uncertainties or decision centres can be included in the model.

The pairwise comparisons are used to obtain local priorities for the various levels of the hierarchy, with respect to the criteria of the immediately higher level.

The process of synthesis includes the local priorities in the priorities of the higher level, giving a complete classification for the alternatives.

The axioms of AHP are:

1. *Reciprocity axiom.* Given any two alternatives, (or subcriteria), $A_i$ and $A_j$, from a set of alternatives A, the decision centre should be able to provide a comparison matrix, comprising elements $a_{ij}$, of these alternatives relative to any criterion $C_j$ from the set of criteria C, in a scale of proportions, confirming reciprocity, such that $a_{ij} = \frac{1}{a_{ji}}, \forall i, j \in A$ (Smith, Bush, & Schmoldt, 1995).
2. *Homogeneity axiom.* The elements subject to pairwise comparison should not be very different. Otherwise there are likely to be elementary errors in the decisions (Forman et al., 1996).
3. *Axiom 3.* The decisions or priorities of the elements of the hierarchy do not depend on the elements of the lower levels.
4. *Axiom 4.* The decision centres should make sure that their deliberations are appropriately represented, so that the results obtained are similar to expectation.

Traditionally, the judgements are given from the lower levels of the hierarchy to the higher, unless it is certain that there is no feedback.

The hierarchies defined by Saaty may be complete or incomplete. In the former case, the criteria analysed are similar for each scenario, for the main objective, or the subcriteria used in the hierarchy are similar with respect to each criterion (Forman & Selly, 2001). In the incomplete hierarchy, isolated subhierarchies are provided for each group of users, but these are integrated into a global hierarchy, by introducing forces related to the overall objective, meaning that each scenario (e.g. each group of users) can include specific objectives.

The relative importance of the alternatives with respect to each decision criterion should be determined in AHP. This is done using language items that include the judgements of the decision group. The language items are generally associated with numerical values to create a scale (Triantaphyllou, Lootsma, Pardalos, & Mann, 1994). The decision group thereby has a discrete set of numerical values to determine the relative importance of criteria and subcriteria. The scale defined by Saaty is shown in Table 1. These values are used because elements with very different characteristics cannot be compared, since it would lead to resolution errors, which would be against the axiom of homogeneity.

*Table 1. Relative importance scale (Saaty, 1980).*

| Intensity of importance | Verbal scale | Meaning |
|---|---|---|
| 1 | Equal importance | The two activities contribute equally to the objectives |
| 3 | One alternative weakly more important than the other | Experience and judgements are slightly favourable to one activity rather than the other |
| 5 | Intermediate importance | Experience and judgements are strongly favourable to one activity rather than the other |
| 7 | Strong importance | One activity is strongly preferred to the other and its dominance has been shown in practice |
| 9 | Absolute importance | The evidence favours one activity rather than the other at the highest level of expression |
| 2, 4, 6, 8 | Intermediate values between adjacent judgements | When compromise values are required |
| Inverse values of the above terms | If activity i has any of the above values (other than zero) when compared with activity j, then j has the inverse value when compared to i. | This is a reasonable assumption when the order in which the comparisons between criteria and alternatives are performed is reversed. |

The comparison matrix of the elements at each level is carried out depending on one or more of the following (Peterson, Silsbee, & Schmoldt, 1994):

- *Importance*: when the objectives are compared with respect to their relative importance.
- *Preference*: when the preference of the alternatives is compared with respect to an objective.
- *Probability*: when uncertainties or scenarios are compared with respect to the probability of their occurrence.

When considering a decision problem with $n$ assessment criteria and $m$ alternatives, applying AHP requires building an $nxn$ comparison matrix between criteria of the same level, and $n$ $mxm$ assessment matrices for the alternatives with respect to each criterion.

The weightings of the different criteria at the same level are denoted by $w_1, w_2, \ldots, w_n$. The pairwise comparison between the criteria $C_i$ and $C_j$ gives the value $a_{ij}$, and determining all the pairwise comparisons between the criteria gives the comparison matrix shown in Equation 1.

$$A = \begin{bmatrix} a_{11}\ a_{12} \cdots a_{1n} \\ a_{21}\ a_{22} \cdots a_{2n} \\ \cdots\ \cdots\ \cdots\ \cdots \\ a_{n1}\ a_{n2} \cdots a_{nn} \end{bmatrix} \tag{1}$$

The elements of matrix $A$ satisfy:

$$a_{ij} = \frac{1}{a_{ij}} \tag{2}$$

If the judgements given in the comparison process between all the criteria are consistent, that is, the decisions made produce coherent judgements, the matrix $A$ is said to be consistent. That is:

$$a_{ik} = a_{ij} * a_{jk} \forall i, j, k \tag{3}$$

When exact measures for the criteria on a scale for making the comparisons are known, a perfectly consistent matrix is obtained, satisfying (Saaty 1980):

$$a_{ij} = \frac{w_i}{w_j} i, j = 1, 2, \ldots, n \tag{4}$$

From a consitent matrix $A$ a normalised matrix $N$ can be obtained, with all columns identical (Taha, 1998).

Equation (4) leads to:

$$a_{ij} \frac{w_j}{w_i} = 1\ i, j = 1, 2, \ldots, n \tag{5}$$

and so:

$$\sum_{j=1}^{n} a_{ij} \frac{w_j}{w_i} = 1\ \ i = 1, 2, \ldots, n \tag{6}$$

or:

$$\sum_{j=1}^{n} a_{ij} w_j = n w_i\ \ i = 1, 2, \ldots, n \tag{7}$$

which is, in matrix form (Saaty, 1980):

$$Aw = nw \tag{8}$$

where $w$ is an eigenvector of $A$ with eigenvalue $n$. That is, since the comparison matrix $A$ is of rank 1, all the eigenvalues of $A$ are zero, except one, with value $n$. The sum of the eigenvalues of a positive matrix is equal to the trace of the matrix, and the non-zero eigenvalue is called the main eigenvalue or ${}_{max}$,

If matrix A is not consistent, suppose ${}_{1},\ldots,{}_{n}$ to be the set of eigenvalues which gives the solution to the previously expressed matrix. The following expression is satisfied:

$$If\ a_{ij} = 1, \forall i \Rightarrow \sum_{i=1}^{n} {}_{i} = n \tag{9}$$

In this case, the relative weight $w_i$, is approximated by the average of the $n$ elements of row i of the normalised matrix $N$.

If $\bar{w}$ is the estimate calculated by the procedure described by Saaty (1980), which consists of writing:

$$\alpha = \sum_{i=1}^{n} w_i \tag{10}$$

and substituting $\bar{w}$ for:

$$\frac{1}{\alpha} w \tag{11}$$

then, from Taha (1998):

$$A\bar{w} = {}_{max}\ \bar{w} \tag{12}$$

where ${}_{max} \geq n$.

The nearer ${}_{max}$ is to $n$, the more consistent the comparison matrix $A$, or the more coherent the judgements will be. Therefore, a factor called the Consistency Index ($CI$) is used as a measure of the consistency of the opinions (Saaty, 1980):

$$CI = \frac{{}_{max} - n}{n-1} \tag{13}$$

The Consistency Ratio ( $CR$ ) is defined as the relation between the Consistency Index and the Random Consistency Index ( $ICR$ ) for a matrix of the same size. This was evaluated by Saaty by generating random comparisons for matrices of sifferent sizes, (Forman & Selly, 2001; Triantaphyllou, & Mann, 1995).

$$CR = \frac{CI}{RCI} \tag{14}$$

The values of $RCI$ are shown in Table 2; these values were obtained from a population of 500 matrices, with $n$ from 1 to 15.

*Table 2. Values of the Random Consistency Index*

| $n$ | 1 | 2 | 3 | 4 | 5 | 6 | 7 | 8 | 9 | 10 | 11 | 12 | 13 | 14 | 15 |
|---|---|---|---|---|---|---|---|---|---|---|---|---|---|---|---|
| ICR | 0 | 0 | 0.58 | 0.90 | 1.12 | 1.24 | 1.32 | 1.41 | 1.45 | 1.49 | 1.51 | 1.48 | 1.56 | 1.57 | 1.59 |

Perfect consistency rarely occurs, and so the judgements obtained from a matrix are considered consistent if $CR$ is less than 10%. A value of $CR$ above 0.10 shows a need to reevaluate the comparisons made, although as mentioned in Forman and Selly, (2001), it is more important to be precise than consistent, which means it may be possible to accept insonsistent judgements.

## MODEL FOR ASSESSING THE WIDENING OF A PREDICTIVE MAINTENANCE STRATEGY

### Decision Criteria

The use of the setting variable Type of Introduction of Predictive Maintenance Programme is recommended (TI). This variable shows whether the PM strategy was originally outsourced or was developed in-house.

The decision criteria considered by the model are structured as follows: Overall state of the PMP, number of critical devices to be included, and ability to put the widening into practice.

Prior to widening the PM strategy, the results already obtained with it should be assessed. The following criteria should be examined:

- *Technical results of introducing a Predictive Maintenance strategy (RES_TEC).* This criterion assesses the diagnostic and analytic capacity shown during the introduction and maintenance of the PMP. This criterion is quantified via the control variable Predictive efficiency (Carnero, 2004), where the possible values are:
- High (A).
- Medium (M).

- Low (B).
- *Organisational results of introducing a Predictive Maintenance strategy (RES_ORG).* The global working of the PM strategy, not related to technical aspects, is examined. The indicator Predictive Audit (Carnero, 2004) gives the value that can be used in the assessed of this widening. This criterion is quantified at three levels:
- High (A).
- Medium (M).
- Low (B).

An unacceptable result in these variables means carrying out improvement actions before widening the predictive programme. It thus means a restriction of this widening. Carrying out improvements can go along with checking the figures given for the previous variables.

As well as these decision criteria, it is considered appropriate to include:

- *Number of additional critical machines at the plant (NUM_CRA).* This variable counts the set of machines still existing in the plant, not previously included in the Predictive Maintenance Programme. This criterion considers the following values:
- Low level (B). When the number of industrial machines is less than 10. This may lead to non-critical machinery being included in the PM strategy, generating additional costs to the Maintenance Department
- Medium level (M). Up to 20 machines.
- High level (A). Over 20 machines.

Finally, the capacity for widening the PM strategy should be assessed. The criteria recommended are these:

- *Acquiring new instruments or using under-used resources (AI).* Consider the values Acceptable (A), and Low (B), meaning suitable inclusion, or lack of additional technical resources in the PM strategy.
- *Acquiring new human resources or using under-used resources (AP).* Consider the values Acceptable (A), and Low (B), meaning suitable inclusion, or lack of additional human resources in the PM strategy.

The alternatives considered are:

- Widening a PM strategy (ALT 1)
- Performing a series of requirements prior to the widening of the PM strategy (ALT 2)
- Not widening the PM strategy (ALT 3).
- Outsourcing the widening of the PM strategy (ALT 4).

Since ALT 2 is not a competitive alternative with respect to ALT 1 and ALT 3, as it is only relevant to the state variables of the PM programme, this alternative is assessed using decision rules. ALT 4 is considered in those cases where the PM programme was originally outsourced through the setting variable TI. If the PM programme is introduced internally, the alternative of outsourcing the widening is not

*Figure 3. Hierarchy of the Assessment model for Widening a PM strategy*

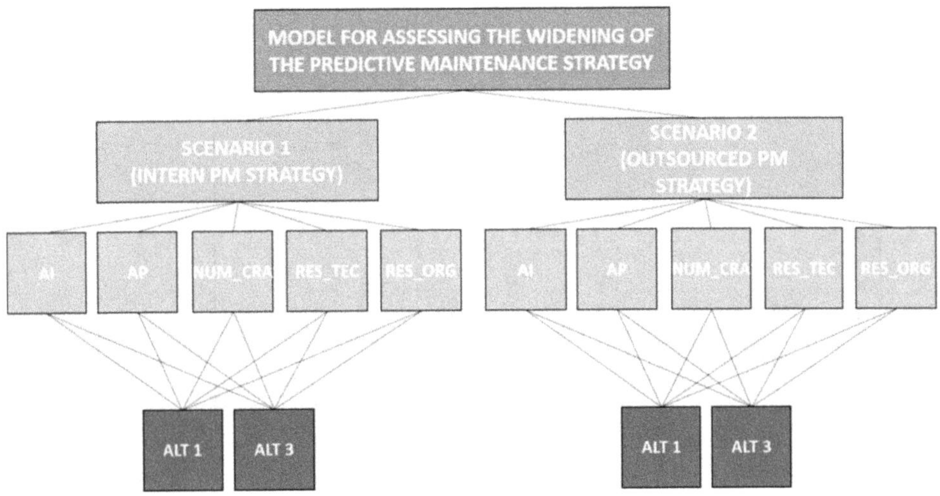


considered, as the outsourced company will use its own procedures and diagnostic techniques and so the results cannot be compared or controlled in the same way that is possible for an in-house PM strategy.

## Eigenvalues

Table 1 shows an example of eigenvalues corresponding to the results of the assessment. A distributive AHP model was used, although it only takes the model chosen in first place. The eigenvalue is used as a probability of the success of the widening. Looking at the different levels of each variable, some variations with possible repetitions are now shown, and each level of the variables is shown by the labels A (High), M (Medium) and B (Low). The eigenvalues shown in Table 3 correspond to an internally introduced PM programme (TI=0), as it will be widened to an outsourced PM strategy by using the decision rules.

## Decision Rules

The following decision rules provide the tools to guarantee the success of the widening.

1. If RES_TEC=A and RES_ORG=B and NUM_CRA=A and AI=A and AP=A then ALT 2.
2. If RES_TEC=B and RES_ORG=A and NUM_CRA=A and AI=B and AP=A then ALT 2.
3. If RES_TEC=B and RES_ORG=A and NUM_CRA=A and AI=A and AP=B then ALT 2.
4. If RES_TEC=A and RES_ORG=B and NUM_CRA=B and AI=A and AP=A then ALT 2.
5. If RES_TEC=A and RES_ORG=B NUM_CRA=A and AI=B and AP=A then ALT 2.
6. If RES_TEC=A and RES_ORG=B NUM_CRA=A and AI=A and AP=B then ALT 2.
7. If RES_TEC=A and RES_ORG=B and NUM_CRA=A and AI=B and AP=B then ALT 2.
8. If RES_TEC=B and RES_ORG=A and NUM_CRA=A and AI=A and AP=A then ALT 2.
9. If RES_TEC=A and RES_ORG=B and NUM_CRA=A and AI=A and AP=A then ALT 2.
10. If RES_TEC=B and RES_ORG=B and NUM_CRA=A and AI=A and AP=A then ALT 2.
11. If RES_TEC=M and RES_ORG=B and NUM_CRA=A and AI=B and AP=A then ALT 2.

*Table 3. Examples of applying the model to different values of the decision criteria*

| Criteria | Value of Variables | Alternative Chosen | Eigenvalue | Index of Overall Inconsistency |
|---|---|---|---|---|
| RES_TEC<br>RES_ORG NUM_<br>CRA AI<br>AP | A<br>A<br>A<br>A<br>A | ALT 1 | 0.900 | 0.0 |
| RES_TEC<br>RES_ORG NUM_<br>CRA AI<br>AP | A<br>A<br>A<br>A<br>B | ALT 1 | 0.774 | 0.0 |
| RES_TEC<br>RES_ORG NUM_<br>CRA AI<br>AP | A<br>A<br>B<br>A<br>A | ALT 1 | 0.662 | 0.0 |
| RES_TEC<br>RES_ORG NUM_<br>CRA AI<br>AP | A<br>B<br>B<br>A<br>A | ALT 1 | 0.591 | 0.0 |
| RES_TEC<br>RES_ORG NUM_<br>CRA AI<br>AP | A<br>M<br>A<br>B<br>A | ALT 1 | 0.760 | 0.0 |
| RES_TEC<br>RES_ORG NUM_<br>CRA AI<br>AP | M<br>A<br>B<br>B<br>B | ALT 3 | 0.635 | 0.0 |
| RES_TEC<br>RES_ORG NUM_<br>CRA AI<br>AP | B<br>M<br>M<br>A<br>B | ALT 3 | 0.502 | 0.0 |
| RES_TEC<br>RES_ORG NUM_<br>CRA AI<br>AP | M<br>B<br>A<br>B<br>B | ALT 3 | 0.370 | 0.0 |
| RES_TEC<br>RES_ORG NUM_<br>CRA AI<br>AP | M<br>M<br>B<br>B<br>B | ALT 3 | 0.649 | 0.0 |
| RES_TEC<br>RES_ORG NUM_<br>CRA AI<br>AP | A<br>B<br>B<br>B<br>B | ALT 3 | 0.662 | 0.0 |

12. If RES_TEC=M and RES_ORG=B and NUM_CRA=A and AI=A and AP=B then ALT 2.
13. If RES_TEC=M and RES_ORG=B and NUM_CRA=A and AI=B and AP=B then ALT 2.
14. If RES_TEC=M and RES_ORG=A and NUM_CRA=A and AI=B and AP=B then ALT 2.
15. If RES_TEC=M and RES_ORG=B and NUM_CRA=A and AI=A and AP=A then ALT 2.
16. If RES_TEC=B and RES_ORG=M and NUM_CRA=M and AI=A AP=A then ALT 2.
17. If RES_TEC=M and RES_ORG=B and NUM_CRA=M and AI=A and AP=B then ALT 2.
18. If RES_TEC=M and RES_ORG=B and NUM_CRA=M AI=B and AP=A then ALT 2.
19. If RES_TEC=M and RES_ORG=B and NUM_CRA=M and AI=B and AP=B then ALT2.
20. If RES_TEC=M and RES_ORG=B and NUM_CRA=M and AI=A and AP=A then ALT2.
21. If RES_TEC=B and RES_ORG=M and NUM_CRA=A and AI=A and AP=B then ALT 2.
22. If RES_TEC=B and RES_ORG=M and NUM_CRA=A and AI=B and AP=A then ALT 2.
23. If RES_TEC=B and RES_ORG=M and NUM_CRA=A and AI=A and AP=A then ALT 2.
24. If RES_TEC=A and RES_ORG=B and NUM_CRA=M and AI=B and AP=B then ALT2.
25. If RES_TEC=B and RES_ORG=B and NUM_CRA=M and AI=A and AP=A then ALT 2.
26. If RES_TEC=B and RES_ORG=A and NUM_CRA=M and AI=B and AP=A then ALT 2.
27. If RES_TEC=B and RES_ORG=A and NUM_CRA=M and AI=A and AP=B then ALT 2.
28. If RES_TEC=A and RES_ORG=B and NUM_CRA=M and AI=B and AP=A then ALT 2.
29. If RES_TEC=A and RES_ORG=B and NUM_CRA=M and AI=A and AP=B then ALT 2.
30. If RES_TEC=B and RES_ORG=A and NUM_CRA=M and AI=A AP=A then ALT 2.

From the previous results, denoted as WIDENING MODEL=ALT X, is evaluated at T1=1. This is done band proposing rules to include ALT 4, which will now be shown. If MEVA gives positive value to the viability of widening, the overall system advises outsourcing, if this was the system used to put PMP in practice.

31. If TI=1 and WIDENING MODEL=ALT 1 then ALT 4.
32. If TI=1 and WIDENING MODEL=ALT 2 then ALT 2.
33. If TI=1 and WIDENING MODEL=ALT 3 then ALT 3.

## Conditional Probabilities for Control of Uncertainty in Decision Making on the Widening of a Predictive Maintenance strategy

When an organisation does not have accurate data on the state of the company, the model may choose an alternative by considering the available data in the databases on other organisations, which have already applied the model and do have that information. There follows a mathematical formulation of the model.

There exist a series of alternatives $A_i$, for evaluation of a PM strategy. The number of alternatives is assumed to be M, that is, i=1, 2, ..., M. There are N decision variable $V_j$, j=1, 2, ..., N. Each of them should have a different value $s_k$, k=1, 2, ..., K.

The probability of choosing alternative $A_i$ is evaluated from the expression:

$$P\left(A_i / \left(\left[V_1(S_k)\right] \cap \left[V_2(S_k)\right] \cap \ldots \cap \left[V_N(S_k)\right]\right)\right)$$

$$\frac{P(A_i)P\left([V_1(s_k)] \cap [V_2(s_k)] \cap \ldots \cap [V_N(s_k)] / A_i\right)}{P\left([V_1(s_k)] \cap [V_2(s_k)] \cap \ldots \cap [V_N(s_k)]\right)}$$

$$\frac{P(A_i)P\left([V_1(s_k)] \cap [V_2(s_k)] \cap \ldots \cap [V_N(s_k) / A_i\right)}{P\left(V_1(s_k)\right)P\left(V_2(s_k)\right)\ldots P\left(V_N(s_k)\right)} \tag{15}$$

It is assumed that the decision variables are independent, a necessary prior step to applying AHP. In the specific case of our problem set, this expression becomes:

$$P\left(A_i / [V_1(s_k)]\right) = \frac{P(A_i)P\left(V_1(s_k) / A_i\right)}{P\left(V_1(s_k)\right)} \tag{16}$$

The previous expression yields the probability that a company will use an alternative $A_i$, when there exist certain values associated with the scenario values.

## Sensitivity Analysis

There now follows an example of a sensitivity analysis of the model described in Figure 4. First comes the analysis of the eigenvalues ALT1=0.522, ALT3=0.478. The x-axis shows the possible weightings that a criterion can have; the current weighting assigned is identified by a red vertical line. The y-axis shows the utility of the alternatives evaluated. As seen from Figure 4 a), an increase in the weighting of the variable RES_TEC, reverses the classification of the alternatives. A similar effect occurs with the variable AP, as shown in Figure 4 b). Modifying the weighting of RES_ORG, NUM_CRA and AI does not change the classification of the alternatives, because alternatives ALT1 and ALT3 do not cross for any value that could possibly be assigned to the weightings of the criteria. However, when the decision rules are included in the model, ALT 2 is recommended, and so the plant should take actions for improvement in order to reach the medium level in this variable. This combination at the initial state of the Model gives the following eigenvalues: ALT1=0.718, ALT3=0.282. With the new eigenvalues, the classification of the alternatives is stable since, as shown in Figure 4 c), there is no reversal for any weighting value assigned to the variable RES_TEC. There is also stability in the classification with the variables RES_ORG, NUM_CRA and AI, as shown in Figure 4 d). For AP, the reversal of the classification occurs for weightings so high that they are not held to be reasonable (Figure 4 e)).

The industrial plant could also have obtained, by applying ALT 2, a level of excellent in ORG_TEC, where the eigenvalues are: ALT1=0.760, ALT3=0.240. The sensitivity analysis shows that the variables RES_TEC, RES_ORG, NUM_CRA and AI do not change the classification of the alternatives when the weightings assigned to them are changed. AP causes reversal but only for weighting values so great that they are not considered acceptable in the analysis. The Model uses the concept of continuous improvement, requiring a given level of excellence before the PM strategy can be widened.

*Figure 4. Sensitivity analysis: a) Reversal of alternatives with modification in the priority of RES_TEC b) Evolution of the classification of alternatives on changing the priority of AP c) Classification of alternatives when modifying the weighting of RES_TEC d) Classification of alternatives on modifying the priorities of RES_ORG, NUM_CRA and AI (from left to right) e) Classification of the alternatives with variation of the weighting of AP.*

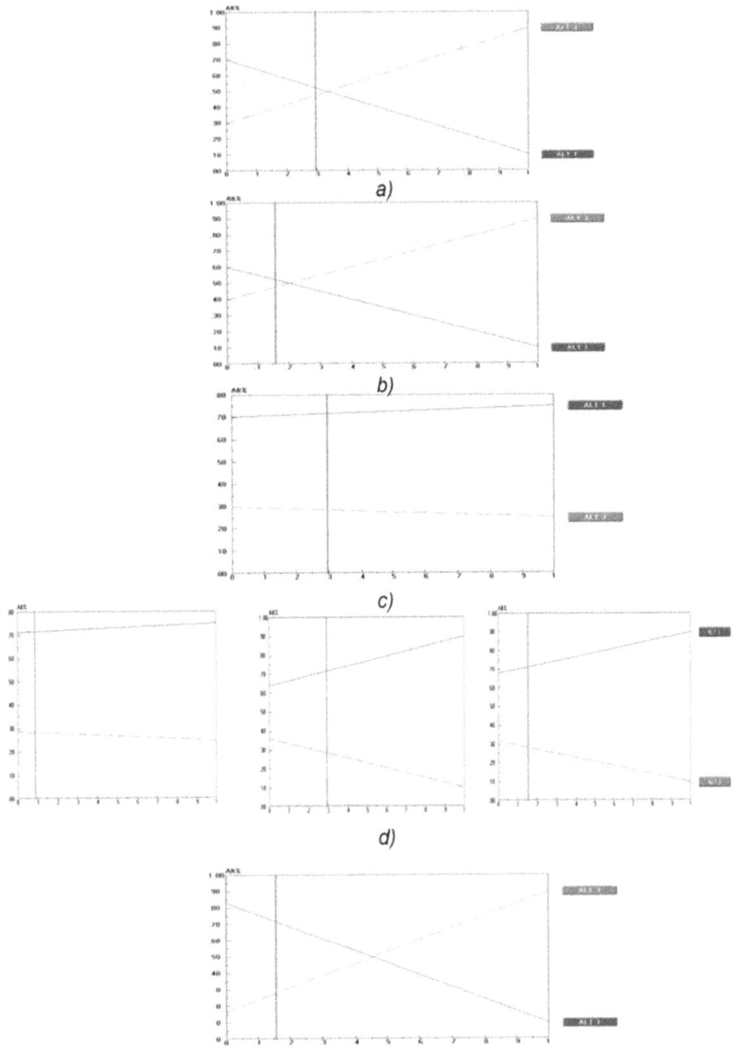


## FUTURE RESEARCH DIRECTIONS

This research could be widened by using fuzzy logic in decision making, either fuzzy AHP or another multicriteria technique such as the fuzzy Technique for Order of Preference by Similarity to Ideal Solution (TOPSIS), and comparing the results.

It would also be useful to use experts in PM strategies as a decision group, rather than a single decision maker, to obtain the judgements, and also to review the Model whether new criteria, relevant to the decision, can be included.

## CONCLUSION

A PM strategy can provide great benefits for an organisation, especially within the scope of Industry 4.0, by making organisations more efficient, and, among other things, increasing customer satisfaction due to improvements in delivery times and product quality, as well as flexibility in production processes by decreasing unproductive time, all of which gives a PM strategy the possibility of leading to competitive advantage for the organisations that use it. This all stems from automatic analysis of a large amount of data from a number of indicators of the condition of machinery. However, introducing a PM strategy is not free from problems that may prevent it from achieving the expected improvements. This Chapter is, therefore, intended to underline the areas which might lead to failure, and which are frequently not considered. These matters are especially relevant to processing industries, but also, and increasingly, in manufacturing companies and service companies such as hospitals, where PM strategies are beginning to be used.

Moreover, matters arising from the introduction and handling of this maintenance strategy are not discussed in the literature, which contributes to a lack of analysis of these problems and their possible solutions. This Chapter therefore characterises the life cycle of PM strategies and describes an assessment model for the widening of this type of programme. There are as yet no decision-making models for this issue, even though it is a common practice to use a PM strategy as a prototype, and subsequently to widen the PM strategy to new machines in the organisation. At times, things such as the possibility of outsourcing the PM strategy are not considered. It is hoped thereby to help in decision making for these strategies, leading to greater success in introducing them and in the expected benefits for the organisations.

## ACKNOWLEDGMENT

This research was supported by the University of Castilla-La Mancha and the European Union through the European Regional Development Fund to the Predictive Analysis Laboratory (PREDILAB) group [2020-GRIN-28770].

## REFERENCES

Bana e Costa, C. A., Carnero, M. C., & Oliveira, M. D. (2012). A multi-criteria model for auditing a Predictive Maintenance Programme. *European Journal of Operational Research*, *217*(2), 381–393. doi:10.1016/j.ejor.2011.09.019

Banks, J. C., Reichard, K. M., Hines, J. A., & Brought, M. S. (2008). Platform degrader analysis for the design and development of vehicle health management systems. *Proceedings of the International Conference on Prognostics and Health Management*, 464. 10.1109/PHM.2008.4711468

Bianchini, A., Pellegrini, M., & Rossi, J. (2019). Maintenance scheduling optimization for industrial centrifugal pumps. *International Journal of System Assurance Engineering and Management*, *10*(4), 848–860. doi:10.100713198-019-00819-4

Cao, Q., Giustozzi, F., Zanni-Merk, C., de Bertrand de Beuvron, F., & Reich, C. (2019). Smart Condition Monitoring for Industry 4.0 Manufacturing Processes: An Ontology-Based Approach. *Cybernetics and Systems, 50*(2), 82–96. Advance online publication. doi:10.1080/01969722.2019.1565118

Carnero, C., & Conde, J. (2000a). Conflictos y utilidades en la relación Gran Empresa-Universidad en proyectos de implantación de programas de mantenimiento predictivo. *Proceedings del 1[er] Workshop of Industrial Organization.*

Carnero, C., & Conde, J. (2000b). Selección de parámetros de control y diagnóstico de fallos en la implantación de programas de Mantenimiento Predictivo. *Proceedings del IV Spanish Congress of Maintenance*, 395-402.

Carnero, C., López-Escobar, C., & Conde, J. (1998). Implantación de programas de mantenimiento predictivo por empresas de servicios técnicos en colaboración. Evaluación mediante indicadores. *Proceeding of VII International Congresso of AEDEM*, 522-527.

Carnero, M. C. (2004). The control of the setting up of a predictive maintenance programme using a system of indicators. *Omega, 32*(1), 57–75. doi:10.1016/j.omega.2003.09.009

Carnero, M. C. (2005). Selection of diagnostic techniques and instrumentation in a predictive maintenance program. A case study. *Decision Support Systems, 38*(4), 539–555. doi:10.1016/j.dss.2003.09.003

Carnero, M. C. (2006). An evaluation system of the setting up of predictive maintenance programmes. *Reliability Engineering & System Safety, 91*(8), 945–963. doi:10.1016/j.ress.2005.09.003

Carnero, M. C. (2016). Fuzzy Multicriteria Model for Selection of Vibration Technology. *Shock and Vibration, 2016*. . doi:10.1155/2016/6723605

Chase, R. B., & Aquilano, N. J. (1978). *Gestión de la producción y dirección de operaciones.* Hispano Europea.

Conde, J., Conde, L. J., & Conde, M. (1999). Maintenance projects evaluation using Fully Bayesian Approach and Logistic Distribution. *Proceedings of International Conference on Safety and Reliability*, 755-759.

Durmus, M. (2019). *Smart Predictive Maintenance: The Key to Industry 4.0.* Retrieved from https://www.aisoma.de/smart-predictive-maintenance-the-key-to-industry-4-0/

Forman, E., & Selly, M. A. (2001). *Decision by Objectives.* World Scientific. doi:10.1142/4281

Ivancevich, J. M., Lorenzi, P., Skinner, S. J., & Crosby, P. B. (1997). *Gestión calidad y competitividad.* McGraw-Hill.

Mobley, R. K. (2002). *An introduction to Predictive Maintenance.* Butterworth-Heinemann.

Peterson, D. L., Silsbee, D. G., & Schmoldt, D. L. (1994). A Case Study of Resources Management Planning with Multiple Objectives and Projects. *Environmental Management, 18*(5), 729–742. doi:10.1007/BF02394636

Rastegari, A. (2017). *Condition Based Maintenance in the manufacturing industry from strategy to implementation* (Thesis). University of New South Wales, Stockholm, Sweden.

Saaty, T. L. (1980). *The Analytic Hierarchy Process*. McGraw Hill.

Shafer, S. M., & Byrd, T. A. (2000). A framework for measuring the effiency of organizational investments in information technology using data envelopment analysis. *Omega*, *28*(2), 125–141. doi:10.1016/S0305-0483(99)00039-0

Shafiee, M. (2015). Maintenance strategy selection problem: An MCDM overview. *Journal of Quality in Maintenance Engineering*, *21*(4), 378–402. doi:10.1108/JQME-09-2013-0063

Smith, R. L., Bush, R. J., & Schmoldt, D. L. (1995). *A Hierarchical Analysis of Bridge Decision Makers*. USDA Forest Service, Timber Bridge Information Resource Center, in cooperation with the Center for Forest Products Marketing, Department of Wood Science and Forest Products, at Virginia Polytechnic Institute and State University.

Taha, H. A. (1998). *Investigación de Operaciones. Una introducción*. Pearson.

Tiddens, W. W., Braaksma, A.J.J., & Tinga, T. (2018). Selecting Suitable Candidates for Predictive Maintenance. International Journal of Prognostics and Health Management, 9(1).

Tiddens, W. W., Braaksma, A. J. J., & Tinga, T. (2015). The Adoption of Prognostic Technologies in Maintenance Decision Making: A Multiple Case Study. *Procedia CIRP*, *38*, 171-176.

Tiddens, W. W., Tinga, T., Braaksma, A. J. J., & Brouwer, O. (2017). The business case for condition-based maintenance: A hybrid (Non-) financial approach. *Proceedings of the 27th European Safety and Reliability Conference*, 821-829.

Triantaphyllou, E., Lootsma, F. A., Pardalos, P. M., & Mann, S. H. (1994). On the evaluation and application of different scales for quantifying pairwise comparisons in Fuzzy Sets. *Journal of Multi-Criteria Decision Analysis*, *3*(3), 133–155. doi:10.1002/mcda.4020030302

Triantaphyllou, E., & Mann, S. H. (1995). Using the analytic hierarchy process for decision making In engineering applications: Some challenges. *International Journal of Industrial Engineering: Applications and Practice*, *2*(1), 35–44.

Veldman, J., Klingenberg, W., & Wortmann, H. (2011). Managing condition based maintenance technology: A multiple case study in the process industry. *Journal of Quality in Maintenance Engineering*, *17*(1), 40–62. doi:10.1108/13552511111116240

## ADDITIONAL READING

Al-Najjar, B. (2012). On establishing cost-effective condition-based maintenance: Exemplified for vibration-based maintenance in case companies. *Journal of Quality in Maintenance Engineering*, *18*(4), 401–416. doi:10.1108/13552511211281561

Carnero, M. C., & Hidalgo, J. C. (2003). Evaluation of condition based maintenance through activity based cost. *Maintenance Journal*, *16*, 54–61.

Peng, Y., Dong, M., & Zuo, M. J. (2010). Current status of machine prognostics in condition-based maintenance: A review. *International Journal of Advanced Manufacturing Technology*, *50*(1-4), 297–313. doi:10.100700170-009-2482-0

Roda, I., Macchi, M., & Fumagalli, L. (2018). The Future of Maintenance Within Industry 4.0: An Empirical Research in Manufacturing. In I. Moon, G. Lee, J. Park, D. Kiritsis, & G. von Cieminski (Eds.), *Advances in Production Management Systems. Smart Manufacturing for Industry 4.0. APMS 2018. IFIP Advances in Information and Communication Technology, Springer, 536, 39-46.* doi:10.1007/978-3-319-99707-0_6

Rosmaini, A., & Kamaruddin, S. H. (2012). An overwiew of time-based and condition-based maintenance in industrial application. *Computers & Industrial Engineering*, *63*(1), 135–149. doi:10.1016/j.cie.2012.02.002

Sharma, N. R., Mishra, A. K., & Agrawal, H. (2019). Assessment and review of maintenance practices in the 4th industrial revolution using the cognitive analytics framework. *Journal of Mines. Metals and Fuels*, *67*(9), 416.

Wang, Q., & Gao, J. (2012). Research and application of risk and condition based maintenance task optimization technology in an oil transfer station. *Journal of Loss Prevention in the Process Industries*, *25*(6), 1018–1027. doi:10.1016/j.jlp.2012.06.002

## KEY TERMS AND DEFINITIONS

**Analytic Hierarchy Process:** Multicriteria model conceived by Tomas L. Saaty in 1980. This gives a complete classification of alternatives using the principles of pairwise comparison, decomposition, and synthesis. Decision centres provide judgements based on a linguistic scale to determine the relative importance of the alternatives with regard to each of the criteria or between the criteria themselves.

**Lubricant Analysis:** A diagnostic technique of a Predictive Maintenance strategy involving the monitoring of a number of physical and chemical parameters of the lubricant used, to obtain information about the state of the machine and diagnose a possible failure.

**Vibration Analysis:** A diagnostic technique of a Predictive Maintenance strategy involving the gathering of vibration data at given points of a rotating or alternating machine. This information is converted to a spectrum by Fourier transform, and then analysed and compared to known patterns, to identify a specific failure.

**Widening and Globalising Stage:** This is a stage of the lifecycle of a Predictive Maintenance strategy in which the amount of machinery to be controlled under the strategy, or its goals, is increased, once the period of return to investment is over. This stage could last 22 months or more, depending on the characteristics of the introduction, the machines chosen, and the business. The goals established for the PM strategy can thus be achieved some three years from the start of introduction.

# Chapter 9
# A Hybrid Asset-Based IT Risk Management Framework

**Baris Cimen**
*Department of Management Information Systems, Bogazici University, Turkey*

**Meltem Mutluturk**
*Department of Management Information Systems, Bogazici University, Turkey*

**Esra Kocak**
https://orcid.org/0000-0002-4808-830X
*Department of Management Information Systems, Bogazici University, Turkey*

**Bilgin Metin**
https://orcid.org/0000-0002-5828-9770
*Department of Management Information Systems, Bogazici University, Turkey*

## ABSTRACT

*Information security has become one of the most important responsibilities of all organizations due to increasing cyber threats. Attackers take advantage of systems vulnerabilities; therefore, system administrators should be aware of potential threats to take necessary actions to protect their organizations and stakeholders. At this point, a risk assessment is needed to discover possible threats for vulnerable systems of the organization and to implement strategies for the business goals. This study proposes a hybrid risk management framework using both qualitative and quantitative methods to analyze risk within organizations and reduce them with practical countermeasures. Based on this framework, case studies have been carried out considering three hypothetical companies identifying possible information security risks, and these risks have been reduced to an acceptable level by applying the proposed risk analysis methodology.*

DOI: 10.4018/978-1-7998-3246-1.ch009

## INTRODUCTION

Cyber-attacks have confounded information technology infrastructure of organizations and posed significant threats to their valuable information resources that have been valued as extremely important assets in cyberspace. Business managers need to be savvy to possible threats, aware of and prepared for both internal and external elements in order to manage cybersecurity risks. Hence, risk analysis procedure assists managers through evaluating possible threats or risks imposed on organizations, measuring the probability and impact of those risks and finally implementing strategies that create additional value to business operations by means of protecting their most precious information assets. Although, the most established risk analysis methodologies give great attention to technical risks, in recent, business organizations need risk analysis that incorporates social and organizational elements of complex systems with technical aspects in order to correctly evaluate and manage those risks. Within this context, the main objective of this paper is to address information risk management concepts considering possible threats and important organizational assets.

Drawing upon BS 7799 (Biery, 2006) information security risk management guidelines, in addition, by contributing additional possible threats, three hypothetical companies have been examined in terms of their business and organizational priorities. To that end, a strategic framework was proposed in order to identify critical business operations and relevant threats along with asset valuation for these three companies. Finally, a summary of evaluations and solutions were proposed to the identified threats for these companies. This study proposes a hybrid risk management framework via using both qualitative and quantitative methods to analyze risks within three hypothetical companies with the aim of incorporating social and organizational aspects alongside technical ones to overcome incompetencies of methods that previous studies did not. The remainder of the paper is structured as follows. The extant literature on cybersecurity risk analysis and management was reviewed and followed by presenting the proposed risk management framework. The paper concludes by providing a summary of the results and an overview of solutions and evaluations.

## BACKGROUND

Business Impact Analysis (BIA) has been defined as the process of conducting risk and gap assessment along with the implementation of global security standards considering people, process and technology in an organization (Sikdar, 2017). Business landscape renders a need for BIA within the context of business continuity management plan as a result of the latest trends such as globalization, e-commerce, enterprise resource planning, outsourcing business operations and legal and regulatory norms (Sikdar, 2017). Moreover, companies have developed BIA programs to cope with crisis events such as technological failures, natural events and deliberate disasters (Păunescu et al., 2018). While performing BIA, it is crucial to think in an organizational context and to understand relationships and dependencies throughout the enterprise including customers, business partners, stakeholders, supply chain vendors along with legal and regulatory institutions (Sikdar, 2017). According to Bjerga and Aven (2016), BIA estimates the possible damages that an enterprise might suffer without taking into consideration recovery methods. Although BIA and risk analysis have been often treated as separate activities, they are linked to each other by basic premises such as, what can happen, what will be affected and what are the resulting effects and impact (Hiles, 2002). Within this context, risk analysis refers to the study of evaluating possible threats and risks. The main objective of a risk analysis is to understand the existing environment and identify risks through analysis of several types of information such as security requirements and objectives, system and network architecture and infrastructure, physical assets, operating systems, data repository, etc. (Schmittling, 2010). According to Patel et al. (2008) risk assessment methods can be either qualitative or quantitative. While qualitative methodologies have used a common approach based on threat, vulnerability, and probability through generally using verbal hazard scales such as low, medium, high (Ghazouani et al., 2014), quantitative risk analysis methods have incorporated into what the literature refers to as probabilistic risk assessment (PRA) including mathematical procedures for risk evaluation (Henriques de Gusmão et al., 2018). Information systems (IS) risk literature contains a compilation of various risk models and conceptualization of risk factors along with risk components. On the other hand, one of the early studies by Rainer (1991) proposed a risk analysis process that uses a combination of qualitative and quantitative methodologies including steps for identifying IT assets, assigning value to these assets, stating possible threats to IT assets along with vulnerability of them to potential threats, and finally determining IT risk exposure for the organization. On the other hand, Alter and Sherer (2004) presented a broadly adaptable model of system-related risks including risk management, possible outcomes and their probabilities, impacts on other systems, and resulting financial gains and losses. Another study by Karabacak and Sogukpinar (2005) adopted a survey-based quantitative approach to analyze the security risks of information technologies by using the method named as Information Security Risk Analysis Method (ISRAM) consisting of seven steps. Moreover, they conducted a case study to exemplify these steps to analyze the risk arising from computer viruses. Furthermore, Sun, Srivastava, and Mock (2006) developed an alternative methodology for the risk analysis of information security systems (ISS) by adopting the Dempster-Shafer theory. They presented a cost-benefit analysis of control measures to reduce information security risk under belief functions. Moreover, Khambhammettu et al. (2013) described a framework for risk assessment including four different approaches to conducting a threat assessment. Although this study contributed, it didn't cover unexpected threats such as insider threats. Another study by Ghazouani et al. (2014) presented an overview of the previous risk methodologies through conducting comparative analysis and then proposed an integrated qualitative approach based on ISO 27005 along with a mathematical formulation of risk. They recommended developing an alternate IS risk management tool by using COBIT standards. Furthermore, Roldan-Molina et al. (2017) compiled different risk assessment tools for information and communication technology (ICT) infrastructure and in the cybersecurity domain.

Recently, Pate-Cornell et al. (2018) conducted a quantitative risk analysis and presented a probabilistic risk analysis framework for cybersecurity in an organization. In this study, three illustrative examples were described to exemplify decision support by quantifying the cyber risk and the risk reduction benefits of a spectrum of countermeasures such as encryption, reduction of the number of entry points, or modification of the operation software (Paté-Cornell et al., 2018). Another recent study by De Gusmao et al. (2018) adopted a fault tree analysis (FTA) approach to determine the vulnerability of cybersecurity and identify potential consequences of cyber-attacks. In this study, they proposed a structured model to measure cyber-attack scenarios that consider the risk of financial losses and analysis of restoration time analysis via fuzzy theory decision (Henriques de Gusmão et al., 2018).

Many methods have been incorporated into the field such as monte-carlo simulation as this method is used in cases with high uncertainty such as finance. In this setting, such a method can come in handy as risk assessment of information security is a complex structure with many parameters and a change in one project on its outcome and probability. Therefore, Monte-Carlo can simulate many possible outcomes according to the given parameters.

Another method frequently used within this field of research is Bayesian Networks (Holm et al., 2005; van der Veeken et al., 2016). This method can help with the likelihood estimations of risk assessment. Other methods used are Agent-based Simulation used in studies such as that of Weimer et al. (2016) and Niazi and Hussein (2011), which can simulate different agents (human or other) given their characteristics as well as their interactions, the game theory approach, fuzzy decision models and social network analysis such as has been used in the study of Yin and Chen (2012).

This paper proposed a framework for how to perform a risk analysis for data centers within enterprises by adopting a hybrid risk analysis methodology including both qualitative and quantitative norms. Based on this framework, case studies have been carried out considering three hypothetical companies identifying possible information security risks, and these risks have been reduced to an acceptable level by applying the proposed risk analysis methodology.

## RISK ASSESSMENT METHODS

### Qualitative Methods

#### Conflicting Incentives Risk Analysis (CIRA)

CIRA models risks by way of conflicting incentives among stakeholders. The primary focus of this method is the stakeholders' actions and perceived outcomes of these actions. This method is similar to game theory and can be applied to the scenario analysis of human-agent interaction and incentives (Rajbhandari, 2013).

#### CORAS

This method is model-based used to conduct a risk analysis. It is a 7-step method for information security risk analysis using its own language built on UML. CORAS models threat scenarios to assets. Although the risk analysis uses qualitative values, quantification can be accomplished (Den Braber et al., 2007).

#### CCTA Risk Analysis Management Method (CRAMM)

CRAMM first identifies and evaluates assets subsequently assessing the threats and vulnerabilities towards these assets and lastly combines the risk estimation and assessment into a single analysis. This method also proposes a risk management method (Yazar, 2002).

#### OCTAVE Allegro

The primary focus of this method is threat profiling assets. This method identifies and profiles assets, identifies threats and vulnerabilities to these assets and provides risk mitigation according to the threat information. This method incorporates threat trees into its process (Caralli et al., 2007).

### Factor Analysis of Information Risk (FAIR)

This method is based on risk assessment and as the name implies, it has 4 pre-defined factors for every loss and probability calculation. This method also provides a way to obtain quantitative results from the analysis (Freund & Jones, 2014).

## Quantitative Methods

### Information Security Risk Analysis Method (ISRAM)

The ISRAM method is a quantitative method that analyses the security risks of information technologies by way of the manager and employee participation. It also complies with the ISO 177999 and ISO 13335 standards. The input is obtained from a discussion among the risk analysis team. From this discussion, the important factors thought to affect the probability and outcome of a breach are found and each factor is given a weight. The outcome of the analysis is a quantitative value that is calculated using a risk formula (Karabacak & Soğukpınar, 2005).

# RISK ANALYSIS FRAMEWORK

Risk Analysis is a process to identify and manage potential problems that may create a threat to key enterprise activities or projects. For a successful Risk Analysis, it is necessary to identify potential threats that may be encountered first and then to estimate the likelihood of these threats. Once the risks are identified, they are analyzed to identify the qualitative and quantitative impact of the risk on the company so that appropriate steps can be taken to mitigate them.

## Risk Analysis Plan

The risk analysis plan includes these definitions and guidelines:

- Threat Identifications
- Risk Assessment
- Risk Mitigation

## Threat Identifications

A more systematic process involves the use of threats checklists and examining possible threats in the company. Some companies have been developing threat checklists based on their experience from past events. These checklists can help the IT team and managers both to identify specific threats in the checklist and to expand team thinking. The past experience of the company's IT consultants and managers and experts in the industry can be important resources to identify the potential risks.

Identifying the sources of threats by category is another method for exploring potential threats. Some examples of categories for potential threats include the following *Table 1*.

*Table 1. Example Thread Categories*

| Thread Category | Extended Categories |
|---|---|
| Technical | Extended categories Requirements, Technology, Interfaces, Performance, Quality, etc. |
| External | Custom, Contract, Market, Snooker, etc. |
| Organizational | Project Dependencies, Logistics, Resources, Budget, etc. |
| Managerial | Planning, Schedule, Estimation, Controlling, Communication, etc. |

## Risk Assessments

Once potential threats are identified, they are assessed based on the likelihood of occurrence of threat and possible losses accompanying the threat. Not all threats are equal. Some threats are more harmful than others, so the cost of a threat incident can vary greatly. After the threats and their possible effects and the likelihood of occurrence are identified, the impact values of the risks can be calculated by using the following formulas:

$$Total\ Impact\ Value\ of\ Risk = Busines\ Impact\ of\ Asset \times \left(Vulnerability\ level \times Likelihood\ of\ Threat\right)$$

$$Threat\ Impact = Vulnerability\ Level \times Likelihood\ of\ Threat$$

The Business Impact, Likelihood of Threat and Vulnerability Level can be explained as below:

### Business Impact of an Asset

Information Security Asset Classification is one of the key necessities because it lets a company assess the value of its information assets and to assign resources to protect them. The business impact from the loss of Confidentiality, Integrity and Availability of the Information Asset is considered. Business impact depends on asset value. Business Impact of the asset is taken as between 1 and 2 in our study.

### Likelihood of Threat

Some threats are expected to occur than others because an information asset is extensively exposed in the system or the number of potential attacks or incidents is high. The likelihood of a threat shows the probability of a potential threat for a vulnerability. Likely of threat can be derived from annual reports on cyber incident frequency. The likelihood percentages and corresponding likelihood values used in risk calculation is given in *Table 2*. In this study, the most significant 35 vulnerability sources are chosen given in *Table 3*. Also related countermeasures and/or controls for these vulnerability sources given in *Table 4*.

Our methodology depends on choosing a proper threat likelihood value in *Table 2* for each vulnerability source for the risk calculation. In our methodology, five of the threat likelihood for vulnerability controls are chosen as Very High, nine of them are accepted as high, nine of them are selected as Medium, and 12 of them are chosen as Low likelihood.

*Table 2. Classification of threat likelihood*

| Likelihood (Percentage range) | Corresponding Likelihood Value |
|---|---|
| Low likelihood – $(0\% < x < 30\%)$ | 1 |
| Medium likelihood – $(30\% \leq x < 60\%)$ | 4 |
| High likelihood – $(60\% \leq x < 80\%)$ | 8 |
| Very High likelihood – $(80\% \leq x \leq 100\%)$ | 16 |

## Vulnerability Level

A vulnerability is a weakness that can be exploited by possible threats such as data breaches, unexpected system failures or natural disasters. In this study 35 potential vulnerability sources are given as shown in Table 3. In Table 4, the related controls and/or countermeasures are presented as courses of actions to reduce these vulnerabilities. An adopted course of action corresponds to a vulnerability level between 1 and 5.

## Threat Impact

The Threat Impact indicates how seriously a potential threat affects a specific vulnerability source in the total system risk. Threat Impact is related with both the security controls and the likelihood of a threat. It shows which vulnerability source is much more exposed to threats for the information asset.

The following list shows the degrees of the Threat Impact given in *Figure 1*.

1. Informational
2. Low
3. Medium
4. High
5. Critical

Having criteria to determine high impact risks can help narrow the focus on a few critical risks that require mitigation. For example, suppose that high-impact risks could cause $100,000 or more losses. Only a few potential risk events meet these criteria. These are the critical few potential risk events that the IT experts and managers should focus on when developing risk mitigation or management plan. Risk assessment is about developing an understanding of which potential risks have the greatest possibility of occurring and can have the most significant negative impact. In *Table 3* some potential threat sources are given.

After defining the potential threat/vulnerability sources table, the total impact value of risk can be represented by *Figure 2*.

*Figure 1. Threat Impact Degree*

| Likelihood: Percentage | Value | Vulnerability Level 1 | 2 | 3 | 4 | 5 |
|---|---|---|---|---|---|---|
| (1-29)% | 1 | Ideal | Informational | Informational | Low | Medium |
| (30-59)% | 4 | Informational | Informational | Low | Medium | High |
| (60-79)% | 8 | Informational | Low | Medium | High | Critical |
| (80-100)% | 16 | Low | Medium | High | Critical | Critical |

## Risk Mitigation

Risk Mitigation is to take reasonable action to minimize the amount of the possible loss suffered. The ways of managing them are investigated. A risk mitigation plan is developed to reduce the impact of a risk. The risks may be mitigated by the following ways:

- Risk avoidance
- Risk sharing
- Risk reduction
- Risk transfer

Each of these abatement techniques can be an effective means of reducing the risks assessed. The risk mitigation plan covers risk mitigation approaches for all identified risk events and actions that IT professionals and managers will take to mitigate these risks.

### Risk Avoidance

Risk avoidance refers to preferring lower-risk alternatives to high-risk ones. It can usually be costlier.

### Risk Sharing

Risk-sharing involves partnering with others to share responsibility for the risk activities. Many organizations often try to reduce political, legal, labor, and other risk types by developing a joint venture with another company.

### Risk Reduction

Risk reduction is the investment to reduce the risk in a company. For example, by making necessary expenditures for a situation that is considered as high risk, the risk can be reduced to an acceptable level.

*Table 3. Potential Vulnerability Sources*

| 1 | Physical Location |
|---|---|
| 2 | UPS Availability |
| 3 | Generator Availability |
| 4 | Environmental ThreatSense Prevention Management System Availability |
| 5 | Internet Connection |
| 6 | Where data is stored |
| 7 | The environment in which backups are stored |
| 8 | Fire Prevention |
| 9 | Maintenance Agreement |
| 10 | Patch Tracking |
| 11 | Can it be replaced easily? (Replaceability) |
| 12 | Network protection (Firewall, IPS, WAF, DDOS, DNS Firewall, …….) |
| 13 | Maintenance Firms Access |
| 14 | Qualification of Authorized Personnel |
| 15 | Backup frequency |
| 16 | Clustering (Hardware / Software) |
| 17 | Server Virtualization |
| 18 | Log Monitoring |
| 19 | Response to Detected Incidents |
| 20 | Network Segmentation |
| 21 | The malicious software detection system |
| 22 | Air Conditioner Availability |
| 23 | Technical team coordination |
| 24 | Insurance for cyber threats |
| 25 | Hardware / Software support team availability (per week) |
| 26 | Cybersecurity Vulnerability Assessment |
| 27 | TEMPEST |
| 28 | Root / Administrator Password Change Frequency |
| 29 | Password Delivery Method |
| 30 | Data Loss Prevention Application |
| 31 | Physical strength of the structure |
| 32 | Compliance with industrial standards |
| 33 | Budget for replacement |
| 34 | Protection of Sensitive Data |
| 35 | Access control of Sensitive Data |

*Figure 2. Total Impact Value of the Risk*

## Risk Transfer

Risk transfer is a risk reduction method that shifts the risk from the company to another party. The purchase of insurance on certain items is a risk transfer method. The risk is transferred from the company to the insurance company.

# RISK ASSESSMENT CASE STUDY OF ABC HOLDING

In this paper, risk analysis of three companies' data centers is uncovered through case studies.

## ABC Holding

ABC holding is a multinational company that mainly produces construction materials. The holding has investments in paint production, fast food and cleaning sectors. The holding has more than 5,000 employees.

All three companies below belong to ABC Holding. All of them basically have air conditioning systems, UPS, Environmental ThreatSense prevention management system, high-speed symmetrical internet connection, fire prevention, patch tracking system, clustering, and log monitoring in their data centers.

### Company A (Headquarters of ABC Holding)

The headquarters of ABC holding facility has a 30 square meter room in the building which is used as a data center. In addition to the properties listed above, there are power generators, high-speed internet connection with a failover link and SLA agreement upon 99.99% availability, firewall, Intrusion Prevention System (IPS), and network-based attack monitoring facilities in the data center. The data center has 5 cabins and 42 physical servers. The data center door is locked and only 5 people who are responsible for the data center have the physical key. Many services, especially an Enterprise Resource Planning (ERP) system and e-mail and Customer Relationship Management (CRM) applications are hosted in this data center.

### Company B (Branch Office #1 of ABC Holding)

Company B is the branch office of the ABC holding. It is responsible for the middle east region. It operates in a 500 square meter office in a plaza with 25 employees. The information technology infrastructure of the branch office is provided from the 16 square meter data center in the office. The data center has a total of 50 physical server computers in 7 server cabins. In addition to the properties listed above, there are RAID, Firewall, Network-based attacks monitoring system, and DLP software in the data center. The file storing and sharing services, regional information databases, e-mail servers and other service applications are hosted from this data center. The entrance to the data center is made by RFID cards. Only 5 people in charge of servers are allowed to access the data center.

### Company C (Branch Office #2 of ABC Holding)

Company C is another branch office of ABC holding. It is responsible for the Europe region. It operates in a 200 square meter office in a trade center with 10 employees. The information technology infrastructure of the branch office is provided from the 10 square meter data center in the office. The data center has a total of 15 physical server computers in 2 server cabins. In addition to the properties listed above, there are RAID, maintenance agreement, Firewall, WAF, IPS, and Malicious software detection system in the data center. File storing and sharing services, regional information databases, e-mail servers and other service applications are hosted from this data center. Entrance to data centers are made by fingerprint readers. Only 2 people in charge of servers are allowed to access the data center.

## Risk Analysis Plan

In this step, the data centers of these 3 companies are examined by experts who have worked in the IT sector for many years and 35 sources of potential threats are found, as shown in *Table 3*.

## Threat Assessment Case Study

In this section, the identified threats of the three companies are assessed based on the likelihood of occurrence of the threat and possible impact of the threat.

In *Table 4*, threat assessments and calculated risk impact values are shown.

## Risk Mitigation

After determining the value of the identified risks, this step attempts to find ways to manage them. First, the minimum and maximum values of the risk assessment table are calculated and the overall risk ranges according to these values are determined. According to *Table 4*, the minimum impact value of risk is 200, and the maximum is 2000. In *Table 5*, overall risk ranges are demonstrated from informational to critical based on the minimum and the maximum values.

According to *Table 3*, the total risk impact values of A, B, and C companies are 1132, 1687, 350 respectively. This means that the risk range of company A is moderate, company B is critical while company C is informational. In the next step, it is necessary to look at what needs to be done to reduce

*Table 4. Threat Assessment Table*

| The Vulnerability Level: 1 (Lowest), 5 (Highest)<br>Likelihood Value of Threat: 1 (Lowest), 16 (Highest)<br>Business Impact: 1 (Lowest), 2 (Highest)<br>Total Impact Value of Risks: max. 2000 points, min. 200 points | | | COMPANY ASSETS | | |
|---|---|---|---|---|---|
| | | | A | B | C |
| | | | Business Impact of the asset | Business Impact of the asset | Business Impact of the asset |
| | | | 2 | 1.75 | 1.25 |
| # | Vulnerability Level | Likelihood Value of Threat | Threat Impact | | |
| 1 | **Physical Location** | | | | |
| | Fingerprint or Retina Reader Entry Section two way authentication (1) Card Entry Section (2) - Key Section / Only Officials can enter (3) - Switching Department / Other staff can enter or Keyless Section (4) - in Outdoor Environment Service (5) | 1 | 3 | 2 | 1 |
| 2 | **UPS Availability** | | | | |
| | More than one UPS (1) Only one UPS (3), Not available (5) | 16 | 16 | 80 | 16 |
| 3 | **Generator** | | | | |
| | Available (1), Not available (5) | 1 | 1 | 5 | 1 |
| 4 | **Environmental ThreatSense prevention management (e.g., long term power failure, pollution, chemicals, liquid leakage, moisture, smoke, heat vs.)** | | | | |
| | Excellent (1) Good (2) Normal (3), Poor (4), Worst (5) | 1 | 2 | 4 | 2 |
| 5 | **Internet Connection** | | | | |
| | SLA and Failover Link (1), SLA or Failover Link (3), Neither SLA nor Failover Link (5) | 1 | 1 | 5 | 3 |
| 6 | **Where data is stored** | | | | |
| | Both SAN/NAS and cloud (1), SAN/NAS or cloud (2), A virtualized server with RAID (3), A virtualized server (4), Server without virtualization (5) | 8 | 40 | 40 | 8 |
| 7 | **Environment in which cold backups are stored** | | | | |
| | Separate Building/Private Section/Cloud/Datacenter (1), Separate building (3), In the same building/ Fireproof safe (4), In the same building (5) | 8 | 16 | 40 | 8 |
| 8 | **Fire Prevention** | | | | |
| | Alarm / Auto Fire Suppression (1), Manual extinguishing / Fire extinguisher (3), No Measure (5) | 4 | 12 | 12 | 12 |
| 9 | **Maintenance Agreement** | | | | |
| | Yes (1), No (5) | 1 | 5 | 5 | 1 |
| 10 | **Patch Tracking Is** | | | | |
| | Automatic but try first (1), Automatic (2), Periodic / Warning As it grows (3), Not Followed (5) | 16 | 48 | 80 | 16 |
| 11 | **Can it be replaced easily?** | | | | |
| | Can be purchased with money (1), available for purchase, very expensive (3), information / prestige will be lost even if purchased (5) | 1 | 5 | 5 | 3 |
| 12 | **Network protection (Firewall, IPS, WAF, DDOS, DNS Firewall, …….)** | | | | |
| | 3+ assets (1), 2 assets (3), 1 (5) | 16 | 48 | 80 | 16 |
| 13 | **Maintenance firms Access with** | | | | |
| | VPN with two way authentication (1), VPN (2), rlogin (3), 3rd party desktop sharing apps (4), without VPN (5) | 4 | 12 | 16 | 4 |
| 14 | **Qualification of Authorized Personnel** | | | | |
| | Advanced Level (1), Intermediate Level (3), Not sufficient (5) | 4 | 12 | 16 | 4 |
| 15 | **Backup frequency** | | | | |
| | Daily (1), Weekly (2), Monthly (3), Once in 3 Month s (4), Yearly (5) | 8 | 24 | 32 | 8 |
| 16 | **Clustering (Hardware / Software)** | | | | |
| | Two or more hardware with load balancer active/active (1), Two hardware one active one backup active passive (3), No cluster (5) | 8 | 8 | 40 | 8 |
| 17 | **Server Virtualization** | | | | |
| | Yes (1), No (5) | 16 | 16 | 80 | 16 |
| 18 | **Log Monitoring** | | | | |
| | SIEM and Real-time Alarms (1), Logging and log backup (3), No log monitoring (5) | 4 | 12 | 20 | 4 |
| 19 | **Response to Detected Incidents** | | | | |
| | in Real-time (1) Within 30 mins (2) Within 2 hours (3) Within 6 hours (4), uncertain (5) | 4 | 16 | 20 | 8 |
| 20 | **Network Segmentation** | | | | |
| | Firewall between VLANs (1), Servers in DMZ (2), VLAN (3), Not available (5) | 1 | 1 | 5 | 5 |
| 21 | **Malicious software detection system** | | | | |
| | Both EDR and corporate anti-virus (1), Corporate anti-virus (3), Personal anti-virus (4), Not available (5) | 8 | 40 | 40 | 8 |
| 22 | **Air Conditioner Availability** | | | | |
| | Available with back up (1), Only one (3), Not available (5) | 4 | 4 | 4 | 4 |
| 23 | **Technical team coordination** | | | | |
| | Excellent (1), Good (2), Normal (3), Poor (4), Worst (5) | 4 | 16 | 20 | 4 |
| 24 | **Insurance for cyber threats** | | | | |
| | Available (1), Not available (5) | 4 | 20 | 20 | 4 |

*continued on following page*

*Table 4. Continued*

| The Vulnerability Level: 1 (Lowest), 5 (Highest)<br>Likelihood Value of Threat: 1 (Lowest), 16 (Highest)<br>Business Impact: 1 (Lowest), 2 (Highest)<br>Total Impact Value of Risks: max. 2000 points, min. 200 points | | | COMPANY ASSETS | | |
|---|---|---|---|---|---|
| | | | A | B | C |
| | | | Business Impact of the asset | Business Impact of the asset | Business Impact of the asset |
| | | | 2 | 1.75 | 1.25 |
| # | Vulnerability Level | Likelihood Value of Threat | Threat Impact | | |
| 1 | Physical Location | | | | |
| 25 | **Hardware / Software support team availability (per week)** | | | | |
| | 168 hours (1), 168 - 84 hours (2), 84 - 42 hours (3), 42- 16 (4), <16 (5) | 4 | 12 | 20 | 4 |
| 26 | **Cybersecurity Vulnerability Assessment** | | | | |
| | Weekly (1), Monthly (2), Yearly (3), Never (5) | 1 | 3 | 5 | 2 |
| 27 | **TEMPEST** | | | | |
| | Available (1), Not available (5) | 1 | 5 | 5 | 1 |
| 28 | **Root / Administrator Password Change Frequency** | | | | |
| | Every day (1), Every week (2), Every 3-4 months (3), Yearly (4), When required or never (5) | 8 | 32 | 40 | 16 |
| 29 | **Password Delivery Method** | | | | |
| | Hard to solve (1), Haphazard, easy to solve password (3), More than one-person, unprotected passwords (5) | 8 | 24 | 40 | 24 |
| 30 | **Data Loss Prevention Application** | | | | |
| | DLP policy and software data classification (1), Data classification software (3), Manual data classification (4), Not available (5) | 8 | 40 | 40 | 40 |
| 31 | **Physical strength of the structure (building)** | | | | |
| | Excellent (1), Good (2), Normal (3), Poor (4), Worst (5) | 1 | 2 | 5 | 3 |
| 32 | **Compliance with industrial standards** | | | | |
| | Yes (1), No (5) | 1 | 1 | 5 | 1 |
| 33 | **Budget for replacement** | | | | |
| | Available (1), Not available (5) | 1 | 5 | 5 | 1 |
| 34 | **Protection of Sensitive Data** | | | | |
| | Tokenization Encryption with key management (1), Tokenization Encryption without key management (3), No Tokenization Encryption (5) | 16 | 48 | 80 | 16 |
| 35 | **Access control of Sensitive Data** | | | | |
| | Both blocking, masking (1), Masking (2), Blocking (3), No access control (5) | 8 | 16 | 40 | 8 |
| | **TOTAL RISK IMPACT VALUE** | | 1132 | 1687 | 350 |

the risk range of companies A and B to an acceptable range (low or informational). The following ways can be used to reduce risks:

- Eliminating the risk source
- Lowering the likelihood of risk occurrence
- Lowering the impact of the risk

In the specified case, eliminating the risk source and lowering the likelihood of the risk occurrence are not possible, the third way should be used. As the likelihood of the threat, which is planned to be reduced, increases, the impact on the overall risk value of the company also increases. *Table 5* shows the

*Table 5. Overall Risk Ranges Table*

| Risk Range | Risk Level |
|---|---|
| **(1640-2000)** | Critical |
| **(1280-1640)** | High |
| **(920-1280)** | Moderate |
| **(560-920)** | Low |
| **(200-560)** | Informational |

*Table 6. Threat Assessment Table (after mitigation)*

| The Vulnerability Level: 1 (Lowest), 5 (Highest)<br>Likelihood Value of Threat: 1 (Lowest), 16 (Highest)<br>Business Impact: 1 (Lowest), 2 (Highest)<br>Total Impact Value of Risks: max. 2000 points, min. 200 points | | | COMPANY ASSETS | | |
|---|---|---|---|---|---|
| | | | A | B | C |
| | | | Business Impact of the asset | Business Impact of the asset | Business Impact of the asset |
| | | | 2 | 1.75 | 1.25 |
| # | Vulnerability Level | Likelihood Value of Threat | Threat Impacts | | |
| 1 | Physical Location | | | | |
| | Fingerprint or Retina Reader Entry Section two way authentication (1) Card Entry Section (2) - Key Section / Only Officials can enter (3) - Switching Department / Other staff can enter or Keyless Section (4) - in Outdoor Environment Service (5) | 1 | 3 | 2 | 1 |
| 2 | **UPS Availability** | | | | |
| | More than one UPS (1) Only one UPS (3), Not available (5) | 16 | 16 | **16*** | 16 |
| 3 | **Generator** | | | | |
| | Available (1), Not available (5) | 1 | 1 | 5 | 1 |
| 4 | **Environmental ThreatSense prevention management (e.g., long term power failure, pollution, chemicals, liquid leakage, moisture, smoke, heat vs.)** | | | | |
| | Excellent (1) Good (2) Normal (3), Poor (4), Worst (5) | 1 | 2 | 4 | 2 |
| 5 | **Internet Connection** | | | | |
| | SLA and Failover Link (1), SLA or Failover Link (3), Neither SLA nor Failover Link (5) | 1 | 1 | 5 | 3 |
| 6 | **Where data is stored** | | | | |
| | Both SAN/NAS and cloud (1), SAN/NAS or cloud (2), A virtualized server with RAID (3), A virtualized server (4), Server without virtualization (5) | 8 | **8*** | **8*** | 8 |
| 7 | **Environment in which cold backups are stored** | | | | |
| | Separate Building/Private Section/Cloud/Datacenter (1), Separate building (3), In the same building/ Fireproof safe (4), In the same building (5) | 8 | **8*** | **8*** | 8 |
| 8 | **Fire Prevention** | | | | |
| | Alarm / Auto Fire Suppression (1), Manual extinguishing / Fire extinguisher (3), No Measure (5) | 4 | 12 | 12 | 12 |
| 9 | **Maintenance Agreement** | | | | |
| | Yes (1), No (5) | 1 | 5 | 5 | 1 |
| 10 | **Patch Tracking Is** | | | | |
| | Automatic but try first (1), Automatic (2), Periodic / Warning As it grows (3), Not Followed (5) | 16 | **16*** | **16*** | 16 |
| 11 | **Can it be replaced easily?** | | | | |
| | Can be purchased with money (1), available for purchase, very expensive (3), information / prestige will be lost even if purchased (5) | 1 | 5 | 5 | 3 |
| 12 | **Network protection (Firewall, IPS, WAF, DDOS, DNS Firewall, .......)** | | | | |
| | 3+ assets (1), 2 assets (3), 1 (5) | 16 | **16*** | **16*** | 16 |
| 13 | **Maintenance firms Access with** | | | | |
| | VPN with two way authentication (1), VPN (2), rlogin (3), 3rd party desktop sharing apps (4), without VPN (5) | 4 | 12 | 16 | 4 |
| 14 | **Qualification of Authorized Personnel** | | | | |
| | Advanced Level (1), Intermediate Level (3), Not sufficient (5) | 4 | 12 | 16 | 4 |
| 15 | **Backup frequency** | | | | |
| | Daily (1), Weekly (2), Monthly (3), Once in 3 Month s (4), Yearly (5) | 8 | **8*** | **8*** | 8 |
| 16 | **Clustering (Hardware / Software)** | | | | |
| | Two or more hardware with load balancer active/active (1), Two hardware one active one backup active passive (3), No cluster (5) | 8 | 8 | **8*** | 8 |
| 17 | **Server Virtualization** | | | | |
| | Yes (1), No (5) | 16 | 16 | **16*** | 16 |
| 18 | **Log Monitoring** | | | | |
| | SIEM and Real-time Alarms (1), Logging and log backup (3), No log monitoring (5) | 4 | 12 | 20 | 4 |
| 19 | **Response to Detected Incidents** | | | | |
| | in Real-time (1) Within 30 mins (2) Within 2 hours (3) Within 6 hours (4), uncertain (5) | 4 | 16 | 20 | 8 |
| 20 | **Network Segmentation** | | | | |
| | Firewall between VLANs (1), Servers in DMZ (2), VLAN (3), Not available (5) | 1 | 1 | 5 | 5 |
| 21 | **Malicious software detection system** | | | | |
| | Both EDR and corporate anti-virus (1), Corporate anti-virus (3), Personal anti-virus (4), Not available (5) | 8 | **8*** | **8*** | 8 |
| 22 | **Air Conditioner Availability** | | | | |
| | Available with back up (1), Only one (3), Not available (5) | 4 | 4 | 4 | 4 |
| 23 | **Technical team coordination** | | | | |
| | Excellent (1), Good (2), Normal (3), Poor (4), Worst (5) | 4 | 16 | 20 | 4 |
| 24 | **Insurance for cyber threats** | | | | |
| | Available (1), Not available (5) | 4 | 20 | 20 | 4 |

*continued on following page*

*Table 6. Continued*

| The Vulnerability Level: 1 (Lowest), 5 (Highest)<br>Likelihood Value of Threat: 1 (Lowest), 16 (Highest)<br>Business Impact: 1 (Lowest), 2 (Highest)<br>Total Impact Value of Risks: max. 2000 points, min. 200 points | | | COMPANY ASSETS | | |
|---|---|---|---|---|---|
| | | | A | B | C |
| | | | Business Impact of the asset | Business Impact of the asset | Business Impact of the asset |
| | | | 2 | 1.75 | 1.25 |
| # | Vulnerability Level | Likelihood Value of Threat | Threat Impacts | | |
| 1 | Physical Location | | | | |
| 25 | **Hardware / Software support team availability (per week)** | | | | |
| | 168 hours (1), 168 - 84 hours (2), 84 - 42 hours (3), 42- 16 (4), <16 (5) | 4 | 12 | 20 | 4 |
| 26 | **Cybersecurity Vulnerability Assessment** | | | | |
| | Weekly (1), Monthly (2), Yearly (3), Never (5) | 1 | 3 | 5 | 2 |
| 27 | **TEMPEST** | | | | |
| | Available (1), Not available (5) | 1 | 5 | 5 | 1 |
| 28 | **Root / Administrator Password Change Frequency** | | | | |
| | Every day (1), Every week (2), Every 3-4 months (3), Yearly (4), When required or never (5) | 8 | **8*** | **8*** | **8*** |
| 29 | **Password Delivery Method** | | | | |
| | Hard to solve (1), Haphazard, easy to solve password (3), More than one-person, unprotected passwords (5) | 8 | **8*** | **8*** | **8*** |
| 30 | **Data Loss Prevention Application** | | | | |
| | DLP policy and software data classification (1), Data classification software (3), Manual data classification (4), Not available (5) | 8 | **8*** | **8*** | **8*** |
| 31 | **Physical strength of the structure (building)** | | | | |
| | Excellent (1), Good (2), Normal (3), Poor (4), Worst (5) | 1 | 2 | 5 | 3 |
| 32 | **Compliance with industrial standards** | | | | |
| | Yes (1), No (5) | 1 | 1 | 5 | 1 |
| 33 | **Budget for replacement** | | | | |
| | Available (1), Not available (5) | 1 | 5 | 5 | 1 |
| 34 | **Protection of Sensitive Data** | | | | |
| | Tokenization Encryption with key management (1), Tokenization Encryption without key management (3), No Tokenization Encryption (5) | 16 | **16*** | **16*** | 16 |
| 35 | **Access control of Sensitive Data** | | | | |
| | Both blocking, masking (1), Masking (2), Blocking (3), No access control (5) | 8 | **8*** | **8*** | 8 |
| | **TOTAL RISK IMPACT VALUE** | | **620** | **679** | **280** |
| *indicates mitigated threats and re-calculated impact values of risks | | | | | |

threat assessments and re-calculated risk impact values of the three companies after the risk mitigation process has been conducted.

As shown in *Table 6*, the overall risk impact values of companies A and B have been reduced to a low level and the risk range of company C has also decreased.

## CONCLUSION

In this paper, three hypothetical companies have been taken into consideration by identifying their possible information security risks and these risks have been reduced to an acceptable level through deploying constructed risk analysis methodology. The proposed method is a hybrid method, which is a combination of qualitative and quantitative parts. In the quantitative part, survey results and risk tables were used to analyze information security risks. In this manner, the proposed method is simple and does not include mathematical procedures or software tools. In the qualitative part, subjective evaluations are made based on IT experts' experiences in this research area.

The main advantage of the proposed framework is to provide guidance for companies that will perform information security risk analysis for their data centers. Furthermore, this study contributes to the existing literature on how to perform risk analysis for data centers. The framework proposes an information risk analysis within an organizational context and from a business-oriented perspective. In this manner, the framework has gratified both organizational and technical requirements regarding trends in the recent business landscape.

## LIMITATIONS

The data and information provided in this study were created by experts working in the IT industry for many years. Each industry can have its own data security challenges and risks. For this reason, it is aimed to present an inclusive case study as much as possible. Also, the mitigation step did not take cost into account, therefore mitigation strategies may change to incorporate cost into the equation.

## REFERENCES

Alter, S., & Sherer, S. A. (2004). A General, But Readily Adaptable Model of Information System Risk. *Communications of the Association for Information Systems*, *14*(1). Advance online publication. doi:10.17705/1CAIS.01401

Biery, K. J. (2006). Aligning an information risk management approach to BS 7799-3:2005. SANS Reading Room.

Bjerga, T., & Aven, T. (2016). Some perspectives on risk management: A security case study from the oil and gas industry. *Proceedings of the Institution of Mechanical Engineers, Part O: Journal of Risk and Reliability*, *230*(5), 512–520. doi:10.1177/1748006X16654589

Caralli, R. A., Stevens, J. F., Young, L. R., & Wilson, W. R. (2007). *Introducing octave allegro: Improving the information security risk assessment process* (No. CMU/SEI-2007-TR-012).

Den Braber, F., Hogganvik, I., Lund, M. S., Stølen, K., & Vraalsen, F. (2007). Model-based security analysis in seven steps—A guided tour to the CORAS method. *BT Technology Journal*, *25*(1), 101–117. doi:10.100710550-007-0013-9

Freund, J., & Jones, J. (2014). *Measuring and managing information risk: a FAIR approach*. Butterworth-Heinemann.

Ghazouani, M., Faris, S., Medromi, H., & Sayouti, A. (2014). Information Security Risk Assessment A Practical Approach with a Mathematical Formulation of Risk. *International Journal of Computers and Applications*, *103*(8), 36–42. doi:10.5120/18097-9155

Henriques de Gusmão, A. P., Mendonça Silva, M., Poleto, T., Camara e Silva, L., & Cabral Seixas Costa, A. P. (2018). Cybersecurity risk analysis model using fault tree analysis and fuzzy decision theory. *International Journal of Information Management*, *43*, 248–260. doi:10.1016/j.ijinfomgt.2018.08.008

Hiles, A. (2002). *Enterprise Risk Assessment and Business Impact Analysis: Best Practices*. Rothstein Associates Inc.

Holm, H., Korman, M., & Ekstedt, M. (2015). A bayesian network model for likelihood estimations of acquirement of critical software vulnerabilities and exploits. *Information and Software Technology*, *58*, 304–318. doi:10.1016/j.infsof.2014.07.001

ISO/IEC 27005:2008, “Information technology -- Security techniques-Information security risk management

Karabacak, B., & Sogukpinar, I. (2005). ISRAM: Information security risk analysis method. *Computers & Security*, *24*(2), 147–159. doi:10.1016/j.cose.2004.07.004

Khambhammettu, H., Boulares, S., Adi, K., & Logrippo, L. (2013). A framework for risk assessment in access control systems. *Computers and Security, 39*(A), 86–103.

Niazi, M. A., & Hussain, A. (2011). A novel agent-based simulation framework for sensing in complex adaptive environments. *IEEE Sensors Journal*, *11*(2), 404–412. doi:10.1109/JSEN.2010.2068044

Paté-Cornell, M. E., Kuypers, M., Smith, M., & Keller, P. (2018). Cyber Risk Management for Critical Infrastructure: A Risk Analysis Model and Three Case Studies. *Risk Analysis*, *38*(2), 226–241. doi:10.1111/risa.12844 PMID:28679022

Patel, S. C., Graham, J. H., & Ralston, P. A. S. (2008). Quantitatively assessing the vulnerability of critical information systems: A new method for evaluating security enhancements. *International Journal of Information Management*, *28*(6), 483–491. doi:10.1016/j.ijinfomgt.2008.01.009

Păunescu, C., Popescu, M. C., & Blid, L. (2018). Business impact analysis for business continuity: Evidence from Romanian enterprises on critical functions. *Management and Marketing*, *13*(3), 1035–1050. doi:10.2478/mmcks-2018-0021

Rainer, R. K. Jr, Snyder, C. A., & Carr, H. H. (1991). Risk analysis for information technology. *Journal of Management Information Systems*, *8*(1), 129–147. doi:10.1080/07421222.1991.11517914

Rajbhandari, L. (2013). *Risk Analysis Using “Conflicting Incentives” as an alternative notion of Risk* (Ph.D. thesis). Gjøvik University College.

Roldán-Molina, G., Almache-Cueva, M., Silva-Rabadão, C., Yevseyeva, I., & Basto-Fernandes, V. (2017). A Comparison of Cybersecurity Risk Analysis Tools. *Procedia Computer Science*, *121*, 568–575. doi:10.1016/j.procs.2017.11.075

Schmittling, R. A. M. (2010). Performing a Security Risk Assessment. *ISACA Journal, 1*, 1–7.

Sikdar, P. (2017). Practitioner’s Guide to Business Impact Analysis. Practitioner’s Guide to Business Impact Analysis. doi:10.1201/9781315187884

Sun, L., Srivastava, R. P., & Mock, T. J. (2006). An information systems security risk assessment model under the Dempster-Shafer theory of belief functions. *Journal of Management Information Systems*, *22*(4), 109–142. doi:10.2753/MIS0742-1222220405

Van der Veeken, P., Van Schooten, S., Shinde, R., Dunnewind, M., & van den Berg, J. (2016, December). Applying Bayesian game theory to analyse cyber risks of bank transaction systems. In *2016 International Conference on Computing, Analytics and Security Trends (CAST)* (pp. 84-89). IEEE. 10.1109/CAST.2016.7914945

Weimer, C. W., Miller, J. O., & Hill, R. R. (2016, December). Agent-based modeling: An introduction and primer. In *Winter Simulation Conference (WSC)*, (pp. 65-79). IEEE.

Yazar, Z. (2002). A qualitative risk analysis and management tool–CRAMM. *SANS InfoSec Reading Room White Paper, 11*, 12-32.

Yin, Q., & Chen, Q. (2012, January). A Social Network Analysis Platform for Organizational Risk Analysis—ORA. In *2012 Second International Conference on Intelligent System Design and Engineering Application (ISDEA)*, (pp. 760-763). IEEE. 10.1109/ISdea.2012.546

## KEY TERMS AND DEFINITIONS

**Residual Risk:** It is the risk that cannot be reduced to the acceptable level after risk processing.

**Risk:** It is a measure of how open assets are against potential threats.

**Risk Analysis:** Evaluation of possible threats and possible risks.

**Risk Processing:** As a result of risk analysis, it is a package/plan of measures to reduce the risks to an acceptable level and maintain this level.

**Threat:** Potential hazard that can partially or totally interrupt the operation of systems, processes and other information system components.

**Threat Impact:** The threat Impact indicates how seriously a potential threat affects a vulnerability in the system.

**Vulnerability:** Weakness leading to a threat.

# Chapter 10
# Fuzzy Greedy Search:
## An Algorithmic Approach for Combinatorial Optimisation

**Kaveh Sheibani**
*ORLab Analytics, Canada*

## ABSTRACT

*In recent years, there has been a growth of interest in the development of systematic search methods for solving problems in operational research and artificial intelligence. This chapter introduces a new idea for the integration of approaches for hard combinatorial optimisation problems. The proposed methodology evaluates objects in a way that combines fuzzy reasoning with a greedy mechanism. In other words, a fuzzy solution space is exploited using greedy methods. This seems to be superior to the standard greedy version. The chapter consists of two main parts. The first part focuses on description of the theory and mathematics of the so-called fuzzy greedy evaluation concept. The second part demonstrates through computational experiments the effectiveness and efficiency of the proposed concept for hard combinatorial optimisation problems.*

## INTRODUCTION

In recent years, there has been a growth of interest in the development of systematic search methods for solving problems in operational research and artificial intelligence. Metaheuristics that are employed as strategies in optimisation are a fairly young research field. They are approaches that organise an interaction between solution improvement procedures and higher-level tactics in order to create a process capable of escaping from premature local optima and performing a robust search of a solution space. A metaheuristic can be viewed as a generic approach, for a type of hard optimisation problem. It is applicable to a wide set of different optimisation problems, with relatively few modifications needed to apply it to a specific problem. A much newer area of research is the hybridisation of metaheuristics. It has become evident that a skilled combination of general ideas from different methods can provide an efficient performance and high flexibility.

DOI: 10.4018/978-1-7998-3246-1.ch010

The use of search techniques on a solution space are central to the design of metaheuristics. Indeed, adopting a robust search technique significantly improves the overall performance. In this chapter, we introduce a new idea for the integration of approaches for hard combinatorial optimisation problems. The proposed methodology evaluates objects in a way that combines fuzzy reasoning with a greedy mechanism. In other words, we exploit a fuzzy solution space (fuzzy set) using greedy methods. Our methodology also attempts to adapt its knowledge from previous experiments, thereby improving the exploration of the promising areas of the search space. The effectiveness and efficiency of this so-called fuzzy greedy evaluation concept are investigated for hard combinatorial optimisation problems.

The chapter consists of two main parts. The first part focuses on description of the theory and mathematics of the so-called fuzzy greedy evaluation concept. The second part demonstrates through computational experiments, the effectiveness and efficiency of the proposed concept for hard combinatorial optimisation problems. The text contains an extensive bibliography, which covers many relevant books and significant papers.

## FUZZY GREEDY EVALUATION

### Introduction

In recent decades there has been a growth of interest in methods for finding optimal solutions to a class of problems called combinatorial optimization. The subject is very wide and many books and articles have been published on its various aspects. Typical examples of combinatorial optimization problems are the travelling salesman (Lawler, 1985; Gutin and Punnen, 2002; Gutin, 2013), variants of the assignment problem (Cela *et al*, 2014), scheduling problems (Pinedo *et al*, 2015), the set covering (Beasley, 1990; Boschetti and Maniezzo, 2015), and vehicle routing problems (Coelho and Laporte, 2015). Combinatorial problems are normally easy to describe but difficult to solve (Korte and Vygen, 2012). The foundations for the theory of computational complexity are to be found in Cook's (1971) seminal paper. In his paper, Cook attempted to classify problems in practice as easy or hard. A problem is called easy if an algorithm can be developed which solves the problem to optimality in a polynomial-time (Du and Pardalos, 2005). A problem is called hard or intractable if such efficient algorithms do not exist or the solution cannot be found within a reasonable computational time bound.

Due to the practical importance of combinatorial optimization problems, many methods have been developed for solving them (Pulleyblank (2014); Martello and Ries (2015); Nishi *et al*, (2017)). These methods can be classified as either exact or approximate. Exact methods guarantee to find an optimal solution in a bounded amount of time. Of course, for those combinatorial optimization problems which belong to the class NP-hard (Garey and Johnson, 1979), exact methods need an exponential amount of time. Thus, approximation methods which usually called heuristics, are often considered to be the only practical tools available to solve hard combinatorial optimization problems.

This section overviews a new idea, so-called fuzzy greedy evaluation concept for the integration of approaches for hard combinatorial optimization problems. The section also provides a comprehensive discussion on mathematics of the proposed methodology. In this presentation, the proposed method can be seen as a generic heuristic for the integration of approaches to hard combinatorial optimization problems. Furthermore, the current work shows another application of the proposed method with different objectives in addition to the previously published literature on the same subject. For example, it shows that how different adaptations of the proposed method can be yielded to different results.

The section consists of two main parts. The first part focuses on description of the theory and mathematics of the so-called fuzzy greedy evaluation concept. The second part demonstrates computational examples of the proposed concept for hard combinatorial optimization problems. The concluding remarks contain some suggestions for further research.

## Fuzzy Sets

In many real-world problems fuzzy sets allow us to represent vague concepts expressed in natural language. The membership function of a fuzzy set $A$ can be denoted by $\mu_A: X \rightarrow [0,1]$. Each fuzzy set should be uniquely defined by one particular membership function. Consider a fuzzy set where membership function is defined in Equation (1). This is one of the general formulae of a parameterized family of membership functions described in Klir and Yuan (1995).

$$\mu_A(x) = \frac{1}{1+\rho(x-r)^2} \tag{1}$$

This fuzzy set expresses, in a particular form, the general concept of a class of real numbers that are close to $r$. When the non-negative parameter $\rho$ increases, the graph of $\mu_A(x)$ becomes narrower. The function has the following properties: $\mu_A(r) = 1$ and $\mu_A(x) < 1$ for all $x\ ^1\ r$. More detailed discussions of fuzzy sets and their properties can be found in (Klir and Yuan, 1995; Zimmermann, 2001; Wang and Klir, 2013).

## Greedy Algorithms

The greedy method is an approach to algorithm design which is particularly suited for solving certain optimization problems. Such algorithms make choices according to what appears to be the best choice at the moment, without any consideration of any previous choices or future consequences. Greedy algorithms do not necessarily yield optimal solutions, but for some problems they do.

The greedy paradigm is often used in solving combinatorial optimization problems (Cook *et al*, 1998). This phenomenon can be explained by the fact that, although it is widely assumed that greedy algorithms rarely produce optimal solutions, they often provide solutions that are significantly better than the worst case (Vince, 2002; Curtis, 2003; Bang-Jensen *et al*, 2004). Hence, a natural question to pose is: can we characterize the type of optimization problems for which greedy algorithms produce optimal solutions? In a sense, the theory of matroids (Whitney, 1935; Edmonds, 1971), and later greedoids (Korte and Lovasz, 1981; Korte and Lovasz, 1983; Korte *et al*, 1991) are useful in determining when the greedy method yields optimal solutions. Vince (2002) put greedy algorithms into a simple framework, which explains when they do not work for all linear objective functions. Curtis (2003) addressed principles for the classification of those greedy algorithms that do provide optimal solutions to optimization problems. In extreme cases, for certain cases of optimization problems, a greedy algorithm may produce the unique worst possible solution (Gutin and Yeo, 2002; Gutin *et al*, 2002). Bang-Jensen *et al* (2004) characterized certain cases in which the greedy algorithm fails.

Because of the simplicity and versatility of greedy algorithms, they are frequently applied in many various areas such as combinatorial optimization problems. A variety of applications is addressed in (Cormen *et al*, 1990; Curtis 2003). For example, the greedy algorithm gives optimal solutions for Dijkstra's algorithm for shortest path (Dijkstra, 1959) and minimal spanning tree (Kruskal, 1956; Prim, 1957). More detailed discussions on the subject can be found in (Horowitz and Sahni, 1978; Papadimitriou and Steiglitz, 1982; Wilson, 1985; Grimaldi, 1989; Cook *et al*, 1998).

## Fuzzy Greedy Evaluation

Consider combinatorial optimization problems, each of which is associated with a discrete solution space $X$, a feasible space $S$ with the property $S \subseteq X$ that is defined by the problem constraints, and an objective function $f: X \rightarrow \mathfrak{R}$. In the case of minimization, the aim is to find a feasible solution $\mathbf{x}^* \in S$ such that $f(\mathbf{x}^*) \leq f(\mathbf{x}), \forall \mathbf{x} \in S$, where, $\mathbf{x} = (x_1,\ldots,x_n)$ is a vector of decision variables (solution). A cost function $c(x)$, $\forall x \in \{x_1,\ldots,x_n\}$ is defined for each specific problem. This function represents the incremental increase in the value of the objective function due to the incorporation of an element '$x$' into the solution under construction. We refer to $c(x)$ as a greedy evaluation function. Let the set of candidate elements be the set of all such elements that can be incorporated into the partial solution under construction without causing infeasibility. Therefore, the selection of the next element '$x$' for incorporation into the partial solution under construction is determined by the evaluation of all candidate elements according to our greedy evaluation function.

In the standard greedy approach, the greedy evaluation function denotes a degree of priority for incorporating the corresponding element '$x$' into the solution under construction. We now describe an alternative approach. We will treat the set $X$ as a fuzzy set with a well-defined membership function $\mu(x)$, the form of which is given by Equation (2).

$$\mu(x) = \frac{1}{1+\lambda^2\left(\left(\frac{1-\lambda}{\lambda}\right)x-\theta\right)^2} \tag{2}$$

The variable $x \in X$ corresponds to one of the variables in the definition of our combinatorial optimization problem. The parameter $\theta$ is a basic measure for evaluating the priority to be assigned to variable $x$. The parameter $\lambda$ is a tuning parameter that is chosen by experimentation such that $0 \leq \lambda < 1$. This parameter will be seen to play an important part in the proposed approach.

The evaluation function in Equation (2) has the following properties: $\mu(\lambda\theta / (1 - \lambda)) = 1$ and $0 < \mu(x) < 1$ for all $x \neq \lambda\theta / (1 - \lambda)$. This implies that $x$ closest numerically to $\lambda\theta / (1 - \lambda)$ should be given higher priority. Let $x_{min}$ and $x_{max}$ be the smallest and the largest values of $x$, respectively. We define a small enough value $\lambda_{min} = x_{min} / (x_{min} + \theta)$ for $\theta > 0$ and any $\lambda \leq \lambda_{min}$, for which inequality $x \geq \lambda_{min}\theta / (1 - \lambda_{min})$ holds. We also define a big enough value $\lambda_{max} = x_{max} / (x_{max} + \theta)$ for $\theta > 0$ and any $\lambda \geq \lambda_{max}$, for which inequality $x \leq \lambda_{max}\theta / (1 - \lambda_{max})$ holds.

Equation (2) is a modification of Equation (1), which is one of the general formulas of the families of fuzzy membership functions. We will refer to this function as a fuzzy greedy evaluation function, and it will replace the role of the greedy evaluation function in determining the degree of priority assigned to an element '$x$'. Figure 1 shows examples of the proposed fuzzy greedy evaluation function on different values of $\lambda$.

To convey the nature of the concept, we give a simple example. Suppose we want to build a list of a given set of integers $X = \{x_1,\ldots,x_n\}$ so that the numbers are ordered in terms of their closeness to a number which is equal to one quarter of the mean value of the numbers, say $\theta/4$, where $\theta = (x_1+\ldots+x_n)/n$. We will set the tuning parameter $\lambda$ in Equation (2) to 0.2; for this $\lambda$ value, the equation shows that $\mu(x)$ is maximized when $x$ takes the value $\theta/4$. Clearly, if element 1 has a value $x_1$ and element 2 has a value $x_2$ and if $|x_1 - \theta/4| < |x_2 - \theta/4|$, then element 1 will be of higher priority than element 2. Thus, it is only necessary to sort the list of candidate elements by descending order of their membership grades, corresponding to $\mu(x)$. Much more will be said below about the role of this parameter and how its value should be set in the context of particular combinatorial optimization problems.

## Alternative Fuzzy Membership Functions

As we have mentioned in sections above, the fuzzy membership functions can have very different shapes of their graphs, in the context of a particular application. We can also mention the following membership functions, which are described in Klir and Yuan (1995).

$$A_1(x) = \begin{cases} \rho(x-r)+1, & x \in [r-1/\rho, r] \\ \rho(r-x)+1, & x \in [r, r+1/\rho] \\ 0 & \text{otherwise} \end{cases} \tag{3}$$

$$A_2(x) = \begin{cases} (1+\cos(\rho\pi(x-r)))/2, & x \in [r-1/\rho, r+1/\rho] \\ 0 & \textit{otherwise} \end{cases} \tag{4}$$

$$A_3(x) = e^{-|\rho(x-r)|} \tag{5}$$

It should be noted that the shape parameter ρ depends on the domain of the membership function for the problem considered, which may be eliminated for combinatorial optimisation problems.

## Fuzzy Greedy Search

The fuzzy set theory has been applied in an attempt to answer this significant question in context of the greedy algorithms: what is the best choice at present? We propose that the fuzzy greedy evaluation function $\mu(x)$ in Equation (2) can be employed to identify a suitable choice. We introduce a new algorithm for evaluating the candidate elements by using $\mu(x)$ at each stage of the classical greedy algorithm. We have discussed the greedy parameter $\lambda$, which needs to be tuned. As a simple procedure, the algorithm can be invoked for a fixed number of times for any given problem instance. This number, say $t$, depends

*Figure 1. Distribution of $\mu(x)$ with $\theta$ = 10, and $\lambda$ = 0.4, 0.5 and 0.6.*

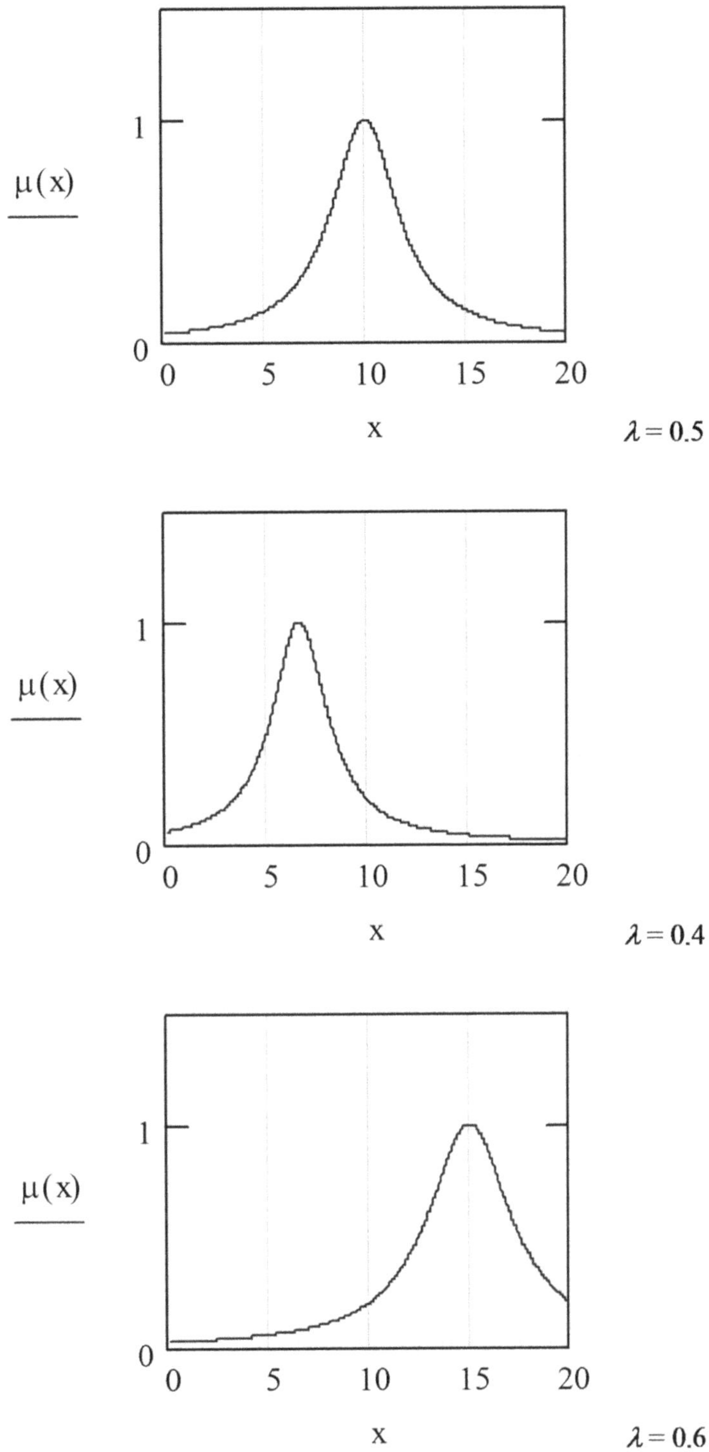

on whether we decide to let $\lambda$ take all possible values to $k = 2, 3, 4,\ldots$ decimal places over the range from $\lambda_{min}$ to $\lambda_{max}$. We will call these procedures as the fuzzy greedy search heuristic. The steps are as follows:

**Algorithm 1.** Fuzzy Greedy Search Heuristic.

```
WHILE (λmin ≤ λ up to k decimal places ≤ λmax) DO
BEGIN
        WHILE (X is not empty) DO
        BEGIN
                        1.      Choose an {x} which maximizes μ(x);
                        2.      X = X - {x};
                        3.      Incorporate {x} to build a solution;
        END
        Keep current best_solution;
END
RETURN best_solution;
```

Suppose $X = \{1,2,\ldots,10\}$. List the elements in order of their closeness to the mean of the elements of $X$. Examine the performance of the fuzzy greedy algorithm using different values of $\lambda$ equal to $\lambda_{min}$, 0.4, 0.5, 0.6 and $\lambda_{max}$.

Let $\theta = 11/2$ as the mean of the elements of $X$. From Equation (2), it is obvious that a value of $\lambda = 0.5$ will give $\mu(\theta) = 1$ so that using Algorithm (1) will give an optimal list. $\lambda_{min}$ is equal to 2/13. $\lambda_{max}$ is equal to 20/31. Using the fuzzy greedy search algorithm we get the ordered lists shown in Table 1.

*Table 1. Performance of the fuzzy greedy algorithm using different values of $\lambda$.*

| $\lambda = \lambda_{min}$ | $\lambda = 0.4$ | $\lambda = 0.5$ | $\lambda = 0.6$ | $\lambda = \lambda_{max}$ |
|---|---|---|---|---|
| 1 | 4 | 5 | 8 | 10 |
| 2 | 3 | 6 | 9 | 9 |
| 3 | 5 | 4 | 7 | 8 |
| 4 | 2 | 7 | 10 | 7 |
| 5 | 6 | 3 | 6 | 6 |
| 6 | 1 | 8 | 5 | 5 |
| 7 | 7 | 2 | 4 | 4 |
| 8 | 8 | 9 | 3 | 3 |
| 9 | 9 | 1 | 2 | 2 |
| 10 | 10 | 10 | 1 | 1 |

In this trivial example, it was intuitively clear what the optimal choice of $\lambda$ should be. However, for the combinatorial optimization problems, this will not be the case; suitable values of $\lambda$ obtain by experimentation. The computational complexity of the fuzzy greedy algorithm is the same as that of the straightforward greedy algorithm. This can be viewed as a transformation of the exchange sort algorithm known as bubble sort. The computational complexity of the algorithm is $O(n^2)$ when bubble sort is used. This can be improved to $O(n \log n)$ if *merge sort* is used. More detailed and comprehensive discussions of sorting methods can be found in (Horowitz and Sahni, 1978; Neapolitan and Naimipour, 2003).

## FUZZY GREEDY HEURISTIC AND FLOW-SHOP SCHEDULING

### Introduction

The flow-shop scheduling problem can be stated as follows: there are *n* jobs to be processed by *m* machines in an identical sequence on each machine. The usual objective is to minimize the completion time of the last job to leave the system, commonly termed the makespan ($C_{max}$). The flow-shop scheduling problem with $m > 2$ belongs to the class of combinatorial optimization problems known to be NP-hard in the strong sense (Garey *et al*, 1976). Hence, approximation methods are generally considered to solve these problems. There are many variations of the flow-shop scheduling problem (Baker, 1974). In this section, the permutation flow-shop scheduling problem (PFSP) with makespan criterion is considered. In the PFSP after completing a job on one machine, the job is processed on the next machine or joins a queue if the machine is busy. All queues are assumed to operate under the first-in-first-out (FIFO) discipline. It means that a job cannot pass another job while waiting in a queue. Therefore, clearly there are *n*! possible schedules. Many heuristics have been proposed for this problem since Johnson's (1954) pioneering work, including Page (1961), Palmer (1965), Campbell *et al* (1970) and Nawaz *et al* (1983). The collections of survey papers (Rajendran, 1995; Framinan *et al*, 2004; Ruiz and Maroto, 2005; Stafford *et al*, 2005) and books (French, 1982; Sule, 1997; Pinedo *et al*, 2015) summarize various developments in the subject. Here, we consider some of those methods that are most relevant to our discussion. Rajendran (1993) developed a heuristic for the total flowtime minimization criterion, similar to the NEH heuristic, which is based on arranging the jobs according to a particular method of ranking. Woo and Yim (1998) developed the heuristic for solving problems with the mean flowtime minimization criterion similar to the NEH heuristic but evaluating the partial sequences corresponding to the lowest flowtime. Davoud Pour (2001) proposed a procedure for makespan minimization based on an insertion method and the idea of job exchanging. Nagano and Moccellin (2002) proposed a heuristic to minimize the makespan, in which the initial list of jobs was assumed to be the final sequence obtained by the NEH heuristic, and then they applied the NEH's insertion technique to construct a new sequence. They also modified the first step of the NEH heuristic by using a lower bound on the total waiting time for each job. Allahverdi and Aldowaisan (2002) developed a set of heuristics for minimizing total completion time. Their method constructs an initial sequence by applying the method of Woo and Yim (1998), and finally the obtained sequence is improved by a general pairwise interchange until it reaches a local optimum. Framinan and Leisten (2003) proposed a heuristic based on an adaptation of the NEH heuristic for flowtime minimization and applying a general pairwise interchange to improve the quality of the partial sequence obtained in the NEH's insertion phase. Framinan *et al* (2003), in their comprehensive study for producing an efficient initial order of the jobs for the NEH heuristic with multiple objectives, tried 177 distinct methods for ranking the jobs in the first phase of the NEH heuristic, of which the original NEH was just one. For makespan minimization, they concluded that the original NEH heuristic was the best method. This research thus confirmed the practical importance of the NEH heuristic for the PFSP with makespan criterion. Zheng and Yamashiro (2010) proposed a novel quantum differential evolutionary algorithm by merging particular evolutionary and local search strategies for permutation flow-shop scheduling problem minimizing the makespan and the total flowtime. Mirabi (2014) proposed a novel hybrid genetic algorithm using a modified approach to generate the initial population by an improved iterated swap procedure for the sequence-dependent permutation flow-shop scheduling problem minimizing the makespan. Liu *et al* (2017) utilized the skewness of the processing times to construct a priority rule to the NEH heuristic for permutation flow-shop scheduling with the makespan criterion. Sioud and Gagné (2018) proposed an enhanced migrating bird optimization algorithm for the permutation flow-shop problem with sequence dependent setup times minimizing the makespan.

In this section, a polynomial-time integrated heuristic for the PFSP with makespan minimization as the objective has been proposed. The section consists of two main parts. The first part focuses on description of the theory and mathematics of the proposed fuzzy greedy heuristic. The second part demonstrates computational examples of the proposed method using benchmark problems. The concluding remarks contain some suggestions for further research.

## The Integrated Heuristic

The proposed integrated fuzzy greedy heuristic (IFGH) consists of two phases: arranging the jobs in priority order and then constructing a sequence by a simple job insertion principle similar to the NEH heuristic. The priority of the jobs is determined according to Equation (2). Where, $x$ is a generic variable associated with the data defining a particular combinatorial optimization problem. The parameter $\theta$ is a basic measure for evaluating the priority to be assigned to $x$. The parameter $\lambda$ is a tuning parameter such that $0 \leq \lambda < 1$ to adjust $\theta$. Here, we represent $x$ with $x_j$ for each job $j$ through Equations (6) and (7), corresponding to strategies TPT (Total Processing Times), and CVPT (Coefficient of Processing Times), respectively. The parameter $\theta$ can be seen as the average of $x_j$ values, through Equation (8).

$$x_j = \sum_{i=1}^{m} P_{ij} \tag{6}$$

$$x_j = \left( \frac{m\sum_{i=1}^{m} P_{ij}^2 - \left(\sum_{i=1}^{m} P_{ij}\right)^2}{m(m-1)} \right)^{1/2} \times \left( \frac{m}{\sum_{i=1}^{m} P_{ij}} \right) \tag{7}$$

$$\theta = \frac{1}{n}\sum_{j=1}^{n} x_j \tag{8}$$

## The IFGH heuristic

1. Calculate $\mu(x_j)$ for each job $j$.
2. Arrange the jobs by descending order of $\mu(x_j)$.
3. Select the next job and insert it in all possible positions in the partial sequence and keep the best one (i.e. minimum makespan) as the current partial sequence.
4. Repeat step 3 until all jobs are scheduled corresponding to each strategy.
5. Repeat steps 1 to 4 for all preferred $\lambda$ values for both strategies TPT and CVPT and return the best solution.

The heuristic has the same computational complexity as the NEH heuristic and so it can be implemented as $O(n^2m)$. The heuristic needs to be invoked for a fixed number of times for any given problem instance. This number, say $t$, depends on whether we decide to let $\lambda$ take all possible values to 2, 3, 4,... decimal places over the range from $\lambda_{min}$ to $\lambda_{max}$ corresponding to each strategy. The choice of $t$ is, therefore, independent of both $m$ and $n$. Hence, our experimental results have a computation time that is $t$ times that of the NEH heuristic. In this sense, our approach is of course more computationally expensive than that of the NEH heuristic. However, because $t$ is a fixed number that is independent of $m$ and $n$, the theoretical computational complexity of the approach remains unaffected by the value of $t$. We would argue that the additional computational expense incurred by this approach, compared to the NEH heuristic, is justified by the significant reduction obtained in the average error. Figure 2 exemplifies the effect of different values of $\lambda$ on the computational performance of the proposed heuristic.

*Figure 2. IFGH on a PFSP instance.*

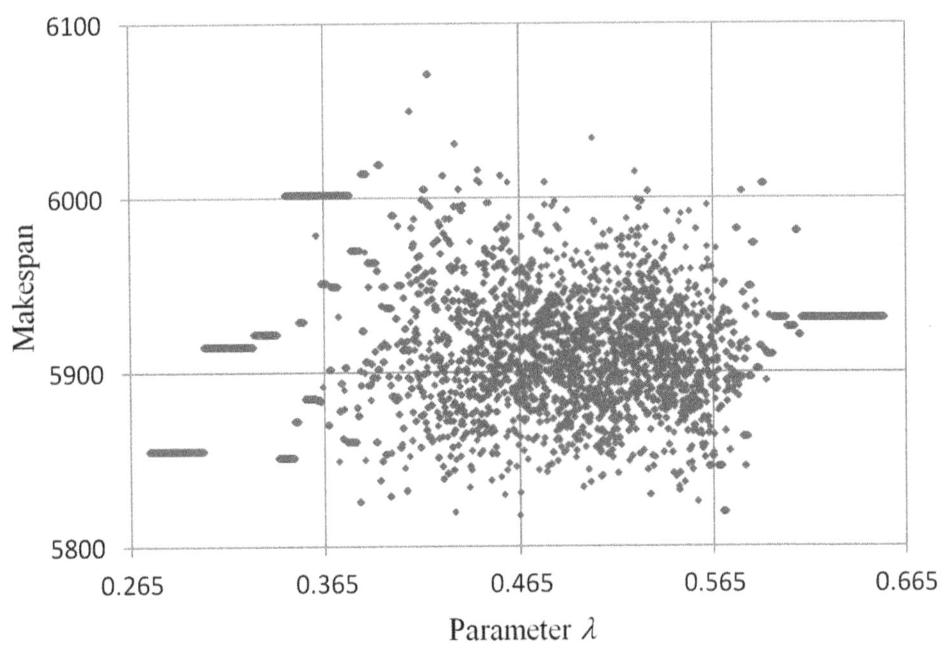

It might be supposed that, instead of using $t$ different $\lambda$ values for the proposed heuristic, a possible alternative approach that is equally expensive computationally would be to repeat the NEH heuristic $t$ times, in each case using a different randomly generated initial job order. In fact this is unlikely to produce an improvement. Framinan *et al* (2003) showed that such randomization does not, in general, improve the NEH heuristic. Randomization was included in 177 different systematic approaches to modifying the NEH heuristic. None of these 177 approaches gave an improvement over the NEH heuristic. This is in direct contrast to the heuristic proposed in this section.

We introduced the tuning parameter $\lambda$ to obtain a good performance of the proposed heuristic. The efficiency of the heuristic depends greatly on the choice of the parameter $\lambda$ in an effective range. If the range is too small, the probability that it includes the best $\lambda$ value will be low. If it is too large, the algorithm may waste computational resources. In this experimentation, we evaluate the results obtained for all values of $\lambda$ between $\lambda_{min}$ and $\lambda_{max}$ up to four decimal places (i.e. with increments of $\lambda$ equal to $10^{-4}$) for each instance of the benchmark problems considered. Where there are several such $\lambda$ values, we select those that correspond to a minimum makespan. We refer to these selected values as $\lambda^*$. Figure 3 shows the distribution of $\lambda^*$ values as a histogram. We found that setting $\lambda$ up to 4 decimal places is sufficiently extensive for the benchmark instances considered (Sheibani, 2005; Sheibani 2010).

*Figure 3. The distribution of $\lambda^*$ values as a histogram.*

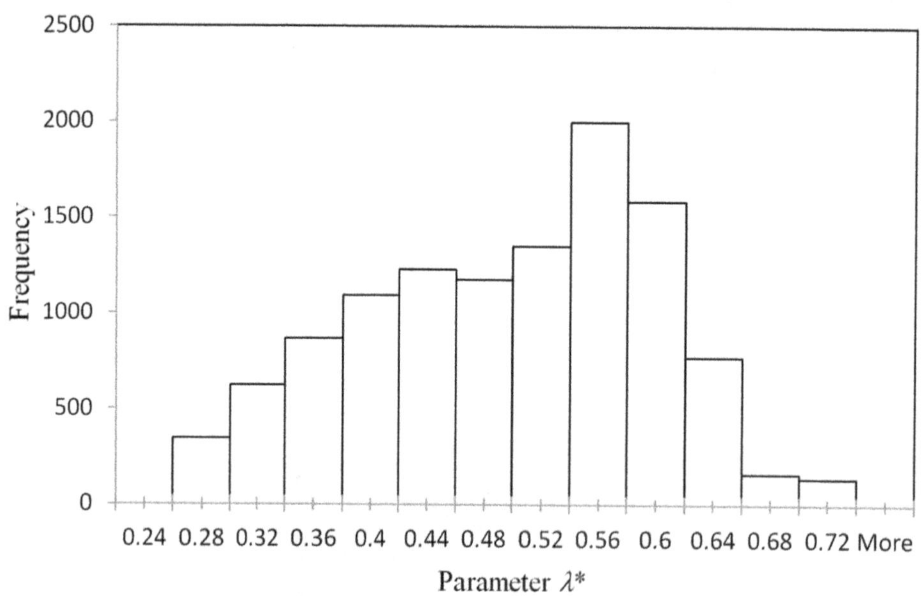


The statistical analyses show that the range of the values of $\lambda^*$ is between 0.2459 and 0.6936 with the standard deviation 0.1024, the average 0.4754, the median 0.4898, and the mode 0.5389, so it is easy to observe that the values of $\lambda^*$ have an asymmetric distribution, with skew –0.2691.

The proposed heuristic was implemented in C++. The test problems are all from an extensive set of Taillard (1993) standard benchmark instances. They were chosen in order to compare our results with those obtained by others. The solution quality is measured by the percentage deviation of the obtained solution from the best-known solution. In Table 2, we compared the performance of the IFGH heuristic

*Table 2. A comparison of the IFGH heuristic.*

| Problem size (Job × Machine) | Problem | NEH | FGH(CVPT) | FGH(TPT) | IFGH |
|---|---|---|---|---|---|
| 20 × 5 | ta001–010 | 2.81 | 1.21 | 0.88 | 0.72 |
| 20 × 10 | ta011–020 | 4.58 | 1.91 | 1.97 | 1.66 |
| 20 × 20 | ta021–030 | 3.60 | 1.82 | 1.94 | 1.72 |
| 50 × 5 | ta031–040 | 1.09 | 0.35 | 0.18 | 0.16 |
| 50 × 10 | ta041–050 | 6.00 | 3.11 | 3.03 | 2.91 |
| 50 × 20 | ta051–060 | 6.09 | 4.47 | 4.15 | 4.11 |
| 100 × 5 | ta061–070 | 0.49 | 0.15 | 0.16 | 0.09 |
| 100 × 10 | ta071–080 | 2.22 | 1.19 | 1.10 | 1.02 |
| 100 × 20 | ta081–090 | 5.61 | 4.15 | 3.99 | 3.94 |
| 200 × 10 | ta091–100 | 1.24 | 0.82 | 0.65 | 0.65 |
| 200 × 20 | ta101–110 | 4.56 | 3.45 | 3.27 | 3.26 |
| 500 × 20 | ta111–120 | 2.23 | 1.79 | 1.61 | 1.61 |
| **Average** | | **3.38** | **2.04** | **1.91** | **1.82** |

with its original adaptations using strategies TPT and CVPT, as well as the NEH heuristic, which has been commonly regarded as one of the most effective heuristics for PFSP in the literature.

The results in Table 2 show that the integration of proposed TPT and CVPT strategies significantly improve relative performance over the NEH heuristic for all the problems of varying sizes in terms of the percentage error obtained.

We recall that for every instance, there exists a value of $\lambda$ for which the makespan obtained is less than or equal to the heuristic's greedy versions ($\lambda < \lambda_{min}$ or $\lambda > \lambda_{max}$). In a sense, the heuristic can be said to dominate the greedy versions. It should be noted that in a particular case where the sum of the processing times of each job is the same on each machine, certain priority rule-based scheduling heuristics for PFSP would assign the same priority to every job, and so in this situation, all possible permutations of the jobs would be equally likely to be selected by those heuristics. This implies practical significance of the proposed integrated heuristic using CVPT. More detailed discussion on the fuzzy greedy heuristics can be found in (Sheibani 2004-2019).

## FUZZY GREEDY SEARCH AND JOB-SHOP PROBLEM

### Introduction

The effectiveness and efficiency of the proposed fuzzy greedy search heuristic are demonstrated on job-shop scheduling problem (JSP) as one of the most challenging classical sequencing problems in the areas of operational research and management science. In general form the JSP can be stated as follows. There are $n$ jobs to be processed by a finite set of $m$ machines on which each job visits each machine at most once in its own predetermined route. There are numerous distinct scheduling goals (Mellor,

1966). One of the usual objectives is to minimise the mean total completion time (MTCT) of the jobs. The JSP belongs to the class of combinatorial optimization problems known to be NP-hard (Garey *et al*, 1976). Exact methods have an exponential time bound and have successfully solved only very small size instances. Hence, approximation methods are generally considered to be the only practical way to solve most real-life problems. Many heuristics have been proposed for the problem. Most of them are based on priority dispatching rules (Panwalker and Iskander, 1977; Haupt, 1989). The collections of survey papers (Pinson, 1995; Holthaus and Rajendran, 1997; Jain and Meeran, 1999) and books (French, 1982; Pinedo *et al*, 2015) summarise various developments in the subject.

The proposed heuristic consists of two phases: building a candidate list of operations in priority order and then constructing a schedule. The candidate list is formed from all possible unscheduled operations, each of which can be incorporated into the partial schedule under construction without causing infeasibility. The priority of the operations in the list is determined according to Equation (2). Here, $x$ is a generic variable associated with the data defining a particular instance of the JSP. The parameter $\theta$ is a basic measure for evaluating the priority to be assigned to $x$. The parameter $\lambda$ is a tuning parameter that is chosen by experimentation such that $0 \le \lambda < 1$ to adjust $\theta$.

Let $P_{i,j}$ be the processing time of job $j$ on machine $i$. Here, $x$ is represented with $x_{i,j}$ corresponding to operation $(i,j)$, and the parameter $\theta$ is set to constants, as specified in Table 3.

*Table 3. Strategies S1, S2, and S3.*

| *Strategy* | $x_{i,j}$ | $\theta$ |
|---|---|---|
| S1 | $P_{i,j} / \sum_{i=1}^{m} P_{i,j}$ | $1/m$ |
| S2 | $\left(\sum_{i=1}^{m} P_{i,j}\right) - P_{i,j}$ | $(1/n)(1-1/m)\sum_{j=1}^{n}\sum_{i=1}^{m} P_{i,j}$ |
| S3 | $P_{i,j}$ | $(1/mn)\sum_{j=1}^{n}\sum_{i=1}^{m} P_{i,j}$ |

The proposed evaluation function has the following properties: $\mu(\lambda\theta/(1-\lambda))=1$ and $0<\mu(x_{i,j})<1$ for all $x_{i,j} \neq \lambda\theta/(1-\lambda)$. This implies that scheduling the job with $x_{i,j}$ closest numerically to $\lambda\theta/(1-\lambda)$ should be given higher priority. The steps of the heuristic are as follows:

***The FGH heuristic for JSP***

WHILE ($\lambda_{min} \le \lambda$ up to k decimal places $\le \lambda_{max}$) DO

BEGIN

1. Calculate $\mu(x_{i,j})$ for each operation;
2. Arrange the candidate list of unscheduled operations;
3. Select the next operation with the highest priority from the candidate list and insert it at the earliest available feasible time slot on the schedule under construction;
4. Repeat steps 2 and 3 until all operations are scheduled;
5. Keep current best_solution;

END

RETURN best_solution;

The computational complexity of the proposed heuristic is the same as that of the corresponding greedy algorithm described above. The heuristic needs to be invoked for a fixed number of times for any given problem instance. This number, say $t$, depends on whether we decide to let $\lambda$ take all possible values to $k = 2, 3, 4,\ldots$ decimal places over the range from $\lambda_{min}$ to $\lambda_{max}$. The choice of $t$ is, therefore, independent of both $m$ and $n$. However, because $t$ is a fixed number that is independent of $m$ and $n$, the theoretical computational complexity of the approach remains unaffected by the value of $t$. For more details on methodology we refer the interested reader to (Sheibani, 2005 and 2010). Figure 4 exemplifies the effect of different values of λ on the computational performance of the proposed heuristic.

*Figure 4. Performance of FGH(S2) on a JSP instance using of different λ values.*

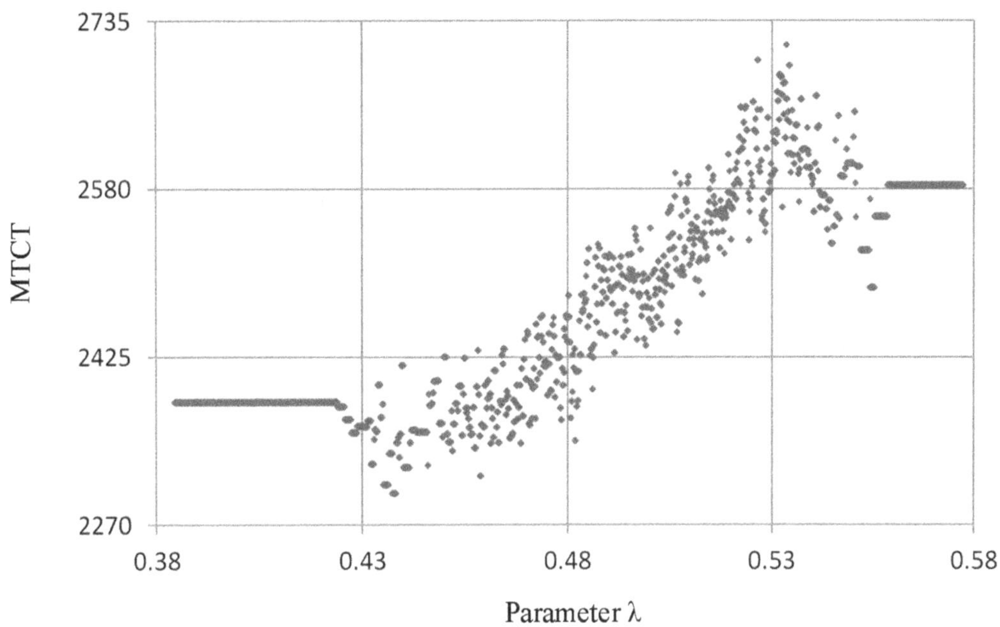


## Experimental Results

The proposed heuristic was implemented in C++. The test problems are all from an extensive set of Taillard's (1993) standard benchmark instances. Three adaptations of the heuristic are presented as FGH(S1), FGH(S2), and FGH(S3) corresponding to strategies S1, S2, and S3.

We introduced the tuning parameter λ to achieve a good performance of the proposed heuristic. The efficiency of the FGH heuristic depends greatly on the choice of the parameter λ in an effective range. If the range is too small, the probability that it includes the best λ value will be low. If it is too large, the algorithm may waste computational resources and so the waiting time for an improvement could be too long. In this experimentation, we evaluate the results obtained for all values of λ between $\lambda_{min}$ and $\lambda_{max}$ up to four decimal places (i.e. with increments of λ equal to $10^{-4}$) for each instance of the benchmark problems considered. Where there are several such $\lambda$ values, we select one that corresponds to a minimum MTCT. We refer to this selected value as $\lambda^*$. Figure 5 exemplifies the distribution of these $\lambda$ values as a histogram.

*Figure 5. The distribution of $\lambda^*$ values as a histogram using FGH(S2).*

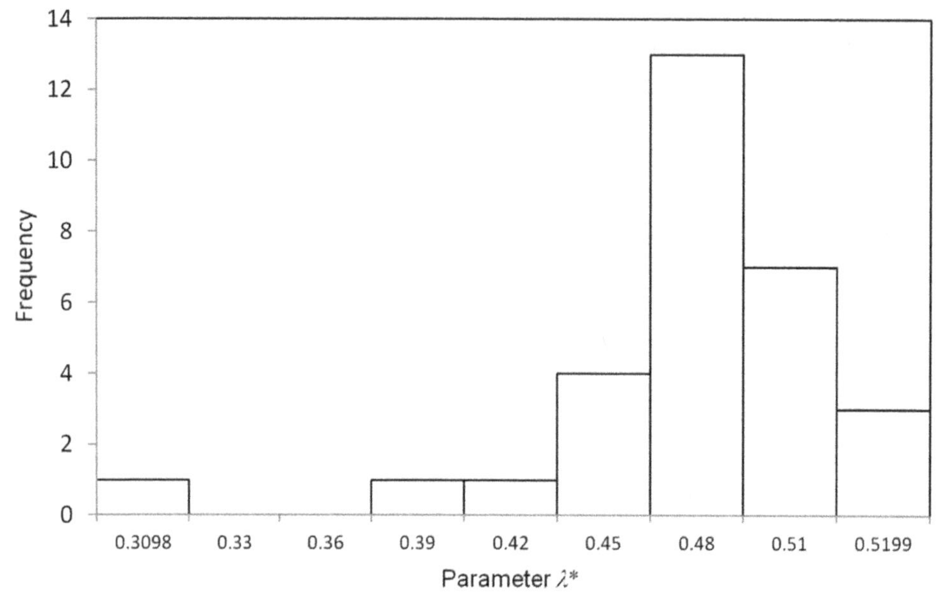


The performance of the proposed FGH heuristic in comparison with two classical dispatching rules: SPT (shortest processing time first), which has been commonly regarded as one of the most effective heuristics for this problem (Holthaus and Rajendran, 1997); and LPT (longest processing time first), is reported in Table 4. We investigated the performance of the proposed heuristic when the fuzzy part is turned off. In this form, we reduce all the adaptations of FGH to corresponding greedy versions by taking either minimum or maximum values of $x_{i,j}$. We refer to these as GHS(shortest strategy first) and GHL(longest strategy first). It is obvious that $\mu(x)$ becomes an increasing function by setting the parameter $\lambda$ to a big enough value ($\lambda \geq \lambda_{max}$). Thus, FGH(S3) arranges the operations by descending order of $x_{i,j}$. This is equivalent to the SPT rule. It is also clear that by setting the parameter $\theta$ to zero and/or setting the parameter $\lambda$ to a small enough value ($\lambda \leq \lambda_{min}$), $\mu(x)$ becomes a decreasing function, so that FGH(S3) arranges the operations by ascending order of $x_{i,j}$. This is equivalent to the LPT rule.

The results show that the fuzzy adaptations significantly improve relative performance over greedy versions for all of the problems of varying sizes in terms of the solution obtained. It is easy to observe that FGH(S2) is superior to all of the methods for all of the problems of varying sizes in terms of the average solution obtained.

*Table 4. A comparison of different addaptations of the FGH heuristic with classical dispatching rules SPT and LPT.*

| Problem Size (Job × Machine) | Problem | FGH(S1) | GHS(S1) | GHL(S1) | FGH(S2) | GHS(S2) | GHL(S2) | FGH(S3) | SPT | LPT |
|---|---|---|---|---|---|---|---|---|---|---|
| 15 × 15 | ta01 | 1306.27 | 1357.27 | 1306.27 | 1245.47 | 1319.47 | 1378.33 | 1244.53 | 1244.53 | 1388.93 |
| | ta02 | 1274.87 | 1392.00 | 1290.47 | 1245.73 | 1304.93 | 1316.40 | 1244.60 | 1399.53 | 1273.53 |
| | ta03 | 1300.60 | 1393.40 | 1361.20 | 1207.07 | 1348.80 | 1321.87 | 1279.00 | 1279.00 | 1337.47 |
| | ta04 | 1268.33 | 1343.67 | 1366.80 | 1188.93 | 1226.73 | 1287.40 | 1242.80 | 1302.73 | 1401.80 |
| | ta05 | 1267.93 | 1359.07 | 1398.00 | 1261.00 | 1356.07 | 1349.80 | 1287.07 | 1440.13 | 1418.40 |
| | ta06 | 1276.67 | 1328.80 | 1380.67 | 1228.93 | 1282.20 | 1352.73 | 1282.67 | 1314.07 | 1349.20 |
| | ta07 | 1330.80 | 1381.33 | 1353.47 | 1250.40 | 1303.00 | 1398.87 | 1287.33 | 1346.60 | 1341.13 |
| | ta08 | 1274.07 | 1290.87 | 1274.07 | 1197.00 | 1298.93 | 1303.60 | 1248.87 | 1331.93 | 1321.07 |
| | ta09 | 1334.13 | 1471.93 | 1449.47 | 1323.13 | 1354.67 | 1525.80 | 1331.27 | 1429.93 | 1480.27 |
| | ta10 | 1318.33 | 1394.40 | 1401.00 | 1261.53 | 1283.60 | 1429.80 | 1248.87 | 1373.47 | 1405.67 |
| | **Average** | **1295.20** | **1371.27** | **1358.14** | **1240.92** | **1307.84** | **1366.46** | **1269.70** | **1346.19** | **1371.75** |
| 30 × 20 | ta41 | 2190.03 | 2291.07 | 2332.20 | 2067.80 | 2178.00 | 2192.53 | 2261.57 | 2304.80 | 2303.07 |
| | ta42 | 2151.63 | 2285.03 | 2179.53 | 2048.43 | 2121.10 | 2236.30 | 2154.47 | 2277.43 | 2154.47 |
| | ta43 | 2078.83 | 2190.47 | 2155.80 | 1945.13 | 2039.33 | 2146.07 | 2102.27 | 2204.43 | 2199.87 |
| | ta44 | 2171.43 | 2223.20 | 2264.77 | 2002.70 | 2068.87 | 2245.83 | 2125.20 | 2160.00 | 2387.43 |
| | ta45 | 2179.93 | 2250.33 | 2258.53 | 2057.77 | 2121.63 | 2265.90 | 2095.20 | 2160.03 | 2300.10 |
| | ta46 | 2210.73 | 2288.03 | 2267.60 | 2101.50 | 2197.63 | 2282.00 | 2195.50 | 2394.83 | 2329.00 |
| | ta47 | 2122.70 | 2198.67 | 2171.57 | 2002.93 | 2081.80 | 2211.50 | 2117.80 | 2180.77 | 2261.60 |
| | ta48 | 2066.80 | 2221.37 | 2146.40 | 1986.50 | 2063.73 | 2118.60 | 2068.87 | 2158.83 | 2175.43 |
| | ta49 | 2136.07 | 2231.40 | 2267.10 | 2007.53 | 2017.27 | 2194.60 | 2109.37 | 2147.03 | 2181.23 |
| | ta50 | 2117.53 | 2302.23 | 2217.73 | 2047.73 | 2100.47 | 2194.17 | 2173.87 | 2173.87 | 2275.60 |
| | **Average** | **2142.57** | **2248.18** | **2226.12** | **2026.80** | **2098.98** | **2208.75** | **2140.41** | **2216.20** | **2256.78** |
| 50 × 15 | ta51 | 2611.24 | 2722.92 | 2660.72 | 2299.46 | 2383.66 | 2583.54 | 2544.60 | 2624.86 | 2693.10 |
| | ta52 | 2618.76 | 2686.30 | 2641.14 | 2264.74 | 2366.20 | 2542.22 | 2597.04 | 2672.10 | 2743.44 |
| | ta53 | 2473.98 | 2671.16 | 2558.14 | 2268.60 | 2282.26 | 2591.80 | 2452.36 | 2579.16 | 2553.98 |
| | ta54 | 2512.02 | 2641.28 | 2645.70 | 2348.12 | 2397.56 | 2501.04 | 2556.14 | 2670.04 | 2610.84 |
| | ta55 | 2577.10 | 2688.26 | 2676.06 | 2287.04 | 2398.76 | 2553.68 | 2483.74 | 2589.78 | 2683.36 |
| | ta56 | 2610.84 | 2633.52 | 2666.24 | 2293.54 | 2293.54 | 2612.70 | 2550.12 | 2633.54 | 2716.16 |
| | ta57 | 2642.16 | 2711.32 | 2834.96 | 2328.04 | 2465.52 | 2574.66 | 2605.22 | 2714.88 | 2753.92 |
| | ta58 | 2667.56 | 2771.18 | 2769.10 | 2445.58 | 2500.10 | 2664.34 | 2656.52 | 2798.26 | 2755.96 |
| | ta59 | 2547.10 | 2679.36 | 2711.58 | 2265.58 | 2332.70 | 2516.56 | 2509.52 | 2591.58 | 2619.40 |
| | ta60 | 2628.76 | 2776.06 | 2660.52 | 2309.48 | 2384.52 | 2541.58 | 2591.12 | 2750.82 | 2639.60 |
| | **Average** | **2588.95** | **2698.14** | **2682.42** | **2311.02** | **2380.48** | **2568.21** | **2554.64** | **2662.50** | **2676.98** |

## CONCLUDING REMARKS

This chapter proposed a new integrated polynomial-time fuzzy greedy heuristic for combinatorial optimization problems with specific reference to the permutation flow-shop and job-shop scheduling problems. We examined the effectiveness and the efficiency of the proposed heuristic on a wide range of benchmark problems of varying sizes. The developed heuristic gives a significantly improved performance relative to the well-known NEH heuristic, which has dominated the field for many years.

The proposed method has the potential for application to other combinatorial optimization problems, if a suitable evaluation measure can be properly defined based on global information for a given problem solution. As it has been shown, technically the proposed heuristic dominates to its greedy version. Thus, it should be highly expected that the proposed method would have a particular capability in the integration of greedy methods for a certain combinatorial optimization problem.

For future research we believe that the following topics are potentially useful: (1) developing efficient adaptations of the proposed heuristic, (2) extending the proposed method to other variations or objectives, and (3) developing efficient methods using the fuzzy greedy evaluation concept in other areas of optimization.

## REFERENCES


Allahverdi, A., & Aldowaisan, T. (2002). New heuristic to minimize total completion time in m-machine flowshops. *International Journal of Production Economics*, *11*(1), 77–83. doi:10.1016/S0925-5273(01)00228-6

Baker, K. R. (1974). *Introduction to Sequencing and Scheduling*. John Wiley & Sons.

Ballis, A., Karapetis, F., & Ballis, T. (2017). Towards the Implementation of Optimal Train Loading Plan in the Athens-Thessaloniki Freight Services. *International Journal of Operations Research and Information Systems*, *8*(2), 78–92. doi:10.4018/IJORIS.2017040104

Bang-Jensen, J., Gutin, G., & Yeo, A. (2004). When the greedy algorithm fails. *Discrete Optimization*, *1*(2), 121–127. doi:10.1016/j.disopt.2004.03.007

Beasley, J. E. (1990). An algorithm for set covering problems. *European Journal of Operational Research*, *31*(1), 85–93. doi:10.1016/0377-2217(87)90141-X

Boschetti, M., & Maniezzo, V. (2015). A set covering based matheuristic for a real-world city logistics problem. *International Transactions in Operational Research*, *22*(1), 169–195. doi:10.1111/itor.12110

Campbell, H. G., Dudek, R. A., & Smith, M. L. (1970). A heuristic algorithm for the n-job m-machine sequencing problem. *Mngt Sci*, *16*(10), B630–B637. doi:10.1287/mnsc.16.10.B630

Cela, E., Deineko, V. G., & Woeginger, G. J. (2014). Linearizable special cases of the QAP. *Journal of Combinatorial Optimization*, 1–11.

Coelho, L. C., & Laporte, G. (2015). Classification, models and exact algorithms for multi-compartment delivery problems. *European Journal of Operational Research*, *242*(3), 854–864. doi:10.1016/j.ejor.2014.10.059

Cook, S. A. (1971). The complexity of theorem-proving procedures. *Proceedings of the 3rd Annual ACM symposium on theory of computing*, 151-158. 10.1145/800157.805047

Cook, W. J., Cunningham, W. H., Pulleyblank, W. R., & Schrijver, A. (1998). *Combinatorial Optimization*. John Wiley & Sons.

Cormen, T., Leiserson, C. E., & Rivest, R. L. (1990). *Introduction to Algorithms*. The MIT Press.

Curtis, S. A. (2003). The classification of greedy algorithms. *Science of Computer Programming*, *49*(1-3), 125–157. doi:10.1016/j.scico.2003.09.001

Davoud Pour, H. (2001). A new heuristic for the n-job, m-machine flow-shop problem. *Production Planning and Control*, *12*(7), 648–653. doi:10.1080/09537280152582995

Dijkstra, E. W. (1959). A note on two problems in connexion with graphs. *Numerische Mathematik*, *1*(1), 269–271. doi:10.1007/BF01386390

Du, D. Z., & Pardalos, P. M. (Eds.). (2005). *Handbook of Combinatorial Optimization*. Springer-Verlag. doi:10.1007/b102533

Edmonds, J. (1971). Matroids and the greedy algorithm. *Mathematical Programming*, *1*(1), 126–136. doi:10.1007/BF01584082

Framinan, J. M., Gupta, J. N. D., & Leisten, R. (2004). A review and classification of heuristics for permutation flow-shop schedul-ing with makespan objective. *J Opl Res Soc*, *55*(12), 1–13. doi:10.1057/palgrave.jors.2601784

Framinan, J. M., & Leisten, R. (2003). An efficient constructive heuristic for flowtime minimisation in permutation flow shops. *Omega Int J Mngt Sci*, *31*(4), 311–317. doi:10.1016/S0305-0483(03)00047-1

Framinan, J. M., Leisten, R., & Rajendran, C. (2003). Different initial sequences for the heuristic of Nawaz, Enscore, Ham to mini-mize makespan, idletime or flow-time in the static permutation flowshop sequencing problem. *International Journal of Production Research*, *41*(1), 121–148. doi:10.1080/00207540210161650

French, S. (1982). *Sequencing and scheduling: an introduction to the mathematics of the job-shop*. Ellis Horwood.

Garey, M. R., & Johnson, D. S. (1979). *Computers and Intractability a Guide to the Theory of NP-Completeness*. WH Freeman & Company.

Garey, M. R., Johnson, D. S., & Sethi, R. (1976). The Complexity of Flowshop and Job-shop Scheduling. *Mathematics of Operations Research*, *1*(2), 117–129. doi:10.1287/moor.1.2.117

Garey, M. R., Johnson, D. S., & Sethi, R. (1976). The complexity of flowshop and jobshop scheduling. *Mathematics of Operations Research*, *1*(2), 117–129. doi:10.1287/moor.1.2.117

Grimaldi, R. P. (1989). *Discrete and Combinatorial Mathematics*. Addison-Wesley.

Gutin, G. (2013). Section 4.6 traveling salesman problems. Handbook of Graph Theory, 336–359.

Gutin, G., & Punnen, A. P. (Eds.). (2002). *The Traveling Salesman Problem and Its Variations*. Kluwer Academic Publishers.

Gutin, G., & Punnen, A. P. (Eds.). (2002). *The traveling salesman problem and its variations*. Kluwer Academic Publishers.

Gutin, G., & Yeo, A. (2002). Anti-matroids. *Operations Research Letters, 30*(2), 97–99. doi:10.1016/S0167-6377(02)00117-7

Gutin, G., Yeo, A., & Zverovich, A. (2002). *Traveling salesman should not be greedy: domination*. Academic Press.

Haupt, R. (1989). A Survey of Priority-Rule Based Scheduling. *OR-Spektrum, 11*(1), 3–16. doi:10.1007/BF01721162

Holthaus, O., & Rajendran, C. (1997). New Dispatching Rules for Scheduling in a Job Shop – An Experimental Study. *Advanced Manufacturing Technology, 13*(2), 148–153. doi:10.1007/BF01225761

Horowitz, E., & Sahni, S. (1978). *Fundamentals of Computer Algorithms*. Computer Science Press.

Jain, A. S., & Meeran, S. (1999). Deterministic Job-Shop Scheduling: Past, Present, and Future. *European Journal of Operational Research, 113*(2), 390–434. doi:10.1016/S0377-2217(98)00113-1

Johnson, S. M. (1954). Optimal two-and-three-stage production schedules with set-up times included. *Naval Research Logistics, 1*(1), 61–68. doi:10.1002/nav.3800010110

Klir, G. J., & Yuan, B. (1995). *Fuzzy Sets and Fuzzy Logic: Theory and Applications*. Prentice Hall.

Korte, B., & Lovasz, L. (1981). Mathematical structures underlying greedy algorithms. Lecture Notes in Computer Science, 117, 205–209. doi:10.1007/3-540-10854-8_22

Korte, B., & Lovasz, L. (1983). Structure properties of greedoids. *Combinatorica, 3*(3-4), 359–374. doi:10.1007/BF02579192

Korte, B., Lovasz, L., & Schrader, R. (1991). *Greedoids*. Springer-Verlag. doi:10.1007/978-3-642-58191-5

Korte, B., & Vygen, J. (2012). *Combinatorial optimization: theory and algorithms. Algorithms and Combinatorics* (Vol. 21). Springer-Verlag Berlin Heidelberg. doi:10.1007/978-3-642-24488-9

Kruskal, J. B. (1956). On the shortest spanning subtree and the traveling salesman problem. *Proceedings of the American Mathematical Society, 7*(1), 48–50. doi:10.1090/S0002-9939-1956-0078686-7

Lawler, E. L. (1985). *The Traveling Salesman Problem: A Guided Tour of Combinatorial Optimization*. John Wiley & Sons.

Liu, W., Jin, Y., & Price, M. (2017). A new improved NEH heuristic for permutation flowshop scheduling problems. *International Journal of Production Economics, 193*, 21–30. doi:10.1016/j.ijpe.2017.06.026

Martello, S., & Ries, B. (2015). Advances in Combinatorial Optimization. *Discrete Applied Mathematics, 196*(C), 1–3. doi:10.1016/j.dam.2015.08.011

Mellor, P. (1966). A Review of Job Shop Scheduling. *Operational Research Quarterly, 17*(2), 161–171. doi:10.1057/jors.1966.24

Mirabi, M. (2014). A novel hybrid genetic algorithm to solve the sequence-dependent permutation flow-shop scheduling problem. *International Journal of Advanced Manufacturing Technology, 71*(1-4), 429–437. doi:10.100700170-013-5489-5

Nagano, M. S., & Moccellin, J. V. (2002). A high quality solution constructive heuristic for flow shop sequencing. *J Opl Res Soc, 53*(12), 1374–1379. doi:10.1057/palgrave.jors.2601466

Nawaz, M., Enscore, E. E. Jr, & Ham, I. (1983). A heuristic algorithm for the m-machine, n-job flow-shop sequencing problem. *Omega International Journal of Management Science, 11*(1), 91–95. doi:10.1016/0305-0483(83)90088-9

Neapolitan, R. E., & Naimipour, K. (2003). *Foundations of Algorithms Using C++ Pseudo-code*. Jones & Bartlett Publishers.

Nishi, T., Ohno, A., Inuiguchi, M., Takahashi, S., & Ueda, K. (2017). A Combined column generation and heuristics for railway short-term rolling stock planning with regular inspection constraints. *Computers & Operations Research, 81*, 14–25. doi:10.1016/j.cor.2016.11.025

Page, E. S. (1961). An approach to the scheduling of jobs on machines. *Journal of the Royal Statistical Society. Series A (General), 23*, 484–492.

Palmer, D. S. (1965). Sequencing jobs through a multi-stage process in the minimum total time—A quick method of obtaining a near optimum. *Opns Res Quart, 16*(1), 101–107. doi:10.1057/jors.1965.8

Panwalker, S. S., & Iskander, W. (1977). A Survey of Scheduling Rules. *Operations Research, 25*(1), 45–61. doi:10.1287/opre.25.1.45

Papadimitriou, C. D., & Steiglitz, K. (1982). *Combinatorial Optimization: Algorithms and Complexity*. Prentice-Hall.

Pinedo, M., Zacharias, C., & Zhu, N. (2015). Scheduling in the service industries: An overview. *Journal of Systems Science and Systems Engineering, 24*(1), 1–48. doi:10.100711518-015-5266-0

Pinson, E. (1995). The Job Shop Scheduling Problem: A Concise Survey and Some Recent Developments. In P. Chretienne, E. G. Coff-man Jr, J. K. Lenstra, & Z. Liu (Eds.), *Scheduling Theory and its Applications* (pp. 277–293). John Wiley & Sons.

Prim, R. C. (1957). Shortest connection networks and some generalizations. *The Bell System Technical Journal, 36*(6), 1389–1401. doi:10.1002/j.1538-7305.1957.tb01515.x

Pulleyblank, W. R. (Ed.). (2014). *Progress in combinatorial optimization*. Academic Press Canada.

Rajendran, C. (1993). Heuristic algorithm for scheduling in a flowshop to minimize total flowtime. *International Journal of Production Economics, 29*(1), 65–73. doi:10.1016/0925-5273(93)90024-F

Rajendran, C. (1995). Heuristics for scheduling in flowshop with multiple objectives. *Eur J Opl Res, 82*(3), 540–555. doi:10.1016/0377-2217(93)E0212-G

Ruiz, R., & Maroto, C. (2005). A comprehensive review and evaluation of permutation flowshop heuristics. *Eur J Opl Res, 165*(2), 479–494. doi:10.1016/j.ejor.2004.04.017

Sheibani, K. (2004). An adaptive greedy genetic algorithm using fuzzy reasoning for the travelling salesman problem. *Proc. Hybrid Metaheuristics, 16th European Conf. on Artificial Intelligence*, 31-49.

Sheibani, K. (2005). *Fuzzy Greedy Evaluation in Search, Optimisation, and Learning* (PhD thesis). London Metropolitan University, London, UK.

Sheibani, K. (2006). *Fuzzy greedy search and job-shop problem. Proc. the UK Planning & Scheduling Special Interest Group.*

Sheibani, K. (2008). Fuzzy Greedy Search in Combinatorial Optimisation. Tadbir Institute for Operational Research, Systems Design, and Financial Services.

Sheibani, K. (2010). A fuzzy greedy heuristic for permutation flow-shop scheduling. *The Journal of the Operational Research Society*, *61*(5), 813–818. doi:10.1057/jors.2008.194

Sheibani, K. (2010). An evolutionary approach using fuzzy greedy initialization to permutation flow-shop scheduling with the makespan criterion. *Proc. IEEE Intl. Conf. on Industrial Engineering and Engineering Management*, 638-641. 10.1109/ICIEEM.2010.5646535

Sheibani, K. (2010). The fuzzy greedy search in combinatorial optimization with specific reference to the travelling salesman problem. *Proc. IEEE Intl. Conf. on Industrial Engineering and Engineering Management*, 1367-1370. 10.1109/IEEM.2010.5674393

Sheibani, K. (2011). A hybrid metaheuristic using fuzzy greedy search operator for combinatorial optimization with specific reference to the travelling salesman problem. *Iranian Journal of Operations Research*, *2*(2), 63–71.

Sheibani, K. (2013). A constructive heuristic for permutation flow-shop scheduling with the makespan criterion: in a particular case where the sum of the processing times of each job is the same on each machine. *Lecture Notes in Management Science, 5*, 18-24.

Sheibani, K. (2017). Fuzzy greedy search heuristic for combinatorial optimization with specific application to machine scheduling. *International Journal of Management Science and Engineering Management*, *12*(2), 148–153. doi:10.1080/17509653.2016.1172993

Sheibani, K. (2018). A very fast heuristic for combinatorial optimization with specific application to priority rule sequencing in operations management: Fuzzy greedy search. *International Journal of Operations Research and Information Systems*, *9*(3), 77–89. doi:10.4018/IJORIS.2018070104

Sheibani, K. (2019). An integrated heuristic for machine sequencing with specific reference to the permutation flow-shop scheduling problem. *International Journal of Strategic Engineering*, *2*(2), 1–8. doi:10.4018/IJoSE.2019070101

Sioud, A., & Gagné, C. (2018). Enhanced migrating birds optimization algorithm for the permutation flow shop problem with sequence dependent setup times. *European Journal of Operational Research*, *264*(1), 66–73. doi:10.1016/j.ejor.2017.06.027

Stafford, E. F. Jr, Tseng, F. T., & Gupta, J. N. D. (2005). Comparative evaluation of MILP flowshop models. *The Journal of the Operational Research Society*, *56*(1), 88–101. doi:10.1057/palgrave.jors.2601805

Sule, D. R. (1997). *Industrial Scheduling*. PWS Publishing Company.

Taillard, E. (1993). Benchmarks for Basic Scheduling Problems. *European Journal of Operational Research*, *64*(2), 278–285. doi:10.1016/0377-2217(93)90182-M

Tucker, A. (1995). *Applied Combinatorics*. John Wiley & Sons.

Vince, A. (2002). A framework for the greedy algorithm. *Discrete Applied Mathematics*, *121*(1-3), 247–260. doi:10.1016/S0166-218X(01)00362-6

Wang, Z., & Klir, G. (2013). *Fuzzy measure theory*. Springer Science & Business Media.

Whitney, H. (1935). On the abstract properties of linear dependence. *American Journal of Mathematics*, *57*(3), 509–533. doi:10.2307/2371182

Wilson, R. J. (1985). *Introduction to Graph Theory* (3rd ed.). Longman.

Woo, D. S., & Yim, D. S. (1998). A heuristic algorithm for mean flowtime objective in flowshop scheduling. *Comput Opl Res*, *25*(3), 175–182. doi:10.1016/S0305-0548(97)00050-6

Zheng, T., & Yamashiro, M. (2010). Solving flow shop scheduling problems by quantum differential evolutionary algorithm. *International Journal of Advanced Manufacturing Technology*, *49*(5-8), 643–662. doi:10.100700170-009-2438-4

Zimmermann, H. J. (2001). *Fuzzy Set Theory and its Applications* (4th ed.). Springer-Verlag. doi:10.1007/978-94-010-0646-0

# Section 3

# Chapter 11
# 55000 Implementation Proposal Integrated With a Risk Management System for a Water Utility Case

**Pablo Zahera**
*University of Seville, Spain*

**Vicente Gonzalez-Prida**
*Universidad Nacional de Educación a Distancia, Spain & University of Seville, Spain*

## ABSTRACT

*The main objective of this chapter is to suggest a methodology for the application of ISO 55000 for a water utility in order to obtain all the benefits of asset management from the starting point of application. In addition, risk is an important part of this methodology, so it also complies with the clauses of ISO 31000. The methodology consists of six steps to be carried out: plan (1), implement (2), risk management (3), monitor (4), analyze (5) and make decisions and improvements (6). The application of this methodology is an iterative process in which the information obtained is going to be used in the previous and subsequent steps so that its benefits are greater as time goes by. At each point in the guide, the actions to be taken in compliance with the clauses and sub-clauses of ISO 55001 will be proposed along with some documents so that companies have a clearer idea of how to proceed.*

## INTRODUCTION

This chapter comes from the Final Master Thesis of Pablo Zahera Bunes, industrial engineering master student, in collaboration with Vicente González-Prida, being a synthesis of the main content of this work.

The main objective of the chapter is to suggest a methodology for the application of ISO 55000 (ISO 55000, 2014), taking as an example a water utility because, due to the kind of activity it performs, the benefits of asset management are obvious, as well as the use of the author's professional experience.

DOI: 10.4018/978-1-7998-3246-1.ch011

The methodology followed consists of the application of a PDCA cycle (Plan Do Check Action) suggested by the International Cupper Association (ICA, 2015), complying with each clause of the standard and highlighting the actions to be taken in the water utility case.

The methodology is composed of 6 steps: Plan (1), implement (2), risk management (3), monitor (4), analyze (5) and decision-making and improvement (6) where:

- Step 1 is the asset management planning and it is the P of PDCA cycle.
- Step 2 and step 3 are the development of the system as the D of PDCA cycle.
- Step 4 and step 5 represents the verification process, where the asset life cycle is analyzed, being the C of the cycle.
- Step 6 consists on the improvement action for the system and it refers to the A in PDCA cycle.

As it shows in the following figure:

*Figure 1. Asset management system methodology. Steps of application.*
*Source: Self made*

The application of this methodology is an iterative process where the information obtained is going to be used in previous and subsequent steps so its benefits will be greater as time goes by.

At each point in the guide, the measures to be taken in compliance with the clauses and sub-clauses of ISO 55001 (ISO 55001, 2014) will be suggested along with some documents to give companies a clearer idea of how to proceed.

## BACKGROUND

The water utilities carry out their activity in a very strict regulatory context, which pursues the efficient management of the resource due to its scarcity, as well as the continuity and quality in the service of this one, all this in constant balance with the environment. In addition, in contrast to other kinds of companies, their profit is regulated, so controlling the cost of operation and optimizing it with respect to the best performance of their assets and minimum risk associated is fundamental.

ISO 55001 (ISO 55001, 2014) certification by water utilities is a mark of assurance for their regulators, helping the company to get the best value from its assets, liberating money for more infrastructure improvements and keeping average household bills down.

## 1. PLAN THE ORGANIZATION'S MANAGEMENT

This is the first logical step to be carried out during implementation. It is also the longest and most important part because it will allow the next steps to be done properly. It includes clauses 4 and 5 of ISO 55000 (ISO 55000, 2014).

### Organizational Context

The main purpose is to analyze the utility's context and to determine the status of the existing asset management system, if any. For example, to review whether there are asset maintenance plans, whether there is an asset monitoring system in place, to determine the budget for asset management, whether there are asset performance objectives and their achievement.

In addition, the utility will have to decide whether to start with a generic asset management system based on ISO 55000 (ISO 55000, 2014), with the intention of full compliance, or to use the existing asset management plan and reverse engineer it to meet the requirements of the standard.

Both options will have its own resources needs and costs, and will be delivered on different time scales.

#### Understanding the Organization and its Context

The asset manager must review the context relating to the following areas of knowledge within the utility that may affect both directly and indirectly the asset management:

- Legality, all the regulations related to the safety of water consumption, the guarantee of its supply, respect for the environment, the employees' occupational safety, etc.

- Technology, everything related to the physical assets of the utility such as pumps, valves, electrical equipment, pipes, control systems, measuring instruments, data acquisition systems, among others.
- Geography, issues such as the size of the wastewater and drinking water treatment infrastructure, the orography of the land where the utility's assets are located, the utility's supply capacity, the number of consumers, the distribution of neighborhoods, towns, and other population centers, among others.
- Economy, reviewing the expenses incurred in the operation, as well as the history of capital investments made, reviewing if there are documented protocols or policies to justify and approve expenses, if the utility innovates, if there are outsourced services from third parties, among others.
- Social, attending topics such as access to drinking water, equality, diversity, job security, if there are education programs about the proper use of resources from the government, water consumption data, among others.
- Organization of the utility, topics such as hierarchical structure and organization chart, company culture, whether there are documented policies (payment, third-party hiring process, investment, quality, supplier management, commercial, etc.), whether employee training is carried out, among others.

## Stakeholder Requirements

Asset managers must know which stakeholder requirements are. It is convenient to document stakeholder's needs, expectations and general data.

Some examples of stakeholders in the utility case are suppliers (hydraulics, automation, electrical equipment, energy, etc.), customers, lobbyists, shareholders, funding bodies, employees, subcontractors, neighbors close to the facilities, other utilities (gas and electricity), government, insurance companies, unions, owners, environmental agencies, etc.

In the stakeholder management, conflicts of interest among stakeholders are common, so it is strongly recommended to use techniques such as the influence matrix to have the best decision-making capacity to deal with these conflicts.

The matrix consist of quantifying the impact and the power of influence of each stakeholder. In order to do this, the asset managers has to evaluate three core questions:

- How much does the asset management project affect them?
- How much influence do they have over the asset management project?
- What is important to the stakeholder?
- How could the stakeholder help in the asset management project?
- How could the stakeholder block the project?

After that, the asset managers must determine the strategy to be followed to involve the stakeholder. As an example, the building of a wastewater treatment plant nearby a neighborhood will affect the neighbors, so the project will have an impact on their lives, and the neighbors will be able to denounce the utility and stop the work. The asset managers will have to deal with the neighbors by offering them solutions, keeping them informed by holding regular meetings and proposing alternatives to their complaints.

## Determine the Scope of the Asset Management System

The asset management system shall be established in combination with other management systems of the utility, such as quality, environmental or risk management systems. In addition, it must be aligned with the objectives and strategic plan of the utility. Therefore, the boundaries where the systems may conflict should be defined. For example, in case the water utility is certified in environmental management with ISO 14001, the strategy of its asset management system must be aligned with that of ISO 14001.

Another important issue in determining the scope is choosing which assets will be included within the asset management system. Depending on the resources of the utility and the strength targeted, the scope will vary. Since the budget is limited, the asset managers will focus on critical assets, which will be addressed in the following points.

## Leadership

The main objective of this point is the commitment and buy-in at senior levels. Seniors must understand the purpose of asset management, and their commitment will be necessary in order to implement measures and operate and improve the system.

For example, if there is a broken valve, the required investment will have to be reviewed and accepted by senior management even though it is not their main role at work. Therefore, they must know the importance of the valve for daily operation and how will affect the utility's objectives.

In addition, senior management has a duty to ensure that the utility's strategic asset management plan is aligned with the strategic plan. This means that any incident, investment or change in the management of assets will influence the rest of the management systems that make up the strategic plan, so they must be able to manage it.

### Policy

The asset policy will be integrated with other management systems such as ISO 9001 (quality), ISO 11000 (collaboration), ISO 14000 (environment), ISO 27000 (safety) and ISO 31000 (risk).

This point will generate a short commitment statement of supplying enough resources and an appropriate organizational structure for the asset management system, signed by senior management. This document will have to compulsory include two elements:

- A commitment of compliance with laws or regulation and legislative requirements.
- A commitment for a continuous review and improvement of the asset management system.

Furthermore, the policy may include some basic principles of asset management such as health and safety as a goal, investments based on life cycle costs, minimizing environmental impact, etc. In addition, it is allowed to talk about long-term objectives or some stakeholder requirements.

### Organizational Roles, Responsibilities and Authorities

The asset managers should assign work team member's roles. Especially in the outsourced services, very common in the water utilities, in order to avoid any interference.

This point will generate personnel requirements sheets for each employee. It may include name, position, functions, experience, formation, aptitudes, etc.

The workforce within the water industry and utilities is getting older, and some of the knowledge and value they generate from experience will be missed. Therefore, procedures should be implemented to avoid this, such as internship programs for students or recording this knowledge through webinars, e-learnings, documents, etc.

## Criteria and Definition of the Asset Portfolio. Scope of the Asset Management System

This point strengthens the following methodology's steps by the introduction of important concepts in asset management. It will be important during the beginning of the application guide.

### Sort of Assets

One of the key points of the ISO 55001 is the concept of asset, which traditionally has been associated to physical asset. In this case, assets are distinguished in four families:

- Human assets, with related topics such as motivation, roles and responsibilities, knowledge, leadership, experience, among others.
- Information assets, like information systems, data collection, regulation, information storage, among others.
- Intangible assets, such as organizational reputation, image, social impact, among others.
- Physical assets, which are machines, equipment, facilities, buildings, furniture, vehicles, commodities, tools, among others.

However, in the utility case, physical assets are the most remarkable.

### Asset Record

In order to gather useful information to develop the context, it is very important to find a procedure about how to record asset information. The ISO 14224 (ISO 14224, 2016) proposes a hierarchical way to do. It consists of nine levels where the number 1 to 5 are high-level ones and the number 5 to 9 are levels, which take into consideration equipment. The objective consists of mapping the utility's assets, allowing its correct classification.

As the table shows, the different hierarchical levels of the assets belonging to the entrance of a pumping station can be mapped in order to know at all times where to place the assets and to know the importance of them.

Through this procedure, the asset managers can get a general idea about the system. It will be part of the asset information to be documented along with drawings, diagrams, data sheets, etc.

*Table 1. Asset classification*

| Level | 1 | 2 | 3 | 4 | 5 | 6 |
|---|---|---|---|---|---|---|
| Industry | Water | | | | | |
| Business category | | Sanitation | | | | |
| Installation category | | | Disposal system | | | |
| | | | Treatment plants | | | |
| | | | Collection systems | | | |
| Plant or unit category | | | | Sewage network | | |
| | | | | Pumping station | | |
| Section or system | | | | | Incoming sewer | |
| Equipment (class or unit) | | | | | | Pipes |
| | | | | | | Well |
| | | | | | | Inlet gate valve |

Source: (EPA, 2016)

## Status of Assets

Especially in water utilities, knowing the assets' condition is crucial and must be continuously recorded and analyzed. In the initial steps, the asset managers should at least do a check for urgent asset repair needs.

There is a lot of information on tools for inspecting the physical state of the assets such as X-ray, infrared, microscope, GIS systems, modeling software, among others. However, the objective of this point is to assign a status mark to each asset and record it, following the same criteria, as shown in the following table:

## Critical Assets

At early stages of the system's implementation, it is desirable that the entire workforce involved with the assets during their life cycle, can discuss and document what the most important assets are.

Efficient resource management is key for utilities, so it is important to know which assets require more attention and resources, in terms of likelihood and/or severity of failure. The experience of the professionals is an important basis to start classifying them.

Two contexts may be possible and will depend on the utility's resources; on the one hand, the water utility has historical data on the likelihood of asset failure. On the other hand, it does not have any documented information about this.

There are some techniques for determining critical assets such as the criticality matrix or Pareto diagrams. Therefore, these techniques, along with regular meetings among experts, can help the asset managers to keep improving its knowledge of the utility context.

To conclude, with this information the asset managers will be able to establish some plans for those assets, in terms of maintenance or operation without having to wait to have a more implemented asset management system.

*Table 2. Quantification of the asset status*

| Level | Clasification | Action | Description |
|---|---|---|---|
| 1 | Very good | Nothing is required Good condition. | Good condition.<br>There is no evidence of damage.<br>May include repaired assets where the repair is as good as the original. |
| 2 | Good | Monitor for changes | Deterioration can affect performance.<br>Includes most repaired assets. |
| 3 | Moderate | Consider specialist' s evaluation | Clearly needs some attention, but it still works.<br>The structure needs to be repaired.<br>Includes assets repaired when the repair is damaged. |
| 4 | Poor | Receive evaluation from specialists | Whether it does not work or works poorly because of deterioration.<br>The integrity of the structure is in question. |
| 5 | Very poor | Replace or repair | Needs urgent attention |

Source: (New Zealand Water and Wastes Association, 2008)

## Level of Service

Asset managers must establish service levels for the assets and the organization as a whole, from which a distinction can be made:

- All those related to customers and business objectives, such as quality, function, capacity or supply reliability. Such as the levels of pollution generated, the volume of water supplied or accidents at the workplace.
- All those related to technical characteristics such as operation (flow and pressure), maintenance, rehabilitation activities or the replacement and improvement of assets.

As an example of setting the level of service of a pumping station:

*Table 3. Level of service parameters of a pumping station*

| Pump station level of service | | | | |
|---|---|---|---|---|
| **Performance variable** | **Measure** | **Acceptable range** | **Current** | **Goal** |
| Stench | Number of complaints/year | 0 | 2 | 0 |
| Spillages | Gallons/spillage | 0-5000 | 56000 | 2000 |
| Pumping | Pumped percentage | 100%-99.89% | 99.68% | 100% |
| Reliability variable | Measure | Acceptable range | Current | Goal |
| SCADA | Leaks/year | 0-2 | 7 | 2 |
| | Duration (h) | 0-10 | 72 | 8 |
| Power | Leaks/year | 0-1 | 1 | 1 |
| | Duration (h) | 0-3 | 7 | 2.5 |

Source: (EPA, 2016)

## Failure Modes

Recording failure modes is probably one of the most important points, because it makes it easier for water utilities to avoid the failure, or at least, to find out the primary cause that has conducted to the failure. In general, there will be information collected in the utility, but it is advisable to go to the asset register and see, with the operators and the more experienced workforce, if there are failure modes that are not documented. In case it is not, the utility could start to register as follows:

*Table 4. Failure modes examples of an electric engine*

| **Failure modes associated with different failures of the electric engine of a booster pump** | |
|---|---|
| **Failure A: Over-consumption protection activates the circuit breaker** | |
| Failure modes | Poor switch calibration |
| | Broken or burnt winding |
| | Bearings in poor condition |
| | Imbalances among phases |
| | The engine gets hot because the fan has broken. |
| Failure B: Engine does not rotate | |
| Failure modes | Broken or burnt winding |
| | Defective power cable connection terminal |
| | Engine power failure (no electric current) |

Source: Failure modes examples of an electric engine

# 2. DEVELOP THE ASSET MANAGEMENT OBJECTIVES

## Strategic Asset Management Plan (SAMP)

The asset managers must be able to set and document for monitoring the asset management objectives, which will build the strategic asset management plan.

Once the context has been developed, the asset managers should address the following questions:

- Where does the utility expect to be in the period under consideration?
- What are the objectives and associated risks in the period under consideration?
- How can the utility achieve them?
- How can the utility measure the objectives?

## Asset Management Goals

There are two kinds of objectives. On the one hand, general ones are set at a high-level and refer to all departments and management systems within the utility. On the other hand, specific objectives are related to the asset management system.

Below, there are several kinds of objectives classified according to their nature. In addition, each objective includes an example of how can the specific objectives help to the compliance of the general ones in the water utility case:

- Political objectives, which are related to the organization's philosophy and image. For example, in the utility case, a general objective could be to improve the organizational image by reducing odors in the purifying facility, which could be a specific asset management objective.
- Economic objectives, which are influenced by either external indicators (taxes) or internal ones (employees' work efficiency). One example could be to reduce organization's costs as a general objective, by reducing asset life cycle cost, as specific asset management objective.
- Social objectives, which are all related with people. For example, in order to improve the reliability of water supplying to clients by replacing old assets without the correct level of service (specific asset management objective).
- Technological objectives, all objectives that result in an improvement in the efficiency of the utility. They usually are related to assets. For example, to improve the asset monitoring (general), through the investment in new asset monitoring solutions such as SCADAs (specific).
- Legal objectives, all related to the law and standards compliance. For example, to fulfill with workplace hazard prevention law (general), through the classification of hazard zones inside a facility (specific).
- Environmental objectives, all related to the environment and sustainability. For example, to reduce the utility environmental impact (general), through the reduction of the electrical energy consumption (specific).
- Others, related to risk management, internal communication, learning programs, etc.

### Asset Management Goals Development

After analyzing the context, the asset managers can establish some asset management plans in order to reach all asset management objectives. All those asset management objectives will conform a strategy or plan, which will contribute to achieve organizational objectives. In this way, the asset management system aligns with the rest of management systems, and with the utility's business strategy.

As it shows in the following figure:

## Support

### Competence and Awareness

One of the most problematic points of ISO 55000 is to consider people as an asset. People carry out the tasks of asset management through their knowledge, expertise, experience, and motivation, ability to work in a team, attitude and alienation with the objectives of the utility, which has an impact on the business results. Therefore, the utility must ensure either that people have all these attributes or identify the gap between what they have and what the utility expects, and act to reduce it.

With this in mind, the asset managers must organize learning and training sessions conducted by asset management experts and trainers. In addition, including people from outsourced companies.

*Figure 2. Decomposition of the utility's general objective into specific asset management*
*Source: Modified from (Román, Padilla & Uceda, 2018)*

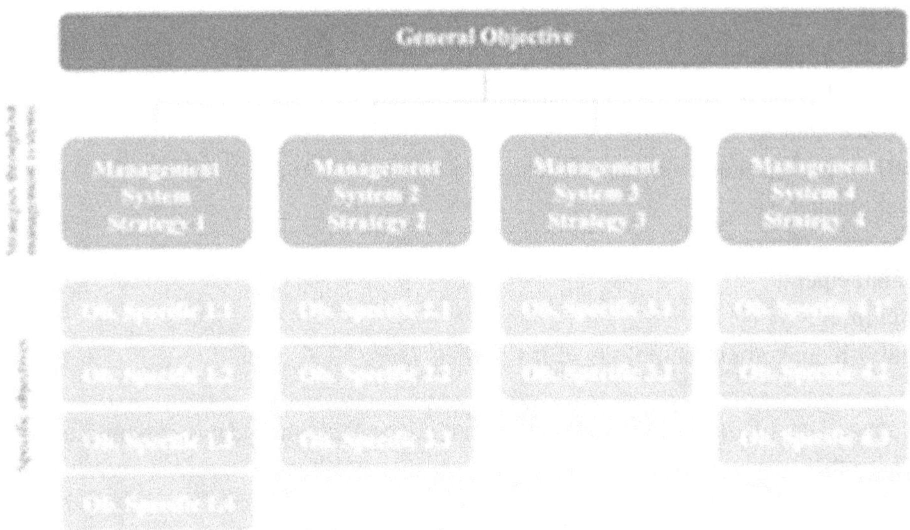


Part of these trainings should focus on employee awareness. For example, in the case of the water utility, a pipe welder should know which is the consequence of his work is, the environmental impact of a failure, the economic impact of leakages, etc. In this way, employees will be aware of the importance of their work and will feel an important part of the utility, which is important in the alienation with the utility objectives.

## Communication Plan

All stakeholders should be informed on the issues that are relevant to them. There will be some, such as organizational performance, that apply to all of them, or others that apply to only some of them. The asset managers will have to be careful about information overload or deficit.

At this point, the utility's communication plan should be reviewed and, if necessary, modify it. The content should be structured as follows (Navarro, 2012):

a) Study of the context/environment, where it is necessary to see the kind of internal communication that exists and to analyze the utility context, which has been dealt with in point 1 of the methodology.
b) Communication objectives, where it is necessary to define what the purpose of the communication plan is. It is part of the strategic plan for asset management, which has been dealt with in point 2 of the methodology.
c) Stakeholders, are those groups of people analyzed in 1.1.2 who are the receivers of the information.
d) The message, which is the information related to the asset management system to be communicated.
e) The strategy, which will consist of explaining how to communicate.
f) The communication actions, explaining how the communication objectives will be achieved.
g) Control and monitoring, defining indicators that, after carrying out periodic measures, will allow corrective actions to be taken and, in this way, improve the communication plan.

In asset management, the information must refers to:

- Planning and strategy.
- Process and procedures.
- Technical and physical properties of assets.
- Maintenance management.
- Information management.
- Financial management.
- Resource management.
- Risk management.
- Audit planning.
- Third party management (suppliers, third parties, etc.).

There are several channels to send information, but the content should have an unique format. Similar to the following table:

*Table 5. Communication message content*

<table>
<tr><td>MAIN MESSAGE</td><td colspan="2"></td></tr>
<tr><td colspan="3"></td></tr>
<tr><td>SENDER</td><td>RECEIVER</td><td>SHIPPING DATE</td></tr>
<tr><td></td><td></td><td></td></tr>
<tr><td>ISSUANCE RESPONSIBLE</td><td colspan="2">CONTENT</td></tr>
<tr><td></td><td colspan="2"></td></tr>
</table>

Source: Self made

# 3. RISK MANAGEMENT

## Asset Management Risk Assesment

This point encompasses the identification, analysis and assesment of risks in asset management.

### Risk Identification

Once identified what are the critical assets and the asset management objectives, the asset managers should determine the associated risks in order to understand the source, the impact and the likelihood that they occur and then, manage to mitigate and reduce them into a controlled level.

During this stage, the most applicable risk analysis techniques are:

- Brainstorming
- Structured or semi-structured interviews
- Delphi

- Checklists
- Preliminary hazard analysis
- HAZOP
- HACCP
- Environmental risk assessments
- SWIFT
- Scenario analysis
- Failure modes and effects analysis
- Cause and effect analysis
- Human reliability analysis
- Reliability – centered
- Risk Matrix

The choice of which one to use will depend on the following factors

- Resources and capacities.
- Nature and degree of uncertainty.
- Complexity.

In the initial steps of implementation, where there are fewer resources to determine the most important risks, techniques such as "brainstorming" among experts will be sufficient, as it makes low use of resources and has low uncertainty and complexity.

When the system is more mature, a cause and effect analysis can be used to determine risks that are not so evident, which can provide quantitative results with greater resource use and complexity.

The final risk list should be checked for consistency with the utility's business plans and should be the starting point for their evaluation.

## Risk Analysis

The risk analysis consists in the evaluation of two aspects, its consequence and the likelihood that it will happen.

### Consequence Analysis

Based on either, the experience of asset operators and other related professionals, or the different prediction software that the market offers, the asset managers should analyze the consequence of asset failure from the following perspectives:

- Asset reparation and substitution costs.
- Impact on safety and quality in water consumption.
- Damage to properties located in areas close to the utility's facilities.
- Hydraulic capacity and reliability of water supply.
- Environmental impact on events such as floods or leaks at pipes.
- Water interruption and weather forecast.

- Stakeholders' interests.

## Probability Estimation

The probability of failure can be determined by statistical or physical approaches since both allow predicting the deterioration of network components. However, the physical approach is not recommended due to the fact that the physical mechanisms that cause breakage of assets are too much complex to determine (Espín, 2014)

Therefore, the statistical approach, which is based on historical information about past failures to determine a pattern of rupture, will be used for the asset decision-making and for the controls to be implemented.

At this point, all the information from asset failures that have occurred must be documented and reported to asset managers. This process can be manual, if the utility does not have enough resources, or automatic, where there is a solution of monitoring system and prediction software implemented that reports the information to a database.

Therefore, asset managers should conduct a statistical study of the likelihood of asset failure, especially for critical assets, being the likelihood one of the two variables that determine criticality. At the same time, this point of risk analysis can help the utility better understand the critical asset and establish plans to reduce high failure rates.

### Control Measures

Once the risk is identified and evaluated, the asset managers should provide control measures for each kind of risk. There will be people and departments responsible of each control and all of them will use documents, policies, processes and procedures to control and mitigate risks.

The types of controls and the issues to be controlled in the utility are very diverse. As an example, in the process of purchasing assets at any level, there are risks of corruption, because, if the purchasing responsible has any conflict of interest with the supplier, and is making a profit due to his position in the utility, this can mean an economic loss for the utility. In addition, it could suppose risk of lack of quality in the purchased assets.

One of the possible control measures would be to establish a model of segregation of duties in the purchasing processes in order to avoid a single person deciding and approving payments, which should be documented in a purchasing policy.

Once the asset managers have identified the risks and assessed the level of control that each of them has, it is convenient to map all this information. This will result in 3 important documents for risk management:

- Risk Map, this document should include a matrix that maps the different departments or professionals whose activity is related to each identified risk. Something similar to what is shown below:

However, it is also can consist of a not too complex matrix showing the risks, the probability of their occurrence, their severity and, if desired, the actions to address and mitigate them.

*Table 6. Risk Matrix*

| | | Risk 1 | Risk 2 | Risk 3 |
|---|---|---|---|---|
| Dpt 1 | Responsible 1 | x | | |
| Dpt 2 | Responsible 2 | | | x |
| Dpt 3 | Responsible 3 | x | | |

Source: Self made

- Risk and Control Matrix, this document should inform which risks are controlled with each control. Something similar as it shows below:

However, it is recommended that this document incorporate the description of the control, the document that supports it (if any), the kind of control (reactive, corrective, detective, and preventive) and the person responsible.

*Table 7. Control matrix*

| | Risk 1 | Risk 2 | Risk 3 |
|---|---|---|---|
| Control 1 | x | | |
| Control 2 | | x | x |
| Control 3 | | | |

Source: Self made

- Coverage Map, it shows who and what department is responsible of each control activity. Something similar as it shows below:

These documents provide a very interesting control and management tool for identifying business activities, associating them with risks differentiated by type and level and with the exogenous and endogenous factors related to these risks. All this allows the utility to have an integral risk management system.

*Table 8. Coverage Map*

| | | Dpt 1 | Dpt 2 | Dpt 3 |
|---|---|---|---|---|
| Control 1 | Activity 1.1 | x | | |
| Control 2 | Activity 2.1 | | | x |
| | Activity 2.2 | x | | |
| | Activity 2.3 | | x | |
| Control 3 | Activity 3.1 | x | | |

Source: Self Made

## 4. MONITOR THE OPERATION

At this point, the system is ready to be used and will serve to evaluate and optimize the proposed management plans. Frequently, it will be the application of the system itself, which will determine whether the plans are adequate, insufficient or wrong in an iterative process. Even the objectives will be susceptible to change, either because more important problems have arisen on which it will be necessary to set new objectives to solve them, or simply because the direction of the utility has changed.

### Performance Evaluation

Within the utility case, their priority is to be reliable in the delivery of resources and reasonable in their bills, so the cost of efficiency is fundamental. In addition, the amount of information they handle related to the different assets (500 assets on average per organization), the geographical dispersion of the assets and the number of actors involved, make digitalization a necessary resource.

#### Monitoring

The evaluation of the assets performance in the public utilities through their monitoring, will allow them to know at every moment their physical condition, in order to plan as well as possible all the tasks of rehabilitation and replacement.

Within this context, monitoring and data acquisition systems (SCADA) and hydraulic simulation models are very useful in meeting these information requirements. Thus, the risk is controlled.

#### Definition of Indicators

All the information collected through monitoring should be analyzed later, which will consist of a comparison with the indicators that provide information on different parameters related to risk, cost and performance on assets. The characteristics of these are as follows:

- Be aligned with the utility's strategic objectives.
- Be balanced and measurable, which means that the variable described must be quantifiable in terms of degree, frequency or quantity.
- Be implemented at all levels of the organization.
- Be understandable, which means that they must be easily recognized by all those who use them.

Below is a table with examples of the different types of key indicators for water utilities.

### Audit

In order to carry out the asset management system's audit, the asset managers will have to decide:

- The scope, which will depend on the associated risks with each audited activity.
- The auditors, and ensure their independence from the audited processes.
- What kind of tools to use, and how they will be used (plans)

*Table 9. Indicators of water utilities*

| 4.1.2.1 Key Performance Indicators | 4.1.2.2 Cost Indicators | 4.1.2.3 Risk Indicators |
|---|---|---|
| Some examples of this type are:<br>-MTBF (Mean Time Between Failures)<br>- MTTR (Mean Time to Repair Fault)<br>- OEE (Overall Equipment Efficiency)<br>- LTIFR (Lost Time Injury Frequency Rate)<br>- Availability (hours without interruption)<br>- Failure rate /critical equipment<br>- Severity of faults<br>- Average time between repairs<br>- Maintainability<br>- Frequency of failures or repairs<br>- Technical losses /Units or total losses/km of network<br>- Reliability<br>- Hours dedicated to the management of available assets/equipment hours | Some examples of this type are:<br>- Total maintenance cost /total asset value<br>- Actual cost (sum of repair costs with restated cost (average acquisition cost - average depreciation value * (1- average age of asset))<br>- Cost of operation/marketed unit<br>- Maintenance costs/Estimated asset replacement costs<br>- Cost of maintenance /Cost of operation<br>- Operating cost/Asset base value<br>- Expected CAPEX value/ Actual capex value | Some examples of this type are:<br>- Degree of exposure/ number of faults<br>- Probability of failure/ cost of maintenance<br>- Investments in training/cost of accidents and incidents<br>- Estimated company exposure costs/ investment in maintenance and operation<br>- Rate of return per maintenance project<br>- Number of warnings from the regulatory body/marketed unit<br>- Cost of fines/ maintenance investments |

Source: (Modified from ISO 31010, 2019)

- How to structure, and store information.

At the beginning of the audit it is essential that all documents generated throughout the application of this methodology remain documented and based on the information they contain, corrective and improvement actions can be carried out.

The audit will generate an audit report, which will include the findings detected and should be delivered to senior management. Once delivered, the results will be analyzed, reviewing the achievement of the planned objectives and the gaps that remain to be covered. Finally, an action plan will be drawn up to resolve the non-conformities detected and improve the asset management system.

## 5. ASSET MANAGEMENT SYSTEM ANALYSIS

In water utilities, the life cycle of the assets has the following phases:

- Asset purchase.
- Asset commissioning.
- Asset operation.
- Cessation of the asset's activity.
- Asset decommissioning.

At this point, the water utility should already have a continuous monitoring system, a system for recording and evaluating incidents, accidents and failures, a study of risks and controls documented, and an organizational context that is properly documented.

The asset managers must then also establish a strategic asset maintenance plan that makes use of all the information collected throughout the implementation of the previous points of the methodology and an analysis of the life cycle costs of the assets that will lead to an investment plan.

## Asset Maintenance Program

Once the asset managers know the most critical assets, it must choose which the best maintenance strategy is for each, always taking into consideration the economic constraints.

*Table 10. Evaluation of indicators for different types of asset maintenance*

| Kind of maintenance | | MTBF | MTTR | Efficiency | OEE | LTIFR |
|---|---|---|---|---|---|---|
| 1 | Corrective | low | low | low | low | low |
| 2 | Preventive | medium | medium | low | medium | medium |
| 3 | Predictive | high | high | high | high | high |
| 4 | Proactive | high | high | high | high | high |

Source: (Espín, 2018)

The asset managers could decide based on the following indicators:

1. Their use can be justified in cases where intervention in the event of plant failure is quick and the costs and risks associated with the failure are minimal.
2. The problem is that money and time are lost in doing maintenance that is not necessary, so it is not very effective against the breakdown neither. This takes a long time to be repaired. However, the only type of breakdown in which it is effective is that related to the age of the asset.
3. This maintenance eliminates the cause of the failure, but as a disadvantage, it has a higher initial cost due to the fact that it requires continuous monitoring and a statistical model based on risk.
4. It is only used in the aerospace industry and a few other high-tech industries. In the water industry, it does not make sense.

## Asset Life Cycle Costs

The asset managers should know all the costs in the asset life cycle in order to establish budget and investment control mechanisms. The following table shows the different costs in the asset life cycle:

Once the costs are known, it is advisable to carry out an analysis of the investments required to ensure the functioning and quality of the asset management system.

It should not be forgotten that asset management plans seek to maintain as far as possible the value of the infrastructure and the quality of the service, so the main investments will be made in terms of renovation, maintenance and modernization of the system.

From the above, investment plans, policies and procedures should emerge, involving all departments and always in accordance with the general interests of the organization and not only of the asset management system.

*Table 11. Associated costs throughout the different stages of the asset life cycle.*

| | |
|---|---|
| Initial investment costs | Some costs at this stage:<br>- Engineering (Schematics, draws, design etc.).<br>- Supplies.<br>- Tender process.<br>- Administration costs.<br>- Inspection and testing.<br>- Spare parts inventory.<br>- Trainings. |
| Installation and commissioning costs | Some costs at this stage are:<br>- Foundations, design, preparation, reinforcement, etc.<br>- Adjustments and connection of components.<br>- Electrical switchgear.<br>- Hydraulic equipment.<br>- Heating systems.<br>- Connection of auxiliary systems.<br>- Supplies.<br>- Tools.<br>- External services. |
| Operation and maintenance costs | According to the EPA, operation and maintenance costs are five, ten and even twenty times greater than initial investment costs (USEPA, 2006). Some costs at this stage are:<br>- Unavailability.<br>- Operational costs.<br>- Maintenance costs.<br>- Replacement costs (spare parts).<br>- Effects on safety and the environment (with emphasis on energy consumption).<br>For example, a pump initial price is a small expense in comparison with the rest of costs in its life cycle, where the energy expenses are very high. However, some studies have shown that utilities can save a 50% of energy consumption through equipment changes and new control systems. (HI, Europump & OIT US, 2001). |
| Decommissioning and removal costs | The costs related to the readjustment to the local environment must be included. |
| Replacement costs | Although the literature on asset management does not include replacement costs within the life cycle of the assets, these are very important to determine because they allow the asset managers to know the economically<br>optimal moment to decommission the asset and, therefore, the type of maintenance to be carried out.<br>A physical asset must be replaced when the following contingencies occur:<br>- Insufficiency.<br>- High maintenance costs.<br>- Obsolescence.<br>At the same time, for its replacement, it must be contemplated:<br>- The planning horizon, that is, the shorter the period of time in which the replacement of the asset is evaluated, the more precise the analysis will be.<br>- The availability of capital to make the investment.<br>New equipment may be similar to installed equipment (when new) or superior. Inside a water distribution system, in pipes, some of the situations that give place to the causes aforementioned are:<br>- Structural deterioration.<br>- Hydraulic damage<br>- Quality deterioration. |

Source: Self made

## 6. CONTINUOUS IMPROVEMENT

The organization must permanently improve the asset management system so that the number of non-conformities found in the audits is reduced, with the objective of reaching zero.

For this purpose, all points in the guide should be constantly reviewed to detect where and how it can be improved. All the areas covered by this methodology will be enhanced in such a way that there will be a more robust system with a lower number of nonconformities.

## FUTURE RESEARCH DIRECTIONS

This chapter aims to take a step forward in the application of ISO 55001, and to facilitate its understanding using a water utility as an example. However, the structure of this methodology still serves any company that wants to comply with this family of standards.

As the scope is limited, and the topic very broad and generic, not as much depth has been put into each step of the methodology to maintain a balance between all steps.

This allows future editors the possibility of deepening each step of the methodology, as this chapter has provided a solid foundation, which also clarifies the order of magnitude of the application. Future research could address issues such as

- The problems that may arise in the implementation of different management systems combined
- Risk management, where traditionally only probability and impact have been considered, although there are other variables on which risk also depends, such as the speed with which it can appear.
- Decision-making with respect to assets and their life cycle with artificial intelligence, large data and the Internet of Things.

## CONCLUSION

Although some companies were already PAS 55 accredited, it is only since the launch of the ISO 5500 family of standards that companies have begun to notice the competitive advantage that this accreditation provides.

There are many benefits that this standard brings, such as the alignment of the company's vision, better financial results, correct risk management, better service and/or product regardless of the activity carried out in the organization, ensuring regulatory compliance, as well as improving reputation and sustainability.

In conclusion, this text will allow any company to have a starting point from which it is possible to proceed to achieve certification in ISO 55001.

## REFERENCES

Espín, P. (2014). *Sistema de gestión integral para empresas de abastecimiento de agua*. Universidad Politécnica de Cartagena.

Hydraulic Institute (HI), Europump, and the US Department of Energy's Office of Industrial Technologies (OIT). (2001). *Pump life cycle costs: A guide to LCC analysis for pumping systems*. Author.

International Copper Association (ICA). (2015). *Guía para la aplicación de la Norma NMX ISO 55001.* Naucalpán de Juarez.

ISO 55000. (2014). *Asset Management — Overview, principles and terminology.*

ISO 55001. (2014). *Asset Management — Management systems — Requirements.*

ISO 14224. (2016). *Petroleum, petrochemical and natural gas industries — Collection and exchange of reliability and maintenance data for equipment.*

ISO. (2018). *(International Organization for Standardization) 31000.* Risk Management — Guidelines.

ISO 31010. (2019). *Risk Management — Risk assesment techniques.*

New Zealand Water and Wastes Association. (2008). *Visual assessment of utility assets.* Author.

Román, M., Padilla, D., & Uceda, K. (2018). Implementación del Plan Estratégico de Gestión de Activos en Red de Energía del Perú. Academic Press.

USEPA GHD. (2006). *Asset Management Training Workshops.* Author.

## ADDITIONAL READING

Fuenmayor, E. (2016). *Análisis de reemplazo de un activo físico.* Retrieved from: https://predictiva21.com/analisis-de-reemplazo-de-un-activo-fisico/

Gebremedhin, S. (2017). *IT Asset Valuation, Risk Assessment and Control Implementation Model.* Retrieved from: https://www.isaca.org/resources/isaca-journal/issues/2017/volume-3/it-asset-valuation-risk-assessment-and-control-implementation-model

Glasson, J & Mannisto, Vesa. (2016). *Implementation guide for an ISO 55001 asset management system. A practical approach for the roads sector.* Conference of European Directors of Roads.

Harvey, R. (2015). *An Introduction to Asset Management Tools Municipal Water, Wastewater and Stormwater Systems.* University of Guelph.

Holte, K. (2010). *Use and Collection of data in Gemini VA in Asset Management. Norwergian University of Science and Technology.* Norwey.

MacGillivray, B. H., Hamilton, P. D., Strutt, J. E., & Pollard, S. J. T. (2007). *Risk analysis strategies in the water utility sector: an inventory of applications for better and more credible decision-making.* University of Cranfield.

Medina, R. (2016). *Tesis de grado, maestría gestión de activos.* Retrieved from https://www.linkedin.com/pulse/la-gestion-de-activos-su-historia-y-definiciones-para-medina-cmrp/

Popovic, A. (2016). *Stakeholder Analysis for Integrated Water Resources Manager. A MENA case study.* The Regional Environmental Center.

Rosés, F. (2002). *Risk management. Una nueva forma de asegurar el éxito empresarial.* ACV Ediciones.

## KEY TERMS AND DEFINITIONS

**Asset:** Item, thing, or entity that has potential or actual value to an organization.

**Asset Manager:** Qualified person(s) with the responsibility of managing the utility's assets by applying the ISO 55000 set of standards.

**Context:** To establish the context means to define the external and internal parameters that organizations must consider when they manage risk.

**Critical Asset:** Having potential to significantly impact on the achievement of the organization's goals.

**Implementation:** The act of implementing or putting into effect; fulfillment.

**Life Cycle:** Stages involved in the management of an asset.

**Management System:** Set of interrelated or interacting elements of an organization to establish policies, objectives, and processes to achieve those objectives.

**Methodology:** A system of broad principles or rules from which specific methods or procedures may be derived to interpret or solve different problems within the scope of a particular discipline.

**Monitoring:** Supervising activities in progress to ensure they are on-course and on-schedule in meeting the objectives and performance targets.

**Performance:** The accomplishment of a given task measured against preset known standards of accuracy, completeness, cost, and speed.

**Risk:** Effect of uncertainty on objectives.

**Utility:** Large firm that owns and/or operates facilities used for generation and transmission or distribution of electricity, gas, or water to general public.

# Chapter 12
# A Strategy for the Control of Uncertainty and WIP:
## A Case Study Based on Lean Practices

**Pablo Viveros Gunckel**
*Universidad Técnica Federico Santa María, Chile*

**Rodrigo Mena Bustos**
*Universidad Técnica Federico Santa María, Chile*

**Diego Figueroa Ramírez**
*Universidad Técnica Federico Santa María, Chile*

**Fredy Kristjanpoller**
https://orcid.org/0000-0001-8970-9371
*Universidad Técnica Federico Santa María, Chile*

**Vicente Gonzalez-Prida**
https://orcid.org/0000-0001-5257-8224
*University of Seville, Spain & UNED, Spain*

## ABSTRACT

*High works-in-process (WIP) levels usually involve significant losses for companies as they trigger dispatching delays, product shortages, increases in immobilized capital, and greater storage costs, among others. This study provides a strategy for gradually decreasing the WIP levels associated with the production line of a pharmaceutical company located in Latin America. The methodology used is based on Lean practices and, more specifically, organized into phases: the diagnosis of the current situation, information processing, mapping out the current situation through value stream mapping, analysis and determination of key factors for uncertainty control, and improvement proposals. The main tools used are value stream mapping, 5-whys, single-minute exchange of dies, and total productive manufacturing.*

DOI: 10.4018/978-1-7998-3246-1.ch012

## I. INTRODUCTION

There are many different types of production systems used by manufacturing companies. These systems can be classified by the type of production process: Continuous, Batch, or Combined (Schenk, Wirth, & Müller, 2010). Generally, the processes that produce by batch have inventories for each stage of the production process and those unfinished products are called Works-In-Process (WIP). This alludes to the fact that batch production has begun but remains incomplete.

The uncertainty and variability that occurs during the different stages of the production process have been studied as have their direct impact on the expected performance of the entire process (Nannapaneni, Mahadevan, & Rachuri, 2016). In batch production systems, uncertainty translates into higher quantities of WIP, which generates costs through increasing immobilized capital and the loss of space, among others (Zandieh, Joreir-Ahmadi, & Fadaei-Rafsanjani, 2017).

Given the above, it is necessary to have a means of controlling and measuring this uncertainty. One way to address this problem is by utilizing Lean Manufacturing (LM) methods, which originated from the strategy that, in order to improve quality, waste and delivery time must be reduced, among other factors (Ghizoni Pereira & Luz Tortorella, 2018). Several authors have studied different tools that make use of LM and their effects on operational and economic performance (Negrão, Godinho Filho, & Almeida Marodin, 2016). Over the course of the last 30 years, LM has led to the development of some of the most popular and effective tools that improve operational efficiency (Möldner, Garza-Reyes, & Kumarc, 2018).

In this way, this study offers a strategy aimed at gradually reducing WIP through the use of LM tools.

The proposed method used in this study is based on the integration of five different phases: the diagnosis of the current situation, information processing, mapping out the current situation through Value Stream Mapping (VSM), analysis and the determination of key factors for uncertainty control and improvement proposals. In addition, an applied case study was performed at a glass factory owned by a Latin American pharmaceutical company.

Finally, the structure of this article is as follows: In Section 2, the theoretical framework of the different tools used is described; In Section 3, the proposed method is detailed; In Section 4, the case study that was performed is described; and in Section 5, the results of the case study and the study's conclusions are given.

## II. REVIEW OF THE MAIN TECHNIQUES TO BE USED AS PART OF THE PROPOSED METHOD:

In LM methodology, there are various tools that serve five different purposes (De Oliveira, Oliveira Sousa, & de Campos, 2019):

- Maximum Resource Availability: Total Productive Maintenance (TPM) and Overall Equipment Efficiently (OEE).
- Maximum Quality: Total Quality Maintenance (TQM).
- Maximum Production Flow: Single-Minute Exchange of Dies (SMED), Poka-Yoke and Cellular Manufacturing.
- Minimum Inventory: Just in Time (JIT) and Kanban.
- Decision-Making: VSM, Kaizen, 5S, Visual Management, 5-Whys.

The most commonly used of these tools are detailed below:

- Kanban: "Kanban is the name of a control and programming system that is synchronized with a production system based on flashcards (Kanban means "visual cards" in Japanese)" (Hernández & Vizán, 2013). Basically, this flashcard system helps machine operators have a global vision of the production process so that they only produce what is necessary by following the JIT-style thinking that the company wants to utilize when implementing this strategy.
- Single-Minute Exchange of Dies (SMED): This is a tool that seeks to reduce machine preparation times (set-up) per unit while also including the time needed to change formats.
- Total Production Maintenance (TPM): "It is a set of techniques aimed at eliminating machine breakdowns through the participation and motivation of all employees. The fundamental idea is that improvement and the improved conservation of production assets are everyone's responsibility, from factory managers down to the assistants of machine operators" (Hernández & Vizán, 2013).
- Poka -Yoke: Also known as a fail-safe system, this is a tool that is applied so that processes are performed and products are created in the correct manner.
- 5S: "The 5S tool correspons to a systematic application based on the principles of order and cleanliness at a work station". (Hernández & Vizán, 2013).
- The 5-Whys analysis: This tool seeks to find the root of a given problem. The idea is simple in that it involves asking the question "Why?" several times in order to separate the symptoms from the causes of a problem (Marathamuthu & Jebaraj, 2015).

One of the most popular tools for decision-making that has great results when comparing an option to its alternatives (Flow Chart, EPK, BPMN, Spaghetti-Diagram, Sankey-Diagram, SCOR and Gunther Method) is Value Stream Mapping, or VSM (Bauer, Ganschar, & Gerlach, 2014). VSM is defined as a tool of continuous improvement that helps workers visualize the complete production process, especially the flow of material and the flow of information that occurs during a given production process. The flow of value, or value stream, is defined as the combination of all the different activities that add value to a product and those that do not add value to a product from the initial input to the delivery of the product to the consumer (Singh, Garg, & Sharma, 2011). The final objective of VSM is to remove, to the extent that it is possible, the waste (Muda) that was identified in the analyzed activities and processes (Rath, 2008). As a result, the VSM tool greatly assists in uncovering and quantifying key factors that affect operational performance. In addition, some authors such as (Brunt, 2000) and (Saboo, Garza-Reyes, Er, & Kumar, 2014) have argued that the current effectiveness of VSM is the ability to develop a value stream that describes the ideal state of a process. The future state of VSM allows companies to look towards the future and understand how to best manage the value stream.

Finally, the decision to use the VSM tool in this study is justified because VSM is a first level tool for analyzing a company's value stream quickly and easily and also identifies areas where improvements can be made (Chaneski, 2014). Likewise, one of the main benefits of VSM is that it allows improvements to be made at the inventory level and for the lead time of products to be evaluated, which was necessary for the company that participatd in this study. It is worth mentioning that a review of 91 jobs in the area of manufacturing showed that the improvements at the inventory level and in lead time were an average of 70% and 56%, respectively (Knoll, Reinhart, & Prüglmeier, 2019).

## III. PROPOSED METHOD:

The method proposed in this study aims to determine the key variables that result in a high level of uncertainty in the production process and thus high levels of WIP.

This is performed in the hope that the resulting proposals focus on the root of a given problem and use the quantitive information pertaining to the process as a foundation. The main tool utilized was VSM of the LM method.

To accomplish this, 5 phases are considered to perform this work: the diagnosis of the current situation, information processing, the creation of a VSM diagram for the current situation, analysis and determination of key factors and the definition and evaluation of improvement proposals. These 5 phases are divided into two main stages: inicial diagnosis and analysis and evaluation and proposals. The schematics of the proposed method of this study are shown in *Figure 1*.

Phase 1: Diagnosis of the current situation:

Depending on the amount of available information, both conducting semi-structured interviews and reviewing company documents are recommended (Díaz-Bravo, Torruco García, Martínez Hernández, & Varela Ruiz, 2013). This fulfills the objective of quantifying the current situation of the production process. During this phase, the first two steps are performed: process mapping and the acquistion of quantitative and qualitative information.

Phase 2: Information Processing:

In this phase, the information obtained from interviews and other existing sources of company information are processed. Next, different insights and a hypothesis explaining which key factors result in a higher quantity of WIP are determined. The use of a data pre-processing technique (Weixiong, 2011) and a data cleaning technique (Runkler, 2012) is recommnded. The steps performed during this phase are steps 3 and 4, as shown in Figure 1. The indicators calculated in order to utilize VSM are as follows: Cycle Time (C/T), Value Added Time (VA), Lead Time(L/T) and Takt Time. In addition, calculating overall equipment effectiveness (OEE) of the production assets is recommended.

Phase 3: Creation of a VSM diagram for the current situation:

In this phase, steps 5 and 6 should be performed. For step 5, the family of products to which the VSM tool will be applied must be selected. For this selection, it is important to consider aspects pertaining to finance, efficiency and strategy as well as collapses in stock.

Also for step 5, it is best to conduct simulations of production scenarios for the selection of a family of products. On the other hand, for step 6, it is suggested that the VSM diagram for the current situation is created using the method laid out by the authors *(Adwait, Saily, Jayant, & Vivek, 2018)* and using the information obtained during phase 2.

Phase 4: Analysis and determination of key factors:

In this phase, steps 7 and 8 are performed. In step 7, an analysis of all that was developed during the previous phase from the operational and managerial perspective is conducted. Next, in step 8, the 5-Whys tool is used to look for and identify the key factors that generate uncertainty in the production process and which then result in a high WIP level. The final objective is to find the root cause of the problem.

Phase 5: Definition and evaluation of improvement proposals:

In this phase, steps 9, 10, 11 and 12 are performed. First, it is recommended that improvement proposals for the production process be defined and evaluated, taking into consideration the key factors identified during the previous phase and by LM tools. Next, it is recommended that a VSM diagram for the future state be created, which will depict the consequences of implementing the improvement proposals in graphic form that is easily understood by management. Finally, in considering the company's situation, culture and the available proposals, a strategy for the implementation of the improvement proposals should be developed, where impact is maximized while the amount of resources needed to create that impact is minimized. In order to do this, it is best to evaluate the proposals from the economic perspective as well as their operational impact in addition to the ease and speed of their implementation.

## IV. CASE STUDY: APPLICATION OF THE PROPOSED METHOD AT A GLASS FACTORY OWNED BY A PHARMACEUTICAL COMPANY:

### 4.1. Introduction and the Context of Application:

The production process of the pharmaceutical company that participated in this study is divided into two sections: the creation of products packaged in glass and the creation of products packaged in plastic. This study focuses on the first production process. For the glass factory, there are two types of containers: vials and bottles. The first contains a single dosis and the second contains multiple doses. In addition, each packaging format is characterized by the size of the container. Glass vials are availble in the following sizes: 2, 3, 5 and 10 mL y glass bottles are availble in: 5, 10 and 20 mL. Finally, the format of product packaging can be either Aggregate or Blister, with the difference arising from the fact that the second format (Blister) includes an additional process which is applied during the blister packaging of vials.

The final products are the result of the following processes. First, the preparation and mixing process is carried out, where chemical components and water are mixed in reactors, leading to the creation of a medicinal product. Then, the filling process is performed, during which medicine is deposited in glass vials. Then, the products are sterilized by high temperatures. Next, products enter the review process, where machines are used to verify that the medicine does not contain any foreign particles or pieces of glass that could harm the consumer. Subsequently, a sealing test occurs and, if a product is to be placed in blister packaging, it must pass through the blister packaging process. Finally, the vials are packed into boxes and sent to the Finished Products Warehouse.

As a result, the main problem facing this company has to do with the high level of uncertainty that takes place following the filling of vials or bottles. This uncertainty means that the total production time

*Figure 1. The proposed method used in this investigation.*

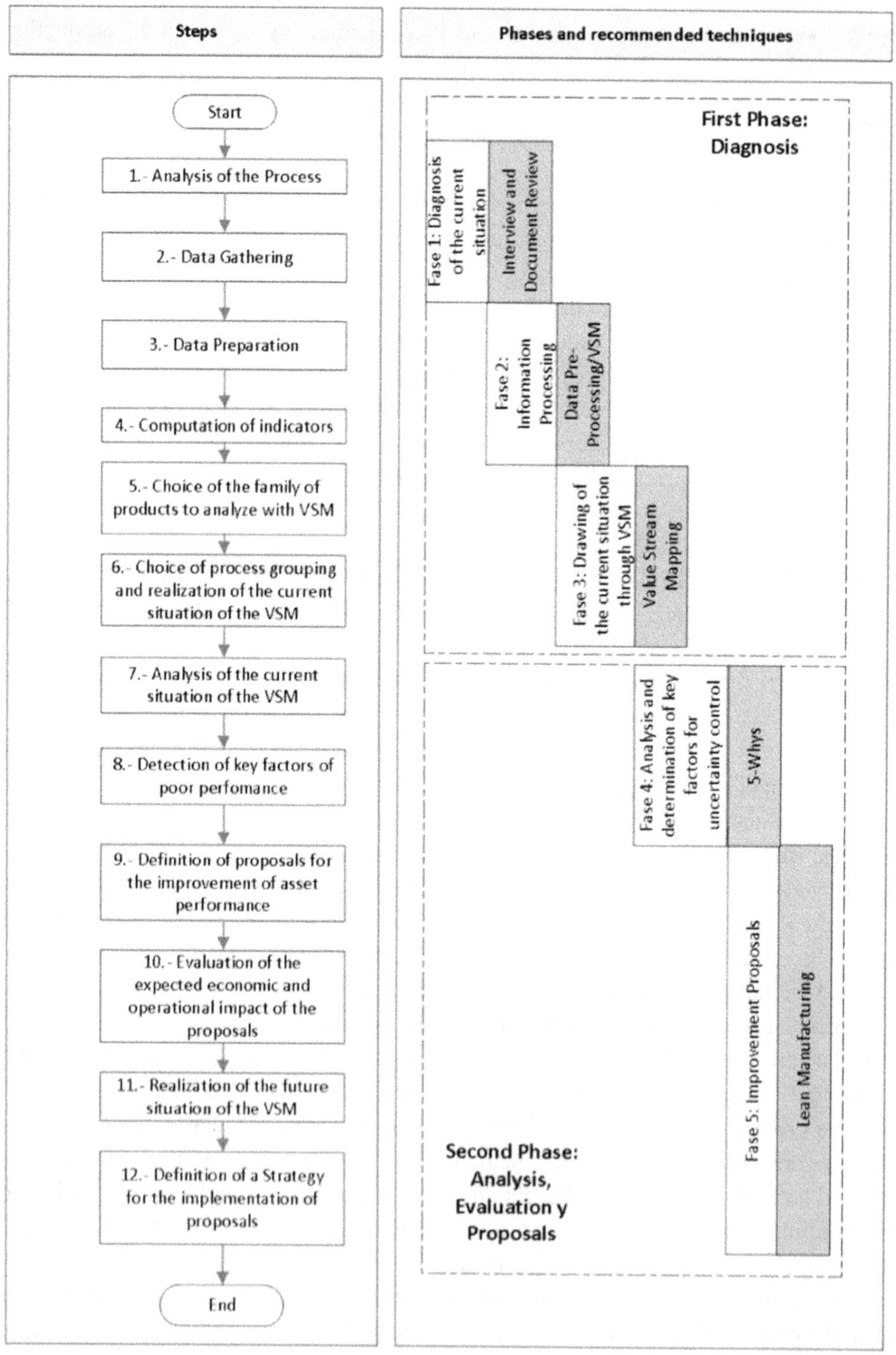

for a pharmaceutical product is highly variable, resulting in an approximate WIP of $2,500,000\left[\frac{u}{month}\right]$, which accounts for $30\%$ of monthly production.

In addition, due to the existence of high WIP levels, some production batches can take up to 2 months to plan. In the case of a collapse in stock, delivery delays and fines may result.

In this context, the main objective of this study is to propose a strategy for gradually reducing the WIP level by identifying key variables that result in uncertainty during the production process.

## 4.2. Application of the Method:

The objective of the proposed method is to deliver a guide for the identification, analysis and control of uncertainty in the production process by using LM tools. In this section, a case study is described whereby the 5 phases of the proposed method and their corresponding steps were applied to a glass factory owned by a Latin American pharmaceutical company.

### 4.2.1. Phase 1: Diagnosis of the Current Situation.

In this phase, the use of interviews and document reviews allowed for the identification of different information sources available to the company and which quantify the current situation of the production area.

Step 1: Process Mapping. The production plan has five main processes: Preparation and mixing, filling, sterilization, review and blistering/packaging. It should be noted that the primary process is the filling of vials and that the subsequent processes are considered secondary and as part of the packaging area.

Step 2: Acquire Information. In this step, it was determined that the company did not have quantitative information regarding production asset performance in the packaging area, only physical asset information.

Given that the information on asset performance is physical, it was decided to collect performance information for a three month period in order to obtain a representative sample of asset performance. In addition, the number of personnel needed to conduct each process and the number of available hours (shifts contracted) for the different processes in the packaging area were obtained. Nominal speeds and process restrictions were also determined. It should be noted that the company does not currently keep records of asset failures. In addition, it was determined that asset maintenance is corrective. Therefore, it was established that, without performing a large-scale study, it is impossible to obtain accurate quantitative information on this phenomenon and, therefore, this aspect of the assessment will only be done with only qualitative information.

Finally, the lack of quantitative information on the process and the null value for performance indicators in the packaging zone means that the factor managers do not know the performance of the production plant and, therefore, do not know the level of uncertainty present in the production process.

### 4.2.2. Phase 2: Information Processing.

Step 3: Preparing the information. In this step, following the method for data pre-processing, outliers of the acquired information were eliminated. This decision was made because there was enough information to make a correct analysis of the problem. The outliers were determined after taking into consideration the data that possessed a production capacity greater than the nominal asset speed and those with a value close to 0, according to the company information acquired during the shifts when another area validates an asset and, by extension, the asset is not working.

Step 4: Calculation of performance indicators. The average real capacity, the % of use and performance of the quality check machines used during the review process and the blister packaging machines in the packaging area were calculated. For the blister packaging machines, the calculation regarding machine use was separated by line and by machine, where the calculation by line refers to the asset as well as packaging while the calculation by machine refers only to the asset.
During this phase, it was possible to determine that there was a discontinuity in the review and blister packaging processes, with the greatest discontinuity occurring in line number 2 of the blister packaging area. This is due to the fact that five people need to be working on this line to prevent problems. However, whether due to absenteeism or an incorrect headcount, only two people normally worked on this line. This lack of personnel meant that the workers had to divide shift work into two segments, with the first segment involving only work with the machine and with second dedicated only to packaging. This resulted in low levels of % asset use, since this only occurred during half of the shift. This discontinutiy is easily seen upon reviewing *Table 2*. In addition, the % quality for each quality check machine was determined and the results are shown in *Table 1*. Likewise, it was determined that the absenteeism and the lack of personnel on the production lines greatly impacted factory performance and, therefore, was a key factor influencing uncertainty in the production process at this company.

Finally, the capacity of the packaging area was determined, which had previously been unknown to the company.

*Table 1. Performance Indicators obtained for the quality check machines.*

| Quality Check Machines | % Utilization | % Perfomance | % Quality |
|---|---|---|---|
| **821** | 80% | 47% | 99% |
| **614** | 48% | 44% | 95% |
| **613** | 48% | 44% | 90% |

### 4.2.3. Phase 3: Creation of a VSM Diagram for the Current Situation

Step 5: Choosing the family of products to analyze using VSM. The family of products chosen were blister packaged products with a size of 10 ml. This choice was made because, based on the projections made, the packaging area will not be able to produce the demand for this product at current performance levels, which will then result in a backorder of $574,522[u]$ for the year 2019.

*Table 2. Performance indicators obtained for the blister packaging area.*

| Blisters | % Line Utilization | % Asset Utilization | Perfomance |
|---|---|---|---|
| 1 | 82% | 74% | 54% |
| 2 | 91% | 56% | 44% |
| 3 | 80% | 64% | 49% |

*Figure 2. VSM diagram of the current situation for a family of 10 mL blister packaging products.*

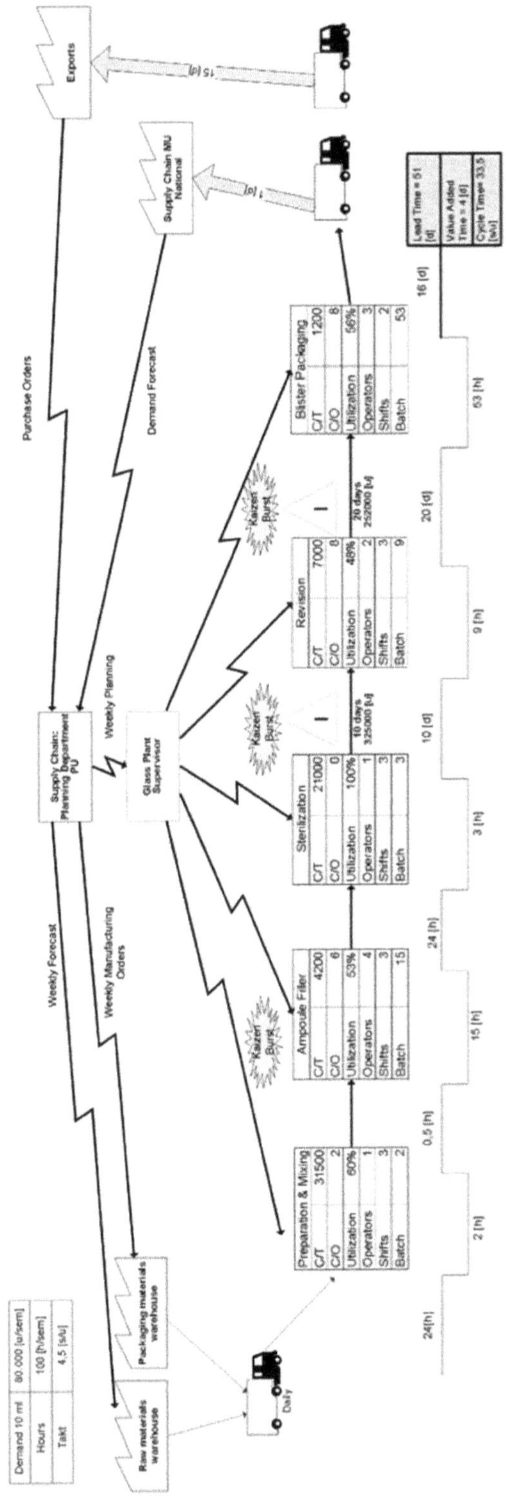

Step 6: Selecting process groupings and creating a VSM diagram for the current situation. First, the decision was made to aggregate the Blistering and Packaging processes into only one activity for the purposes of creating the VSM diagram. This was done because these activities both are characterized as having a continuous work flow and also share the same production capacity.

In addition, BPT transfer activities and the release of products were added to the dispatch of finished products given that those two activities take less than 6 hours to complete and, therefore, are not worth considering separately for the purposes of this study. Finally, the VSM diagram for the glass factory was created and the following indicators were obtained for this family of products and its respective production process: L/T, VA, C/T and Takt Time.

The results of this phase can be seen in *Figure 2.*

### 4.2.4. Phase 4: Analysis and Determination of key Factors.

Step 7: Analysis of the current situation. With the creation of a VSM diagram for the current situation, the three main problem areas that create uncertainy in the production process were identified:

1. WIP after sterilization and before the review process.
2. WIP after review and before blister packaging.
3. Long periods of delay during format changes during different process stages: Filling, Review-Packaging and Blistering-Packaging.

In addition, it was determined that value was being added only 8% of the time during the production process. It was also calculated that the total L/T of this family of products was 51 days. This long L/T time, when compared to other products created at the plant, is mainly due to the low real capacity of the machines.

Step 8: Identification of the key factors controlling uncertainty. By using the VSM diagram, interviews and the 5-Whys tool determined the key factors creating uncertainty on the production lines and thus the variables that must be controlled in order to reduce the high levels of WIP. The underlying factors of this problem were the same for both the review lines and the blister packaging lines, with the difference based only on the magnitude of each factor:

- Low Capacity on Review lines and Blister packaging lines:
  - Process Discontinuity:
    § Absenteeism.
    § Incorrect headcount.
  - High Format Changing times and a high frequency of failures:
    § No process standardization.
    § Poorly informed machine operators

In the case of blister packaging lines, the factor that most affected low performance was due to process discontinuity. For the review lines, the most influential factor was high format changing time. In addition, the existing levels of WIP between the sterilization and review processes were mainly due to the fact that these processes are not performed on a line after the products are reviewed due to the low number of operators available. Thus, in order to know the number of personnel missing for optimal performance, it was decided to perform an overhaul of the entire glass factory. It was then determined that 28 additional employees were needed. These results can be seen in *Annex 2.*

### 4.2.5. Phase 5: Definition and Evaluation of Improvement Proposals.

Step 9: Definition of the proposals to improve performance and decrease uncertainty. Considering the factors identified in the previous phase, the following solutions were proposed:

- Increase personnel count by two in line 2 of the blister packaging zone, but for the first shift.
- Increase personnel by twenty-three people in the packaging zone (considering what was already suggested in the first proposal).
- Process standardization and the application of SMED and TPM in the filling and packaging zones.

Step 10: Evaluation of the economic and operational impact expected from the proposals. With respect to the first proposal, it was determined that the proposed change would increase production capacity by $1800\left(\frac{u}{h}\right)$ and, given that this capacity would allow the factory to meet its yearly demand for this particular product, the economic benefit would be $40,000\left(\frac{\$}{year}\right)$.

With respect to the second proposal, which involved increasing the number of personnel by twenty-three in the packaging area, the proposed changes were projected to eliminate the existing WIP in this area, resulting in an expected economic benefit of $60,000\left(\frac{\$}{year}\right)$, which also considers the fact that fines for delivery delays for finished products would be eliminated.

A projection of factory capacity was also conducted following the implementation of the first two proposals and it determined that the factory would be able to meet its sales forecast for this year.

Regarding process standardization and the application of SMED and TPM, it was determined that it would be possible to improve the speed of formatting changes by $10-20\%$ and by $15\%$ for machine maintenance. By applying just SMED, the expected economic benefit would be between $[48,000-96,000]\left(\frac{\$}{year}\right)$ if applied to one filling machine, one quality check machine and one blister packaging machine, which would result in an increase in the factory's total productive capacity.

Step 11: Creation of a VSM diagram for the expected future situation. Using the information regarding the proposals and the expected results of their implementation, a VSM diagram for the future state of the glass production factory was created and is shown in *Figure 3.*

It was determined that, by implementing the suggested proposals, WIP for this family of products would be completely eliminated in the packaging area. It would also decrease WIP by $80\%$ after the sterilization process and prior to the review process. Finally, it would also reduce Lead Time by an estimated $66\%$.

Step 12: Definition of a strategy for the implementation of the proposals. Over the short term, it was recommended that two employees be added to blister packaging line number 2 during the first shift.

*Figure 3. VSM diagram of the expected future situation for a family of 10 mL blister packaging products.*

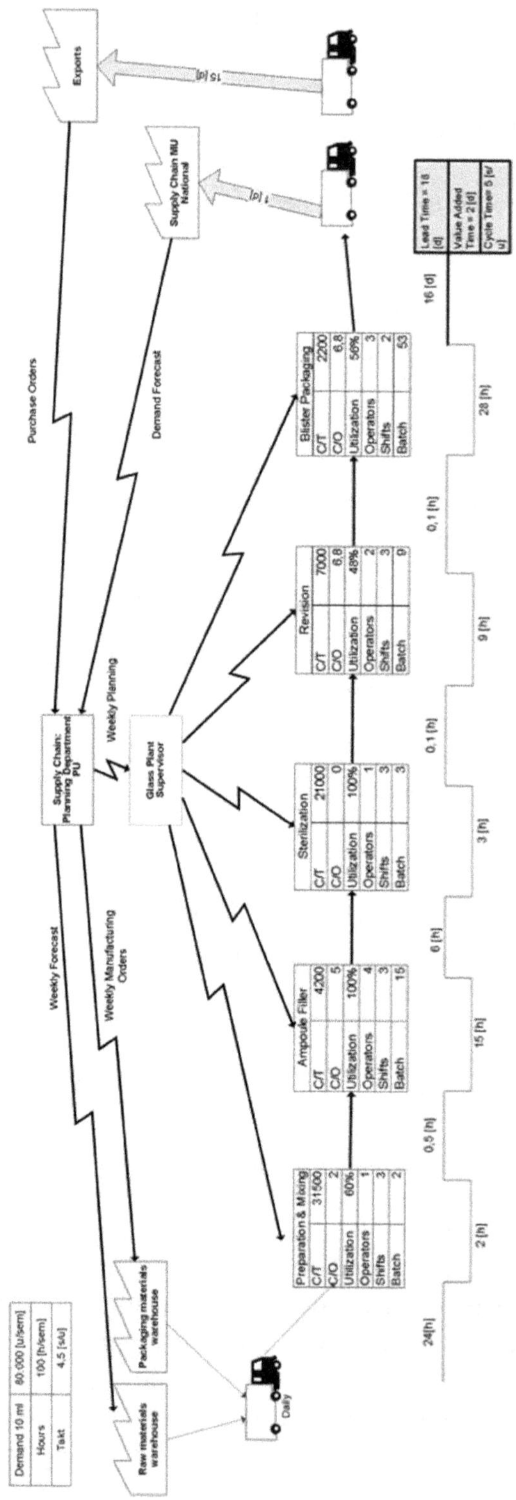

Subsequently, it was recommended that the format changing process be standardized and that maintenance be performed for the different assets of the production process.

Over the medium to long term, it was recommended to increase the number of personnel in the packaging area with more employees who would be distributed over the three shifts and in different areas of the factory. Finally, it was proposed that the company apply the LM, SMED and TPM techniques, with the objective of further reducing total production time, given that it is possible to both reduce format changing times and increase asset availability.

Finally, the implementation of these proposals implies an improvement in the control of uncertainty in the production process. This is because, by improving overall factory performance, the variability in production time will be reduced and, therefore, factory managers will have more consistent results regarding the production time of the different products manufactured at the plant.

## V. RESULTS AND CONCLUSION

A high level of Works-In-Progress (WIP) at a company creates uncertainty in the production process that results in great losses for the company, such as collapses in stock, delivery delays and immobilized capital, among others. Therefore, it is essential to have tools available that can control uncertainty and high levels of WIP.

In this way, the strategy proposed in this article seeks to gradually reduce the levels of WIP experienced by the company that participated in the case study through the use of Lean Manufacturing (LM) tools. The main LM tools used were a combination Value Stream Mapping (VSM) diagrams and the 5-Whys method, which were used to determine the root cause of the identified problem. The importance of VSM is due to the fact that it provides an overview of the entire flow of value of the company, from the initial input of material to the delivery of the final product. Thus, this makes it easy to determine in which area or during which process the problem occurs, thereby allowing it to be more easily investigated and mitigated by using both quantitative and qualitative information.

It is worth mentioning that, by conducting this case study, it was determined that LM methodology provides extremely useful tools for the identification and subsequent control of uncertainty in the production process, specifically the VSM tool. VSM is very useful in situations where a company has little information on asset performance and does not know which factors create uncertainty in the production process.

Also, during in the case study, the key factors of uncertainty in relation to WIP were determined and were mainly due to the poor management of production lines. Most were specifically related to an incorrect count of the number of people needed to operate the blister packaging lines continously.

With respect to the proposals presented to the company, it was determined that the overall WIP elimination strategy will generate an economic benefit of between $\left[100,000-148,000\right]\left[\$/\,year\right]$ as well as a 66% decrease in Lead Time and a considerable increase in the capacity of production lines.

It is worth pointing out that the application of the strategy proposed in this article will largely depend on company leadership and its culture. This is because those variables are fundamental to ensuring that the application of the proposals be sustainable over time and that a change in company mentality takes place that favors the continued improvement of the factory's operation.

## REFERENCES

Adwait, D., Saily, K., Jayant, G., & Vivek, K. (2018). Design and evaluation of a Lean Manufacturing framework using Value Stream Mapping (VSM) for a plastic bag manufacturing unit. Materials Today: Proceedings, 5(2), 7668–7677. doi:10.1016/j.matpr.2017.11.442

Bauer, W., Ganschar, O., & Gerlach, S. (2014). Development of a Method for Visualization and Evaluation of Production. *Proceedings of the 47th CIRP Conference on Manufacturing*, 17, 481-486.

Braglia, M., Carmignani, G., & Zammori, F. (2006). A new value stream mapping approach for complex production systems. *International Journal of Production Research, 44*(18-19), 3929–3959. doi:10.1080/00207540600690545

Brunt, D. (2000). From current state to future state: Mapping the steel to component supply chain. *International Journal of Logistics: Research and Applications, 3*(3), 259–271. doi:10.1080/713682765

Chaneski, W. (2014). Revisiting The Value Stream. Obtenido de Modern Machine Shop: https://www.mmsonline.com/columns/revisiting-the-value-stream

De Oliveira, R. I., Oliveira Sousa, S., & de Campos, F. (2019). Lean manufacturing implementation: Bibliometric analysis 2007–2018. *International Journal of Advanced Manufacturing Technology, 101*(1-4), 979–988. doi:10.1007/s00170-018-2965-y

Díaz-Bravo, L., Torruco García, U., Martínez Hernández, M., & Varela Ruiz, M. (2013). La entrevista, recurso flexible y dinámico. *Investigación en Educación Médica, 2*(7), 162–167. doi:10.1016/S2007-5057(13)72706-6

Ghizoni Pereira, L., & Luz Tortorella, G. (2018). A Literature Review on Lean Manufacturing in Small Manufacturing Companies. In J. Davin (Ed.), *Progress in Lean Manufacturing (págs* (pp. 39–68). Springer., doi:10.1007/978-3-319-73648-8_3.

Hernández, J., & Vizán, A. (2013). Lean Manufacturing Conceptos, Técnicas e implantación. Obtenido de https://www.eoi.es/es/savia/publicaciones/20730/lean-manufacturing-concepto-tecnicas-e-implantacion

Knoll, D., Reinhart, G., & Prüglmeier, M. (2019). Enabling value stream mapping for internal logistics using multidimensional process mining. *Expert Systems with Applications, 124*, 130–142. doi:10.1016/j.eswa.2019.01.026

Marathamuthu, S., & Jebaraj, S. (2015). The use of 5-WHYs technique to eliminate OEE's speed loss in a manufacturing firm. *Journal of Quality in Maintenance Engineering, 21*(4), 419–435. doi:10.1108/JQME-09-2013-0062

Möldner, A. K., Garza-Reyes, J. A., & Kumarc, V. (2018). Exploring lean manufacturing practices' influence on process innovation performance. *Journal of Business Research*, •••, 1–17. doi:10.1016/j.jbusres.2018.09.002

Nannapaneni, S., Mahadevan, S., & Rachuri, S. (2016). Performance evaluation of a manufacturing process under uncertaintyusing Bayesian networks. *Journal of Cleaner Production, 113*, 947–959. doi:10.1016/j.jclepro.2015.12.003

Negrão, L., Godinho Filho, M., & Almeida Marodin, G. (2016). Lean practices and their effect on performance: A literature review. *Production Planning and Control*, *28*(1), 33–56. doi:10.1080/09537287.2016.1231853

OECD. (2015). Health at a Glance 2015. Obtenido de OECD ilibrary: https://www.oecd-ilibrary.org/social-issues-migration-health/health-at-a-glance-2015_health_glance-2015-en

Rath, F. (2008). Tools for developing a quality management program: Proactive tools (process mapping, Value stream mapping, fault tree analysis and failure mode and effects analysis). *International Journal of Radiation Oncology, Biology, Physics*, *71*(1), 187–190. doi:10.1016/j.ijrobp.2007.07.2385 PubMed

Runkler, T. (2012). Data Preprocessing. In T. Runkler (Ed.), *Data Analytics* (pp. 21–33). Springer Vieweg., doi:10.1007/978-3-8348-2589-6_3.

Saboo, A., Garza-Reyes, J., Er, A., & Kumar, V. (2014). A VSM improvement-based approach for lean operations in an Indian manufacturing SME. Int. J. Lean Enterprise Research, 1(1), 41–58. doi:10.1504/IJLER.2014.062281

Schenk, M., Wirth, S., & Müller, E. (2010). Factory Planning Manual. Springer. doi:10.1007/978-3-642-03635-4

Singh, B., Garg, S., & Sharma, S. (2011). Value stream mapping: Literature review and implications for Indian Industry. *International Journal of Advanced Manufacturing Technology*, *53*(5-8), 799–808. doi:10.1007/s00170-010-2860-7

Toyota. (2019). A production system based on the philosophy of achieving the complete elimination of all waste in pursuit of the most efficient methods. Obtenido de Toyota: https://global.toyota/en/company/vision-and-philosophy/production-system

Weixiong, Z. (2011). Computer Science & Engineering Mckelvey School of Engineering. Obtenido de https://www.cse.wustl.edu: http://iasri.res.in/ebook/win_school_aa/notes/Data_Preprocessing.pdf

Womack, J., & Jones, T. D. (2003). *Lean Thinking*. Simon & Schuster, Inc.

Zandieh, M., Joreir-Ahmadi, M., & Fadaei-Rafsanjani, A. (2017). Buffer allocation problem and preventive maintenance planningin non-homogenous unreliable production lines. *Springer London*, *91*(5-8), 201–211.

# Chapter 13
# Use of Mathematical Models in a Mechanical Metal Industry to Improve Production Planning and Control

**Leoni Pentiado Godoy**
*University Federal de Santa Maria, Brazil*

**Wagner Pietrobelli Bueno**
*Universidade Federal do Rio Grande, Brazil*

**Tais Pentiado Godoy**
https://orcid.org/0000-0001-5021-5315
*Universidade Federal de Santa Maria, Brazil*

**Clandia Gomes**
*Universidade Federal de Santa Maria, Brazil*

**Maria Carolina Martins Rodrigues**
https://orcid.org/0000-0003-2575-8611
*Algarve University, Portugal*

**Luciana Aparecida Barbieri da Rosa**
https://orcid.org/0000-0001-9240-0236
*Universidade Federal de Santa Maria, Brazil*

## ABSTRACT

*This chapter aims to propose an improvement in decision making in the planning sector and production control (PPC) with application of a mathematical model. In the methodology, the qualitative approach was used because the linguistic codifications are interpreted and characterized by a case study applying a questionnaire to the managers of the company of the metal mechanic sector. In this context,*

DOI: 10.4018/978-1-7998-3246-1.ch013

*six constructs were structured as a proposal for performance improvement, being composed of costs, management, inspection, processes, and capacity. The chapter reports the main results achieved during fuzzy sets application, obtaining a better result compared to FAHP in which there were certain oscillations between the percentage of constructs. The construct prioritized by managers and specialists was the cost construct, reaching 38.60%, being advantageous for the industry when the cost is placed in order of manufacture (subconstruct), followed by the prioritized management construct with 28.50%.*

## 1. INTRODUCTION

Due to the competitive market within the manufacturing sector, organizations need to have efficient and effective planning and process management so that decisions made in the business environment are able to obtain as much information as possible (PRAJOGO et al., 2018). So, the entire productive operational network is integrated, enabling the study of the organization concept by associating the characteristics of the competitive market of the metalworking industries (LONGENECKER, et al., 2013, MUSETTI; ESCRIVÃO FILHO, 2016). Organizations are made up of open systems and are therefore affected directly and indirectly by competition in their operating environment (factory floor). In this sense, it is necessary to guide, monitor and evaluate the decision-making processes in the management of planning and production control (PPC) lays the organization's survival in close competition markets, production control (MUSETTI ESCRIVÃO FILHO, 2016).

Any environment where resources can be find, mainly financial, it will be transformed into a tangible good of service for the consumers, will carry with it the challenge and the necessity to reduce the risks in the decision-making process (BRANDÃO et al., 2017). According to Skolud; Krenczyk; Davidrajuh, (2016), the poor performance of PPC has been one of the main causes that have led companies around the world to terminate the business, or file for bankruptcy.

The improvement process, focuses on business identification and analysis, including quality improvement, waste removal and maintenance of the achieved improvements (AQLAN, AL-FANDI, 2018). However, there is a need for a methodology to identify the sectors of the PCP with the best opportunities, as well as to prioritize and select projects for improvement and allocate resources to the selected projects (SOLIMAN; SAURIN; ANZANELLO, 2018). In order to prioritize and propose an improvement in the decision making in the sector (PPC), methodologies were used to help managers in decision making presenting more accuracy in data analysis.

For Rodriguez, Coast e Do Carmo (2013), the modeling of decision problems, from the multiple criteria perspective, has received important diffusion in the environment of Production Engineering. Salomon, Marins e Duduch, (2009) made use of the AHP method to model the supplier selection process in situations that have multiple decisions. They showed the AHP method applied to the selection of suppliers of new equipment for two assembly lines of an auto parts factory. The model considered the same set of criteria for the two lines and allowed to incorporate aspects of multi-decision modeling to the problem, initially presented as two independent models.

Therefore, one of the used methods was the Fuzzy-Analytic Hierarchy Process (FAHP). Chang (1996) developed the method, which was seeking to rank the factors considered most important, in aa certain situation. FAHP is a method that allows to approach phenomena of uncertainties and inaccuracies, considering a mathematical model that fits for decision processes of complex and nebulous environments (CHANG, 1996; SOUZA; BIANCHI; BOENTE, 2015). The other model used was the Fuzzy Sets (fuzzy sets), which contributes to simulations and modeling. Fuzzy Sets is a logic introduced by Zadeh (1965), proposing as concept a theory capable of understanding human reasoning by assigning information which is only necessary for certain multivalued decision situations (WANG; XIONG; JIANG, 2014).

In this way, the research is guided by the following question: What are the relevant attributes contributing to the productive decision-making processes in companies with manufacturing on demand?

The study focused on the preference for small companies in the metalworking sector with production projections on demand, being justified, because there are restrictions that contribute to analysis and management models of planning and production control (MUSETTI FILHO, 2016). The article aims to offer as a contribution, to scholars and new entrepreneurs, efficient processes for evaluating projects in PPC. Thus, the study is applied to the managers of a metalworking company, located in the northern region of RS.

## 2 THEORETICAL REFERENCE

Here the theoretical items are presented to support the proposed work, relating projects of Planning and Production Control, PPC; Fuzzy-Analytic Hierarchy Process, FAHP and Fuzzy Sets.

### 2.1 Production Planning and Control Projects (PPCP) in the Metalworking Sector

Organizations willing to stay in competitive markets are of the utmost importance the need to improve the interactive and innovative way of projects (GONDIM, 2011). Thus, obtaining information from the competitive market, facilitates managers to choose more appropriate methods, according to different situations (NORO et al., 2012; ABDEL-MAKSOUD et al., 2015). Metalworking is one of the most representative industrial sectors in the economy of Rio Grande do Sul, with 37.6% of industrial GDP, according to the Federation of Industries of Rio Grande do Sul (FIERGS, 2015).

For the realization of the project in PPC it is necessary to seek information and analyze the economic and technical components, which are characterized in the product, projected demand, marketing channels, sales price and other similar procedural factors according to each company (WOILER; MATHIAS, 2010; KHAMAKSORN, 2016). Thus, as in many manufacturing sectors, metalworking is no different, as all search for models of projects capable of supplying the deficits not expected by supply and demand in the market.

Thus, the complex metalworking sector encompasses numerous sectors of activity in projects in Brazil, being a very relevant area and prone to economy. The metalworking sector of Rio Grande do Sul is in second place in terms of concentration of industrial companies, representing 12.5% of the national total. The concentration is higher in the Metropolitan Region of Porto Alegre with 44.3%, then in the Northeast with 26.4%, last in the Northwest with 15.2%, according to data from the Federation of Industries of the State of Rio Grande do Sul (FIERGS, 2015). Rio Grande do Sul is the second-best market in machinery and equipment with 24%, followed by metal products 18%, and metallurgy with 14% (FIERGS, 2015).

Therefore, in view of the above, the guarantee of good results is linked to the good Planning and Production Control, of the entire production process. Considering this, it becomes possible to act correctly, even if there are irregularities, process failures, and this practice also allows the reduction of operating costs. The PPC comes to support the management in the decision making, and the biggest problems of production, is to manage the planned means and not the production goals.

In the 1980s, scholars made the first scientific FAHP appearances. Specialists known as Van Laarhoven e Pedrycz (1983), proposed the hybrid integration of fuzzy logic methodologies. It is logical to draw Boolean configurations, where there are propositions of judgments made of linguistic terms for the scale values between [0,1]. This connection is affected only with the application of the Analytic Hierarchy Process methodology known as AHP. The purpose of the studies carried out by the AHP methodology is the hierarchy of the actions that are in the foreground, those of the background are to choose the order of decision making. These measures should be consistent with the Consistency Index (CI), which evaluates the degree of inconsistency of the matrix performed in pairs by Equation 1 (ZARE et al., 2016).

$$CI = \frac{|\text{»}\max - N|}{N-1} \tag{1}$$

Subsequently, the Consistency Ratio (CR) is performed allowing to consider the inconsistency, according to the judgments made, through Equation 2, which should be <0.10.

*Table 1. Conversion Scale*

| Scale | Evaluation- AHP | Reverse-AHP | Evaluation-*Fuzzy* | Inversa-*Fuzzy* |
|---|---|---|---|---|
| **Extremely preferred** | 1 | 1 | (1;1;3) | (1/3;1;1) |
| **Very strongly prederred** | 3 | 1/3 | (1;3;5) | (1/5;1/3;1) |
| **Strongly preferred** | 5 | 1/5 | (3;5;7) | (1/7;1/5;1/3) |
| **Moderately preferred** | 7 | 1/7 | (5;7;9) | (1/9;1/7;1/5) |
| **Equally preferred** | 9 | 1/9 | (7;9;9) | (1/9;1/9;1/7) |
| **Intermediate values** | 2,4,6, e 8 | ½, ¼, 1/6, 1/8 | | |

Source: SAATY (2008); KAHRAMAN, (2006).

$$CR = \frac{CI}{IR} \tag{2}$$

The Fuzzy-AHP theory is a variation of the traditional AHP (Analytic Hierarchy Process) approach proposed by Saaty (1987). Approach AHP uses integers to perform the pairwise comparison of criteria, and the Fuzzy-AHP method employs fuzzy numbers for this purpose. Some authors consider this approach more appropriate because peer-to-peer comparisons are fuzzy in nature (CHAN; KWONG; DYLON, 2012). This method is considered one of the most effective when involving decision making and the need for greater objectivity in responding within complex environments (GOMES et al., 2004).

First the AHP scales (odd values) are organized and later transformed to a fuzzy scale, as shown in Table 1.

To model this type of uncertainty (preferably human), fuzzy sets can be incorporated into the AHP pair comparison. Therefore, the FAHP approach allows for a more accurate and robust description of the decision process. Zadeh introduced the fuzzy set theory, to rationalize the uncertainty associated with imprecision, analogously to human thought. The scale recommended by Saaty (1991), shown in Table 01, part 1 to 9, with 1 meaning the indifference of importance of one criterion in relation to the other, until the extreme importance of one criterion over another, with intermediate stages of importance between these levels 1 and 9, which are the intermediate values shown (2, 4, 6, and 8).

In addition, disregarding the comparisons between the criteria themselves, which represent 1 on the scale, only half of the comparisons need to be made, as the other half consists of the reciprocal comparisons in the matrix, which are the reciprocal values (when looking for a condition of compromise between two definitions). These were compared, as shown in Table 01. It is important to note that the most important elements of comparison were used as an integer value of the scale, and the least important, as the inverse of this unit. Through the values presented by the author, the manager makes his judgment prioritizing the analyzed variables. This judgment takes place before its perception, thus being subjectively. The prioritized criteria and sub-criteria is the result of the judgment made by the manager.

Therefore, Van Laarhoven e Pedrycz (1983) suggested that, in order to justify the applicability of FAHP, the four sequential steps should be elaborated as follows (KAHRAMAN, 2006; ZARE et al., 2016). Steps to perform FAHP calculations (KAHRAMAN, 2006):

1) Identify the main criteria and sub-criteria;
2) Structure the AHP model based on the main criteria and sub-criterium;
3) To determine the weights of the main criteria established;
4) Classification of the matrix for decision making.

Following this, four processes are proposed by Kahraman (2006), oriented by Chang's extension research. Two mathematical foundations were developed to simplify FAHP calculations. Equations 3 and 4 show the equations suggested by Chang (1996)

$$1, \text{if } mi \geq mj \qquad 0, \boldsymbol{if\ lj \geq li} \tag{3}$$

$$\frac{lj - ui}{(mi - ui) - (mj - lj)} \tag{4}$$

Authors Chen, Wang e Lin, (2015), Zhao, Tang e Ning, (2016) define FAHP as a suitable study instrument to solve decision-making problems in environments composed of multiple uncertainties, with the emergence of the best organizational strategies.

## 2.3 Fuzzy Sets

Fuzzy logic was introduced in mids-1965 by Lofti Zadeh, where the need to calculate the uncertainties was evident, since Fuzzy theory states that within a set gaps exist and must be considered, this lead to a more concrete analysis (ZADEH, 1965; KLIR; YUAN, 1995; HUANG et al., 2015). For example, in a highly competitive environment, which is the case of the company studied, it is necessary that the decisions taken become more and more precise.

The contribution of the Fuzzy or Fuzzy Sets establishes behavior modelling of the constructs in relation to the variables that occur in the market (SELLITTO; WALTER, 2006; YAMAKAWA; MIGUEL; AOKI, 2014). When elaborating probabilities of possible scenarios according to the simulated combinations, we used for the input variables corresponding weights between 0 and 1 (ZADEH, 1965).

The Fuzzy Sets, are characterized by mathematical expressions, where Zadeh (1965) defines basic guidelines of the sets, stating that:

Let X be the object environment, where any element of X is denoted by x. Thus, X = {x};

The Fuzzy set A in X is qualified by a membership function fA (x) in which each point of a real number is associated in the interval of [0,1], in which the value of fA (x) in x represents the 'degree of pertinence' of x in A;

The closer the value of fA (x) is to that of the stipulated set, the greater the degree of pertinence. Zadeh (1983) defined quantifiers of the first type as quantifiers used to represent absolute quantities (using Fuzzy numbers). The quantifiers of the second type are used to represent relative quantities (defined by Fuzzy numbers in [0,1]). In such an approach, there is the possibility of an alternative not fully true, or even fully false, with the possibility of presenting this interval scale with this variation (LIMA JUNIOR, 2013, MASSA et al., 2015). In relation to the rule that directs the data so that they can vary between extremes 0 and 1, becoming a multivalued logic, according to Equations 5 and 6 (KAHRAMAN, 2008; MASSA et al., 2015).

$$f(x) = \begin{cases} 1 \; if, and\ only\ if, x \in a \\ 0 \; if, and\ only\ if, x \notin a \end{cases} \tag{5}$$

$$\mu(x) = \begin{cases} 1 \; if, and\ only\ if, x \in a \\ 0 \; if, and\ only\ if, x \notin a \\ 0 \le \mu(x) \le 1 \; if, \; x\ belongs\ to\ partially, a \end{cases} \tag{6}$$

Thus, the concept is to work with as much information as possible, of bivariate or multivalence, by means of uncertainty methods, being explained with middle term expressions: hot medium, more or less cold, a little distant among others (KAR, PANI, 2014, DRAGOVIĆ et al., 2015).

The fuzzification is composed by triangular and trapezoidal Fuzzy. The proposition between these two methods of search for results, is the relationship between the model variables and Fuzzy regions. Equations 7 and 8 respectively represent the two fuzzification methods. A triangular Fuzzy value is computed if its membership function is according to Equation 7:

$$ÆA(x) = \begin{cases} 0, if\ x \le a \\ \dfrac{x-a}{u-a}, \text{if} \quad a < x \le u \\ \dfrac{x-b}{u-b}, if\ u \le x < b \\ 0, if\ x \ge b \end{cases} \tag{7}$$

It is a trapezoidal value and mentioned if its a membership function is presented according to Equation 8.

$$\varphi A(x) = \begin{cases} \dfrac{x-a}{b-a}, if \quad a \le x < b \\ 1, if\ b \le x \le c \\ \dfrac{d-x}{d-c}, if \quad c < x \le d \\ 0, otherwise \end{cases} \tag{8}$$

The equation that corresponds to the defuzzification is the center of gravity, once it provides the mean of the areas of all the corresponding figures of the Fuzzy set. Thus, the equations for this Fuzzy domain are 9 and 10, and so, any real number, which in a way can represent the Fuzzy set B (quoted in the equations) can be considered a defuzzification of B.

## 3. METHOD OF RESEARCH

To be considered as scientific research, the present study followed a script of methodological procedures. For Marconi e Lakatos, (2010) the method is the set of rational and systematic activities, which more safely allows to reach the general objective of the research, tracing the path to be followed by the researcher (MARCONI, LAKATOS, 2010, DEMO, 2013). The objective of this research is to propose an improvement in decision making in the planning and production control (PPC) sector, with the application of a mathematical modeling. Therefore, it is necessary to define which constructs, should be allocated to the project management in PPC, as it is a company with production characteristics under request.

Aiming to provide a perceptibility of the points resulting from this evaluation, the research contemplates the following steps: introduction, bibliographical research, methodology, results and discussion, and finally the conclusions. In the bibliographic research, the topics related to the Planning and Production Control (PPC) projects in the metalworking sector, multi-criteria decision methods such as Fuzzy-AHP and Fuzzy sets were based. The perspective of this stage was to understand the fundamental definitions of project approaches, inserting management, process analysis and systems modification to ensure greater efficiency and effectiveness in production.

Thus, the production process of the parts was thoroughly analyzed in order to obtain the most accurate data possible. The diagnosis stage was carried out, with the observation of the production cycle of the pieces, as well as interviews with the managers and operators, in order to collect the necessary information to justify the present study. It is evidenced that this stage lasted about 120 days, during which the research was being carried out in the company.

The Fuzzy-AHP modeling step is the definition of the criteria and alternatives, evidenced during the investigation. The reasons given by Saaty (2008), regarding the comparison judgment at par. Given the combination of the proportions through the scales, one can observe the status of each alternative in relation to the system used. The order of the definition of the criteria and alternatives is in accordance with the structure elaboration of the problem in evidence. The limitations of the research are found in the collected data, as gathered information was from the company's management body with a focus group method.

Justification for this research is the fact that the elements or variables that lead to the studied company were searched, but did not reach the productivity and ideal quality, as well as verified which variables would be prioritized to improve company's performance. The research was designed as a case study in which the researcher involved must understand the approaches to be more effective to investigate problems, answering objective questions of the research, providing practical and theoretical contributions. (YIN, 2015, MIGUEL, et al., 2012).

To collect data from the research, was through a questionnaire with related questions, with five alterative responses which are: equal importance, not important, very important, most important and extremely important. Following that, the data will be collected through primary and secondary sources, documents made available by the company and semi-structured interviews. Cervo e Bervian (2002) state that conducting a survey through questionnaires, it is necessary to establish objective criteria, with questions that lead to the answer in a simple and easy to understand.

The related criteria for this research were established based on the literature and the sub-criteria were selected based on a pilot test in the company, located in the northern region of RS. According to the test we selected the subcategories for each matrix of the criteria. To do so, the six criteria are: costs, management, storage, inspection, processes and capacity. The questionnaire was applied to four managers using the focal group technique. In order to analyze the information collected in the research, we used software such as Excel and Assistant to perform FAHP and Infuzzy calculations to make the simulations and proof of the proposed performance improvement of planning, planning and production control.

## 4. ANALYSIS AND DISCUSSION OF RESULTS

The FAHP methodology together with the Fuzzy Sets seek to refine the variables that involve the business field, based on the multi-criteria method, ratifying which factors can be understood as a secure decision making.

### 4.1 Fuzzy-Analytic Hierarchy Process (FAHP)

Adopting the FAHP method, Table 02 presents the percentage of importance for each construct, contemplated by the managers and the relevance of the subcategories for each matrix-construct. According to Saaty (1991), for the matrices to be adequate and satisfactory for research, the Consistency Ratio values, RC <0.10. For the general matrix of the criteria the RC obtained 0.0598, thus, the proposed model is appropriate to follow the research. The criteria of each individual construct obtained in the cost (Cr)

*Table 2. Structure of Manufacturing Competitiveness*

| Terms Theoretical | Management in Decision-Making Processes Manufacturing industry (100%) | |
|---|---|---|
| **Constructs** | **Concepts** | **Relative Importance "ai"** |
| Cost (38,60%) | Automate batch transport (11.01%)<br>Invest in infrastructure (21.58%)<br>Invest in the machinery sector (11.21%)<br>Investment in a team (25.43%)<br>Control costs by order of manufacturing (30.77%) | 4,25%<br>8,33%<br>4,33%<br>9,81%<br>11,88% |
| Management (28,50%) | Execution of information systems as MRP (20.38%)<br>Differentiated solutions (21.81%)<br>Application of environmental management (2.88%)<br>Training of personnel (23.27%)<br>Leadership development (31.60%) | 5,81%<br>6,22%<br>0,82%<br>6,63%<br>9,03% |
| Estorage (20,90%) | Sales volume versus capacity (20.17%)<br>Security stock (0.76%)<br>Innovate with automation (35.68%)<br>Preventive maintenance (20.91%)<br>Corrective maintenance (21.48%) | 4,21%<br>0,16%<br>7,46%<br>4,37%<br>4,70% |
| Inspection (6,80%) | Focal inspection (25.60%)<br>Use of indicators (0,20%)<br>Chronoanalysis performance (4.50%)<br>Adjustments to fluctuations (23.50%)<br>Inspection of productive capacity (13.60%)<br>Ergonomic application (30.90%)<br>Investment in intellectual capital (1.70%) | 1,74%<br>0,01%<br>0,31%<br>1,60%<br>0,92%<br>2,10%<br>0,12% |
| Processes (4,80%) | Application of manufacturing cell (27.00%)<br>Use of positional manufacturing cell (13.60%)<br>Use of operational control (18.00%)<br>Use of departmentalization (23.00%)<br>Relevance on sustainability within the company (10.30%)<br>Use of information systems (5.60%)<br>Innovate in the reformulation of new processes (2,50%) | 1,30%<br>0,65%<br>0,87%<br>1,10%<br>0,49%<br>0,27%<br>0,12% |
| Capacity (0,30%) | Global process control (34.90%)<br>Process control isolated (6.00%)<br>Characteristic layout with company services (28,00%)<br>Setups by product (20.90%)<br>Use of statistical controls (3.10%) | 0,11%<br>0,02%<br>0,08%<br>0,02%<br>0,07% |
| | Total | 100% |

Source: Prepared by the authors, (2019).

0,0775; in management (Cr) 0.0811; on inspection (Cr) 0.0669; in the procedures (Cr) 0.0516; in storage (Cr) 0.0398 and capacity (Cr) 0.0743.

The arguments explained in Table 02, stands out for the managers, the construct, "cost" obtained the most importance in the study with 38.60%. This factor materializes because the companies need before even starting their activities, to combine prices together with their customers as raw material, logistics, inventory, personnel and machines (MUSETTI; ESCRIVÃO-FILHO, 2016).

From the managers' point of view, carrying out projects in companies with custom ordering features is more cost effective, but it is complex compared to industries that work with mass production. The capacity presented 0.30% being second in the choice of the managers. This percentage occurs due to the characteristics of the industry, it does not need know-how to carry out emergency manufacturing

*Table 3. Hierarchical Structure for each Sub-construct*

| Language Variables | Numerical variables | | | |
|---|---|---|---|---|
| | Máx % | Current % | Gap % | Order |
| Manufacturing cell application | 1,30 | 0,00 | 1,30 | 17 |
| Use of positional manufacturing cell | 0,65 | 0,04 | 0,61 | 22 |
| Operational control capability | 0,87 | 0,11 | 0,76 | 20 |
| Use of departmentalization | 1,10 | 0,18 | 0,92 | 18 |
| Relevance of sustainability theme within the company | 0,49 | 0,04 | 0,45 | 23 |
| Use of information systems | 0,27 | 0,03 | 0,24 | 25 |
| Innovate in reformulating new processes | 0,12 | 0,00 | 0,12 | 27 |
| Focal inspection | 1,74 | 0,00 | 1,74 | 15 |
| Use of indicators | 0,01 | 0,03 | (0,02) | 35 |
| Performing chronoanalysis | 0,31 | 0,02 | 0,29 | 24 |
| Adjustments to fluctuation | 1,60 | 0,08 | 1,52 | 16 |
| Inspection of productive capacity | 0,92 | 0,10 | 0,82 | 19 |
| Ergonomic application | 2,10 | 0,07 | 2,03 | 14 |
| Investment in intellectual capital | 0,12 | 0,00 | 0,12 | 28 |
| Sales volume vs. capacity | 4,21 | 0,01 | 4,20 | 11 |
| Safety stock | 0,16 | 0,01 | 0,15 | 26 |
| Innovate with Automation | 7,46 | 0,00 | 7,46 | 5 |
| Preventive maintenance | 4,37 | 0,03 | 4,37 | 10 |
| Corrective maintenance | 4,70 | 0,09 | 4,61 | 9 |
| Global Process Control | 0,11 | 0,22 | (0,11) | 33 |
| Process control isolated | 0,02 | 0,00 | 0,02 | 29 |
| Characteristic layout with company services | 0,08 | 0,06 | 0,02 | 30 |
| Characteristic layout by product | 0,02 | 0,01 | 0,01 | 31 |
| Product Setups | 0,07 | 0,06 | 0,01 | 32 |
| Use of statistical controls | 0,00 | 0,01 | (0,01) | 34 |
| Automate batch transport | 4,25 | 0,08 | 4,17 | 12 |
| Investing in infrastructure | 8,33 | 0,19 | 8,14 | 4 |
| Adherence to technologies | 4,33 | 0,40 | 3,93 | 13 |
| Team investment | 9,81 | 0,86 | 8,95 | 2 |
| Control costs by manufacturing order | 11,88 | 1,79 | 10,09 | 1 |
| Execution of information systems as MRP | 5,81 | 0,39 | 5,42 | 8 |
| Differentiated solutions | 6,22 | 0,10 | 6,12 | 7 |
| Application of environmental management | 0,82 | 0,08 | 0,74 | 21 |
| Training | 6,63 | 0,28 | 6,35 | 6 |
| Leadership developmen | 9,03 | 0,63 | 8,40 | 3 |

Source: Prepared by the authors, (2019).

activities. The construct capacity is likely to undergo changes in its importance if it is applied in other organizations. Three constructs (capacity, costs, management) have a constant linearity in their importance weightings according to the experts.

In Table 03, the specific hierarchy for each subcategory of the 35 survey questions is found. The purpose of this analysis is to simplify decision-making for efforts in PCP projects. It is perceived that simplification within the manufacturing sector is important, but the use of FAHP shows which constructs stand out most for the PCP to apply efforts and control. Thus, managers are directed to the main action to be contemplated when planning an activity. First, we begin with the cost-to-manufacturing process, followed by team investment, leadership development, and investment in infrastructure.

Therefore, in this analysis it was observed that the infrastructure is among the four most expressive. This fact stands out according to the managers, due to the evolution of the market under the order to order. In many negotiations, in which customers demand that their product remain in the premises of the company for an odd period of days, and to use them later.

## 4.2 Degree of Fuzzy Relevance

The contribution of the Fuzzy or Fuzzy Sets is to model the behavior of the constructs in relation to the variables that occur in the market (YAMAKAWA; MIGUEL; AOKI, 2014). Performing probabilities of possible results according to the simulated combinations using for the input variables corresponding weights between 0 and 1 (ZADEH, 1965). Table 04 provides the input values stipulated according to the researchers analyze.

In consideration of the input aspects, the output values are composed of weights between 0 and 100 as decisive factors for the defuzzification of the triangular methods Equation 7 and trapezoid Equation 8. As a risky action plan (20; 40; 55); plan of action to be analyzed (40, 60, 70, 80) and acceptable action plan (75, 85, 100).

## 4.3 Simulations FUZZY

Fuzzy logic (diffuse or nebulous) makes the results similar to human thinking, modeling the qualitative sense of the answers to make decisions that are closer to the correct one (VAN LAARRHOVEN; PEDRYCZ, 1983). Figure 02 shows the history of 76 simulations, which were performed based on the values added according to Table 03. These combinations performed by the researchers, in which the best defuzzification point for the proposed method was sought.

Thus, the graph of Fig. 01, is a result of the center of gravity method of Equations 09 and 10. In which it presents the defuzzification obtained of 56.7%. This number will inform managers what is most pertinent to do, whether to follow the actual FAHP project or the simulated Fuzzy Sets project.

In this context, it is important to highlight Table 2, which presented the results of the FAHP constructs, and the cost presented 38.60%; the management of 28.50%; stocking 20.90%; inspection 6.80%; processes 4.80%; capacity 0.30%. It can be observed that, the cost and management obtained a defuzzified value of 58.99%. Therefore, these are the simulations of degree of capacity among all the constructs. Based on the FAHP values, making new adjustments with real answers from managers, we find the approach of the new proposal, which obtained an improvement upto 63.66%, undergoing significant changes in the weights of some constructs.

*Table 4. Input variables*

| **Aggregation of values for degree of relevance in Costs** | | |
|---|---|---|
| Equal importance | Triangular | (0; 0,1; 0,2) |
| Little important | Triangular | (0,1; 0,35; 0,4) |
| Very important | Trapezoidal | (0,35; 0,5; 0,65; 0,7) |
| Fairly important | Triangular | (0,65; 0,75; 0,85) |
| Extremely important | Triangular | (0,75; 0,90; 1) |
| Aggregation of values for degree of pertinence in Capacity | | |
| Equal importance | Triangular | (0,11; 0,13; 0,32) |
| Little important | Triangular | (0,3; 0,43; 0,48) |
| Very important | Trapezoidal | (0,43; 0,54; 0,59; 0,6) |
| Fairly important | Triangular | (0,59; 0,68; 0,7) |
| Extremely important | Triangular | (0,7; 0,89; 0,99) |
| Aggregation of values for degree of relevance in Management | | |
| Equal importância | Triangular | (0; 0,18; 0,23) |
| Little importante | Triangular | (0,18; 0,3; 0,38) |
| Muito importante | Trapezoidal | (0,3; 0,45; 0,52; 0,63) |
| Bastante importante | Triangular | (0,52; 0,68; 0,76) |
| Extremamente importante | Triangular | (0,68; 0,85; 1) |
| Aggregation of values for degree of pertinence in Storage | | |
| Equal importance | Triangular | (0,1; 0,2; 0,3) |
| Little important | Triangular | (0,2; 0,4; 0,5) |
| Very important | Trapezoidal | (0,4; 0,55; 0,6; 0,65) |
| Fairly important | Triangular | (0,6; 0,7; 0,8) |
| Extremely important | Triangular | (0,75; 0,85; 1) |
| Aggregation of values for degree of pertinence in Processes | | |
| Equal importance | Triangular | (0; 0,12; 0,18) |
| Little important | Triangular | (0,15; 0,23; 0,28) |
| Very important | Trapezoidal | (0,23; 0,29; 0,33; 0,38) |
| Fairly important | Triangular | (0,3; 0,54; 0,65) |
| Extremely important | Triangular | (0,54; 0,89; 1) |
| Aggregation of values for degree of pertinence in Inspection | | |
| Equal importance | Triangular | (0,11; 0,18; 0,23) |
| Little important | Triangular | (0,17; 0,25; 0,29) |
| Very important | Trapezoidal | (0,24; 0,31; 0,38; 0,42) |
| Fairly important | Triangular | (0,33; 0,69; 0,77) |
| Extremely important | Triangular | (0,7; 0,85; 1) |

Source: Prepared by the authors, (2017)

*Figure 1. Graph of Fuzzy Sets simulation*
*Source: Prepared by the authors (2019).*

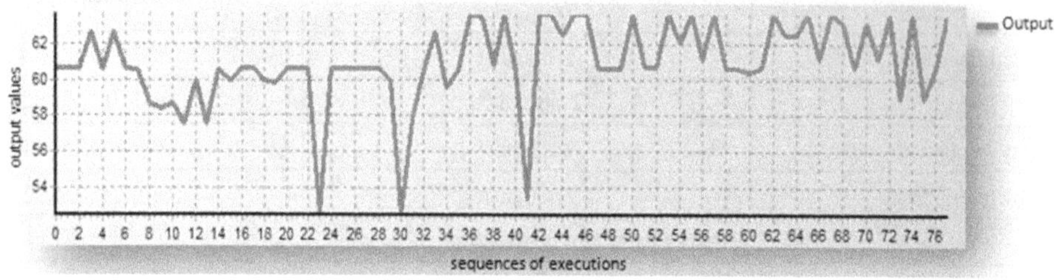


*Figure 2. Defuzzification*
*Source: Prepared by the authors (2019).*

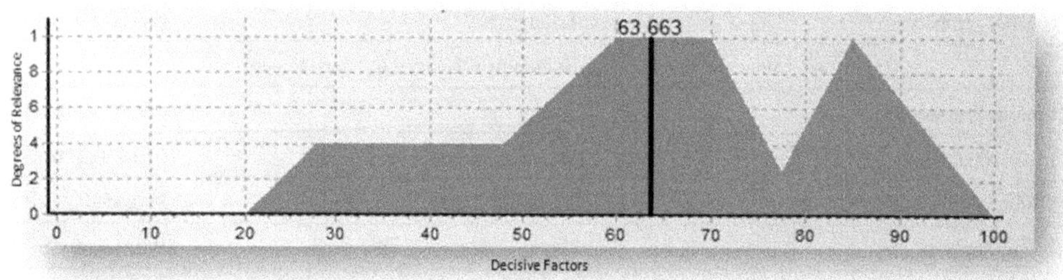


Figure 2 shows the result of the center of gravity method. Thus, the defuzzification is obtained reaching 63.66%, informing managers what is more important to be done, whether following the actual FAHP project or the Fuzzy sets modeled project.

Evaluating the mathematical proposals, Table 05 presents two decisions for the PPC. Note that there was a dissension of values for the inspection construct, in which the actual project was 21% and climbed to 35% with the simulation. So being, the inspection construct can be considered a synonym of control in efficient production.

*Table 5. Evaluation of the Project in Management*

| Control Parameter | Real FAHP | Simulated Fuzzy Sets |
|---|---|---|
| Process | 4,80% | 3% |
| Storage | 6,80% | 5% |
| Inspection | 21% | 35% |
| Capacity | 0,30% | 0% |
| Costs | 38,60% | 36% |
| Management | 28,50% | 21% |
| Total | 100% | 100% |
| Degree of Project Accuracy | 58,99% | 63,66% |

Source: Prepared by the authors (2019.

After performing the simulations versus the real plan, the percentage of the FAHP project for the Fuzzy Sets resulted in a 4.67% difference. The two methods confirmed the importance of these constructs for the metalworking sector. As a solution to operational problems, the manager obtains two ways of assisting their decision making at the time applied to the PPC through FAHP or Fuzzy Sets. The sector that needs to add more focus in the company, and what needs less control for certain activities in the company. Identifying which sector needs greater exposure coming from managers, and which sector needs less control for certain activities. These linguistic values can come by analyzing each construct for each sub-construct and emphasizing what must be changed internally in the industry.

## FINAL CONSIDERATIONS

The main contribution of the research was the use of the Fuzzy-AHP and Fuzzy Sets methods, which contribute to decision-making, using the constructs and sub-constructs relevant to the metallurgical industries. Approach, carried out through multi-criteria bases like pilot tests, applying questionnaires to managers and supervisors specialized in production. Mathematical models were used to emerge a new philosophical approach, of which paths are necessary and fundamental, for the fulfillment of a new management in decision making in the metallurgical sector.

Therefore, this study is portrayed in a linguistic way, and the techniques used such as FAHP and Fuzzy Sets resulted in reasoning projections aimed at reliable productions. In favor of this logical approach, allocating values for linguistic terms and using these techniques, the dimensions of the multivalued decision output responses remain with an acceptable degree of reliability, and the dimensions of the multivalued decision output responses remain with a degree of acceptable reliability for Saaty <0.10. It is important to observe the contribution of the FAHP in the treatment of uncertainties, because, in these types of works, empirical knowledge of specialists is used. Fuzzy Logic has an important contribution to the processing and integration of data from the metallurgical industry.

The first construct prioritized by managers and specialists was cost, reaching 38.60%, being advantageous for the industry when the cost is placed in a manufacturing order (sub-construction), followed by management with 28.50%. The constructs costs and management have a constant linearity in their importance weights according to the experts, as they admit that the set of these variables can represent good results if applied in production.

Storage with 20.90% of the specialists' preference, the others of lesser importance. So, in management, the preferred sub-criteria was "Training of personnel" (23.27%) and "Leadership development" (31.60%), being the leader responsible for facilitating, monitoring and evaluating the performance of a team. In this case, managers believe that, if there is training for the development of leaders, the impact on the productivity of the sector will be imminent, that is, these sub-constructs cited become the main coefficients to start a PPC consolidation.

The leader's primary role is the link established with the company's goals and the aspirations of employees. The leader's job is to inspire capacity, ethics and commitment to seek increasingly satisfactory results in your organization. In the "Personnel training", the fundamental role of this sub-construct is the development of the employee, which can be specific knowledge of the area or increase productivity, performance, motivation and updating to new technologies. A team that undergoes personal training, in addition to increasing skills, acquires motivation, self-confidence and self-esteem.

Most of the problems in decision making are in human subjectivity and in the vagueness of the real world. Numerous methods and tools are used to approach the uncertainties of future events in the industrial sector, which go through limitations to constantly improve the management system. An uncertain and competitive business environment causes industry professionals to seek flexible approaches to manufacturing processes. Competitive industrial professionals adopt various strategies for invariant market situations, where constant changes in design and demand rarely occur. The practical contribution made in this study is to propose better knowledge of these methods applied in production.

Finally, several case studies can be carried out to assess the acceptability and usability by professionals of methods such as those developed in this article to support the evaluation of improvements in decision making in the production planning and control (PPC) sector, applying new models, such as diffuse TOPSIS and sensitivity analysis.

The objective of this work is to propose an improvement in the decision making in the planning and production control (PPC) sector with the application of a mathematical modeling. Conducting pilot tests, applying questionnaires to managers and, using mathematical models to emerge a new philosophical proposal, what are the necessary steps and, fundamental, for the fulfillment of a good management in PPC.

The basis of this research was the use of the FAHP and Fuzzy Sets method, which contributed to the decision making, using the constructs and sub-construct relevant to the metalworking industries. The construct prioritized by managers and specialists was the cost construct, reaching 38.60%, being advantageous for the industry when the cost is placed in order of manufacture (sub-construct). Performing manufacturing order costs is highly complex for management. Second is "investment in teamwork" with 25.43%.

The third preference of managers and specialists in the cost sub-construct was "investment in infrastructure 21.58%, followed by the others of minor importance. In this conclusion, the managers believe, that if we get trained people and developed leaders, the impact on the productive sector will be eminent, that is, these two sub-constructs become the main coefficients for beginning a PPC consolidation.

The results obtained in the application of the methods can perceive the quality of the application because it is a mathematical model using techniques that involve linguistic purposes such as the PPC area. The management construct with 28.50% preference of managers and experts. It occurs due to the sub-construct, for example, "development leaders" with 31.60%, being that the leader is responsible for facilitating, monitoring and evaluating the performance of a team. The key role of the leader is the connecting link makes with the company's goals and the aspirations of employees. The leader's job is to inspire ability, ethics, and commitment so that leaders can seek increasingly satisfying results in their organization. The "Training of personnel" with 23.27% the fundamental role of this sub-construct is the development of the employee, who may be specific knowledge of the area or to enhance productivity, performance, motivation and update to new technologies. A team that undergoes personal training in addition to having the skills increased acquires motivation, self-confidence and self-esteem.

The constructs costs and management have a constant linearity in their importance weights according to the experts, because, managers admit that the set of these variables can represent good results if applied in the production. The application of the Fuzzy Sets simulations, which were proposed, obtained a greater veracity in relation to the real data obtained in the study compared to FAHP (Godoy et al., 2018), being efficient in the aid of planning and production control management, fact that occurs due to the variables inserted in the research for mathematical modeling Fuzzy Sets.

It is understandable that the structure evaluated by adjacent constructs and concepts can undergo changes depending on the objectivity of each industry. All aspects presented and evaluated in the research are important to replicate the study in other industries of the same sector, in favor of improving consistency indices. The FAHP method had moderately high weights averaging 0.0751, but acceptable for Saaty <0.10.

It is assumed that, due to the moderate inconsistency found in the research, it can be a striking fact to improve the Fuzzy Sets simulations. The application of the Fuzzy Sets simulations obtained greater veracity in relation to the data obtained in the study compared to the FAHP, being efficient in the aid, for planning management and production control.

Finally, several case studies could be carried out to evaluate the acceptability and usability by professionals of methods such as those developed in this article to support the evaluation of improvement in decision making in the planning and production control (PPC) sector applying new models such as Fuzzy TOPSIS and sensitivity analysis.

## Thanks

This study was funded by the Scholarship Program of the National Council for Scientific and Technological Development (CNPq). In the State of Rio Grande do Sul, the student of the Graduate Program in Production Engineering of the Federal University of Santa Maria at the Technology Center. Our compliments.

## REFERENCES

Abdel-Maksoud, A., Elbanna, S., Mahama, H., & Pollanen, R. (2015). The use of performance information in strategic decision making in public organizations. *International Journal of Public Sector Management, 28*(7), 528–549.

Aqlan, F., & Al-Fandi, L. (2018). Prioritizing process improvement initiatives in manufacturing environments. *International Journal of Production Economics, 196*, 261–268. Advance online publication. doi:10.1016/j.ijpe.2017.12.004

Brandão, A., Belisário, G. C., Tomazini, I. M., & Gomes, K. Q. (2017). Análise bibliométrica da literatura sobre métodos multicritério de auxílio à tomada de decisão aplicados a problemas de PCP. *Brazilian Journal of Production Engineering, 3*(2), 128–140. http://periodicos.ufes.br/BJPE

Cervo, A. L., & Bervian, P. A. (2002). *Metodologia científica*. Editora Pearson Prentice Hall.

Chan, K. Y., Kwong, C. K., & Dillon, T. S. (2012). *Computational Intelligence Techniques for New Product Design. Studies in Computational Intelligence - 403*. Springer., doi:10.1007/978-3-642-27476-3

Chang, D. Y. (1996). Applications of the extent analysis method on fuzzy AHP. *European Journal of Operational Research, 95*(3), 649–655. doi:10.1016/0377-2217(95)00300-2

Chen, W. C., Wang, L. Y., & Lin, M. C. (2015). A Hybrid MCDM Model for New Product Development: Applied on the Taiwanese LiFePO Industry. Mathematical Problems in Engineering, 2015, 15. doi:10.1155/2015/462929

Demo, P. (2013). *Introdução à metodologia da ciência* (2nd ed.). Atlas.

Dragović, I., Turajlić, N., Pilčević, D., Petrović, B., & Radojević, D. (2015). *A Boolean Consistent Fuzzy Inference System for Diagnosing Diseases and Its application for Determining Peritonitis Likelihood*. Academic Press.

FIERGS. (2015). *Caderno setorial Rio Grande do Sul metal mecânico*. FIERGS.

Gomes, L., Araya, M., & Carignano, C. (2004). *Tomada de Decisões em Cenários Complexos: Introdução aos Métodos Discretos do Apoio Multicritério à Decisão*. Thomson.

Gondim, F. M. (2011). As ações da sustentabilidade empresarial como suporte à gestão de projetos. RESAC, sociedade, administração e contemporaneidade, 1(1).

Huang, S., Chang, J., Leng, G., & Qiang, H. (2015). Integrated index for drought assessment based on variable fuzzy set theory: A case study in the Yellow River basin, China. *Journal of Hydrology (Amsterdam)*. Advance online publication. doi:10.1016/ j.jhydrol.2015.05.032

Kahraman, C. (2008). *Fuzzy Multi-criteria decision making: Theori and applications with recent development*. Springer.

Kahraman, C. T., Ertay, T., & Büyüközkan, G. (2006). A fuzzy optimization model for QFD planning process using analytic network approach. *European Journal of Operational Research*, *171*(2), 390–411. doi:10.1016/j.ejor.2004.09.016

Kar, A. K., & Pani, A. K. (2014). Exploring the importance of different supplier selection criteria. *Management Research Review*, *37*(1), 89–105. doi:10.1108/MRR-10-2012-0230

Khamaksorn, A. (2016). Project Management Knowledge and Skills for the Construction Industry. *International Conference on Civil, Architecture and Sustainable Development (CASD-2016).*

Klir, G., & Yuan, B. (1995). *Fuzzy sets and fuzzy logic* (Vol. 4). Prentice hall.

Lima, F. R. Junior. (2013). Comparação entre os métodos Fuzzy TOPSIS e Fuzzy AHP no apoio à tomada de decisão para seleção de fornecedores (Mestrado em Engenharia de produção). Escola de Engenharia de São Carlos da Universidade de São Paulo, São Carlos.

Longenecker, J. G., Moore, C. W., Petty, J. W., & Palich, L. E. (2013). *Administração de pequenas empresas*. Cencage Learning.

Marconi, M. D. A., & Lakatos, E. M. (2010). *Fundamentos de metodologia científica. 7*. Atlas.

Massa, F. H. D., Cazier, O., Tison, T., & Lallemand, B. (2015). Finite element analysis of frictionless contact problems using fuzzy control Poulinapproach. *Engineering Computations*, *32*(3), 585–606. doi:10.1108/EC-11-2013-0289

Miguel, P. A. C. (Ed.). (2012). *Metodologia de pesquisa em engenharia de produção e gestão de operações*. Elsevier.

Musetti, T. F., & Escrivão Filho, E. (2016). Monitoramento do ambiente organizacional em pequenas empresas: Estudo de casos nos setores Metalmecânica e base tecnológica da região central do estado de São Paulo. *Gestão da Produção Operações e Sistemas*, *11*(2), 139–154.

Noro, G. B., & Amarilho, A. (2012). *Gestão de Projetos Sustentáveis: o Caso América Latina Logística. Simpósio de excelência em Gestão e Tecnologia*. Evento SEGET.

Prajogo, D., Toy, J., Bhattacharya, A., Oke, A., & Cheng, T.C.E. (2018). The relationships between information management, process management and operational performance: Internal and external contexts. International Journal of Production Economics, 199, 95-103. doi: 10.1016 / j.ijpe.2018.02.019

Rodriguez, D. S. S., Costa, H. G., & Do Carmo, L. F. R. R. S. (2013). Métodos de auxílio multicritério à decisão aplicados a problemas de PCP: Mapeamento da produção em periódicos publicados no Brasil. Gestão Produção, São Carlos, 20(1), 134–146. doi:10.1590/S0104-530X2013000100010

Saaty, R. W. (1987). The Analytic Hierarchy Process-what it is and how it is used. Mat/d Modelling, 9(3-5), 161-176. doi:10.1016/0270-0255(87)90473-8

Saaty, T. L. (2008). Decision making with the analytic hierarchy process. *International Journal of Services Sciences*, *1*(1), 83–98. doi:10.1504/IJSSCI.2008.017590

Saaty, T. L. (1991). *Método de Análise Hierárquica. Tradução e Revisão por Wainer da Silveira e Silva.* McGraw-Hill.

Salomon, V. A. P., Marins, F., & Duduch, M. (2009). Tomada de Decisões Múltiplas Aplicada à Seleção de Fornecedores de Equipamentos de uma Linha de Montagem em uma Fábrica de Autopeças. Pesquisa Operacional para o Desenvolvimento, 1(3), 208-217.

Sellitto, M. A., & Walter, C. (2006). Avaliação do desempenho de uma manufatura de equipamentos eletrênicos segundo critérios de competição. *Produção*, *16*(1), 34–47.

Skolud, B., Krenczyk, D., & Davidrajuh, R. (2016). Solving Repetitive Production Planning Problems. An Approach Based on Activity-oriented Petri Nets. In International Conference on European Transnational Education. Springer International Publishing.

Soliman, M., Saurin, T. A., & Anzanello, M. J. (2018). The impacts of lean production on the complexity of socio-technical systems. *International Journal of Production Economics*, *197*, 342–357. doi:10.1016/j.ijpe.2018.01.024

Souza, A. B., Bianchi, J. M. B. B., & Boente, A. N. P. B. (2015). Fuzzyahp: uma ferramenta de gestão Tecnológica para tomada de decisão Muticriterial. Revista científica digital da faetec, 8(1).

Van Laarhoven, P. J. M., & Pedrycz, W. (1983). A fuzzy extension of saaty's priority theory. *Fuzzy Sets and Systems*, *11*(1-3), 229–241. doi:10.1016/S0165-0114(83)80082-7

Wang, B., Xiong, H., & Jiang, C. (2014). A Multicriteria Decision Making Approach Based on Fuzzy Theory and Credibility Mechanism for Logistics Center Location Selection. TheScientificWorldJournal, 2014, 9. doi:10.1155/2014/347619 PubMed

Woiler, S., & Mathias, W. F. (2010). *Projetos. Planejamento, elaboração, análise* (2nd ed.). Editora Atlas.

Yamakawa, E. K., Miguel, P. A. C., & Aoki, A. R. (2014). Aplicação de fuzzyqualityfunctiondeployment para seleção de indicadores de eficiência energética para utilização em um software de gestão de energia. *Science & Engineering Journal, 23*(2), 21–31.

Yin, R. K. (2015). *Estudo de Caso: Planejamento e Métodos* (5th ed.). The Bookman.

Zadeh, L. A. (1965). Fuzzy sets. *Information and Control, 8*(3), 338–353. doi:10.1016/S0019-9958(65)90241-X

Zare, R., Nouri, J., Abdoli, M. A., Atabi, F., & Alavi, M. (2016). The Integrated Fuzzy AHP and Goal Programing Model Based on LCA Results for Industrial Waste Management by Using the Nearest Weighted approximation of FN: Aluminum Industry in Arak, Iran. Advances in Materials Science and Engineering, 2016, 13. doi:10.1155/2016/1359691

Zhao, B., Tang, T., & Ning, B. (2016). Applying Hybrid Decision-Making Method Based on Fuzzy AHP-WOWA Operator for Emergency Alternative Evaluation of Unattended Train Operation Metro System. Mathematical Problems in Engineering, 2016, 12. doi:10.1155/2016/4105079

# Chapter 14
# Implementation of an Integral Methodology for the Simulation of Renewable Energy Scenarios Applied to an Isolated Area

**Elisa Peñalvo-López**
*Universitat Politècnica de València, Spain*

**Francisco Javier Cárcel-Carrasco**
*Universitat Politècnica de València, Spain*

**Angel Pérez-Navarro**
*Universitat Politècnica de València, Spain*

**Elias Hurtado-Pérez**
*Universitat Politècnica de València, Spain*

## ABSTRACT

*This chapter provides an implementation of an integral methodology for the simulation of renewable energy scenarios applied to an isolated village situated in Democratic Republic of Congo. A specific area in Democratic Republic of Congo is selected as a case study in order to validate the methodology. In addition, this chapter includes the experimental validation of the HRES configuration, operating a scalable prototype in LabDER at UPV (Universitat Politècnica de València, Spain). Experimental tests conducted in the laboratory are described and main results and conclusions are discussed.*

DOI: 10.4018/978-1-7998-3246-1.ch014

## 1. INTRODUCTION

Energy Planning (EP) analyses the alternative paths for energy evolution of a region by studying different energy scenarios in three temporal ranges: short (1 to 10 years), medium (10 to approximately 30 years) and long term analysis (more than 30 years). Generally, it begins examining as a reference the actual scenario, named "Business As Usual (BAU)", and its evolution for the time span considered. Then, it compares the results from other different alternative scenarios under the same time span and demand constraints. Each energy scenario involves assessing and matching in an optimal manner the energy sources and their conversion with the energy requirements of different demand sectors (commercial, industrial, residential, etc.). Although, it may seem a simple idea, it becomes a complex problem in which various decisions and criteria converge, together with the existence of complex relationships between the different actors involved in the simulation process: generation, demand, emissions, economics, and technologies (Loken, 2007; Cárcel et al., 2018; Gómez et al., 2018; Peñalvo et al., 2019a; Peñalvo et al., 2019b). Models in energy planning are important in emerging communities and in developed urban areas, since they determine the energy path and goals for the next time period. Precise modelling requires large computational resources, thus a trade-off between exactness and resources needs to be balanced. In order to approximate reality with acceptable computational resources, models are based on certain hypothesis that tackle possible scenarios and casuistry, using estimations and assumptions which may or may not become valid under initial premises but that are unknown at the moment of modelling (Grubb, 1993).

Literature reviews show numerous tools for energy planningbased on HRES dimensioning. The applicability and benefits of these software tools, which facilitate calculations and reduce processing time, is widely demonstrated. Tools allow comparing different alternatives in order to assess the advantages and disadvantages of each solution, assisting in the evaluation of different HRES configurations and energy scenarios within a feasible time frame. Connolly et al. (Connolly, 2010) published a review of 37 computer tools for analysing the integration of renewable energy into various energy systems. Sinha and Chandel (Sinha & Chandel, 2014) provided also a description of different computer programs used for HRES dimensioning but did not focus on the combination of different renewable systems. Delgado-Antillón and Domínguez- Navarro (Delgado, 2018) used a genetic algorithm (GA) to optimize the capacity of a distributed system, minimising the storage capacity and power losses. Giallanza et al. (Giallanza, 2018) studied the probability of seasonal load loss ratio using an iterative method to design a standalone PV-wind-battery system. Rullo et al. (Rullo, 2019) proposed anoff-grid PV-wind-battery system dimensioning model based on active energy management to achieve the lowest operational cost. Dufo-López et al. (Dufo-López, 2016) presented a multi-objective evolutionary algorithm for optimizing off-grid HRESs by minimizing the total NPC, along with the maximization of HDI and job creation. Benavente et al. (Benavente, 2019) optimized the size of a stand-alone PV-battery system for three scenarios in Bolivia. The results provided a reference for the application of off-grid PV-battery systems in rural areas.

Also, Multicriteria Decision Methods applied to renewable energy planning are widely discussed in the literature. Pohekar´s analysis (Pohekar, 2004) (Polatidis, 2006) showed that Multi-Attribute Utility Theory (MAUT) (Wang, 2010)was the most common MCDM method used in energy planning bibliography, including Weighted Product Method (WPM) (Chang, 2001), Analytical Hierarchical Process (AHP)(Saaty, 1987), Analytic Network Process (ANP) (Saaty, 1996), MAUT, fuzzy methods and decision support systems (DSS), among others. The main objective of MADM is to select the alternative that has the highest score according to the set of evaluation criteria.

Revision of the state of the art reveals that great efforts have been devoted to study decentralized planning, electrification of isolated areas and multi-criteria analysis, together with the exploration of projects for electrifying remote communities with renewable energies. It also denoted the noteworthy interest in the different aspects associated to remote electrification of communities: software development for decentralized planning, including HRES; optimization of design system's operation to respond different demand profiles; multi-criteria approaches. However, all these scientific and technical works deal with specific issues of the challenge, but it has not been identified in the literature a holistic methodology for HRES optimization considering decentralized planning and multi-criteria assessment of quantitative and qualitative aspects of the project's context, which are key in the success of real implementation.

Traditionally, remote and non-connected areas do not use energy planning in their energy analysis due to their small size. Electrification plans normally follow simple schemes, implementing simple solutions based on one technology, the most suitable one depending on the available natural resource (solar radiation, wind, type of biomass residues, etc.), (Cherni *et al.,* 2007; Peñalvo, 2017a). These schemes are mainly referred to stand-alone systems based on photovoltaic and/or wind configurations, storing the excess of energy in batteries and using private diesel generators as backup. However, this simple approach based on stand-alone systems do not take into consideration important factors, such as: energy needs of the population, potential flexibility of consumers, expected demand's growth and synergies between different renewable generation systems, which invites to analyze more in detail the different variables and relationships involved in energy planning from a new perspective, considering decentralized energy modeling and customer's participation.

In this chapter it is presented a case study implementing the integrated methodology for the simulation of renewable energy scenarios. The methodology has been applied to an isolated village situated in Central Africa in Republic Democratic of Congo (RDC). The village is a rural community mainly dedicated to agriculture for subsistence.

In the last part of the chapter, the principles for validating models to simulate the electrical behaviour of individual charges are presented, developing techniques of aggregation of these charges with which to estimate their joint physical evolution, and applying algorithms to obtain these groups Load Management strategies, studying the effect that these actions cause from two points of view: user (changes in consumption patterns, in the services obtained), and electricity production (modification of the demand period and optimization of resources). In addition, there have traditionally been greater difficulties in predicting the consumption of residential customers, due to the component of subjectivity that their demand possesses, so it is of interest to have tools that help to manage these loads in an increasing and efficient way

## 2.METHODOLOGY OF ENERGY PLANNING FOR AREAS ISOLATED

Renewable Hybrid Systems are one of the most promising approaches for non-connected areas with significant difficulties to be connected to the electrical distribution network. In addition to this, distributed generation resources also enhance the properly integration of different energy resources in order to maximize the coverage of the demand energy needs and facilitate the operation of such mini grids with specific problematic.

Nowadays, it exists many different software that tackle this issue, such as the popular HOMER or Hybrid2, however, none of them take into consideration the demand flexibility in the system nor a multi-criteria approach, considering qualitative assessments (social and political),. Therefore the benefits associated to these aspects have not been properly analyzed nor assessed.

In order to tackle this issue from a holisticperspective, research objectives should include:

- Energy Planning
- Identification of best techno economic HRES configurations
- Multi-Criteria Analysis
- Ranking of Best HRES solutions

Figure 1A summarizes the Integral Methodology for Isolated Electrification developed in this study in order to fulfill the above mentioned research objectives. Figure 1B presents SIMESEN tool for energy scenarios assessment, which is composed by 3 independent modules (Peñalvo, 2017b; Cárcel *et al.*, 2017): Input data, calculation engine and results output.

*Figure 1. A:Integral Methodology for Isolated Electrification. B: Energy Planning Model SIMESEN*

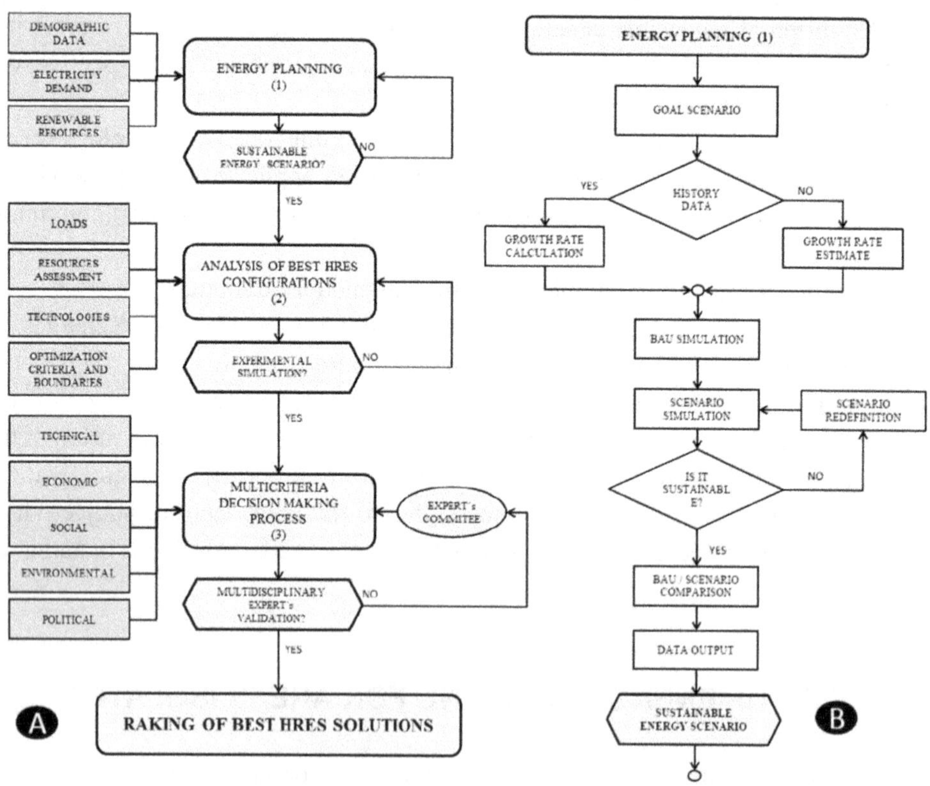


Once the energy planning study is completed, the methodology will address the optimization of the HRES configuration, following the scheme depicted in Figure 2. This stage evaluates the different combinationof renewable and non-renewable systems for meeting the energy demand. It identifies the system configuration of hybrid systems capable to respond to the overall energy needs of the custom-

*Figure 2. Method for identifying HRES Feasible Configurations*

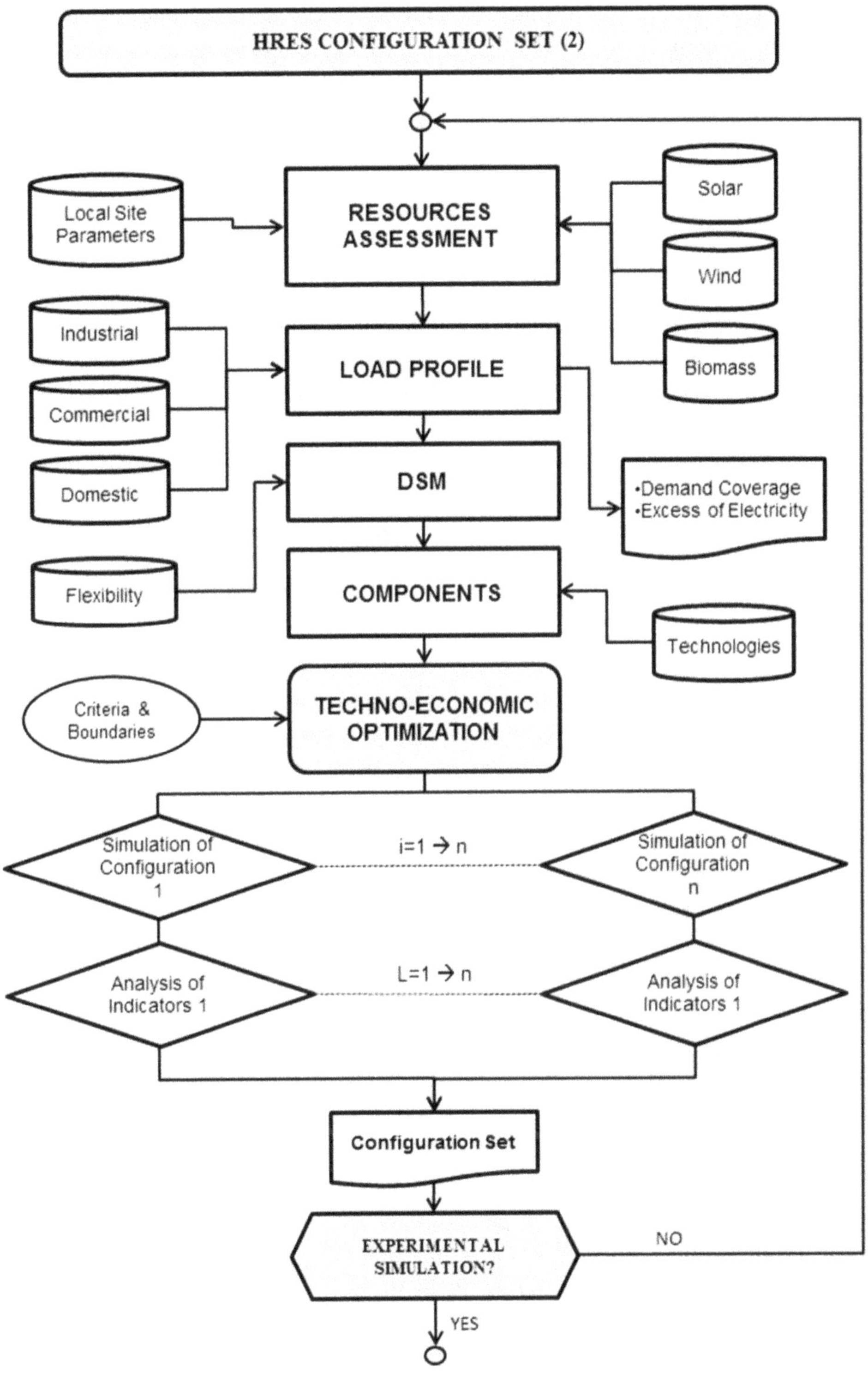

ers. In order to do the evaluation, first it is analysed the demand of the different customers, then natural resources are evaluated to determine the potential for renewable generationsystems. Next, it is identified the available back-up systems such as fossil fuel generator or storage systems. Finally the most promising renewable hybrid configurations are experimentally validated in the lab.Main components of this approach are described in the next paragraphs.

## 3. ENERGY PLANNING

Electricity is necessary for the development of any community. However, electricity supply becomes a challenge for remote and isolated communities. Renewable hybrid energy systems may safely generate electricity for minimum demand requirements without implementing large facilities or networks. Throughout this section it is studied the integration of HRES in the national energy context as a strategy for isolated areas electrification.

In this section, an energy scenario based on the implementation of HRES is explored as a means for providing energy access to a higher percentage of the population and economic development to the country. Then, it is compared with scenario BAU (Business As Usual), representing a continuous tendency of the actual context.

### 3.1. Energy Context of RDC

Republic Democratic of Congo (RDC) is located in the centre of Africa, near the equator. It is the second largest country in Africa in terms of area (2,345,441 km$^2$) withan estimated population of 74,88 million of inhabitants, having a growth rate of 3,1% from 1990 to 2014. Approximately, 75% of the total population lives in rural areas and it is dependent on agriculture, being land cultivation and forestry their major economic activities. RDC owns approximately 52% of total freshwater reserves in Africa and land characterized by dense tropical rainforest in the basin and central west, dry forests and savannahs in the southern part and mountain ecosystems along the eastern border.

RDC is also characterized by its immense natural forest reserves (about 155 million ha) together with a natural rainforest which represents 60% of the forests in Congo Basin and an important carbon storage area. In terms of energy, RDC has a great potential with abundant and varied energy resources: biomass, solar, wind, hydraulic energy, hydrocarbons, including methane gas from Lake Kivu, mineral coal, oil shale, etc. The sustainable development of these energy resources for contributing to economic and social development of the country will certainly be the main agenda of the national and regional Governments.

RDC's energy sector is currently centralized and characterized by a state monopoly. However, it has a high energy potential for distributed energy supply over the country's territory which is almost unexploited. Estimations of RDC's energy potential are around 100 GW of exploitable capacity, of which almost half (44 GW) is concentrated at the Inga site. Despite this extraordinary energy potential, especially for renewable energy generation (mainly hydro based), RDC's national electrification rate is 9% with strong disparities between urban and rural areas (Kinshasa is 44%, while Bandundu is 0,6%) (FAO, n.d.a). Moreover, transport and distribution electricity network are out of date and unable to support actual current consumption, which results in constant power outages. In the places where electricity is essential, such as hospitals, hotels, public buildings, etc., the problem is circumvented by using fossil fuel generators to get some security in electricity supply. However, it exits many remote areas where

*Figure 3. SIMESEN Energy Flow of RDC. Year 2014 (Souce: International Energy Agency)*

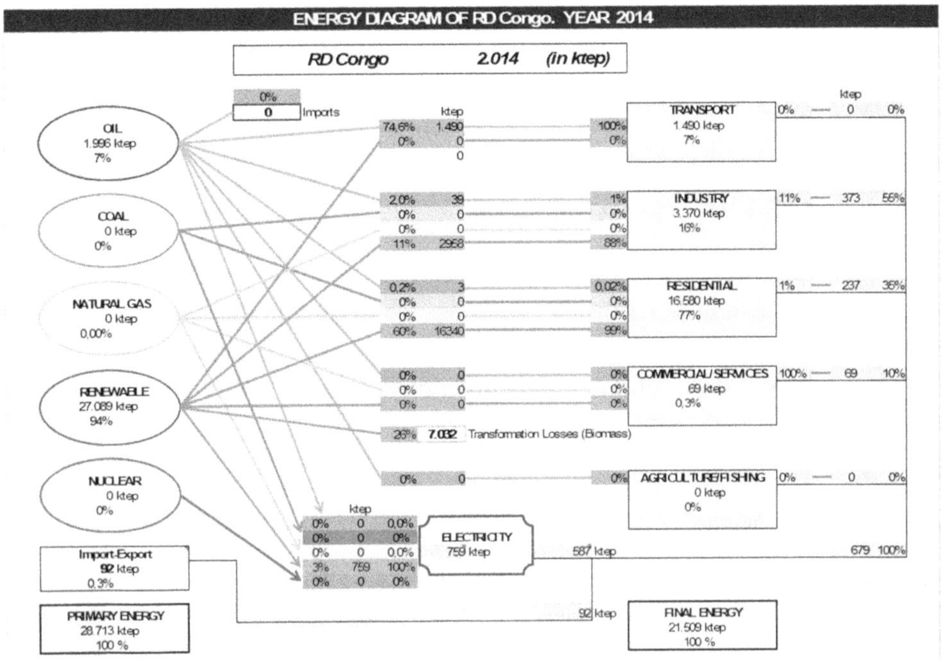


grid extension is costly and simply do not have access to electricity (FAO, n.d.a;Taliotis et.al, 2014). In these cases, renewable hybrid systems may be the best alternative to provide energy access and economic development to the societies, without large initial investment in network extension.

Data from 2014, provided by International Energy Agency (IEA, n.d.a, n.d.b), represented in a Sankey's Energy diagram for RDC shows a national energy supply of 30,04 Mtep in contrast to 28,79 Mtep in 2013, mainly based in three energy sources: oil, biomass and hydro. Produced oil is exported to other countries (1,07Mtep) while oil products such as gasoline and diesel are imported in large quantities (1,78 Mtep), so RDC presents an oil product dependency from the exterior, mainly in transport (1094 ktep). Biomass is the highest energy supply in the country, and it is based on traditional fuels such as firewood, charcoal and waste, which denotes the low level of electrification in the country. Hydro is the only source for electricity generation, together with minor contributions from coal and natural gas in the country.

Total final energy consumption reached 21,51Mtep, an increment of 12,5% from previous year (19,11 Mtep in 2013). Main consuming sector is the residential (74%), followed by industry (15%) and transport (7%), while other forms of energy such as electricity contributes with only 4.1%.

Total primary energy per capita is 0,38tep/capita, remaining far below Africa's average (0,67 tep/capita). Similarly, electricity consumption per capita shows a value of 110 kWh/capita, below Africa's average of 570 kWh/capita, while European Union index reaches 5900 kWh/capita. In contrast, CO2 emissions per capita is 0,06 tCO2/tep in RDC, while Africa's average is about 0,96 tCO2/tep. This highlights the renewable energy potential of RDC.

Gross domestic product (GDP) per person employed was 52,2 billion of USD2010 in 2014, with a total primary energy per GDPppp value of 0,55 tep/GDPppp, 27% higher than Africa's average (0,15 tep/GDPppp).

Figure 3 represents the energy flow diagram of RDC in year 2014, including the origin (energy source) and destination (sector) of energy units. As it may be observed only oil and renewable energies are supplying the country. Country imports are oil products for transport and electricity; while hydro is completely for electricity generation and biomass is mainly used in residential by means of traditional fuels such as wood.

## 3.2 BAU vs HRES Scenario

In this section, two energy scenarios are compared. One is the Business As Usual (BAU) based on a continuous trend where participation of each energy source to the national mix is maintained, together with the electricity import-export balance (table 1). To do so, an historical analysis has been carried out in order to forecast the growth rates of population, GDP, primary energy and final consumption for each sector (table 1).

*Table 1. Growth rates for Scenarios.*

| Economic Sector | Growth Rate (%) |
|---|---|
| Industry | 1,4 |
| Transport | 7,5 |
| Services | 0,5 |
| Residential | 2,7 |
| Agric. / Fish. | 0,1 |

| Other | Growth Rate (%) |
|---|---|
| Population | 3,1 |
| $GDP_{PPP}$ | 0,4 |

The second scenario is the proposed HRES, which is based on increasing the access to electricity by means of renewable hybrid systems and palliating the raise of CO2 emissions introducing biofuels

*Table 2. Energy Indicators for BAU Scenario.*

| | | SCENARIO | BAU | RD Congo | | | |
|---|---|---|---|---|---|---|---|
| **Indicators** | **Units** | **2014** | **2015** | **2020** | **2025** | **2030** | **2035** |
| Population | Million | 74,9 | 77,2 | 89,9 | 104,6 | 121,8 | 141,8 |
| GDPppp | M€$_{2010}$ | 52.200.000 | 52.412.599 | 53.488.650 | 54.586.794 | 55.707.483 | 56.851.180 |
| Consumption of Electricity | TWh | 7,9 | 8,0 | 8,7 | 9,5 | 10,4 | 11,3 |
| $CO_2$ Emissions | Mt | 4,53 | 4,85 | 6,82 | 9,64 | 13,68 | 19,45 |
| Primary Energy (EP) | ktep | 28.713 | 29.520 | 33.978 | 39.265 | 45.586 | 53.214 |
| EP Generated | ktep | 20.057 | 20.558 | 23.264 | 26.344 | 29.852 | 33.848 |
| Import-Export | ktep | 92 | 94 | 102 | 111 | 121 | 132 |
| Generated Electricity | ktep | 587 | 597 | 648 | 706 | 771 | 843 |
| Exterior Dependency | % | 30,15 | 30,36 | 31,53 | 32,91 | 34,51 | 36,39 |
| GDPppp/capita | M€$_{2010}$/inhab | 0,70 | 0,68 | 0,60 | 0,52 | 0,46 | 0,40 |
| TEP/capita | tep/hab | 0,383 | 0,382 | 0,378 | 0,375 | 0,374 | 0,375 |
| TEP/GDPppp | tep/M€$_{2010}$ | 0,55 | 0,56 | 0,64 | 0,72 | 0,82 | 0,94 |
| Electricity/capita | MWh/inhab | 0,11 | 0,10 | 0,10 | 0,09 | 0,09 | 0,08 |
| $CO_2$/TEP | t/tep | 0,16 | 0,16 | 0,20 | 0,25 | 0,30 | 0,37 |
| $CO_2$/GDPppp | t/M€$_{2010}$ | 0,09 | 0,09 | 0,13 | 0,18 | 0,25 | 0,34 |
| $CO_2$/capita | t/inhab | 0,060 | 0,063 | 0,076 | 0,092 | 0,112 | 0,137 |
| Fraction ER in EP* | % | 69,9 | 69,6 | 68,5 | 67,1 | 65,5 | 63,6 |
| Fraction ER in EE* | % | 86,5 | 86,5 | 86,5 | 86,5 | 86,5 | 86,5 |

Fraction ER in EP and EE*: % of renewable energies contribution to primary energy and electrical generation.

in the transport sector. In order to understand the impact of the proposed measures, next it is shown a series of indicators and graphs representing the comparative analyses of both scenarios.

Next it is provided a series of energy indicators (Table 2, 3) for both scenarios and their forecasted evolution within the time period of study (from 2014 to 2035).

*Table 3. Energy Indicators for HRES Scenario.*

| | | SCENARIO | HRES | RD Congo | | | |
|---|---|---|---|---|---|---|---|
| **Indicators** | **Units** | **2014** | **2015** | **2020** | **2025** | **2030** | **2035** |
| Population | Millions | 74,88 | 77,19 | 89,87 | 104,64 | 121,83 | 141,84 |
| GDP | Billion€$_{2010}$ | 52 200 000,00 | 52 412 598,61 | 53 488 650,37 | 54 586 793,92 | 55 707 482,79 | 56 851 179,86 |
| Consumption of Electricity | TWh | 7,90 | 8,07 | 11,64 | 18,84 | 33,48 | 66,74 |
| $CO_2$ Emissions | Mt | 4,53 | 4,40 | 5,94 | 8,53 | 12,37 | 18,22 |
| Primary Energy (EP) | ktep | 28 713,00 | 29 524,03 | 34 675,03 | 41 107,47 | 50 314,52 | 64 915,28 |
| EP Generated | ktep | 20 057,00 | 20 721,46 | 24 155,96 | 28 320,24 | 34 346,50 | 44 194,58 |
| Import-Export | ktep | 92,00 | 93,98 | 135,61 | 219,47 | 390,01 | 777,54 |
| Generated Electricity | ktep | 587,00 | 599,62 | 865,27 | 1 400,29 | 2 488,45 | 4 961,07 |
| Exterior Dependency | % | 30,15 | 29,81 | 30,34 | 31,11 | 31,74 | 31,92 |
| GDP/capita | M€$_{2010}$/inhab | 0,70 | 0,68 | 0,60 | 0,52 | 0,46 | 0,40 |
| TEP/capita | tep/inhab | 0,383 | 0,382 | 0,386 | 0,393 | 0,413 | 0,458 |
| TEP/GDP | tep/M€$_{2010}$ | 0,55 | 0,56 | 0,65 | 0,75 | 0,90 | 1,14 |
| Electricity/capita | MWh/inhab | 0,11 | 0,10 | 0,13 | 0,18 | 0,27 | 0,47 |
| $CO_2$/TEP | t/tep | 0,16 | 0,15 | 0,17 | 0,21 | 0,25 | 0,28 |
| $CO_2$/GDP | t/M€$_{2010}$ | 0,09 | 0,08 | 0,11 | 0,16 | 0,22 | 0,32 |
| $CO_2$/capita | t/inhab | 0,06 | 0,06 | 0,07 | 0,08 | 0,10 | 0,13 |
| Fraction ER in EP* | % | 69,9 | 70,2 | 69,7 | 68,9 | 68,3 | 68,1 |
| Fraction ER in EE* | % | 86,5 | 86,5 | 86,5 | 86,5 | 86,5 | 86,5 |

Analyzing the most important indicators, figure 4A **¡Error! No se encuentra el origen de la referencia.**represents the analysis of the primary energy and CO2 emissions for both scenarios. As it may be observed, scenario HRES shows an increment of total primary energy with respect to BAU. This is due to the increment in distributed electric generation with renewable hybrid systems. CO2 emissions are similar in both scenarios; HRES is slightly lower since this scenario considers a penetration of 10% of biofuels in transport, thus replacing part of the fossil fuel consumption in this sector.

Energy intensity is analyzed using three main energy indicators: total primary energy per capita (TEP/capita), average consumption of electricity per capita (electricity/capita), and total primary energy per gross domestic product per employed person (TEP/GDPppp) (figure 4.B). In this regard, it is observed how HRES increases the electrification rate to 0,58 kWh/capita by 2035, while continuing with actual energy paradigm will quasi maintain electricity rate to actual value.

Exterior dependency is similarly reduced in HRES scenario (figure 4.C). Penetration of renewable hybrid systems in remote areas will increase the percentage of population with electricity access without compromising their energy independence, since HRES are off-grid power plants based on local resources. In the case of continuing with actual scenario (BAU), fossil fuels and electricity dependency will increase along the years.

Electricity generation (figure 5A) is notorious in the HRES scenario to cover a wider range of electricity access in isolated areas, which will also imply higher electricity consumption and economic development of the area. Increasing electricity rate to Africa's actual values of electricity per capita will require a significant effort and compromise from national and regional government to promote HRES as a local and sustainable alternative for rapid energy development in off-grid areas.

Final Energy consumption per economic sector (figure 5B) indicates a growing energy demand in residential segment in HRES with respect to BAU, while Industry and Transport present similar energy

*Figure 4. A: Primary energy consumption and associated $CO_2$ Emissions. B: Energy Intensity. C: Exterior Dependency*

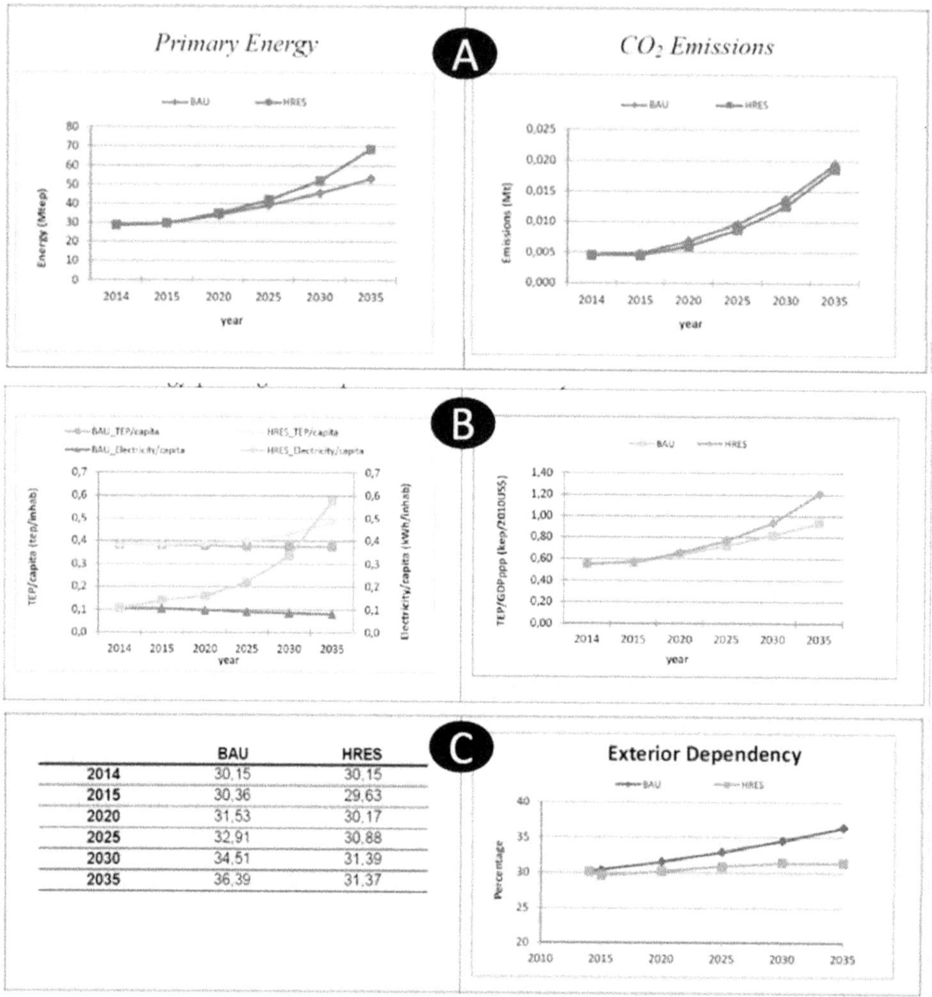


| | BAU | HRES |
|---|---|---|
| 2014 | 30.15 | 30.15 |
| 2015 | 30.36 | 29.63 |
| 2020 | 31.53 | 30.17 |
| 2025 | 32.91 | 30.88 |
| 2030 | 34.51 | 31.39 |
| 2035 | 36.39 | 31.37 |

evolution. HRES scenario presents an increment in Residential energy consumption originated by the electrification process in residential segment, reaching 50 Mtep in 2035.

Analysis of Energy sources (figure 5C) highlights the rapid growth of renewable energies to respond to residential electrical needs in 2035. In both scenarios, production with renewable energies increases significantly, but specially in HRES, reaching the value of around 4,7Mtep in 2035 in contrast with 3,3 Mtep in BAU2035. Oil demand is similar in both scenarios, presenting a value of 6,8 Mtep in BAU2035 versus 5,8 Mtep in HRES2035, this difference is due to decreasing demand of oil products (gasoline and diesel) in transport and replace them with biofuels, increasing biomass requirements in this economic sector.

Emission will evolve similarly in both scenarios (figure 5D). Nevertheless, HRES scenario presents a slight reduction of emissions in transport due to the biofuel penetration in the transport, and an increment in emissions due to the extended used of electricity in the country.

*Figure 5. A: Electricity. B: Energy Demand. C: Energy Sources. D: $CO_2$ Emissions per sector*

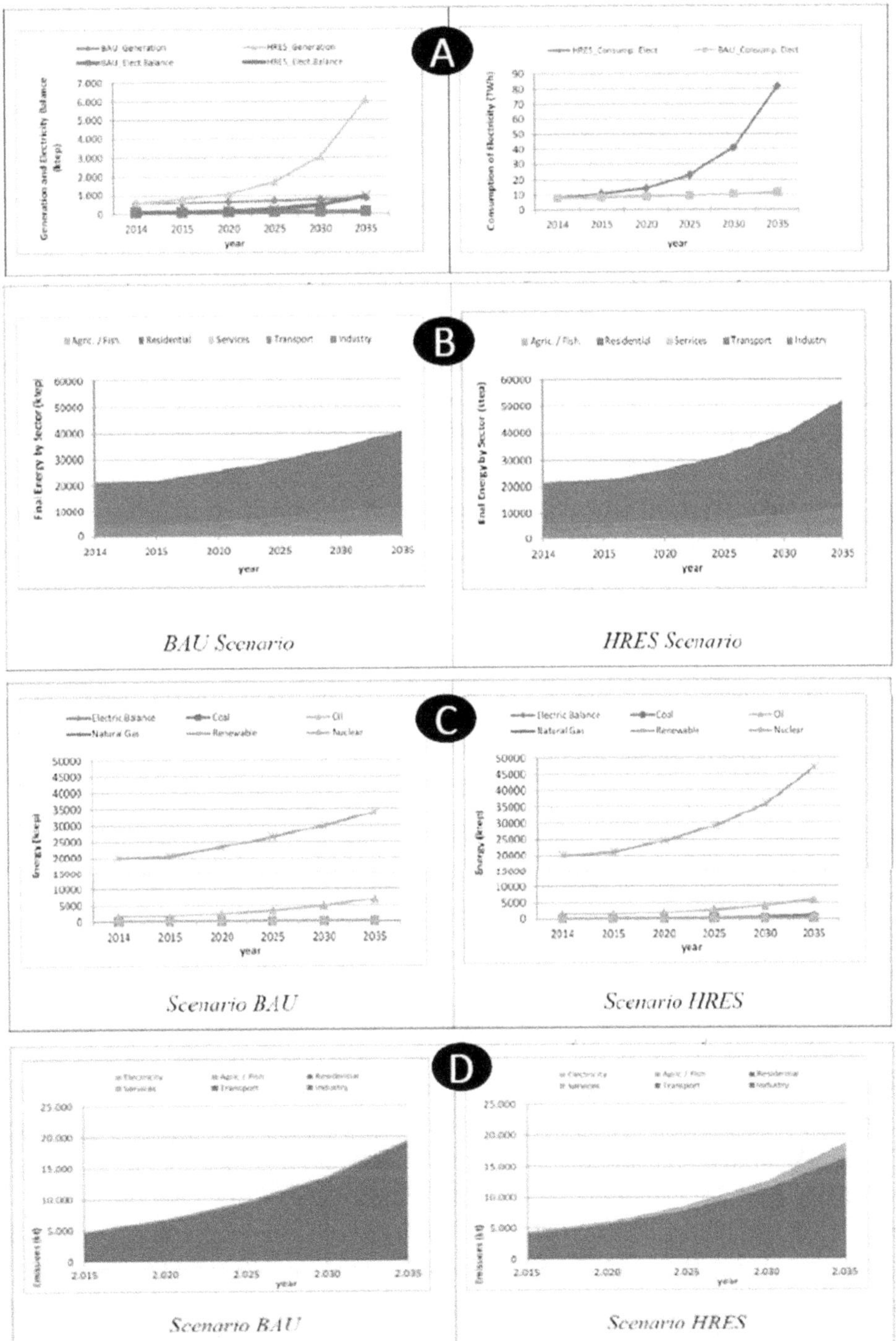

As it may be observed from the analysis, it is feasible to alleviate energy poverty in the country by means of HRES penetration in remote areas where access to electricity is not available. Once it is analyzed the macro impact of the HRES scenario, a micro-energy analysis needs to be conducted in order to verify that HRES configurations will meet the energy requirements of isolated societies in RDC.

## 4. HRES CONFIGURATION SET

Once the macro-perspective of the sustainable energy development has been set, it is analysed its applicability to a small community. To do so, different HRES configurations are studied from a techno-economic perspective in order to identify a set of HRES system architectures that will meet the needs of the village and also, facilitate the energy transformation of the country. The aim is to validate whether the HRES Scenario proposed in the previous section can be implemented at a small scale, identifying the energy systems that will comply with it.

In this section, a series of HRES design schemes are modeled and simulated to study its response to the energy requirements of a remote community in the area of Kinshasa (Republic Democratic of Congo). The objective is to analyze whether the community energy needs are met in a reliable and sustainable manner, while analyzing the economic and technological aspects of it. The study is conducted in several stages, beginning with the estimation of the village´s demand of electricity, to continue with natural resource assessment, analysis of the technologies, and evaluating a set of different HRES configurations with HOMER software package, developed by NREL for the design of micro-power systems.

### 4.1 Demand

The village is composed by a school, a bakery industry, a health clinic and households, and street lighting. Demand of the entire community is organized in common loads, including general lighting to commercial and industrial loads.

Residential demand (figure 6A) mainly refers to household's loads, including illumination, TV and radio. It has been considered an operation period of electricity for lighting from 18:00 to 23:00 hr which coincides with nocturnal activity of the occupants. Lighting is composed byfive lamps of 20 W, two of which are designed to light the bedroom, another two for the living room and kitchen, and the remaining one is used to illuminate the main door. Not all lamps are always turned out ; at midday it is only used one lamp. The radio of 40 W is supposed to be turned on between 7:00 to 12:00 hr and 18:00 to 20:00 hr; and the TV of 80W, considering 5W of the standby, is on from 12:00 to 14:00 hr and during nocturnal activity period. The number of households in the village is 30.

Commercial loads include a school and a health clinic. The schoolis composed by classrooms, a toilet for employees and students, and an office for professors. In terms of electrical loads, it consumes in lighting (indoor and outdoor) and office appliances (figure 6B). External illumination includes 4 lamps of 20W from 18:00 to 22:00 hr, classroom lighting uses20 lamps of 18W from 9:00 to 12:00 hr and 14:00 to17:00 hr, office lighting is equipped with 2 lamps of 15W operating also from 9:00 to 12:00 hr and 14:00 to17:00 hr, and the toilet is illuminated with a lamp of 15W from 9:00 to 12:00 hr. Moreover, the school is equipped with a 60 W computer used from 9:00 to 12:00 hr and 14:00 to17:00 hr, and a 200 W printer operated only two hours a day.

The health clinic is composed of external (6 lamps of 40W) and indoor lighting (10 lamps of 18W), a vaccine freezer, a communication system, a microscope, a printer, a computer and a TV. Due to the fact that is a health clinic, the external lighting is operating all night between 21:00 and 9:00 while the rooms' lighting is on from 8:00 to 22:00 hr (figure 6C). The vaccine freezer is working all day because of the great importance of this service. The communication apparatus' operation is from 8:00 to 11:00 hr and from 13:00 to 16:00 hr. That tool has a power of 50W. A computer of 60W is working from 8:00 to 17:00 hr. The printer of 200W with a standby of 40 W, works from 14:00 to 15:00 hr and the microscope from 8:00 to 11:00 hr. Finally, the TV of 80 W and 5 W of standby, is considered to be turned on between 8:00 and 21:00 hr.

Industrial electric demand is composed by a bakery and a milling machine for bread preparation (figure 6D). The baking machine has a capacity power of 2850 W and it is operated in two work shifts. The first one is in the morning from 5:00 to 8:00 hr while the second shift works from 15:00 to 17:00 hr. Subsequent, milling process begins from 8:00 to 11:00 hr and in the afternoon from 15:00 and 17:00 hr.

*Figure 6. A: Residential Load for the Case Study. B: School Loads. C: Health Centre Loads. D: Industrial Loads*

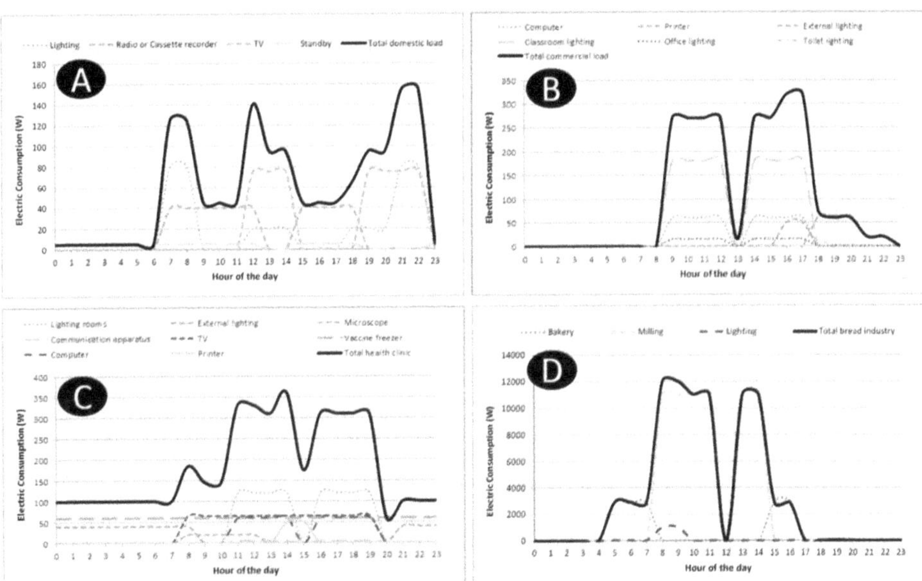


## 4.2 Microgrids and Electric Load Control

It is necessary to ensure the quality, security and reliability of energy supply, supply electricity at the lowest cost long term, and ensure an adequate return (Lujano *et al.*, 2012) of each project compliance with defined rules for the use of energy by the users. For this it is necessary to validate models to simulate the electrical behaviour of single loads (Anderson and Mirheydar, 1992) developing techniques of aggregation of these charges to estimate their joint physical evolution, and apply to these groups, algorithms for Management strategies, studying the effect of these actions from two points of view: user (changes in the patterns of consumption, services obtained), and electricity (modification of the period of demand and optimization of resources) (Dietrich, 2012). In addition, traditionally major difficulties have been found to predict consumption of residential customers, due to the component of subjectivity that has your demand, so it is of interest to have tools that help manage these charges increased and efficient way (Figure 7), (Schneider, n.d.). This interest overlaps the evolution that is currently suffering the electricity market, tending to position users as active elements, for which such programmes could, in principle, be implemented both from the supply side Electric, since demand (Figure 8), (Schneider, n.d.). This problem is acute in renewable systems on island where the generation can be affected by environmental characteristics where there is the possibility of connecting other additional sources and found that guarantee a minimum autonomy for loads priority.

The main objective for the election of a methodology that allows to identify the behavior of a set of elementary charges, should be with a characterization of the loads likely to be added, presenting the criteria for aggregation existing and the main types of groups of cargo that can be generated for energy optimization.

In the light of the elemental models developed for loads and once chosen the methodology of aggregation of elementary charges, the interest focuses on the study and implementation of strategies for control of loads that allow an appropriate modification of the global demand curve, based on the objectives and constraints that are initially marked, and taking into account the instant generation capacity and the State of energy accumulation, when the generation system is on island with media finite production and storage.

Thus, and taking into account the electrical behavior of shaped charges, the problem can be focused from two perspectives: control of loads, and development of new patterns of charge based on the energy

*Figure 7. Cycle of energy efficiency*

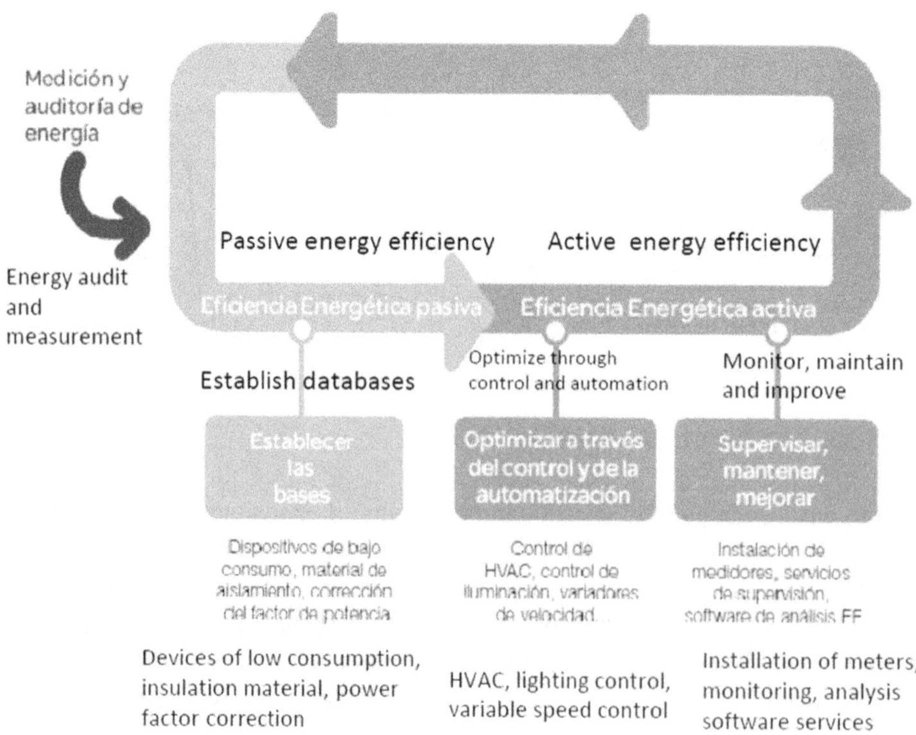

*Figure 8. System optimization of consumption*

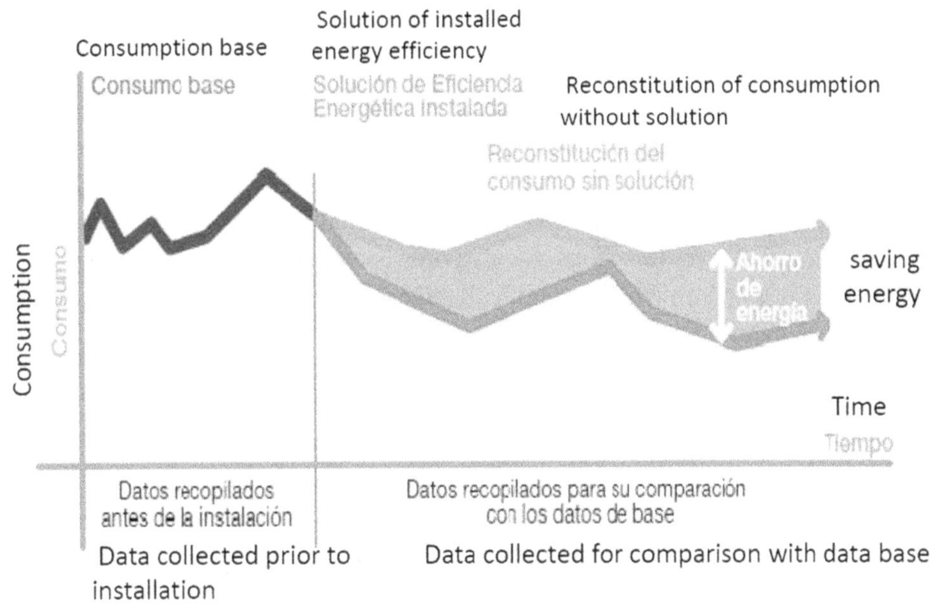


generated and stored Within this context, the power generation (microgrids) are an effective proposal of integration of different sources of renewable energy. The microgrids can work connected to the network or in an isolated manner, so that they offer to the consumer greater reliability of supply to provide different sources of generation.

Increase the reliability of the micro itself, as well as the diagnosis of failures of these is the subject of recent research in the scientific community and industrial. For efficiency the microgrids are evolving systems of predictive maintenance, integrating the fault diagnosis, monitoring by condition (status tracking) and the management of the maintenance in a single system. Similarly, operating systems of consumption or electrical charges that are coupled to renewable systems, affect the conditions of operation or operation of the complete system, making necessary the simulation of the dynamic response of the electrical systems through accurate modeling of components thereof: generators of renewable energy, accumulation of energy with its associated controls systems, and electric consumer loads and protections (load shedding fundamentally), and prioritization of consumption.

For the adequate control of electrical loads fed from the micro-grid system of hybrid renewable, it is necessary to manage the defrosting by automatically disconnecting specific loads. For this, the possibility of carrying out a load shedding must be integrated. The control unit must measure the values of the quantities of electrical and energy, generating the appropriate alarms. Each alarm is associated with a set of activation and non activation and at a setting of deactivation, each of them with timing. The order of non activation must be sent through an output module that activates the appropriate maneuvers (Schneider, n.d.) (Figure 9).

It is necessary to perform an intense research activity to implement renewable hybrid systems and enable the storage of renewable energy to ensure the reliability of these systems, with appropriate control and manoeuvre of consumption to ensure continuous service. With this objective it is necessary to study renewable hybrid systems in the kW range and design and participate in major field experiences of power.

## 4.3. Principles of a System of Monitoring, Control and Discrimination of Electrical Loads

The principles that should build such a system, is the design of a system of monitoring, control and discrimination of electrical charges that allow you to monitor, in real time, each and every one of the variables of the system to assist in a system of generation hybrid renewable energy connected in island, shedding loads / connect electrical loads, depending on the characteristics of the production of energy and the State of charge of the energy storage system, automatically allowing a greater balance between production and consumption, avoiding stops by zeros due to excesses of consumption or to give greater autonomy to the whole of system-wide generator/consumer of energy. All focused on consumers energy systems that operate isolated (on island).

*Figure 9. System non-priority loads*

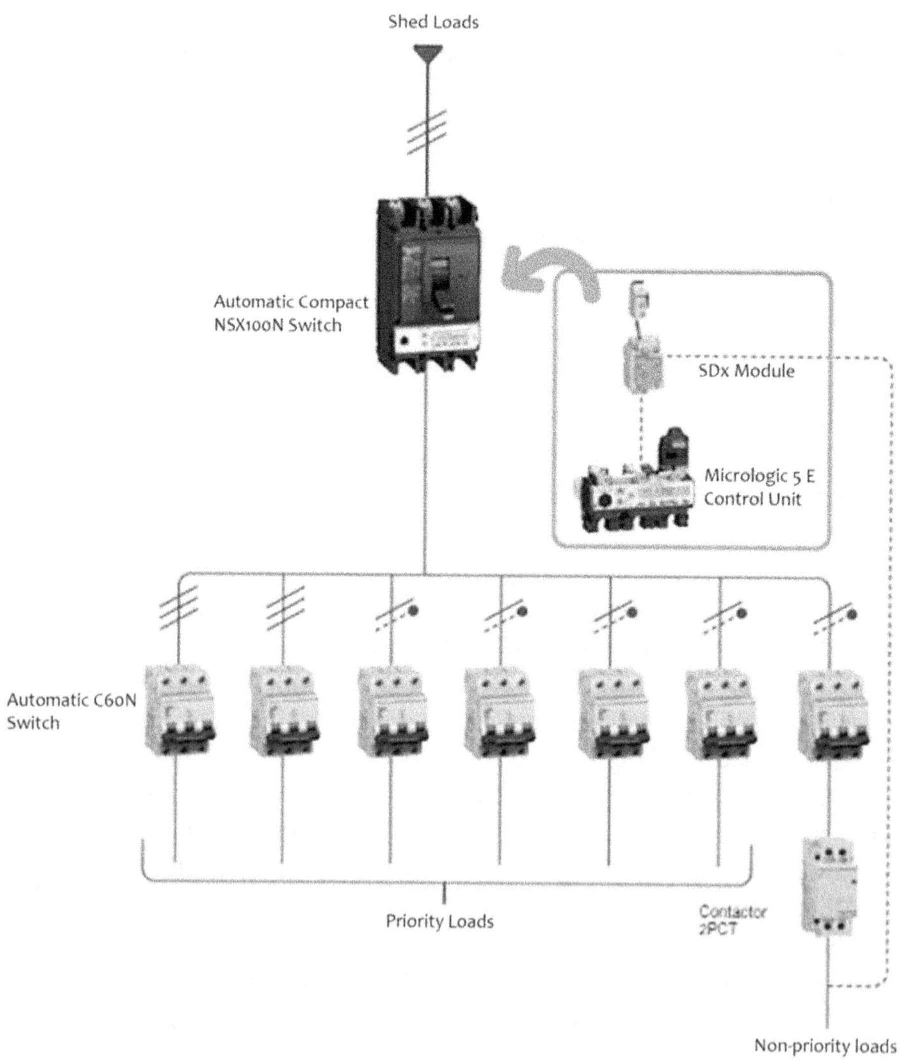

The system is modular in terms of the functions of monitoring, troubleshooting and monitoring of the State offering flexibility in the type of operation of loads depending on the categorization based on their criticality, and the ability to generate energy or store it system hybrid. Generation management tasks will include information of each modular system as soon as their condition and variables such as loads, the powers of generation and demand, using an expert system based on rules that ensure the proper functioning of the System.

To achieve the goal, it should be considered:

- **Sensing**. Should be included in the system all additional sensors to the control that are necessary to complete the main objective.
- **Communications network**. The information collected on each source of energy and consumer items is to be led by network to the central system.
- **Heuristic analysis** of each subsystem of generation/consumption based on expert knowledge and a qualitative diagnosis of physical systems for these subsystems. Heuristic analysis will consist of an analysis detailed of each subsystem generation/consumption recovering empirical information from the experts, justifying it with the technical information of the critical loads and prioritization of disconnection of loads for the implementation of an intelligent system of diagnosis of optimal energy balance. Simulation models for analysis and subsequent calculation of ranges, will be carried out for each case values limits, etc., where main signals for tracking status of the most relevant characteristics are characterized as they can be: storage of medicines, freezers and refrigerators for grocery, systems of pumping of potable water, etc..
- Development of **detailed analytical models** for specific components that allow taking into account the peculiarities of these same subsystems. Certain components or subsystems allow the design of analytical models that allow using diagnostic techniques based on residual analysis. In this way, different generation/consumption systems will be analyzed to check what parts of them are likely to be diagnosed with this technique.
- Development of **models and systems of monitoring of the status of the consumption of loads** based on the forecast production and stored energy. Estimate of the behavior of a particular component or subsystem based on measurements of variables of the same. The algorithms can be as simple as the derivative of the signal to obtain a trend, or as complex as the application of special techniques with Petri nets, and artificial intelligence with Fuzzy logic. Each subsystem will be analyzed to verify algorithms to apply in each case.
- Implementation of an **expert system for management** of the generation/consumption for the mode of operation in parallel of distributed generation. The microgrid must operate with a variable topology to optimize the distribution of energy, guaranteeing the autonomy of the energy needed by the consumption and optimizing the operation of the whole system of generation hybrid renewable.

## 4.4 Resources

Assessing natural resources in the area under study is necessary to determine the capacity for renewable generation. A feasibility analysis of the renewable potential at the specific location highlights solar, wind and biomass as major potential sources, therefore further analysis of these resources is performed.

Solar resource is evaluated using the PVGIS-CMSAF (Huld, *et al.*,2012; Šúri *et al.*,2007) solar radiation database at the location of Kinshasa (4°19'54" South, 15°18'50" East, and elevation: 283 m a.s.l.), resulting an annual average of solar radiation as high as 6,4 kWh/m2/day and a clearness index mean in the order of 0,6. Solar radiation is considerable throughout the year (figure 10) thus significant PV power could be expected without high variations.

*Figure 10. Solar Radiation Profile*

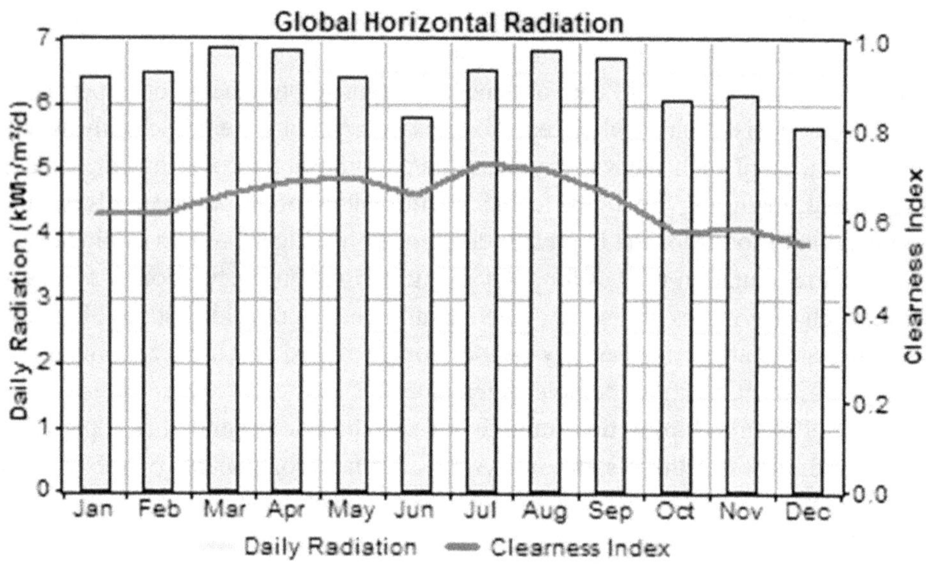


*Figure 11. Monthly wind speed average*

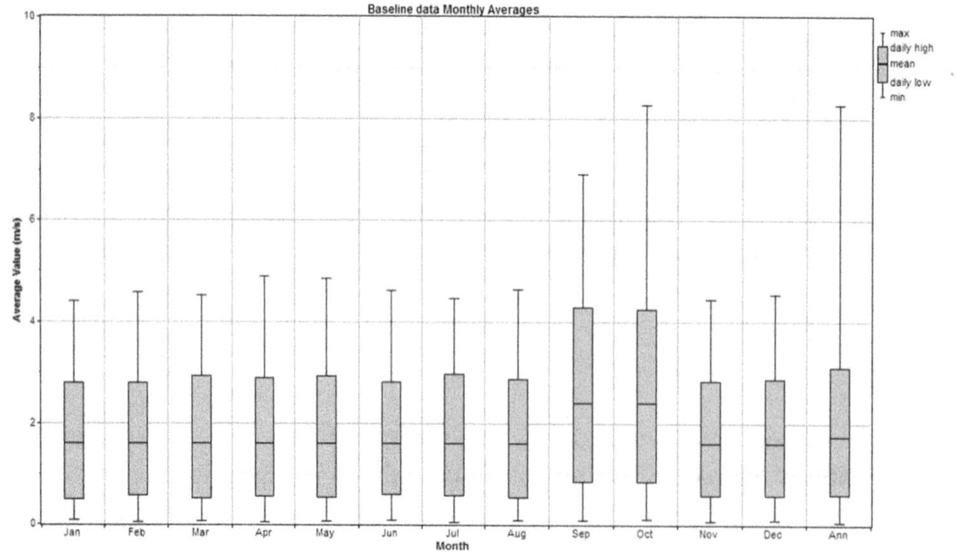

Wind resource is assessed using meteorological data from METEOBLUE, a local weather source (figure 11) Average wind speed throughout the year is 1,8 m/s, except for September and October where average increases to 2,3 m/s. Despite this low average, the wind profile shows daily peaks close to 3 m/s from November to August, and over 4 m/s in September and October, with highest values around 4,5 m/s and over 8 m/s, respectively.

Analysing the daily profile of wind speeds, it is observed similar patterns. The day start with minimum wind speed values, then it intensifies until it reaches its highest record around noon. From this point, wind speed starts decreasing until reaching a minimum value again. Figure 12 represents the daily profile of January, with a minimum of 2 m/s and a 3,8 m/s maximum, while in a windy month like October, the figure represents a minimum wind speed value of 3 m/s and maximum values over 5 m/s.

*Figure 12. Wind speed daily profile*

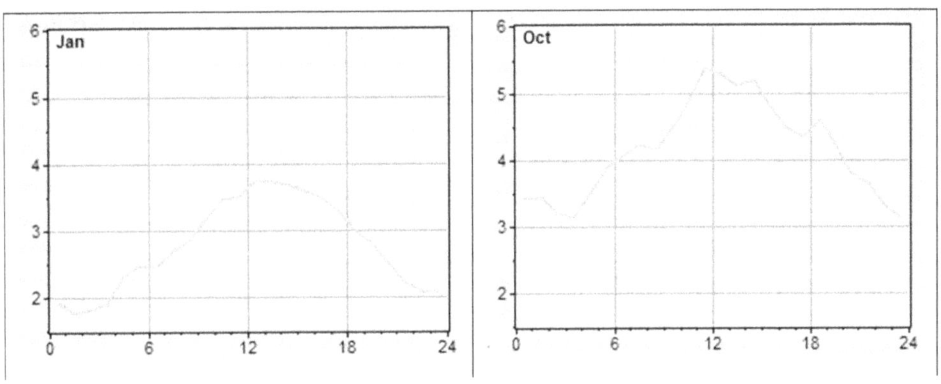


In addition to the wind speed, it is necessary to analyse main wind directions in order to prevent any potential obstacles. Wind rose for this case study shows wind direction mainly from South-West (SW), so special attention should be paid to this direction when installing the wind turbine.

Regarding biomass availability, RDC has 145 million hectares of forest (47% of the total forest resources in Africa) with an annual growth of 100 million tons (35% moisture), (METEOBLUE Weather, n.d.). It is estimated that RDC consumed 54,7 million tons (74,5 million m3) of wood for energy use, mainly for residential energy as firewood and small-scale commercial sectors(FAO, n.d.b.; BMVBS, n.d.). Moreover, reducing the demand for firewood is a key factor for reducing deforestation and ex-

*Table 4.Characterization of biomass forestry species in DRC.*

| Samples | Moisture, (%bh) | Ashes, (%bs) | Density, (g/cm$^3$) | Higher Heating Value (HHV), Wet (MJ/kg) | HHV, Dry (MJ/kg) |
|---|---|---|---|---|---|
| $S_1$: FUMA | 8,31% | 0,6% | 0,434 | 16,14 | 17,60 |
| $S_2$: NTOLA | 13,25% | 0,37% | 0,513 | 16,95 | 19,54 |
| S3: LIFAKI | 13,57% | 2,36% | 0,614 | 15,74 | 18,21 |

(%bh)= % of wet base; (%bs)= % of dry base

haustion of DRC's natural resources (BMVBS, n.d.; Stecher, n.d.), so more efficient forms of biomass energy generation should be implemented and promoted, being gasification a clean alternative. Biomass potential is large, annual forest growing represent an energy equivalency of 28 million of toe (tons of oil equivalent), while agricultural biomass waste is estimated in 0,48 million of toe and it will probably increase over the time as a consequence of general agricultural progress (World Bank, n.d.). Characterization of available biomass has been carried out at the Institute for Energy Engineering (IIE), where three samples of the most representative species were analyzed, obtaining the results detailed in table 4. Despite biomass species present similar characteristics, NTOLA has higher HHV and lower percentage of ashes, being the more convenient alternative for the gasification process.

*Table 5.Gasification Plant Operation Parameters.Source.*

| Parameters | IIE Gasifier | Literature Range Min-Max |
|---|---|---|
| Gas generation per kg of biomass, $Nm^3/kg$ | 2,1 | 2-2,5 |
| Electrical generation per kg of biomass, kW/kg | 0,9 | 0,7-1,5 |
| Input air per gas produced, $m^3/m^3$ | 0,7 | 0,5-0,8 |
| Maximum gas LHV, $kCal/m^3$ | 1595 | 900-1700 |
| Efficiency of the conversion biomass-gas, % | 79% | 65%-85% |
| Efficiency of the conversion gas-electricity using a IC Engine, % | 22% | 18%-32% |

## 4.5 Technologies

This section includes a description of the main renewable energies to be considered in the study. As identified in previous section, main natural resources are solar, wind and biomass, since the area under study did not have access to the Congo River and therefore micro hydro power generation was not available.

Solar PV system is composed by panels in series with no tracking system. Panels are configured as fixed tilted with an optimal slope of -7° at the location 4°19' south (latitude). Estimated losses of the PV system are also considered, such as those related to temperature and low irradiance (12,8%), angular reflectance effects (2,9%), and other losses in wiring, inverter etc. (14.0%). In overall terms, total losses in PV system have been estimated in 27%. Economic modeling includes capital and replacement cost of 1kWPV are 2,3k€ and 1,5k€, respectively, and annual operation and maintenance cost as 0,015 €/kWPV.

Considering average wind speeds in the area, a small-scale wind turbine of 3 kW has been selected. Wind turbine is a DC machine with a hub height of 20 m and 15 years lifetime. Output power is defined by its power curve.

Economic modelling for small wind turbines considers a capital cost of 3,6k€ per installed kWWIND, reduced to 2,3k€ in replacement purchases. Regarding to annual costs in operation and maintenance, this has been estimated in 200 €/year.

Biomass gasifier is modelled based on experimental data obtained at LabDER (Pérez et.al, 2016), the IIE laboratory at UPV, where a gasifier was designed and constructed for small scale validation of the system (table 5). Gasification plant technology is a downdraft fixed bed, fabricated as a mobile unit to facilitate its operation and educational purposes. Lifetime operating hours of the gasifier unit is 20,000 hours considering a minimum load ratio of 90%. Syngas generated in this system is used as an electric power by generation through an internal combustion engine with a conversion efficiency of 22%.

Based on experimental data, initial investment of the gasification plant is 7,7k€/kWGASFwith a replacement cost of 3,8k€/kWGASF, since many of the parts may be re-used. In terms of operation and maintenance, an annual cost of 0,115€/kWGASF is registered. Generation fixed costs associated to the gasification plant are 2,05€/hr while marginal cost is observed as 0,102 €/kWh.

## 4.6 Results

Modelling and simulation of different HRES configurations has been carried out using HOMER as it is widely used in literature (HOMER, n.d.; Bahramara*et al.,* 2016). Detailed data regarding the analysed demand, resources and components are introduced as input data of the model, together with the unitary technology cost (capital, replacement and O&M), the dispatch strategy and model constraints. HOMER uses all this information to simulate all different system configurations, or combinations of components, generating a list of feasible configurations sorted by net present cost, considering the capital, replacement, operation and maintenance, fuel, and interest costs.It simulates the operation of each HRES system, calculating the hourly energy balance. For each of the 8760 hours of the year, HOMER compares the electric demand with the energy supplied by the system and calculates the energy flow to and from each component of the system. Then, it determines whether a configuration meets the electric demand under pre-defined conditions, such as minimum fraction of renewable or demand capacity shortage and estimates the cost of installing and operating the system over the lifetime of the project.

Results of this simulation are presented in table 6, showing a list of possible HRES configurations capable to satisfy the electric requirements of the community.

*Table 6. List of feasible HRES alternatives.*

| PV (kW) | G3 | Dsl (kW) | BGPP (kW) | T-105 | Conv. (kW) | Disp. Strgy | Initial Capital | Operating Cost ($/yr) | Total NPC | COE ($/kWh) | Ren. Frac. | Diesel (L) | Biomass (t) | Dsl (hrs) | BGPP (hrs) |
|---|---|---|---|---|---|---|---|---|---|---|---|---|---|---|---|
| 10 | | 6 | | 40 | 8 | CC | $ 46,275 | 3,942 | $ 88,355 | 0.535 | 0.87 | 995 | | 759 | |
| 10 | 1 | 6 | | 40 | 8 | CC | $ 57,275 | 3,966 | $ 99,612 | 0.603 | 0.89 | 836 | | 645 | |
| 20 | | | | 80 | 8 | CC | $ 66,800 | 3,558 | $ 104,783 | 0.635 | 1.00 | | | | |
| 20 | 1 | | | 80 | 8 | CC | $ 77,800 | 3,933 | $ 119,784 | 0.726 | 1.00 | | | | |
| | | 6 | | 20 | 4 | CC | $ 19,075 | 13,515 | $ 163,341 | 0.989 | 0.00 | 6,301 | | 4,060 | |
| | 1 | 6 | | 20 | 4 | CC | $ 30,075 | 13,298 | $ 172,024 | 1.042 | 0.05 | 6,012 | | 3,938 | |
| 20 | | 6 | | | 8 | CC | $ 58,875 | 12,402 | $ 191,264 | 1.159 | 0.76 | 5,425 | | 5,773 | |
| 20 | 1 | 6 | | | 8 | CC | $ 69,875 | 12,600 | $ 204,379 | 1.238 | 0.77 | 5,342 | | 5,689 | |
| | | | 10 | 40 | 4 | LF | $ 74,980 | 16,335 | $ 249,353 | 1.510 | 1.00 | | 40 | | 3,670 |
| 10 | | | 10 | 20 | 8 | CC | $ 95,780 | 14,557 | $ 251,176 | 1.521 | 1.00 | | 36 | | 3,305 |
| 10 | | 6 | 10 | 20 | 8 | CC | $ 100,655 | 14,248 | $ 252,746 | 1.531 | 0.99 | 246 | 34 | 220 | 3,132 |
| | | 6 | 10 | 20 | 4 | CC | $ 76,655 | 16,997 | $ 258,099 | 1.563 | 0.86 | 1,687 | 35 | 1,369 | 3,165 |
| | 1 | | 10 | 40 | 4 | LF | $ 85,980 | 16,614 | $ 263,328 | 1.595 | 1.00 | | 40 | | 3,658 |
| 10 | 1 | | 10 | 20 | 8 | CC | $ 106,780 | 14,814 | $ 264,914 | 1.605 | 1.00 | | 36 | | 3,284 |
| 10 | 1 | 6 | 10 | 20 | 8 | CC | $ 111,655 | 14,545 | $ 266,915 | 1.617 | 0.99 | 220 | 34 | 200 | 3,132 |
| | 1 | 6 | 10 | 20 | 4 | CC | $ 87,655 | 17,176 | $ 271,005 | 1.642 | 0.87 | 1,612 | 35 | 1,325 | 3,160 |
| | | 6 | 10 | | | CC | $ 68,455 | 24,303 | $ 327,882 | 1.986 | 0.69 | 5,532 | 35 | 5,564 | 3,196 |
| 10 | | 6 | 10 | | 4 | CC | $ 92,455 | 23,178 | $ 339,880 | 2.059 | 0.82 | 4,864 | 34 | 5,186 | 3,132 |
| | 1 | 6 | 10 | | 4 | CC | $ 84,455 | 24,626 | $ 347,335 | 2.104 | 0.70 | 5,476 | 35 | 5,518 | 3,190 |
| 10 | 1 | 6 | 10 | | 4 | CC | $ 103,455 | 23,409 | $ 353,346 | 2.140 | 0.82 | 4,797 | 34 | 5,117 | 3,132 |
| | | 12 | | | | CC | $ 15,750 | 35,777 | $ 397,657 | 2.409 | 0.00 | 16,808 | | 8,760 | |
| | 1 | 12 | | | 4 | CC | $ 31,750 | 35,929 | $ 415,289 | 2.516 | 0.02 | 16,654 | | 8,704 | |
| | | | 10 | | | CC | $ 63,580 | 35,913 | $ 446,944 | 2.707 | 1.00 | | 96 | | 8,760 |
| 20 | | | 10 | | 8 | CC | $ 111,580 | 31,570 | $ 448,584 | 2.717 | 1.00 | | 82 | | 7,509 |
| 20 | 1 | | 10 | | 8 | CC | $ 122,580 | 31,681 | $ 460,772 | 2.791 | 1.00 | | 82 | | 7,446 |
| | 1 | | 10 | | 4 | CC | $ 79,580 | 36,161 | $ 465,588 | 2.820 | 1.00 | | 95 | | 8,708 |

Following a description of the experimental validation of small scaled HRES systems is presented in the next section.

## 4.7 Experimental Validation at Small-Scaled

Any optimization process of a HRES system for a particular application could be improved by an experimental simulation of the selected configuration in order to check its capability to supply the demand profile in a reliable and feasible way. To address this experimental simulation, we are using the LABDER laboratory at IIE-UPV that allows for the simulation of different HRES configurations and its response to different load curves (Pérez et.al, 2016).

Two configuration results are shown here: HRES1, that includes a combination of photovoltaic panels, biomass gasification and batteries bank, dully controlled to supply a typical demand curve in the residential sector with two main peaks, in the order of 4 kW, in the morning and in the evening; and HRES2, a second hybrid system to cover the same demand curve where, in addition to the PV and the biomass gasification plants, a wind generator was included.

To facilitate the experiment, the demand curve was compressed to 3 hours duration in the pre-programmed load, considering this time was long enough to check the capabilities of the two configurations to fulfil the energy requirements.

Total power generated by HRES1 can be expressed by the addition of the contributions from each of the renewable sources:

$$P_{gen}(t) = \sum_i c_i(t) * P_i(t) \tag{1}$$

Where represents the nominal power of each renewable source and the contribution at each moment of that source to the total generated power.

*Figure 13. A: Total generated power and contributions from each element in HRES1. B: Contributions from the different HRES1 elements to the generated power. C Losses in the HRES1. D: Voltage and frequency of the HRES1 delivered power to the load. E: Fluctuation levels for voltage and frequency of the HRES1 power output. F: Power generated and transferred to the load by the HRES2*

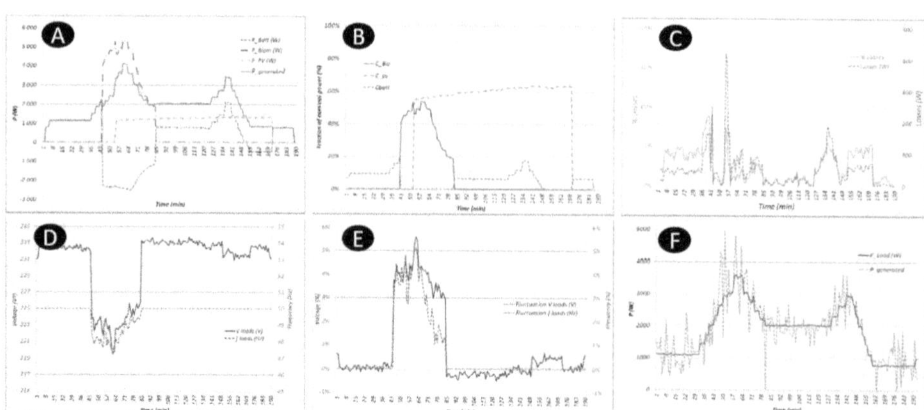

Figure 13A shows the total power generated to cover the demand profile by adding the contribution of the solar panels, the biomass gasifier and the batteries.

As it may be observed in the plot, at the beginning of the experiment, in the absence of solar radiation and with full charged storage bank, the demand is met with the electricity stored in the batteries. When the charging level of the battery bank decreases down to 50%, biomass gasifier starts the generation, covering the demand and recharging the storage bank. Later, PV energy starts, but the gasification plant continues running until battery storage is again full charged, which allows beginning the next cycle of operation under the same conditions of the previous one.

Negative power in the case of batteries indicates that batteries, at that time, are in a charge process taking energy either from the biomass gasifier, in the first part of the discharge, or from the photovoltaic system, during the final part of the discharge. Control system was setup to change to battery charge status when the bank is below 50% of its nominal full charge. The normal solar radiation profile in the experiment was almost constant due to experiment was made around noon. Sunset and sundown were substituted by an on-off control of the PV energy input to the system.

Figure 13B) plots the contributions of each of the elements of the HRES1 to the total generated power. PV panels contributed with around 60% of their nominal power, while biomass gasifier contribution is slightly smaller, in the order of 50%, and battery bank is just working at 20%. So, the system has enough power to supply any extra power demand that could appear.

By comparison of the generated power and the power transmitted to the load is possible to deduce the losses in the system.

$$P_{loss}(t) = P_{gen}(t) - P_{load}(t) \tag{2}$$

and the fraction they represent:

$$f_{loss}(t) = {P_{loss}(t)} \big/ {P_{gen}(t)} \tag{3}$$

These losses are shown at (figure 14), both in absolute and percentage values.

Results indicate that losses are small, 3% on average, and by comparison with (figure 13C), that shows which source is the dominant one at each time, it can be deduced that main losses appear when battery bank is the dominant source or sink, reaching values at that moment in the range 5-10%.

Finally, the quality of the electric power generated by the HRES1 has been monitored by measuring the voltage and frequency of the transmitted power to the load. (figure13D) summarize the results. In the absence of the gasification system, the fluctuation level is almost zero. When the gasifier starts to operate, its associated syngas engine introduces, due to its less stable behaviour, some fluctuation level, in the order of 4%, for both, voltage and frequency, of the output power as detailed at (figure 13E).

In summary, it could be concluded that the assembled hybrid system HRES1, whose design was obtained with the optimization methodology, seems to be able to cover the residential electricity demand profile in a flexible and reliable way.

A second hybrid system to cover the same demand curve was checked in LABDER. In this new system (HRES2), in addition to the PV and the biomass gasification plants, a wind generator was included. A similar behaviour to the previous system in its capability to supply the demand curve was observed, as detailed by the data plotted at (figure 13F)

The main problem with this configuration is the high level of fluctuations in the power production from the wind generator, which was filtered by the battery bank and the gasifier, as can be deduced from figure 14 where contribution of each component of the HRES2 system is detailed. The power transmitted to the load did not present these fluctuations

*Figure 14.Contribution of each energy source to the power output in HRES2.(a: wind generator, b: battery bank, c: biomass gasifier, d: PV panels)*

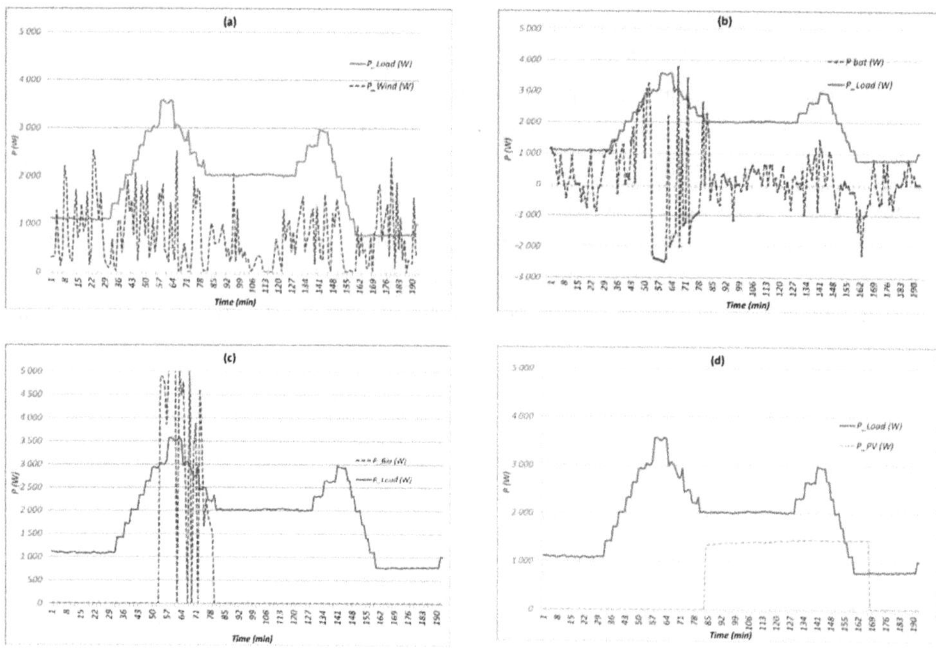


Apart from this fact, the system can cover the demand curve with the same level of reliability than the previous one, as shown at figure 15.

In conclusion, it is important to assure the feasibility of HRES configurations deduced from optimization methodologies, before its implementation on site. To check the expected behaviour of the proposed system an experimental simulation of that system at the minimum meaningful level could be advisable. This test would require the use of experimental installations with an adequate micro grid that includes all the possible renewable sources and storage systems assumed in the configuration.

## 5. MULTI-CRITERIA ASSESSMENT

Finally, a multi-criteria evaluation of the most promising alternatives is carried out in this section. Initially, a series of expert from different disciplines and professions were invited to participate in the assessment. Expert committee was composed by a hres technician, an economist, a social worker, a

*Figure 15.Voltage and frequency of the HRES2 delivered power to the load*

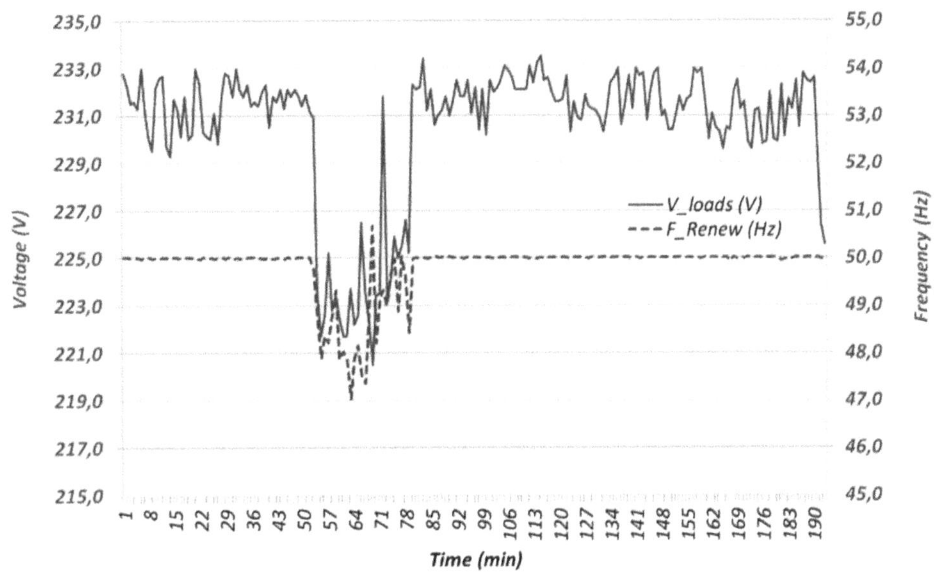


politician and a representative of the community. First decision to be taken was to identify and select the three electrification alternatives that resulted more attractive to them, based on their professional experience. This resulted in three hres systems composed by (table 7):

1- Photovoltaics + Genset + Batteries
2- Photovoltaics + Wind turbine + Batteries
3- Photovoltaics + Biomass Gasifiers+Batteries

*Table 7. Set of HRES configurations.*

| | PV (kW) | G3 | Dsl (kW) | BGPP (kW) | T-105 | Conv. (kW) | Disp. Strgy | Initial Capital | Operating Cost ($/yr) | Total NPC | COE ($/kWh) | Ren. Frac. | Diesel (L) | Biomass (t) | Dsl (hrs) | BGPP (hrs) |
|---|---|---|---|---|---|---|---|---|---|---|---|---|---|---|---|---|
| 1 | 10 | | 6 | | 40 | 8 | CC | $ 46,275 | 3,942 | $ 88,355 | 0.535 | 0.87 | 995 | | 759 | |
| | 10 | 1 | 6 | | 40 | 8 | CC | $ 57,275 | 3,966 | $ 99,612 | 0.603 | 0.89 | 836 | | 645 | |
| | 20 | | | | 80 | 8 | CC | $ 66,800 | 3,558 | $ 104,783 | 0.635 | 1.00 | | | | |
| 2 | 20 | 1 | | | 80 | 8 | CC | $ 77,800 | 3,933 | $ 119,784 | 0.726 | 1.00 | | | | |
| | | | 6 | | 20 | 4 | CC | $ 19,075 | 13,515 | $ 163,341 | 0.989 | 0.00 | 6,301 | | 4,060 | |
| | | 1 | 6 | | 20 | 4 | CC | $ 30,075 | 13,298 | $ 172,024 | 1.042 | 0.05 | 6,012 | | 3,938 | |
| | 20 | | 6 | | | 8 | CC | $ 58,875 | 12,402 | $ 191,264 | 1.159 | 0.76 | 5,425 | | 5,773 | |
| | 20 | 1 | 6 | | | 8 | CC | $ 69,875 | 12,600 | $ 204,379 | 1.238 | 0.77 | 5,342 | | 5,689 | |
| | | | | 10 | 40 | 4 | LF | $ 74,980 | 16,335 | $ 249,353 | 1.510 | 1.00 | | 40 | | 3,670 |
| 3 | 10 | | | 10 | 20 | 8 | CC | $ 95,780 | 14,557 | $ 251,176 | 1.521 | 1.00 | | 36 | | 3,305 |
| | 10 | | 6 | 10 | 20 | 8 | CC | $ 100,655 | 14,248 | $ 252,746 | 1.531 | 0.99 | 246 | 34 | 220 | 3,132 |
| | | | 6 | 10 | 20 | 4 | CC | $ 76,655 | 16,997 | $ 258,099 | 1.563 | 0.86 | 1,687 | 35 | 1,369 | 3,165 |
| | | 1 | | 10 | 40 | 4 | LF | $ 85,980 | 16,614 | $ 263,328 | 1.595 | 1.00 | | 40 | | 3,658 |
| | 10 | 1 | | 10 | 20 | 8 | CC | $ 106,780 | 14,814 | $ 264,914 | 1.605 | 1.00 | | 36 | | 3,284 |
| | 10 | 1 | 6 | 10 | 20 | 8 | CC | $ 111,655 | 14,545 | $ 266,915 | 1.617 | 0.99 | 220 | 34 | 200 | 3,132 |
| | | 1 | 6 | 10 | 20 | 4 | CC | $ 87,655 | 17,176 | $ 271,005 | 1.642 | 0.87 | 1,612 | 35 | 1,325 | 3,160 |
| | | | 6 | 10 | | | CC | $ 68,455 | 24,303 | $ 327,882 | 1.986 | 0.69 | 5,532 | 35 | 5,564 | 3,196 |
| | 10 | | 6 | 10 | | 4 | CC | $ 92,455 | 23,178 | $ 339,880 | 2.059 | 0.82 | 4,864 | 34 | 5,186 | 3,132 |
| | | 1 | 6 | 10 | | 4 | CC | $ 84,455 | 24,626 | $ 347,335 | 2.104 | 0.70 | 5,476 | 35 | 5,518 | 3,190 |
| | 10 | 1 | 6 | 10 | | 4 | CC | $ 103,455 | 23,409 | $ 353,346 | 2.140 | 0.82 | 4,797 | 34 | 5,117 | 3,132 |
| | | | 12 | | | | CC | $ 15,750 | 35,777 | $ 397,657 | 2.409 | 0.00 | 16,808 | | 8,760 | |
| | | 1 | 12 | | | 4 | CC | $ 31,750 | 35,929 | $ 415,289 | 2.516 | 0.02 | 16,654 | | 8,704 | |
| | | | | 10 | | | CC | $ 63,580 | 35,913 | $ 446,944 | 2.707 | 1.00 | | 96 | | 8,760 |
| | 20 | | | 10 | | 8 | CC | $ 111,580 | 31,570 | $ 448,584 | 2.717 | 1.00 | | 82 | | 7,509 |
| | 20 | 1 | | 10 | | 8 | CC | $ 122,580 | 31,681 | $ 460,772 | 2.791 | 1.00 | | 82 | | 7,446 |
| | | 1 | | 10 | | 4 | CC | $ 79,580 | 36,161 | $ 465,588 | 2.820 | 1.00 | | 95 | | 8,708 |

*Figure 16. AHP multi-criteria scheme*

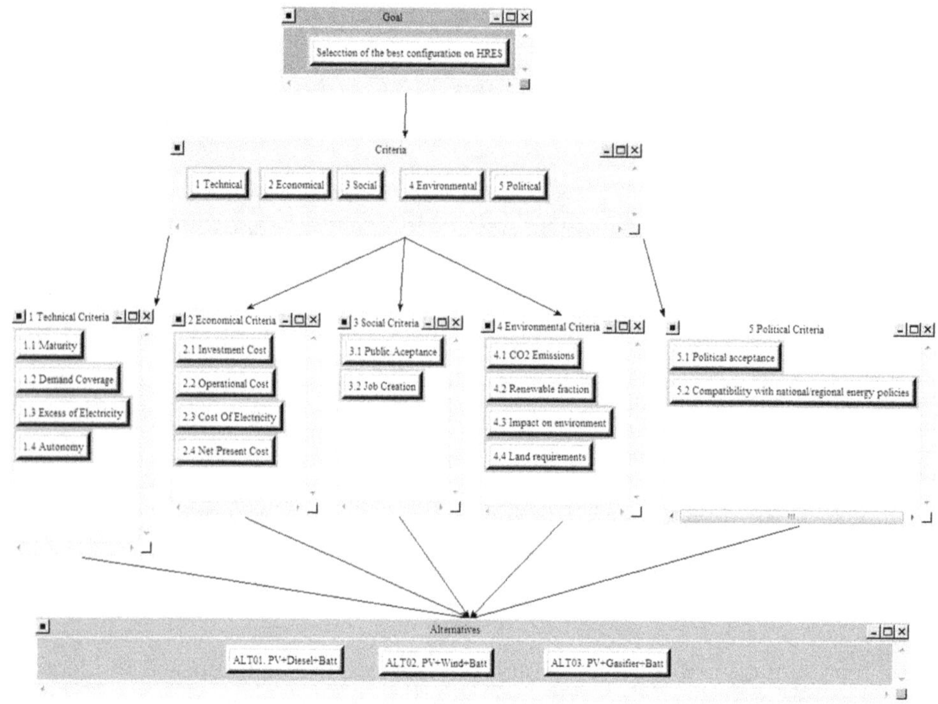


## 5.1 Expert's Validation

Once the goal and criteria of the analysis are well defined and the alternatives are selected, it is carried out the multi-criteria assessment based on the AHP method. Initial step consists on setting the hierarchical structure of the problem (figure 16) and collecting the results of the pair-wise comparisons from the expert's committee, which was performed by means of a questionnaire. In this regard, special attention is provided to the consistency ratio based on the matrix size, since a high inconsistency value will indicate that assessment have been inconsistent and, therefore, the expert committee should reconsider their evaluation.

Initial assessment involves comparing first level alternatives, which results in the following weights:

C1. Technical: 21,3%
C2. Economic: 21,3%
C3. Environmental: 12,3%
C4. Social: 12,3%
C5. Political: 32,9%

Inconsistency in this level is 1,7%, far below from the 10% limit for matrix size of 5. Thus, it is possible to proceedto the next level of pair-wise comparison.

Second level variables to be analyzed involve:

*Table 8.Evaluation of quantitative variables.*

| | x | 1/x | Normalized Weight (1/x) |
|---|---|---|---|
| **ALT1** | 70,000 € | 0.00001429 | 0.38884354 |
| **ALT2** | 119,400 € | 0.00000838 | 0.22802721 |
| **ALT3** | 71,000 € | 0.00001408 | 0.38312925 |
| | **Sum** | 0.00003675 | 1 |

C1. Technical characteristics:
C1.1 Maturity: 27,6%
C1.2 Demand coverage: 19,5%
C1.3 Excess of electricity: 13,8%
C1.4 Autonomy: 39,1%

Inconsistency is equal to 4,5%, below the consistency ratio of 9% for a 4x4 matrix.

C2. Economic Criteria.
C2.1. Initial investment: 27.8%
C2.2. O&M: 16,3%
C2.3. Cost of Electricity: 16,3%
C2.4. Net Present Cost: 39,6

Inconsistency is equal to 2,3%, below the consistency ratio of 9% for a 4x4 matrix.

C3. Social Criteria
C3.1. Public acceptance: 50%

*Table 9.Evaluation of quantitative variables.*

| | | | |
|---|---|---|---|
| **C1.1 Maturity** | 49,3 % | 31,1 % | 19,6 % |
| | | | |
| **C1.3 Excess of electricity** | 87,0 % | 7,8 % | 5,2 % |
| | | | |

*Table 10.Assessment of Economic characteristics in each alternative.*

| | | | |
|---|---|---|---|
| C2.1. Initial investment | 48,1 % | 28,6 % | 23,3 % |
| C2.3. Cost of Electricity | 47,8 % | 35,2 % | 17,0 % |

*Table 11.Assessment of Social characteristics in each alternative.*

| | | | |
|---|---|---|---|
| C3.1. Public acceptance | 31,1 % | 20,6 % | 49,3 % |

C3.2. Job creation: 50%

Inconsistency is equal to 0%, so the consistency ratio condition is fulfilled.

*Table 12.Assessment of Environmental characteristics in each alternative.*

| | | | |
|---|---|---|---|
| C4.1. $CO_2$ emission | 0,0 % | 98,4 % | 1,6% |
| C4.3. Impact on environment | 10,9 % | 34,4 % | 54,7 % |

*Table 13.Assessment of Political characteristics in each alternative.*

| | | | |
|---|---|---|---|
| C5.1. Political acceptance | 41,3 % | 25,9 % | 32,8% |

C4. Environmental Criteria
C4.1. CO2 emission: 28,6%
C4.2. Renewable fraction: 28,6%
C4.3. Impact on environment: 28,6%
C4.4. Land requirements: 14,2%

Inconsistency is equal to 0%, so the consistency ratio condition is fulfilled.

C5. Political Criteria
C5.1. Political acceptance: 50%
C5.2. Compatibility with national/regional energy policies: 50%

Inconsistency is equal to 0%, so the consistency ratio condition is fulfilled.

Finally, alternatives are pondered based on qualitative and quantitative considerations. Qualitative aspects are evaluated by means of a questionnaire, while quantitative variables are assessed using specific

*Table 14.Global weights of the alternatives*

| | |
|---|---|
| **ALT1. PV + Diesel + Batt** | **41,44 %** |
| **ALT3. PV + Gasifier + Batt** | 20,57 % |

values obtained from the above detailed simulations. Normalized weights will be obtained from these values and introduce in the comparison matrix (table 8).

Continuing with the evaluation, next it is assessed each alternative per criteria (technical, economic, social, environmental, and political), obtaining the following weights (table 9, 10, 11, 12, 13).

Finally, global weights for each alternative are calculated according to the following mathematical expression:

$$W_{Global} = \sum\left(W_{C1} * W_{C2} * W_{A}\right) \tag{4}$$

Where:

$W_{C1}$ : are the weighs of the first level analysis
$W_{C2}$ : are the weighs of the second level analysis
$W_{A}$ : are the weighs of the alternatives

Thus, final global weights of the alternatives are ALT1 (41,4%), ALT2 (38,0%) and ALT3 (20,6%), which indicates that the best alternative for this study case is the HRES configuration that includes PV+DieselGen+Batteries.

## 7. CONCLUSION

This chapter presented an implementation of the integral methodology introduced in previous unit. The methodology has been applied to an isolated area near Kinshasa in Republic Democratic of Congo. First it has been analyzed the energy context of the country, and then BAU scenario has been compared with a scenario based of HRES, aiming to increase the electrification rate in the country without compromising their energy independency or the environment. Results from this first analysis showed that a sustainable scenario based on HRES is possible.

Next, it has been modeled and simulated a series of HRES configurations to obtain a set of feasible HRES alternatives for the area of Kinshasa. Once, the promising HRES configurations have been identified, an experimental validation of small scaled systems has been carried out in LabDER, Distributed Energy Resources Laboratory at UPV, concluding that these systems can provide reliable supply of energy to isolated communities.

Finally, a multi-criteria evaluation of the promising HRES alternatives was carried out by an expert's committee, who has analyzed the importance of each variable in comparison with the rest. Final result from this last stage of the methodology is the selection of the most adequate HRES alternative for Kinshasa area, considering not just the economic, environmental and technical aspects, but also the socio-political context of the isolated community.

In renewable generation systems applied to buildings connected in island, requested power, limited character, you must follow a procedure and guidelines for consumption that optimize the service provided.

Electrical loads connected to the renewable system, its priority should be studied from the point of view of the consumer and by models and systems of monitoring of the status of the consumption of loads based on the forecast production and stored energy.

You must be able to operate with a variable topology to optimize the distribution of energy, guaranteeing the autonomy of the energy needed by the consumption and optimizing its use.

Main conclusions extracted from the development of this work are:

- The sustainable development of countries with low electrification rates based on HRES is possible and should be accompanied by national and regional energy directives. Enhancing HRES in remote communities will provide energy access to isolated areas, promoting the economic development of these communities without increasing CO2 emissions.
- HRES modelling and simulation of off-grid HRES configurations may be optimized in a first stage based on the electricity load, climatic data sources, and economics of the power components in which the NPC has to be minimized to select an economic feasible power system. However, in a second stage is crucial to gather the opinion of local experts and local residents.
- The process of decision taken in selecting feasible energy projects is a complex problem that needs to be tackled from different perspectives (technical, economic, social, environmental and political). In this regard, the use of multi-criteria methods is essential for the success and continuity of the project in the future.
- State of the art revealed an important effort carried out by the scientific community to study isolated areas electrification, mainly based on decentralized energy generation, but an integral methodology for the optimization of this electrification based on HRES had not been yet published.
- This integral methodology should optimize the quality of energy services to non-connected areas from three main approaches: technical, economical, and sustainability, assessing the potential for energy generation with renewable energies, evaluating possible hybrid configurations, together with demand side management strategies.
- In addition, the methodology should analyse the available resources with potential for demand response actions for distributed generation and energy storage needs, and integrate qualitative assessment of social and political aspects of the implementation of HRES in the isolated community using the Analytic Hierarchy Process. The methodology has been developed in this thesis and used, as a test case of its applicability and example of the procedure to follow in its application, to a local community located in the Democratic Republic of Congo. Two possible HRES configurations were experimentally simulated in a significant scale in LABDER laboratory to demonstrate its feasibility and reliability behaviour.
- Electrification of rural areas will remain a challenge for governments in developing countries like Republic Democratic of Congo. Comparing renewable energy systems with conventional generation units powered by fuel may not be cost effective yet, but the necessity of environmental protection and global tendency toward more environmental friendly habits will derive in a decrement of fossil fuels and its substitution by renewable energy technologies.

## ACKNOWLEDGMENT

This work has been conducted within the framework of the two research project: "Monitoring, control and discrimination of electrical loads connected to renewable generation hybrid systems in configuration on island", within the programme for the promotion of the scientific research, technological development and innovation in the Comunitat Valenciana [2017/647] funded by the Conselleria de Educación, Investigación, Cultura y Deporte of the GeneralitatValenciana and reference number: GV/2017/023; and research project "Investigación experimental de la integración de sistemas renovables híbridos para un desarrollo sostenible de países emergentes" funded by Universidad Politècnica de València, under grant Primeros Proyectos de Investigación (PAID-06-18), Vicerrectorado de Investigación, Innovación y Transferencia de la Universitat Politècnica de València (UPV).

## REFERENCES

Anderson, P. M., & Mirheydar, M. (1992, May). An Adaptive method for Setting under frequency Load Shedding Relays. *IEEE Transactions on Power Systems*, *7*(2), 647–655. doi:10.1109/59.141770

Bahramara, S., ParsaMoghaddam, M., & Haghifam, M.R. (2016). Optimal planning of hybrid renewable energy systems using HOMER: A review. *Renewable & Sustainable Energy Reviews*, *62*, 609–620.

Benavente, F., Lundblad, A., Campana, P. E., Zhang, Y., Cabrera, S., & Lindbergh, G. (2019). Photovoltaic/battery system sizing for rural electrification in Bolivia: Considering the suppressed demand effect. *Applied Energy*, *235*, 519–528. doi:10.1016/j.apenergy.2018.10.084

BMVBS. (n.d.). Final Report - Global and Regional Spatial Distribution of Biomass Potentials - Status quo and options for specification. Available at: https://www.dbfz.de/web/fileadmin/user_upload/DBFZ_Reports/DBFZ_Report_7.pdf

Cárcel Carrasco, F. J., Peñalvo López, E., & de Murga, G. (2018). Oficinas auto-sostenibles para las agencias de ayuda internacional en zonas geográficas remotas. *Dyna (Bilbao)*, *93*(3), 272–277. doi:10.6036/8507

Chang, Y. H., & Yeh, C. H. (2001). Evaluating airline competitiveness using multi-attribute decision making. *Omega*, *29*(5), 405–415. doi:10.1016/S0305-0483(01)00032-9

Cherni, J., Dayner, I., Henao, F., Jaramillo, P., & Smith, R. (2007). Energy supply for sustainable rural livelihoods. A multi-criteria decision support. *Energy Policy*, *35*(3), 1493–1502. doi:10.1016/j.enpol.2006.03.026

Connolly, D., Lund, H., Mathiesen, B. V., & Leahy, M. (2010). A review of computer tools for analysing the integration of renewable energy into various energy systems. *Applied Energy*, *87*(4), 1059–1082. doi:10.1016/j.apenergy.2009.09.026

Delgado-Antillón, C. P., & Domínguez-NavarroJ, A. (2018). Probabilistic siting and sizing of energy storage systems in distribution power systems based on the islanding feature. *Electric Power Systems Research*, *155*, 225–235. doi:10.1016/j.epsr.2017.10.013

Dufo-López, R., Cristóbal-Monreal, I. R., & Yusta, J. M. (2016, August 1). Optimisation of PV-wind-diesel-battery stand-alone systems to minimise cost and maximise human development index and job creation. *Renewable Energy*, *94*(Supplement C), 280–293. doi:10.1016/j.renene.2016.03.065

FAO. (n.d.a). Democratic Republic of Congo. Bioenergy and Food Security Projects (BEFS). Available at: http://www.fao.org/energy/36342-0d2a826525757566a86e626cfa3de1fb0.pdf

FAO. (n.d.b). State of the World's Forest. Available at: http://www.eip.gov.eg/Upload/Publications/i2000e.pdf

Giallanza, A., Porretto, M., Puma, G. L., & Marannano, G. (2018). A sizing approach for stand-alone hybrid photovoltaic-wind-battery systems: A Sicilian case study. *Journal of Cleaner Production, 199*, 817–830. doi:10.1016/j.jclepro.2018.07.223

Gómez, F., Valcuende, M., Matzarakis, A., & Cárcel Carrasco, J. (2018). Design of natural elements in open spaces of cities with a Mediterranean climate, conditions for comfort and urban ecology. *Environmental Science and Pollution Research International, 25*(26), 26643–26652. doi:10.1007/s11356-018-2736-1 PubMed

Grubb., Edmonds, J., ten Brink, P., & Morrison, M. (1993). The cost of limiting fossil-fuel CO2 emissions: A survey and analysis. Ann Rev Energy Environ, Ann Rev, California., 18(1), 397–478. doi:10.1146/annurev.eg.18.110193.002145

Huld, T., Müller, R., & Gambardella, A. (2012). A new solar radiation database for estimating PV performance in Europe and Africa. Solar Energy, 2012(86), 1803–1815. doi:10.1016/j.solener.2012.03.006

IEA. (n.d.a). Africa Energy Outlook. World Energy Outlook Special Report. OECD/IEA.

IEA. (n.d.b). Country Statistics. https://www.iea.org/statistics/statisticssearch/report/?country=CONGOREP&product=balances&year=2014

Kristin, D. (2012). Demand response in an isolated system with high wind integration. Power Systems. *IEEE Transactions on, 27*(1), 20–29.

Loken, E. (2007). Use of multicriteria decision analysis methods for energy planning problems. *Renewable & Sustainable Energy Reviews, 11*(7), 1584–1595. doi:10.1016/j.rser.2005.11.005

Lujano-Rojas, J. M., Monteiro, C., Dufo-López, R., & Bernal-Agustín, J. L. (2012, June). Optimum residential load management strategy for real time pricing (RTP) demand response programs. *Energy Policy, 45*, 671–679. doi:10.1016/j.enpol.2012.03.019

NREL. (2009). *HOMER the Micro-Power Optimization Model; ver.2.68Beta*. NREL.

Onyeji, I. (2014) Harnessing and Integrating Africa's Renewable Energy Resources. Renewable Energy Integration, 27-38.

Peñalvo López, E. (2017a). Metodología de Evaluación y Optimización de Sistemas Renovables Híbridos para Electrificación de Zonas Aisladas de la Red (Doctoral dissertation). UPV.

Peñalvo-López, E., & Cárcel-Carrasco, F. J. (2019). An Analysis about Learning to Increase Women's Participation and Employment in Europe's Energy Transition: Evidence from the European Project MEnS. *Sustainability, 11*(16), 4345. doi:10.3390/su11164345

Peñalvo-López, E., Cárcel-Carrasco, F. J., Devece, C., & Morcillo, A. I. (2017b). A Methodology for Analysing Sustainability in Energy Scenarios. *Sustainability, 1590*(9), •••. doi:10.3390/su9091590

Peñalvo López, E., Pérez-Navarro, A., Hurtado Pérez, E., & Cárcel Carrasco, J. (2019). Comprehensive Methodology for Sustainable Power Supply in Emerging Countries. *Sustainability*, *11*, 1–22. doi:10.3390u11195398

Pérez-Navarro, A. (2016). Experimental verification of hybrid renewable systems as feasible energy sources. *Renewable Energy*, *86*, 384–391.

Pohekar, S. D., & Ramachandran, M. (2004). Application of multi-criteria decision-making to sustainable energy planning–a review. *Renewable & Sustainable Energy Reviews*, *8*(4), 365–381. doi:10.1016/j.rser.2003.12.007

Polatidis, H. (2006). Selecting an appropriate multi-criteria decision analysis technique for renewable energy planning. Energy Sources Part B Vol. doi:10.1080/009083190881607

Rullo, P., Braccia, L., Luppi, P., Zumoffen, D., & Feroldi, D. (2019). Integration of sizing and energy management based on economic predictive control for standalone hybrid renewable energy systems. *Renewable Energy*, *140*, 436–451. doi:10.1016/j.renene.2019.03.074

Saaty, R. W. (1987). The analytic hierarchy process–what it is and how it is used. *Mathematical Modelling*, *9*(3-5), 161–176. doi:10.1016/0270-0255(87)90473-8

Saaty, R. W. (1996). *Decision-making with dependence and feedback: the analytic network process*. RSW Publications.

Schneider. (n.d.). www.schneiderelectric.com

Sinha, S. S. C., & Chandel, S. S. (2014, April 1). Review of software tools for hybrid renewable energy systems. *Renewable & Sustainable Energy Reviews*, *32*(Supplement C), 192–205. doi:10.1016/j.rser.2014.01.035

Stecher, K., Brosowski, A., & Thrän, D. (n.d.). Biomass Potential in Africa. IRENA. Available at: http://www.irena.org/DocumentDownloads/Publications/IRENA-DBFZ_Biomass%20Potential%20in%20Africa.pdf

Šúri, M., Huld, T. A., Dunlop, E. D., & Ossenbrink, H. A. (2007). Potential of solar electricity generation in the European Union member states and candidate countries. Solar Energy, 2007(81), 1295–1305. doi:10.1016/j.solener.2006.12.007

Taliotis, C., Bazilian, M., Welsch, D. G., & Howells, M. (2014). Grand Inga to power Africa: Hydropower development scenarios to 2035. *Energy Strategy Reviews*, *4*, 1–10.

Wang, M. (2010). The comparison between MAUT and PROMETHEE. IEEE international conference on industrial engineering and engineering management (IEEM), 753–757. doi:10.1016/j.esr.2014.01.001

Weather, M. E. T. E. O. B. L. U. E. (n.d.). Weather history Kinsasa. https://www.meteoblue.com/en/weather/archive/export/kinsasa_dr-congo_8325634

World Bank. (n.d.). Deforestation Trends in the Congo Basin: Reconciling Economic Growth and Forest Protection. Working Paper 5: Wood-based Biomass Energy. Author.

## KEY TERMS AND DEFINITIONS

**Electric Load Control:** A remotely controllable switch that can turn power to a load or appliance on or off. Such a device could also be used to regulate the amount of power that a load can consume.

**Electric Microgrids:** Microgrids are electricity distribution systems containing loads and distributed energy resources, (such as distributed generators, storage devices, or controllable loads) that can be operated in a controlled, coordinated way either while connected to the main power network or while is landed.

**Energy Planning:** Energy planning has a number of different meanings. However, one common meaning of the term is the process of developing long-range policies to help guide the future of a local, national, regional, or even the global energy system.

**Energy Scenarios:** Energy scenarios provide a framework for exploring future energy perspectives, including various combinations of technology options and their implications.

**Renewable Energy:** It is energy produced from sources that do not deplete or can be replenished within a human's lifetime. The most common examples include wind, solar, geothermal, biomass, and hydropower.

**Renewable Hybrid Systems:** A hybrid energy system, or hybrid power, usually consists of two or more renewable energy sources used together to provide increased system efficiency as well as greater balance in energy supply.

# Chapter 15
# An Investigation for CA-Based PageRank Validation in View of Power-Law Distribution of Web Data to Enhance Trustworthiness and Safety for Green Cloud

**Arnab Mitra**
https://orcid.org/0000-0002-7382-089X
*Siksha 'O' Anusandhan (Deemed), India*

## ABSTRACT

*An investigation on cellular automata (CA)-based validation of PageRank with reference to the power-law distribution is presented in this chapter to enhance the trustworthiness and safety of Clouds. Web traffic data for several Clouds were analyzed in view of power-law distribution to explore whether they are natural or manmade. Results obtained for CA-based PageRank validation were compared with Alexa®, which further supported the power-law distribution with some cut-offs, which ensured effectiveness and accuracy for CA-based validation of PageRank. Hence, this exploration helps to enhance the trustworthiness for any computed PageRank, and thus, it helps for the automated decision-making process towards an enhanced trust and safety of Clouds.*

## 1 INTRODUCTION

Automated Decision making has a major impact in today's technology-enhanced life: starting from finance sector (e.g., online loan approval) to assessments in online recruitment scenarios (in "https://ico.org.uk/for-organisations/guide-to-data-protection/guide-to-the-general-data-protection-regulation-gdpr/automated-decision-making-and-profiling/what-is-automated-individual-decision-making-and-

DOI: 10.4018/978-1-7998-3246-1.ch015

profiling/", accessed on September 10, 2019). Thus, one can easily recognize that the impact of WWW (World Wide Web). It is not difficult to realize that on-line resources are playing a crucial role in the development of our knowledge as they offer easy and mobility access to resources. The resources available at several web servers are often considered as Clouds (Mell et al., 2011). Hence, "health information on the World Wide Web is of variable quality, methods are needed to assist consumers to identify health websites" (Griffiths et al., 2005).

In an online knowledge acquisition scenario, a user intended to collect information over the Clouds often requires the help of a search engine to scroll down the available relevant Clouds (web servers). The search engine provides an index of sorted Clouds based on information relevancy with respect to the user's query. The search engine generated index of Clouds is commonly known as PageRank-ed list. By convention, low the index number refers to high the PageRank. Several dynamic and static PageRank computing algorithms have been offered by researchers and practitioners to improve the trustworthiness of computed PageRank (Pasquinelli, 2009; Kundu et al., 2006; Guha et al., 2015). Though, the major emphasis is found to be present towards the computations of PageRank (Pasquinelli, 2009; Kundu et al., 2006; Guha et al., 2015 and 2016), validations of computed PageRank are also found to be present towards the enhanced reliability and safety (Mitra et al., 2015 and 2017; Mitra, 2019). We found that PageRank validation may result to enhance the reliability of computed PageRank and further may enhance web security (Mitra, 2019).

In this regard, we feel that it is not out of scope to mention that advancements in today's Cloud Computing technologies have further transformed the earlier days' Clouds into the Green Clouds (Baliga et al., 2011; Berl et al., 2010; and Patel et al., 2015). Thus, with synchronization with emerging Green Clouds, an emphasis towards energy-based PageRank calculation (Guha et al., 2015) and modeling of low energy consuming PageRank validation (Mitra et al., 2017 and Mitra, 2019) have also been presented by researchers in recent times.

On the other hand, Power-Law is known as a functional relationship among two measures, where a change in one measure consequences in a proportional adjustment in the other, and it is independent of the first size of those measures (i.e., one measure differs as a power of another). Power-Law distributions may be observed in several natural systems; it shows the capability of a system towards self-organizing when it is continually pushed to a disordered state by any "natural perturbations (Barabási 2003)" (Logan et al., 2013). Further, it may be noted that finest diversity, which is acute for resilience, may also be found from the Power-Law patterns (Logan et al., 2013).

For this reason, Power-Law has appealed for certain interest over years, which often leads to the exploration of amazing physical consequences, "and for its appearance in a diverse range of natural and man-made phenomenon" (Clauset et al., 2009). As described in (Clauset et al., 2009), an amount $x$ conforms a Power-Law if it is drawn from a probability distribution $p(x) \propto x^{-\alpha}$, where ± is known as exponent / scaling constant. The exponent / scaling parameter ( $\alpha$ ) typically follows the range $2 < \alpha < 3$, although there might be occasionally exceptions. In run-through, only a few empirical phenomena follow Power-Laws for all values of $x$. Most frequently, the Power-Law is applicable for values greater than some maximum of $x_{min}$. In those scenarios, it is considered that the conclusion of the distribution confirms a Power-Law (Clauset et al., 2009).

Several cases for natural and man-made scenarios were analyzed and presented with reference to Power-Law distribution in (Clauset et al., 2009). Besides different uses of Power-Law distribution targeting Cloud and or IoT were presented by several researchers over time (Girau, 2017; Darwish et al., 2018;

and Xu et al., 2019). Unfortunately, we find PageRank validation was not examined. In our finding, an analysis based on Power-Law distribution of web data related to the validation of PageRank may also help and strengthen the validation process. Thus, it may enhance the reliability and further may result for enhanced web / Cloud security. For this reason, we have carried out an analysis towards Cellular Automata (CA) (Chaudhuri et al., 1997) based PageRank validation in view of Per-Law distribution and have reported the same in our presented research. We found that CA based models potentially fit towards a low cost and low power consuming physical modeling in Cloud environments (Mitra et al., 2015 and 2017; and Mitra, 2019). In this regard it is important to mention that CA is a dynamic modeling tool towards complex problems; it evolves in discrete space and time using CA rule(s). Basic description of CA generated transition diagram is briefly described in following Section 3.

A brief discussion on CA as we found from (Chaudhuri et al., 1997) is described next. Although we request that the reader(s) may go through a book (Chaudhuri et al., 1997) to know more about CA basics and its dynamics.

**CA -** Cells of the CA configuration are set together in form of a lattice, and evolves using CA rule over discrete space and discrete time. CA evolution may be realized over multi-dimensional CA configurations. ECA (Elementary Cellular Automata) is known as simple CA configuration where the CA configuration is achieved with $3$ -cells only at a linear configuration. Brief discussion on ECA configuration is followed next.

**ECA configuration -** ECA configuration is stated as periodic boundary (first and last cell of the CA configuration are connected to each other) or, null boundary (first and last cell of the CA configuration are linked to ground) configuration.

In presented chapter, we have strictly followed null boundary ECA configuration due to its low-cost and easy physical implementation capabilities.

**ECA rule -** Next state of a cell in ECA configuration is achieved based on some functions, which are referred to as ECA rule (total $256$ ECA rules, starting from Rule $0$ to Rule $255$ ; also known as Wolfram Code).

The significant points of our research as presented in this chapter are followed.

i. it presents a theoretical base in the direction for Green Computing of CA based PageRank validation;
ii. it initiates an analysis on PageRank validation with reference to Power-Law distribution;
iii. it helps to validate the results obtained with CA based PageRank validation with reference to the Power-Laws distribution;
iv. it facilitates an autonomous decision-making process about the safety and reliability of a Cloud with reference to the dynamic web traffics.

Rest of the chapter organization is as followed. Past significant works are presented in Section 2; background works are briefly mentioned in Section 3; proposed investigation is presented in Section 4; experimental results along with related analysis are mentioned in Section 5; finally, conclusions are drawn in Section 6.

## 2 RELATED WORKS

As presented in (Mell et al., 2011), Cloud Computing is a highly emerging research area. Further, we found that the information related to the wellbeing of any website (i.e. Cloud) is of variable quality (Griffiths et al., 2005). In our studies we found that there exists a scope for a number of research areas to enhance the existing Clouds. Among several others, enactments in PageRank computation is found to be one of the important and promising research areas in Cloud computing. A number of dynamic and static PageRank algorithms were introduced by researchers to enhance the ranking of Clouds (Pasquinelli, 2009; Kundu et al., 2006; and Guha et al., 2015). A few attempts were also found towards the validation of calculated PageRank (Mitra et al., 2015 and 2017; and Mitra, 2019). In our research we have chosen the validation of calculated PageRank, which is relatively new area of research, and have a great potential to enhance the reliability, safety, and security of Clouds (Mitra et al., 2017; and Mitra, 2019). CA based validation of PageRank is considered here as it is already mentioned in Section 1 that CA is capable towards the physical modeling of low power consuming, low-cost dynamic and complex models (Chaudhuri et al., 1997; Mitra et al., 2015 and 2017; and Mitra, 2019). A brief description on the detailed energy and power consumption by the CA physical components is presented next.

It was presented in (Chaudhuri et al., 1997) that CA based models may be realized with $D$-flip flops (FF) at very low-cost. Hence a brief report on power consumptions by the different types of D-flip flops as presented in (Mitra, 2019) based on (Oskuii, 2004; and Mehta et al., 2011), is followed next. The energy – delay space comparison result for eight different FFs in $0.13\mu m$ $CMOS$ technology determined that the transmission gate (TG) FFs were superior with reference to the power performance, although the pulse triggered FFs were faster with a comparison to the TG FFs. Further a comparative result about the power consumption comparison among '*DT–TG*', '8*T*', and '*MT CMOS*' FFs, confirmed that the maximum power consumptions happened for the '*DT–TG*' FFs, and the least power consumptions were found for the '*MT CMOS*' FFs. The power ingestions for '*DT–TG*' flip-flop at 1.0 *vdd* were 1.20*E*–05 watt (*W*), and at 2.0 *vdd* were 1.20*E*–05 *W*; further, the power consumptions for '*MT CMOS*' flip-flop at 1.0 *vdd* were 1.17*E*–07 *W*, and at $2.0 v$dd were $1.12E-06W$ (in simulation based on the $BSIM\ V390mm$ technology based tanner $EDA$ tool) (Mehta et al., 2011). Thus, it ensured that very low power consuming CA based models are potentially capable for uses towards Green Clouds. Literatures related to CA-based PageRank validation as we found in our studies, are briefly presented in Section 3.

## 3 BACKGROUND

CA based modeling of PageRank validation with $3$-CA cells in null boundary configuration were first introduced in (Mitra et al., 2015). Further, an energy-based modeling of PageRank validation was presented in (Mitra et al., 2017), which was further enhanced into an energy-stability based modeling of PageRank validation (Mitra, 2019). It was examined in (Mitra et al., 2017) that the alteration of PageRank for any Cloud is directly proportional to the alteration of energy for that particular Cloud. Thereafter in (Mitra, 2019), CA based transition diagrams were modeled into signals and it was further explored that for a Cloud the "change in stability index is directly proportional to the change in PageRank" (Mitra, 2019). CA generated transition diagrams towards different scenarios for PageRank validation as described in (Mitra et al., 2015 and 2017; and Mitra, 2019), are once again presented in following Figure 1. Detailed CA based modeling of PageRank validation may be found in (Mitra et al., 2015).

Let us follow the convention to understand the transition diagram as presented in

*Figure 1. Transition diagrams for CA based validation of PageRank in different page traffic scenarios*

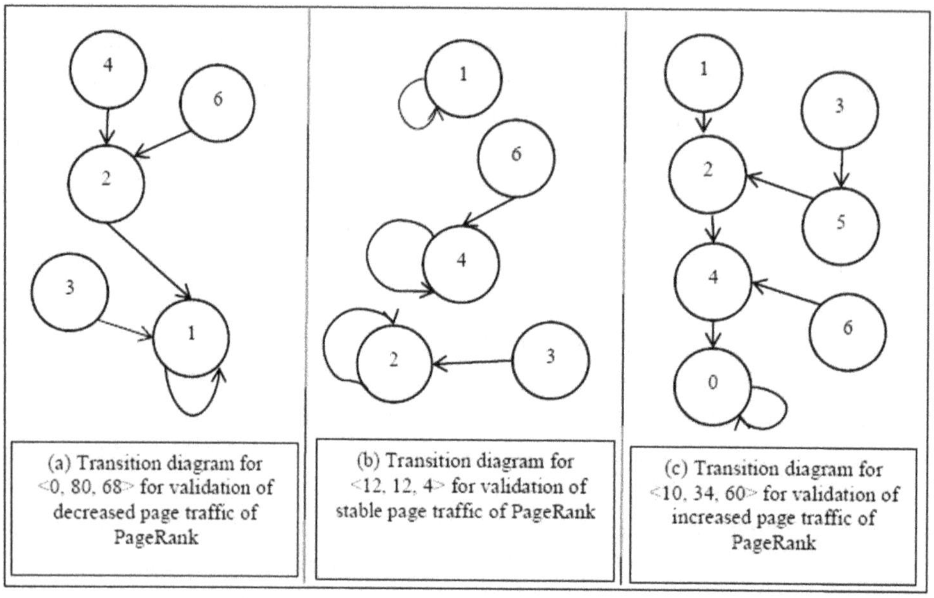


Figure 1. Circles indicate states, transitions are indicated by arrows; directs the collection of CA rules (Wolfram codes), those are in charge of the generation of the transition diagrams (Mitra et al., 2015 and 2017; and Mitra, 2019).

In Figure 1, no states denoting state $0$, state $5$ and state $7$ are present as they were found to be invalid (refer detailed design of (Mitra et al., 2015)) with reference to the validation of PageRank in CA.

The roles of energy and energy-stability for the validation of PageRank were earlier investigated in (Mitra et al., 2017; and Mitra, 2019), and it was concluded that CA-based model in PageRank validation is potentially fit towards Green Cloud. Hence, further energy-based investigation is beyond the scope of this research. Here, the CA based validation of PageRank is investigated through Power-Law distributions to ensure the validation of PageRank for Clouds, which is presented in following Section 3.

## 3 PROPOSED APPROACH

Proposed system flowchart is presented in following Figure 2.

Let us follow the convention for Figure 2: dotted bold line / $N$ - indicates the path followed by being negative (no) and continuous bold line / $Y$ - indicates the path followed by being positive (yes) at each checking (diamond box).

The proposed flowchart as presented in Figure 2 emphasizes on the uses of historical traffic data for any Cloud. Historical data over longer period helps towards the validation for change of PageRank of any Cloud with precision. A final checking is applied based on the Power-Law distributions to validate or, invalidate the change of web traffic for any Cloud. Thus, the additional checking strengthens the trustworthiness of any Cloud, which empowers safe uses and enhanced security. Algorithm 1 was designed to facilitate the proposed system flow.

*Figure 2. Proposed system for Power-Law distribution on web data towards CA-based validation on PageRank*

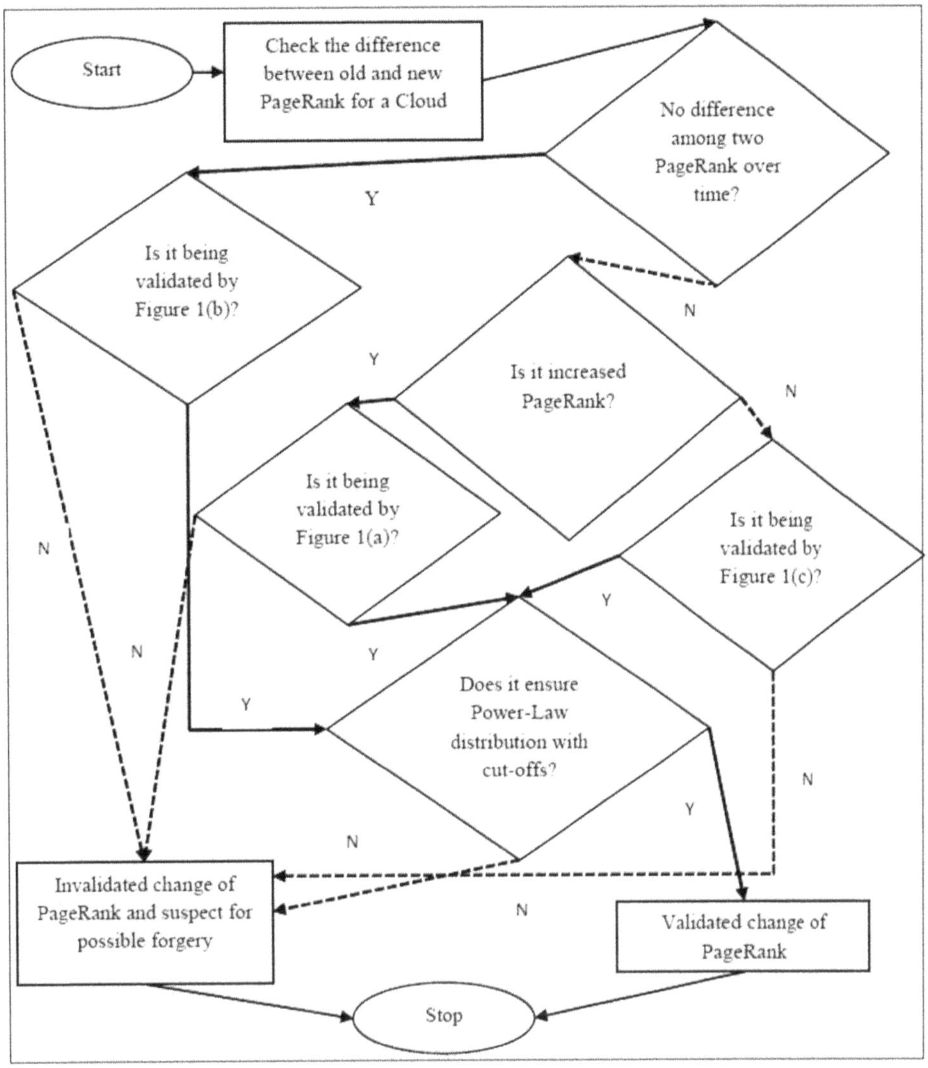


## Algorithm 1. Validation for PageRank changes using Power-Law distribution

Input: web traffic data.
Output: validated change of PageRank.
Step 1. Start
Step 2. Compute change_of_PageRank
Step 3. If change_of_pageRank == 0, follow Step 6 else Step 11
Step 4. Else if change_of_pageRank < 0, follow Step 7 else Step 11
Step 5. Else if change_of_pageRank > 0, follow Step 8 else Step 11
Step 6. If is it supported by CA based design of Figure 1 (b), follow Step 9 else Step 11
Step 7. If is it supported by CA based design of Figure 1 (c), follow Step 9 else Step 11

Step 8. If is it supported by CA based design of Figure 1 (a), follow Step 9 else Step 11
Step 9. If is it supported by Power_Law distribution, follow Step 10, else Step 11
Step 10. Validate change of PageRank
Step 11. Invalidate change of PageRank and suspect for possible manipulation in present PageRank
Step 12. Stop

## 4 RESULT ANALYSIS

Real time traffic data for numerous Clouds were collected and analyzed. Though, several Clouds were considered, only a single instance of a Cloud is presented in following Table 1. Each Cloud in Table 1 was chosen carefully so that the Cloud properly represents each different possible scenario (i.e., increased, decreased and static (no changed) of PageRank change over time (Mitra et al., 2015 and 2017; and Mitra, 2019)). We collected data over span of more than one year. Partial data of Table 1 representing the web traffic data of 2018 for mentioned URL (Uniform Resource Locator), were also used for data analysis in (Mitra, 2019). More detailed description about those traffic data may be found in (Mitra, 2019). With reference to the focus of presented research, we only used a simple form of traffic data. Data shown in Table 1 were obtained with Alexa® (in "https://www.alexa.com/toolbar", accessed on January 27, 2018 and September 12, 2019). Once again please recall that the high values in PageRank indicate low PageRank (Kundu et al., 2006; Pasquinelli, 2009; Guha et al., 20155 and 2016; Mitra et al., 2015 and 2017; and Mitra, 2019). As per suggestions from the PageRank validation model in (Mitra et al., 2019), the following parameters e.g., bounce rate, page-views per day per user, span of visiting (duration), number of in-links and average load time were only considered for monitoring and analyzing the change of web-traffic (if any).

The changes for PageRank as observed in Table 1 are graphically presented in following Figure 3.

In (Clauset et al., 2009), it was examined that the "hits to websites" (Clauset et al., 2009) and "links to websites" (Clauset et al., 2009) support for Power-Law distribution, with cut-offs. Further form our studies (Clauset et al., 2009; and Chaudhuri et al., 1997), we concluded that, the support towards Power-Law distribution leads to an identification as natural phenomenon or man-made (i.e., manipulation). Thus, we conclude that the "hits to websites" (Clauset et al., 2009) and "links to websites" (Clauset et al., 2009) which supports Power-Law distribution with cut-offs, may be considered as natural rather than being manipulated.

The data from Table 1 were further analyzed and compared. It is obvious from Table 2 that the bounce rate (same as hits to the website) and in-links play crucial role towards the stability of PageRank; thus, they also play crucial roles in the validation of PageRank (Mitra et al., 2017; and Mitra, 2019). Hence, we found that, validated PageRank change (in all different scenarios) will also support the Power-Law distribution with cut-offs. The detailed comparison is presented in Table 2.

Thus, from Table 2, it is observed that Power-Law distributions with some cut-offs were supported at the time of validation of the PageRank at each different traffic scenarios.

*Table 1. Traffic data in real time for three different Clouds at three different PageRank scenarios using Alexa®*

| URLs | Data collected on January 27, 2018 (Mitra, 2017) | | Data collected on September 12, 2019 | | Remark |
|---|---|---|---|---|---|
| | *Traffic data* | *PageRank* | *Traffic data* | *PageRank* | |
| "google.co.in" | Total 29.20% bounce rate | 8 | Total 34.8% bounce rate | 42 | Decreased PageRank |
| | On average 10.38-page views per day per user | | On average 7.150-page views per day per user | | |
| | Total 7m01s visit time | | Total 5m02s visit time | | |
| | Total 33889 in-links | | Total 18450 in-links | | |
| | Average load time 1.218s | | Average load time 1.472s | | |
| "youtube.com" | Total 57.90% bounce rate | 2 | Total 55.5% bounce rate | 2 | Stable (no change) PageRank |
| | On average 4.77-page views per day per user | | On average 5.66-page views per day per user | | |
| | Total 8m15s visit time | | Total 8m59s visit time | | |
| | Total 2186717 in-links | | Total 1677382 in-links | | |
| | Average load time 1.009s | | Average load time 2.883s | | |
| "springerlink.com" | Total 67.30%bounce rate | 567 | Total 66.9% bounce rate | 483 | Increased PageRank |
| | On average 2.12-page views per day per user | | On average 2.10-page views per day per user | | |
| | Total 2m17s visit time | | Total 2m29s visit time | | |
| | Total 67575 in-links | | Total 38280 in-links | | |
| | Average load time 1.802s | | Average load time 1.644s | | |

## 5 CONCLUSION

It is observed that the Power-Law distribution with some cut-off was supported at the time of validation of the PageRank at each different traffic scenarios (refer Table 2 of Section 4). Hence, it is concluded that, if the changes (if any) found in PageRank for any Cloud is being capable to be validated by CA-based PageRank validation model, then it will also support the Power-Law distribution with some cut-offs. For this reason, it further ensures that the changes in PageRank (if any) validated by CA-based model indicates that it is natural rather than being caused by an intentional manipulation. Additionally, it is found CA based PageRank validation is an efficient choice towards modeling of Green Cloud (Mitra et al., 2017; and Mitra, 2019). Hence, we found that CA-based PageRank validation process along with an additional analysis using Power-Law distribution is efficient enough to decrease the uncertainty and risk associated with the choices for correct selection between options. Thus, the Power-Law distributions-based analysis using CA based validation of changed PageRank is a green approach and may further capable to automatically ensure the enhanced reliability and safety of Clouds at reduced uncertainty and risk.

*Figure 3. Changed PageRank for different Clouds over time*

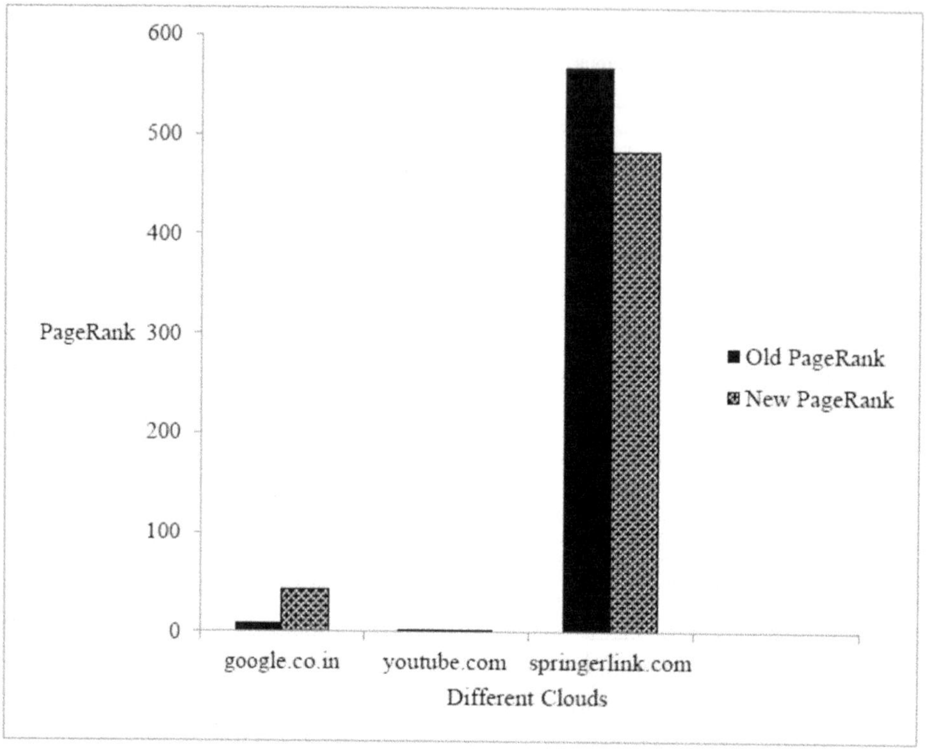


*Table 2. PageRank validation at different traffic scenarios and remarks*

| URLs | Remark on the stability of PageRank as collected from Alexa® (refer Figure 3) | Support for Power-Law distribution | CA based validation for PageRank | Remarks about the validation of PageRank change (if any) |
|---|---|---|---|---|
| "youtube.com" | Stable | Yes, with cut-offs (Clauset et al., 2009). | Yes, it is capable of being validated by transition diagram of Figure 1(b). | No suspicious behavior/ abnormality is observed. Thus, it may be recommended as safe Cloud. |
| "springerlink.com" | Increases in PageRank | Yes, with cut-offs (Clauset et al., 2009). | Yes, it is capable of being validated by transition diagram of Figure 1(c). | No suspicious behavior/ abnormality is observed. Thus, it may be recommended as safe Cloud. |
| "google.co.in" | Decreases in PageRank | Yes, with cut-offs (Clauset et al., 2009). | Yes, it is capable of being validated by transition diagram of Figure 1(a). | No suspicious behavior/ abnormality is observed. Thus, it may be recommended as safe Cloud. |

## ACKNOWLEDGMENT

Author sincerely acknowledges the helpful review comments received from the anonymous reviewers, those have further helped to strengthen the quality and readability of the chapter.

## REFERENCES

Alexa® Internet Tool. (n.d.). https://www.alexa.com/toolbar

Baliga, J., Ayre, R. W., Hinton, K., & Tucker, R. S. (2011). Green cloud computing: Balancing energy in processing, storage, and transport. *Proceedings of the IEEE*, *99*(1), 149–167. doi:10.1109/JPROC.2010.2060451

Berl, A., Gelenbe, E., Di Girolamo, M., Giuliani, G., De Meer, H., Dang, M. Q., & Pentikousis, K. (2010). Energy-efficient cloud computing. *The Computer Journal*, *53*(7), 1045–1051. doi:10.1093/comjnl/bxp080

Chaudhuri, P. P., Chowdhury, D. R., Nandi, S., & Chattopadhyay, S. (1997). *Additive cellular automata: theory and applications* (Vol. 1). John Wiley & Sons.

Clauset, A., Shalizi, C. R., & Newman, M. E. (2009). Power-law distributions in empirical data. *SIAM Review*, *51*(4), 661–703. doi:10.1137/070710111

Darwish, A., & Hassanien, A. E. (2018). Cyber physical systems design, methodology, and integration: The current status and future outlook. *Journal of Ambient Intelligence and Humanized Computing*, *9*(5), 1541–1556. doi:10.100712652-017-0575-4

Girau, R. (2017). *Architectural and Application Solutions for The Cloud Internet of Things* (Doctoral Thesis). University of Cagliari. Retrieved on January 16, 2020 from https://iris.unica.it/retrieve/handle/11584/249555/303095/tesi%20di%20dottorato_Roberto%20Girau.pdf

Griffiths, K. M., Tang, T. T., Hawking, D., & Christensen, H. (2005). Automated assessment of the quality of depression websites. *Journal of Medical Internet Research*, *7*(5), e59. doi:10.2196/jmir.7.5.e59 PMID:16403723

Guha, S. K., Kundu, A., & Dattagupta, R. (2016). Energy Stability in Cloud for Web Page Ranking. In *Proceeding of the International Conference on Information Systems Design and Intelligent Applications*. Springer India. 10.1007/978-81-322-2752-6_61

Guha, S. K., Kundu, A., & Duttagupta, R. (2015). Energy Based Web Page Ranking. *International Journal of Artificial Life Research*, *5*(1), 27–40. doi:10.4018/IJALR.2015010102

Kundu, A., Dutta, R., & Mukhopadhyay, D. (2006, December). An Alternate Way to Rank Hyper-linked Web-Pages. In *9th International Conference on Information Technology (ICIT'06)* (pp. 297-298). IEEE. 10.1109/ICIT.2006.26

Logan, L. H., Karlsson, E. M., Gall, H. E., Park, J., Emery, N., Owens, P., Niyogi, D., & Rao, P. S. C. (2013). Freshwater Wetlands: Balancing Food and Water Security with Resilience of Ecological and Social Systems. In Climate Vulnerability: Understanding and Addressing Threats to Essential Resources (pp. 105-116). Elsevier.

Mehta, K., Arora, N., & Singh, B. P. (2011). Low power efficient D flip flop circuit. In *International Symposium on Devices MEMS, Intelligent Systems & Communication (ISDMISC)* (pp. 16-19). IJCA.

Mell, P., & Grance, T. (2011). *The NIST definition of cloud computing*. NIST.

Mitra, A. (2019). On investigating energy stability for Cellular Automata based PageRank validation model in Green Cloud. *International Journal of Cloud Applications and Computing*, *9*(4), 66–85. doi:10.4018/IJCAC.2019100104

Mitra, A., & Kundu, A. (2015). Page Ranking Validation using Cellular Automata in Cloud. *International Journal of Cloud Applications and Computing*, *5*(3), 1–19. doi:10.4018/IJCAC.2015070101

Mitra, A., & Kundu, A. (2017). Energy Efficient CA based Page Rank Validation Model: A Green Approach in Cloud. *International Journal of Green Computing*, *8*(2), 59–76. doi:10.4018/IJGC.2017070104

Online study material: What is automated decision making and profiling. (n.d.). Accessed from https://ico.org.uk/for-organisations/guide-to-data-protection/guide-to-the-general-data-protection-regulation-gdpr/automated-decision-making-and-profiling/what-is-automated-individual-decision-making-and-profiling/

Oskuii, S. T. (2004). *Comparative study on low-power high-performance flip-flops* (Unpublished Master of Technology Thesis). Linköping University. Retrieved May 15, 2017, from http://www.diva-portal.org/smash/get/diva2:19406/FULLTEXT01.pdf

Pasquinelli, M. (2009). Google's PageRank algorithm: A diagram of cognitive capitalism and the rentier of the common intellect. *Deep search: The politics of search beyond Google*, 152-162.

Patel, Y. S., Mehrotra, N., & Soner, S. (2015, February). Green cloud computing: A review on Green IT areas for cloud computing environment. In *Futuristic Trends on Computational Analysis and Knowledge Management (ABLAZE), 2015 International Conference on* (pp. 327-332). IEEE. 10.1109/ABLAZE.2015.7155006

Xu, L. D., & Duan, L. (2019). Big data for cyber physical systems in industry 4.0: A survey. *Enterprise Information Systems*, *13*(2), 148–169. doi:10.1080/17517575.2018.1442934

# Compilation of References


Abdel-Maksoud, A., Elbanna, S., Mahama, H., & Pollanen, R. (2015). The use of performance information in strategic decision making in public organizations. *International Journal of Public Sector Management*, *28*(7), 528–549.

AckerR. (2020). https://www.infosecurity-magazine.com/blogs/big-data-big-risks/

Ackert, S. P. (2010). Basics of aircraft maintenance programs for financiers—evaluation & insights of commercial aircraft maintenance programs. www. aircraftmonitor. com/uploads/1/5/9/9/15993320/basics_of_aircraft_maintenance_programs_for_ financiers___v1. pdf

Adwait, D., Saily, K., Jayant, G., & Vivek, K. (2018). Design and evaluation of a Lean Manufacturing framework using Value Stream Mapping (VSM) for a plastic bag manufacturing unit. Materials Today: Proceedings, 5(2), 7668–7677. doi:10.1016/j.matpr.2017.11.442

Ain, Q., Iqbal, S., Khan, S., Malik, A., Ahmad, I., & Javaid, N. (2018). IoT Operating System Based Fuzzy Inference System for Home Energy Management System in Smart Buildings. *Sensors (Basel)*, *18*(9), 2802. doi:10.339018092802 PMID:30149631

Alexa® Internet Tool. (n.d.). https://www.alexa.com/toolbar

Allahverdi, A., & Aldowaisan, T. (2002). New heuristic to minimize total completion time in m-machine flowshops. *International Journal of Production Economics*, *11*(1), 77–83. doi:10.1016/S0925-5273(01)00228-6

Almeida, A. T., & Bohoris, G. A. (1995). Decision theory in maintenance decision making. *Journal of Quality in Maintenance Engineering*, *1*(1), 39–45. doi:10.1108/13552519510083138

Al-Najjar, B., & Alsyouf, I. (2003). Selecting the most efficient maintenance approach using fuzzy multiple criteria decision making. *International Journal of Production Economics*, *84*(1), 85–100. doi:10.1016/S0925-5273(02)00380-8

Al-Najjar, B., & Kans, M. (2006). A model to identify relevant data for problem tracing and maintenance cost-effective decisions. A case study. *International Journal of Productivity and Performance Management*, *55*(8), 616–637. doi:10.1108/17410400610710170

Al-Najjar, G. (1999). Economic criteria to select a cost-effective maintenance policy. *Journal of Quality in Maintenance Engineering*, *5*(3), 236–247. doi:10.1108/13552519910282692

Alsyouf, I. (2007). The role of maintenance in improving companies, productivity and profitability. *International Journal of Production Economics*, *105*(1), 70–78. doi:10.1016/j.ijpe.2004.06.057

Alter, S., & Sherer, S. A. (2004). A General, But Readily Adaptable Model of Information System Risk. *Communications of the Association for Information Systems*, *14*(1). Advance online publication. doi:10.17705/1CAIS.01401

Anderson, P. M., & Mirheydar, M. (1992, May). An Adaptive method for Setting under frequency Load Shedding Relays. *IEEE Transactions on Power Systems, 7*(2), 647–655. doi:10.1109/59.141770

Animah, I., & Shafiee, M. (2019). Maintenance strategy selection for critical shipboard machinery systems using a hybrid AHP-PROMETHEE and cost benefit analysis: A case study. *Journal of Marine Engineering and Technology*, 1–12. do i:10.1080/20464177.2019.1572705

Aqlan, F., & Al-Fandi, L. (2018). Prioritizing process improvement initiatives in manufacturing environments. *International Journal of Production Economics, 196*, 261–268. Advance online publication. doi:10.1016/j.ijpe.2017.12.004

Arbonies, A., & Calzada, I. (2004). El poder del conocimiento tácito: Por encima del aprendizaje organizacional. *Intangible Capital, 6*(0), 1–17.

Arens, A., Elder, R., & Beasley, M. (2014). *Auditing and assurance services-An integrated approach; includes coverage of international standards and global auditing issues, in addition to coverage of.* Aufl.

Armstrong, J. S. (Ed.). (2001). *Principles of forecasting: a handbook for researchers and practitioners* (Vol. 30). Springer. doi:10.1007/978-0-306-47630-3

Asuquo, M. P., Wang, J., Zhang, L., & Phylip-Jones, G. (2019). Application of a multiple attribute group decision making (MAGDM) model for selecting appropriate maintenance strategy for marine and offshore machinery operations. *Ocean Engineering, 179*, 246–260. doi:10.1016/j.oceaneng.2019.02.065

Aven, T., & Ylönen, M. (2016). Safety regulations: Implications of the new risk perspectives. *Reliability Engineering & System Safety, 149*, 164–171. doi:10.1016/j.ress.2016.01.007

Azadivar, F., & Shu, V. (1999). Maintenance policy selection for JIT production systems. *International Journal of Production Research, 37*(16), 3725–3738. doi:10.1080/002075499190013

Bäcklund, J., Blom, A. F., & Beevers, C. J. (1982). *Fatigue thresholds, Fundamental and Engineering Applications* (vols. 1-2). Engineering Materials Advisor Services Ltd.

Baez, J. (2007). *Investigación cualitativa*. Esic Editorial.

Bahoque, E., Gómez, O., & Pietrosemoli, L. (2007). Gestión del Conocimiento en la Industria de la Construcción: Estudio de un caso. Revista Venezolana de Gerencia (RVG). *Año, 12*(39), 393–409.

Bahramara, S., ParsaMoghaddam, M., & Haghifam, M.R. (2016). Optimal planning of hybrid renewable energy systems using HOMER: A review. *Renewable & Sustainable Energy Reviews, 62*, 609–620.

Baker, K. R. (1974). *Introduction to Sequencing and Scheduling*. John Wiley & Sons.

Baliga, J., Ayre, R. W., Hinton, K., & Tucker, R. S. (2011). Green cloud computing: Balancing energy in processing, storage, and transport. *Proceedings of the IEEE, 99*(1), 149–167. doi:10.1109/JPROC.2010.2060451

Ballis, A., Karapetis, F., & Ballis, T. (2017). Towards the Implementation of Optimal Train Loading Plan in the Athens-Thessaloniki Freight Services. *International Journal of Operations Research and Information Systems, 8*(2), 78–92. doi:10.4018/IJORIS.2017040104

Bana e Costa, C. A., & Chagas, M. P. (2002). *A career choice problem: An example of how to use MACBETH to build a quantitative value model based on qualitative value judgments*. Working Paper LSEOR 02.53 ISBN 0 7530 1517, The London School of Economics and Political Science.

Bana e Costa, C. A., De Corte, J. M., & Vansnick, J. C. (2006). On the Mathematical Foundation of MACBETH. In J. Figueira, S. Greco & M. Ehrgott (Eds.), Multicriteria Decision analysis: State of the art surveys (pp. 409-444). New York, NY: Springer.

Bana e Costa, C. A., Carnero, M. C., & Oliveira, M. D. (2012). A multi-criteria model for auditing a Predictive Maintenance Programme. *European Journal of Operational Research, 217*(2), 381–393. doi:10.1016/j.ejor.2011.09.019

Bana e Costa, C. A., Da Costa-Lobo, M. L., Ramos, I. A., & Vansnick, J. C. (2002). Multicriteria Approach for Strategic Town Planning: The Case of Barcelos. *International Journal of Management Sciences, 30*(3), 227–242.

Bana e Costa, C. A., De Corte, J. M., & Vansnick, J. C. (2011). MACBETH (Measuring Attractiveness by a Categorical Based Evaluation Technique). In J. J. Cochran, L. A. Cox Jr, P. Keskinocak, J. P. Kharoufeh, & J. C. Smith (Eds.), *Encyclopedia of Operations Research and Management Science*. John Wiley & Sons. doi:10.1002/9780470400531.eorms0970

Bana e Costa, C. A., Nunes da Silva, F., & Vansnick, J. C. (2001). Conflict dissolution in the public sector: A case-study. *European Journal of Operational Research, 130*(2), 388–401. doi:10.1016/S0377-2217(00)00037-0

Bana e Costa, C. A., Oliveira, C. S., & Vieira, V. (2008). Priorization of bridges and tunnels in earthquake risk mitigation using multicriteria decision analysis: Application to Lisbon. *Omega, 36*(3), 442–450. doi:10.1016/j.omega.2006.05.008

Bana e Costa, C. A., & Vansnick, J. C. (1994). MACBETH - An interactive path towards the construction of cardinal value functions. *International Transactions in Operational Research, 1*(4), 489–500. doi:10.1016/0969-6016(94)90010-8

Bana e Costa, C. A., & Vansnick, J. C. (1999). The MACBETH approach: basic ideas, software and an application. In N. Meskens & M. Roubens (Eds.), *Advances in Decision Analysis* (pp. 131–157). Kluwer Academic Publishers. doi:10.1007/978-94-017-0647-6_9

Banegas, J. R., & Artalejo, R. (2000). Popper and the Problem of Induction in Epidemiology. *Revista Espanola de Salud Publica, 74*(4). Advance online publication. doi:10.1590/S1135-57272000000400003 PMID:11031841

Bang-Jensen, J., Gutin, G., & Yeo, A. (2004). When the greedy algorithm fails. *Discrete Optimization, 1*(2), 121–127. doi:10.1016/j.disopt.2004.03.007

Banks, J. C., Reichard, K. M., Hines, J. A., & Brought, M. S. (2008). Platform degrader analysis for the design and development of vehicle health management systems. *Proceedings of the International Conference on Prognostics and Health Management*, 464. 10.1109/PHM.2008.4711468

Baranova, I. V., Parfenenko, Yu. V., & Nenja, V. G. (2011). Complex mathematical model for computer calculation of delivery and heat distribution in pipeline system. *IEEE 6th International Conference on Intelligent Data Acquisition and Advanced Computing Systems: Technology and Applications, IDAACS 2011, 2*, 556–559. 10.1109/IDAACS.2011.6072828

Barlow, R. E., Hunter, L. C., & Proschan, F. (1963). Optimum checking procedures. *Journal of the Society for Industrial and Applied Mathematics, 11*(4), 1078–1095. doi:10.1137/0111080

Barragán, A. (2009). Aproximación a una taxonomía de modelos de gestión del conocimiento. *Intangible Capital, 5*(1), 65–101. doi:10.3926/ic.2009.v5n1.p65-101

Barsom, J. M. (1971). Fatigue-Crack Propagation in Steels of Various Yield Strengths, *Transactions of the ASME Journal of Engineering for Industry, Series B, 93(4).* 10.1115/1.3428061

Barsom, J. M. (1974). *Fatigue Behaviour of Pressure-Vessel Steels. WRC Bulletin, 194.* Welding Research Council.

Barsom, J. M., & Rolfe, S. T. (1977). *Fracture and Fatigue Control in Structures. Applications of Fracture Mechanics.* Prentice-Hall, Inc.

Bashiri, M., Badri, H., & Hejazi, T. H. (2011). Selecting optimum maintenance strategy by fuzzy interactive linear assignment method. *Applied Mathematical Modelling*, *35*(1), 152–164. doi:10.1016/j.apm.2010.05.014

Bauer, W., Ganschar, O., & Gerlach, S. (2014). Development of a Method for Visualization and Evaluation of Production. *Proceedings of the 47th CIRP Conference on Manufacturing*, 17, 481-486.

Beasley, J. E. (1990). An algorithm for set covering problems. *European Journal of Operational Research*, *31*(1), 85–93. doi:10.1016/0377-2217(87)90141-X

Ben Houria, Z., Besbes, M., Elaoud, B., Masmoudi, M., & Masmoudi, F. (2015). Maintenance strategy selection for medical equipments using fuzzy multiple criteria decision making approach. *Proceedings - CIE 45: 2015 International Conference on Computers and Industrial Engineering.*

Ben Houria, Z., Masmoudi, M., Al Hanbali, A., Khatrouch, I., & Masmoudi, F. (2016). Quantitative techniques for medical equipment maintenance management. *European Journal of Industrial Engineering*, *10*(6), 703–723. doi:10.1504/EJIE.2016.081017

Benavente, F., Lundblad, A., Campana, P. E., Zhang, Y., Cabrera, S., & Lindbergh, G. (2019). Photovoltaic/battery system sizing for rural electrification in Bolivia: Considering the suppressed demand effect. *Applied Energy*, *235*, 519–528. doi:10.1016/j.apenergy.2018.10.084

Berl, A., Gelenbe, E., Di Girolamo, M., Giuliani, G., De Meer, H., Dang, M. Q., & Pentikousis, K. (2010). Energy-efficient cloud computing. *The Computer Journal*, *53*(7), 1045–1051. doi:10.1093/comjnl/bxp080

Bertolini, M., & Bevilacqua, M. (2006). A combined goal programming-AHP approach to maintenance selection problem. *Reliability Engineering & System Safety*, *91*(7), 839–848. doi:10.1016/j.ress.2005.08.006

Bevilacqua, M., & Braglia, M. (2000). The analytic hierarchy process applied to maintenance strategy selection. *Reliability Engineering & System Safety*, *70*(1), 71–83. doi:10.1016/S0951-8320(00)00047-8

Bianchini, A., Pellegrini, M., & Rossi, J. (2019). Maintenance scheduling optimization for industrial centrifugal pumps. *International Journal of System Assurance Engineering and Management*, *10*(4), 848–860. doi:10.100713198-019-00819-4

Biery, K. J. (2006). Aligning an information risk management approach to BS 7799-3:2005. SANS Reading Room.

Bjerga, T., & Aven, T. (2016). Some perspectives on risk management: A security case study from the oil and gas industry. *Proceedings of the Institution of Mechanical Engineers, Part O: Journal of Risk and Reliability*, *230*(5), 512–520. doi:10.1177/1748006X16654589

BMVBS. (n.d.). Final Report - Global and Regional Spatial Distribution of Biomass Potentials - Status quo and options for specification. Available at: https://www.dbfz.de/web/fileadmin/user_upload/DBFZ_Reports/DBFZ_Report_7.pdf

Boland, P. J., & Proscan, F. (1982). Periodic Replacement with Increasing Minimal Repair Cost at Failure. *Operations Research*, *30*(6), 1083–1089. doi:10.1287/opre.30.6.1183

Boschetti, M., & Maniezzo, V. (2015). A set covering based matheuristic for a real-world city logistics problem. *International Transactions in Operational Research*, *22*(1), 169–195. doi:10.1111/itor.12110

Braglia, M., Carmignani, G., & Zammori, F. (2006). A new value stream mapping approach for complex production systems. *International Journal of Production Research*, *44*(18-19), 3929–3959. doi:10.1080/00207540600690545

Brandão, A., Belisário, G. C., Tomazini, I. M., & Gomes, K. Q. (2017). Análise bibliométrica da literatura sobre métodos multicritério de auxílio à tomada de decisão aplicados a problemas de PCP. *Brazilian Journal of Production Engineering*, *3*(2), 128–140. http://periodicos.ufes.br/BJPE

British Standard Institute (BSI), British Standard – ISO 31000:2010. (n.d.). *Risk management. Principles and guidelines.* BSI.

British Standard Institute (BSI). British Standard – BS25999-1:2006: Business Continuity Management – Part 1: Code of practice. Available at http://www.bsigroup.com

British Standard Institute (BSI). British Standard – BS25999-2:2007: Business Continuity Management – Part 2: Specification. Available at http://www.bsigroup.com

Broek, D. (1982). *Elementary Engineering Fracture Mechanics.* Kluwer Academic Publishers., doi:10.1007/978-94-011-9055-8

Brunt, D. (2000). From current state to future state: Mapping the steel to component supply chain. *International Journal of Logistics: Research and Applications, 3*(3), 259–271. doi:10.1080/713682765

Bueno, E. (1999). *Gestión del conocimiento y capital intelectual' Experiencias en España.* Comunidad de Madrid- Euroforum Escorial Madrid.

Bueno, E. (2000). *La dirección del conocimiento en el proceso estratégico de la empresa: información, complejidad e imaginación en la espiral del conocimiento.* Euroforum Escorial.

Bueno, E. (2002a). Enfoques principales en Dirección del Conocimiento (Knowledge Management) y tendencias. In R. Hernández (Ed.), *Gestión del Conocimiento: Desarrollos teóricos y aplicaciones.* Ediciones La Coria, Fundación Xavier de Salas.

Bueno, E. (2002b). La sociedad del conocimiento: un nuevo espacio de aprendizaje de las personas y organizaciones. In *La Sociedad del Conocimiento.* Monografía de la Revista Valenciana de Estudios Autonómicos.

Buhari, M., & Adamu, S. (2012). Short-Term Load Forecasting Using Artificial Neural Network. *Proceedings of the International MultiConference of Engineers and Computer Scientist.* Retrieved from http://www.iaeng.org/publication/IMECS2012/IMECS2012_pp83-88.pdf

Bush, R. E., & Bale, C. S. E. (2014). Realising the Social Benefits of District Heating Throught Strategic Planning. *Proceeding of the 14th International Symposium on District Heating and Cooling.* Retrieved from http://eprints.whiterose.ac.uk/81735/

Buttery, L.M. (1978, Jan.). New survey of U.S. Maintenance Costs. Hydrocarbon Processing.

Campbell, H. G., Dudek, R. A., & Smith, M. L. (1970). A heuristic algorithm for the n-job m-machine sequencing problem. *Mngt Sci, 16*(10), B630–B637. doi:10.1287/mnsc.16.10.B630

Canterla, J. P. (2016). Methodology for the Development of a Quality Management System, in accordance with the new ISO 9001:2015 Standard, integrated with a Risk Management System (ISO Standard 31000: 2010), in Companies depicted to Transport of Perishable Goods by Road. Degree project for the master's in industrial management (MOIGE). University of Seville, Spain.

Cao, Q., Giustozzi, F., Zanni-Merk, C., de Bertrand de Beuvron, F., & Reich, C. (2019). Smart Condition Monitoring for Industry 4.0 Manufacturing Processes: An Ontology-Based Approach. *Cybernetics and Systems, 50*(2), 82–96. Advance online publication. doi:10.1080/01969722.2019.1565118

Caralli, R. A., Stevens, J. F., Young, L. R., & Wilson, W. R. (2007). *Introducing octave allegro: Improving the information security risk assessment process* (No. CMU/SEI-2007-TR-012).

Cárcel Carrasco, F. J. (2014b). La gestión del conocimiento en la ingeniería del mantenimiento industrial: Investigación sobre la incidencia en sus actividades estratégicas. Omnia Science. doi:10.3926/oms.197

Cárcel Carrasco, J. (2014d). Planteamiento de un modelo de mantenimiento industrial basado en técnicas de gestión del conocimiento. Omnia Science, 2014. Advance online publication. doi:10.3926/oms.198

Cárcel Carrasco, F. J., & Grau, J. (2012c). Aspectos estratégicos del mantenimiento industrial relativos a la eficiencia energética. *Revista Mantenimiento, 257*, 25–30.

Cárcel Carrasco, J., Peñalvo López, E., & de Murga, G. (2018). Oficinas auto-sostenibles para las agencias de ayuda internacional en zonas geográficas remotas. *Dyna (Bilbao)*, *93*(3), 272–277. doi:10.6036/8507

Cárcel Carrasco, J., Rodríguez Méndez, M., & Carnero Moya, M. C. (2016c). Impact of the knowledge management in maintenance engineering. Effects on industrial production. In *Optimum Decision Making in Asset Management*. IGI Global.

Cárcel-Carrasco, F. J., Pascual-Guillamón, M., Solano García, L., Salas Vicente, F., & Pérez-Puig, M. A. (2019b). Pitting Corrosion in AISI 304 Rolled Stainless Steel Welding at Different Deformation Levels. Applied Sciences (Basel, Switzerland), 9(16), 3265. doi:10.3390/app9163265

Cárcel-Carrasco, F. J., Pascual-Guillamón, M., Salas-Vicente, F., & Donderis-Quiles, V. (2019a). Influence of Heat Treatment in the Microstructure of a Joint of Nodular Graphite Cast Iron when Using the Tungsten Inert Gas Welding Process with Perlitic Grey Cast Iron Rods as Filler Material. *Metals*, *9*(1), 48. doi:10.3390/met9010048

Carcel-Carrasco, F. J., & Rodriguez-Mendez, M. (2013a). *Industrial maintenance and tacit knowledge: An introduction about its incidence*. DYNA Management., doi:10.6036/MN5857

Cárcel-Carrasco, J., Pascual-Guillamón, M., & Pérez-Puig, M. (2016b). Effects of X-rays radiation on AISI 304 stainless steel weldings with AISI 316L filler material: A study of resistance and pitting corrosion behavior. *Metals*, *6*(5), 102. doi:10.3390/met6050102

Cárcel-Carrasco, J., Pérez-Puig, M. A., Pascual-Guillamón, M., & Pascual-Martínez, R. (2016a). An Analysis of the Weldability of Ductile Cast Iron Using Inconel 625 for the Root Weld and Electrodes Coated in 97.6% Nickel for the Filler Welds. *Metals*, *6*(11), 283. doi:10.3390/met6110283

Carman, C. M., & Katlin, J. M. (1966). *Low Cycle fatigue crack propagation of high- strength steels.* ASME paper N° 66-MET.

Carnero, C., & Conde, J. (2000a). Conflictos y utilidades en la relación Gran Empresa-Universidad en proyectos de implantación de programas de mantenimiento predictivo. *Proceedings del 1er Workshop of Industrial Organization.*

Carnero, C., & Conde, J. (2000b). Selección de parámetros de control y diagnóstico de fallos en la implantación de programas de Mantenimiento Predictivo. *Proceedings del IV Spanish Congress of Maintenance*, 395-402.

Carnero, C., López-Escobar, C., & Conde, J. (1998). Implantación de programas de mantenimiento predictivo por empresas de servicios técnicos en colaboración. Evaluación mediante indicadores. *Proceeding of VII International Congresso of AEDEM*, 522-527.

Carnero, M. C. (2016). Fuzzy Multicriteria Model for Selection of Vibration Technology. *Shock and Vibration, 2016.* . doi:10.1155/2016/6723605

Carnero, M. C. (2004). The control of the setting up of a predictive maintenance programme using a system of indicators. *Omega*, *32*(1), 57–75. doi:10.1016/j.omega.2003.09.009

Carnero, M. C. (2005). Selection of diagnostic techniques and instrumentation in a predictive maintenance program. A case study. *Decision Support Systems, 38*(4), 539–555. doi:10.1016/j.dss.2003.09.003

Carnero, M. C. (2006). An evaluation system of the setting up of predictive maintenance programmes. *Reliability Engineering & System Safety, 91*(8), 945–963. doi:10.1016/j.ress.2005.09.003

Carnero, M. C. (2007). Model for the Selection of Predictive Maintenance Techniques. *INFOR, 45*(2), 83–94. doi:10.3138/infor.45.2.83

Carnero, M. C., & Gómez, A. (2016). A multicriteria decision making approach applied to improving maintenance policies in healthcare organizations. *BMC Medical Informatics and Decision Making, 16*(1), 47. Advance online publication. doi:10.118612911-016-0282-7 PMID:27108234

Carnero, M. C., & Gómez, A. (2018). Optimization of Maintenance in Production and Storage Systems for Domestic Water. *Water Resources Management, 32*(1), 359–380. doi:10.100711269-017-1815-2

Carvalho, R., & Lourenço, J. C. (2002). A multicriteria model for assigning new orders to service suppliers. *European Journal of Operational Research, 139*(2), 390–399. doi:10.1016/S0377-2217(01)00367-8

Castanier, B., Grall, A., & Bérenguer, C. (2005). A condition-based maintenance policy with non-periodic inspections for a two-unit series system. *Reliability Engineering & System Safety, 87*(1), 109–120. doi:10.1016/j.ress.2004.04.013

Cavalcante, C. A., & Lopes, R. S. (2015). V. R. S. Multi-criteria model to support the definition of opportunistic maintenance policy: A study in a cogeneration system. *Energy, 80*, 32–40. doi:10.1016/j.energy.2014.11.039

Cegarra, J., & Moya, B. (2003). Orientadores del capital relacional. *Cuadernos Americanos, 16*(26), 79–97.

Cela, E., Deineko, V. G., & Woeginger, G. J. (2014). Linearizable special cases of the QAP. *Journal of Combinatorial Optimization*, 1–11.

Cervo, A. L., & Bervian, P. A. (2002). *Metodologia científica.* Editora Pearson Prentice Hall.

Chai, G., Forsman, T., Gustavsson, F., & Wang, C. (2015). Formation of the grained area in martensitic steel during very high cycle fatigue. *Fatigue & Fracture of Engineering Materials & Structures, 38*(11), 1315–1323. doi:10.1111/ffe.12345

Chai, G., Zhou, N., Ciurea, S., & Andersson, M. (2012). Author. *Scripta Materialia, 66*, 769–772. doi:10.1016/j.scriptamat.2012.02.003

Chaneski, W. (2014). Revisiting The Value Stream. Obtenido de Modern Machine Shop: https://www.mmsonline.com/columns/revisiting-the-value-stream

Chan, F. T. S., & Prakash, A. (2012). Maintenance policy selection in manufacturing firms using the fuzzy MCDM approach. *International Journal of Production Research, 50*(23), 7044–7056. doi:10.1080/00207543.2011.653451

Chang, D. Y. (1996). Applications of the extent analysis method on fuzzy AHP. *European Journal of Operational Research, 95*(3), 649–655. doi:10.1016/0377-2217(95)00300-2

Chang, Y. H., & Yeh, C. H. (2001). Evaluating airline competitiveness using multi-attribute decision making. *Omega, 29*(5), 405–415. doi:10.1016/S0305-0483(01)00032-9

Chan, K. Y., Kwong, C. K., & Dillon, T. S. (2012). *Computational Intelligence Techniques for New Product Design. Studies in Computational Intelligence - 403.* Springer., doi:10.1007/978-3-642-27476-3

Charmaz, K. (2006). Constructing grounded theory. A practical guide through qualitative analysis. *Sage (Atlanta, Ga.).*

*Charpy Impact Test on Metallic Materials- Test Method (V and U notches), Bs EN 10045-1*. (1990). British Standards Publication.

Chase, R. B., & Aquilano, N. J. (1978). *Gestión de la producción y dirección de operaciones*. Hispano Europea.

Chaudhuri, P. P., Chowdhury, D. R., Nandi, S., & Chattopadhyay, S. (1997). *Additive cellular automata: theory and applications* (Vol. 1). John Wiley & Sons.

Chemweno, P., Pintelon, L., Van Horenbeek, A., & Muchiri, P. (2015). Development of a risk assessment selection methodology for asset maintenance decision making: An analytic network process (ANP) approach. *International Journal of Production Economics*, *170*, 663–676. doi:10.1016/j.ijpe.2015.03.017

Chen, W. C., Wang, L. Y., & Lin, M. C. (2015). A Hybrid MCDM Model for New Product Development: Applied on the Taiwanese LiFePO Industry. Mathematical Problems in Engineering, 2015, 15. doi:10.1155/2015/462929

Cheong, K. F. R., & Tsui, E. (2010a). The roles and values of personal knowledge management: An exploratory study. *VINE Journal of Information and Knowledge Management Systems*, *40*(2), 204–227.

Cheong, K. F. R., & Tsui, E. (2010b). Exploring the synergy between business process management and personal knowledge management. *Cutter IT Journal*, *23*(5), 28–33.

Cherni, J., Dayner, I., Henao, F., Jaramillo, P., & Smith, R. (2007). Energy supply for sustainable rural livelihoods. A multi-criteria decision support. *Energy Policy*, *35*(3), 1493–1502. doi:10.1016/j.enpol.2006.03.026

Chong, A. K. W., Mohammed, A. H., Abdullah, M. N., & Rahman, M. S. A. (2019). Maintenance prioritization – a review on factors and methods. *Journal of Facilities Management*, *17*(1), 18–39. doi:10.1108/JFM-11-2017-0058

Chramcov, B., Baláte, Ja., & Princ, M. (2010). Heat Demand Forecasting for Concrete District Heating System. *International Journal of Mathematical Models and Methods in Applied Sciences*, *4*(4), 231–239.

Christer, T. (1981). Ways of assessing and improving maintenance performance. PEMEC'81.

Chua, A. Y. K., & Goh, D. H. (2008). Untying the knot of knowledge management measurement: A study of six public service agencies in Singapore. *Journal of Information Science*, *34*(3), 259–274. doi:10.1177/0165551507084139

Clark, J. R. (1970). *Fatigue Crack Growth Characteristics of Heavy Section. ASTM A533 Grade B Class 1 Steel Weldments. ASME Paper N° 70-PVP 24*. American Society of Mechanical Engineers.

Clark, W. G., & How, J. R. (1974). *August. Fatigue Crack Initiation and Growth Properties Affect the Material Selection and Design Criteria*. Metals Engineering Quarterly.

Clauset, A., Shalizi, C. R., & Newman, M. E. (2009). Power-law distributions in empirical data. *SIAM Review*, *51*(4), 661–703. doi:10.1137/070710111

Clintock, M. C. (1963). On the Plasticity of the Growth of Time Fatigue Cracks. In *Fracture of Solids* (p. 65). Willey-Interscience.

Coates, V., Farooque, M., Klavans, R., Lapid, K., Linstone, H. A., Pistorius, C., & Porter, A. L. (2001). On the future of technological forecasting. *Technological Forecasting and Social Change*, *67*(1), 1–17. doi:10.1016/S0040-1625(00)00122-0

Coelho, L. C., & Laporte, G. (2015). Classification, models and exact algorithms for multi-compartment delivery problems. *European Journal of Operational Research*, *242*(3), 854–864. doi:10.1016/j.ejor.2014.10.059

Colino, J., Martinez, J., & Martínez-Carrasco, F. (2010). La gestión de la innovación en la industria. El caso de la región de Murcia. *Economía Industrial*, *377*, 150–158.

Colino, J., & Riquelme, P. (2000). Estructura industrial y desarrollo tecnológico en la región de Murcia. *Economía Industrial*, *335-336*, 171–183.

Conde, J. (1999). El Mantenimiento efectivo: principios y métodos. Working paper, GIO-0500-UCLM, Ciudad Real.

Conde, J., Conde, L. J., & Conde, M. (1999). Maintenance projects evaluation using Fully Bayesian Approach and Logistic Distribution. *Proceedings of International Conference on Safety and Reliability*, 755-759.

Connolly, D., Lund, H., Mathiesen, B. V., & Leahy, M. (2010). A review of computer tools for analysing the integration of renewable energy into various energy systems. *Applied Energy*, *87*(4), 1059–1082. doi:10.1016/j.apenergy.2009.09.026

Construction specification for fixed offshore structures in North Sea EEMUA 158. (2014). *Engineering Equipment and Materials Users Association, 158.*

Cook, S. A. (1971). The complexity of theorem-proving procedures. *Proceedings of the 3rd Annual ACM symposium on theory of computing*, 151-158. 10.1145/800157.805047

Cook, W. J., Cunningham, W. H., Pulleyblank, W. R., & Schrijver, A. (1998). *Combinatorial Optimization*. John Wiley & Sons.

Cooper, R. J., Silva, H. F., & Trevisan, R. E. (2004). Influencia del precalentamiento en las propiedades de uniones soldadas de acero API 5L-X80 soldadas con alambre tubular autoprotegido. *Rev. Metalurgia-Madrid, 40*(4), 280–270. doi:.2004.v40.i4.275 doi:10.3989/revmetalm

Cormen, T., Leiserson, C. E., & Rivest, R. L. (1990). *Introduction to Algorithms*. The MIT Press.

*COSO*. (2013). *Internal Control - Integrated Framework*. COSO.

Crespo, A. (2004). *Ingeniería de mantenimiento. Técnicas y métodos de aplicación a la fase operativa de los equipos*. AENOR.

Crespo, A., & Sánchez, A. (2002). Models for maintenance optimization: A study for repairable system and finite time periods. *Reliability Engineering & System Safety*, *75*(3), 367–377. doi:10.1016/S0951-8320(01)00131-4

Curtis, S. A. (2003). The classification of greedy algorithms. *Science of Computer Programming*, *49*(1-3), 125–157. doi:10.1016/j.scico.2003.09.001

Darnell, H., & Bert, A. (1978). *The role of maintenance management in achieving industrial efficiency, Maintenance'78*. BMCA.

Darwish, A., & Hassanien, A. E. (2018). Cyber physical systems design, methodology, and integration: The current status and future outlook. *Journal of Ambient Intelligence and Humanized Computing*, *9*(5), 1541–1556. doi:10.100712652-017-0575-4

Dataware Technologies. (1998). Seven Steps to Implementing Knowledge Management in Your Organization. Manual de trabajo.

Davoud Pour, H. (2001). A new heuristic for the n-job, m-machine flow-shop problem. *Production Planning and Control*, *12*(7), 648–653. doi:10.1080/09537280152582995

De Oliveira, R. I., Oliveira Sousa, S., & de Campos, F. (2019). Lean manufacturing implementation: Bibliometric analysis 2007–2018. *International Journal of Advanced Manufacturing Technology*, *101*(1-4), 979–988. doi:10.1007/s00170-018-2965-y

De Quevedo, E., De La Fuente, J., & Delgado, J. (2005). Reputación Corporativa y Creación de Valor. Marco Teórico De Una Relación Circular. *Investigaciones Europeas de Dirección y Economía de la Empresa, 11*(2), 81–97.

Del Moral, A. (2007). *Gestión del Conocimiento.* Thompson Editores.

Delgado-Antillón, C. P., & Domínguez-NavarroJ, A. (2018). Probabilistic siting and sizing of energy storage systems in distribution power systems based on the islanding feature. *Electric Power Systems Research, 155*, 225–235. doi:10.1016/j.epsr.2017.10.013

Demo, P. (2013). *Introdução à metodologia da ciência* (2nd ed.). Atlas.

Den Braber, F., Hogganvik, I., Lund, M. S., Stølen, K., & Vraalsen, F. (2007). Model-based security analysis in seven steps—A guided tour to the CORAS method. *BT Technology Journal, 25*(1), 101–117. doi:10.100710550-007-0013-9

Dhouib, D. (2014). An extension of MACBETH method for a fuzzy environment to analyze alternatives in reverse logistics for automobile tire wastes. *Omega, 42*(1), 25–32. doi:10.1016/j.omega.2013.02.003

Díaz-Bravo, L., Torruco García, U., Martínez Hernández, M., & Varela Ruiz, M. (2013). La entrevista, recurso flexible y dinámico. *Investigación en Educación Médica, 2*(7), 162–167. doi:10.1016/S2007-5057(13)72706-6

Digital Roadmap for District Heating & Cooling. (2019). DHC+ Technology Platform. Retrieved from https://www.euroheat.org/wp-content/uploads/2018/05/Digital-Roadmap_final.pdf

Dijkstra, E. W. (1959). A note on two problems in connexion with graphs. *Numerische Mathematik, 1*(1), 269–271. doi:10.1007/BF01386390

District Heating System Management Guide. (n.d.). Retrieved September 24, 2019, from https://projects.bre.co.uk/DHCAN/pdf/Management.pdf

Donelson, D., Ege, M., & McInnis, J. (2017). Internal Control Weaknesses and Financial Reporting Fraud. *Auditing: A Journal of Practice & Theory, 36*(3), 45-69.

Dorfman, P. (2001). The accidental knowledge manager. *Knowledge Management, 4*(2), 36–41.

Douglas, D. (2004). Grounded theory and the 'And' in entrepreneurship research. *Electronic Journal of Business Research Methods, 2*(2).

Dragović, I., Turajlić, N., Pilčević, D., Petrović, B., & Radojević, D. (2015). *A Boolean Consistent Fuzzy Inference System for Diagnosing Diseases and Its application for Determining Peritonitis Likelihood.* Academic Press.

Du, D. Z., & Pardalos, P. M. (Eds.). (2005). *Handbook of Combinatorial Optimization.* Springer-Verlag. doi:10.1007/b102533

Dufo-López, R., Cristóbal-Monreal, I. R., & Yusta, J. M. (2016, August 1). Optimisation of PV-wind-diesel-battery stand-alone systems to minimise cost and maximise human development index and job creation. *Renewable Energy, 94*(Supplement C), 280–293. doi:10.1016/j.renene.2016.03.065

Durmus, M. (2019). *Smart Predictive Maintenance: The Key to Industry 4.0.* Retrieved from https://www.aisoma.de/smart-predictive-maintenance-the-key-to-industry-4-0/

Dwinanto, I., Siregar, S., & Anitawati, A. (2019). Internal Control Over Financial Reporting, Organizational Complexity and Financial Reporting Quality. Int. *Journal of Economics and Management, 13*(2), 331–342.

Edmonds, J. (1971). Matroids and the greedy algorithm. *Mathematical Programming, 1*(1), 126–136. doi:10.1007/BF01584082

Edvinsson, L., & Malone, M. S. (1997). *Intellectual Capital: Realizing Your Company's True Value by Finding its Hidden Brainpower*. Harper Business.

Eguren Martí, M. L., & Castán Farrero, J. M. (2011). La cadena de valor bajo un enfoque sistémico y sus implicaciones en el análisis de la creación de valor en la empresa. In *XXI Congreso Nacional de la Asociación Científica de Economía y Dirección de la Empresa (ACEDE)*. Barcelona: ACEDE.

Eich, D. (2008). A Grounded Theory of High-Quality Leadership Programs: Perspectives From Student Leadership Development Programs in Higher Education. *Journal of Leadership & Organizational Studies*, *15*(2), 176–187. doi:10.1177/1548051808324099

Eisenhardt, K. (1989). Building theories from case studies research. *Academy of Management Review*, *14*(4), 532–550. doi:10.5465/amr.1989.4308385

El Haddad, M. H., Dowling, N. E., Topper, T. H., & Smith, K. N. (1980). J Integral Application for Short Fatigue Cracks at Notches. *International Journal of Fracture*, *16*(1), 15–30. doi:10.1007/BF00042383

Elber, W. (1971). *The Significance of Crack Closure. Damage Tolerance Aircraft Structure. ASTM STP 486*. American Society for Testing and Materials., doi:10.1520/STP26680S

Elices, M. (1992). *Mecánica De La Fractura Aplicada A Sólidos Elásticos Bidimensionales*. Departamento de Ciencias de Materiales E.T. Superior de Ingenieros de Caminos de Madrid.

Emblemsvag, J., & Tonning, L. (2003). Decision support in selecting maintenance organization. *Journal of Quality in Maintenance Engineering*, *9*(1), 1–24. doi:10.1108/13552510310466765

Emovon, I., Norman, R. A., & Murphy, A. J. (2018). Hybrid MCDM based methodology for selecting the optimum maintenance strategy for ship machinery systems. *Journal of Intelligent Manufacturing*, *29*(3), 519–531. doi:10.100710845-015-1133-6

ENORO Meter & Energy Data Management. (n.d.). Retrieved September 24, 2019, from https://www.enoro.fi/mdmandedm

Espín, P. (2014). *Sistema de gestión integral para empresas de abastecimiento de agua*. Universidad Politécnica de Cartagena.

Euroheat & Power Task Force Customer Installations. (2011). District Heating in Buildings. Retrieved from https://www.buildup.eu/sites/default/files/content/District%20Heating%20in%20buildings_final.pdf

Ewalds, H. L., & Wanhill, R. J. H. (1991). *Fracture Mechanics*. Edward Arnold Edition.

Fang, T., & Lahdelm, R. (2013). Uncertainty Analysis of District Heating State Estimation Model. *Proceedings of the IASTED International Conference Modelling, Identification and Control (MIC 2013)*, 380 – 387. 10.2316/P.2013.794-077

FAO. (n.d.a). Democratic Republic of Congo. Bioenergy and Food Security Projects (BEFS). Available at: http://www.fao.org/energy/36342-0d2a826525757566a86e626cfa3de1fb0.pdf

FAO. (n.d.b). State of the World's Forest. Available at: http://www.eip.gov.eg/Upload/Publications/i2000e.pdf

Febles, J., & Oreja, J. (2008). Factores externos e internos determinantes de la orientación de la cultura estratégica de las empresas. *Investigaciones Europeas de Dirección y Economía de la Empresa*, *14*(1), 13–32. doi:10.1016/S1135-2523(12)60009-4

Feng, M., Li, C., & McVay, S. (2009). Internal control and management guidance. *Journal of Accounting and Economics*, *48*(2-3), 190–209. doi:10.1016/j.jacceco.2009.09.004

FERMA. (2003). *Risk management standards*. Federation of European Risk Management Associations.

Fernandez, E., Avella, L., & Fernandez, M. (2006). Estrategia de Producción (2nd ed.). Madrid, España: Editorial McGraw Hill.

Fernández, J. C., & Canals, X. (2006). Quality standards for medical devices. In Proceedings of the XV Winter Course Standards in Telemedicine CATAI (pp. 133-140). Tenerife.

Ferrada, X., & Serpell, A. (2009). La Gestión del Conocimiento y la Industria de la Construcción. *Revista de la Construcción*, *8*(1), 46–58.

Feyerabend, P. (1975). *Against Method*. Verso.

FIERGS. (2015). *Caderno setorial Rio Grande do Sul metal mecânico*. FIERGS.

Flage, R., & Aven, T. (2015). Emerging risk – conceptual definition and a relation to black swan type of events. *Reliability Engineering & System Safety*, *144*, 61–67. doi:10.1016/j.ress.2015.07.008

Fleming, K. N. (2004). Markov models for evaluating risk-informed in-service inspection strategies for nuclear power plant piping systems. *Reliability Engineering & System Safety*, *83*(1), 27–45. doi:10.1016/j.ress.2003.08.009

Fontoura, M., Hoeflich, S. L., Dias, E. M., & Wee, H. M. (2016). A note on supply chain risk classification: Discussion and proposal. *International Journal of Production Research*, *54*(5), 1568–1569. doi:10.1080/00207543.2015.1067375

Forman, E., & Peniwati, K. (1998). Aggregating individual judgments and priorities with the Analytic Hierarchy Process. *European Journal of Operational Research*, *108*(1), 165–169. doi:10.1016/S0377-2217(97)00244-0

Forman, E., & Selly, M. A. (2001). *Decision by Objectives*. World Scientific. doi:10.1142/4281

Framinan, J. M., Gupta, J. N. D., & Leisten, R. (2004). A review and classification of heuristics for permutation flowshop schedul-ing with makespan objective. *J Opl Res Soc*, *55*(12), 1–13. doi:10.1057/palgrave.jors.2601784

Framinan, J. M., & Leisten, R. (2003). An efficient constructive heuristic for flowtime minimisation in permutation flow shops. *Omega Int J Mngt Sci*, *31*(4), 311–317. doi:10.1016/S0305-0483(03)00047-1

Framinan, J. M., Leisten, R., & Rajendran, C. (2003). Different initial sequences for the heuristic of Nawaz, Enscore, Ham to mini-mize makespan, idletime or flow-time in the static permutation flowshop sequencing problem. *International Journal of Production Research*, *41*(1), 121–148. doi:10.1080/00207540210161650

French, S. (1982). *Sequencing and scheduling: an introduction to the mathematics of the job-shop*. Ellis Horwood.

Freund, J., & Jones, J. (2014). *Measuring and managing information risk: a FAIR approach*. Butterworth-Heinemann.

From Industry 4.0 to Energy 4.0: Future Business Models and Legal Relations. (2016). Retrieved from http://www.germanenergyblog.de/wp-content/uploads/2016/04/160317_Lang_IBE_Bochum_From_Industry_4_0_to_Energy_4_0_fin.pdf

Frost. (1960). Notch Effects and the Critical Alternating Stress Required to Propagate a Crack in an Aluminium Alloy Subjected to Fatigue Loading. *Journal of the Mechanical Engineering Society*, *2*(2), 109-119.

Gabrielaitiene, I., Kacianauskas, R., & Sunden, B. (2003). Thermo-hydraulic Finite Element Modelling of District Heating Network by the Uncoupled Approach. *Journal of Civil Engineering and Management*, *IX*(3), 153–162. doi:10.1080/13923730.2003.10531321

Garey, M. R., & Johnson, D. S. (1979). *Computers and Intractability a Guide to the Theory of NP-Completeness*. WH Freeman & Company.

Garey, M. R., Johnson, D. S., & Sethi, R. (1976). The Complexity of Flowshop and Job-shop Scheduling. *Mathematics of Operations Research*, *1*(2), 117–129. doi:10.1287/moor.1.2.117

Ghazouani, M., Faris, S., Medromi, H., & Sayouti, A. (2014). Information Security Risk Assessment A Practical Approach with a Mathematical Formulation of Risk. *International Journal of Computers and Applications*, *103*(8), 36–42. doi:10.5120/18097-9155

Ghizoni Pereira, L., & Luz Tortorella, G. (2018). A Literature Review on Lean Manufacturing in Small Manufacturing Companies. In J. Davin (Ed.), *Progress in Lean Manufacturing (págs* (pp. 39–68). Springer., doi:10.1007/978-3-319-73648-8_3.

Ghosh, D., & Roy, S. (2009). A decision-making framework for process plant maintenance. *European Journal of Industrial Engineering*, *4*(1), 78–98. doi:10.1504/EJIE.2010.029571

Giallanza, A., Porretto, M., Puma, G. L., & Marannano, G. (2018). A sizing approach for stand-alone hybrid photovoltaic-wind-battery systems: A Sicilian case study. *Journal of Cleaner Production*, *199*, 817–830. doi:10.1016/j.jclepro.2018.07.223

Gibbs, M. T., & Browman, H. I. (2015). Risk assessment and risk management: A primer for marine scientists. *ICES Journal of Marine Science*, *72*(3), 992–996. doi:10.1093/icesjms/fsu232

Girau, R. (2017). *Architectural and Application Solutions for The Cloud Internet of Things* (Doctoral Thesis). University of Cagliari. Retrieved on January 16, 2020 from https://iris.unica.it/retrieve/handle/11584/249555/303095/tesi%20di%20dottorato_Roberto%20Girau.pdf

Glaser, B. G., & Strauss, A. L. (1967). *The discovery of grounded theory*. Aldine deGruyter.

Gomes, L., Araya, M., & Carignano, C. (2004). *Tomada de Decisões em Cenários Complexos: Introdução aos Métodos Discretos do Apoio Multicritério à Decisão*. Thomson.

Gómez de León, F. C., & Ruiz, J. J. (2006). Maintenance strategy based on a multicriterion classification of equipments. *Reliability Engineering & System Safety*, *91*(4), 444–451. doi:10.1016/j.ress.2005.03.001

Gómez, F., Valcuende, M., Matzarakis, A., & Cárcel Carrasco, J. (2018). Design of natural elements in open spaces of cities with a Mediterranean climate, conditions for comfort and urban ecology. *Environmental Science and Pollution Research International*, *25*(26), 26643–26652. doi:10.1007/s11356-018-2736-1 PubMed

Gómez, J. F., & Crespo, A. (2009). Framework for implementation of maintenance management in distribution network service providers. *Reliability Engineering & System Safety*, *94*(10), 1639–1649. doi:10.1016/j.ress.2009.04.003

Gondim, F. M. (2011). As ações da sustentabilidade empresarial como suporte à gestão de projetos. RESAC, sociedade, administração e contemporaneidade, 1(1).

González-Palma, R., Suárez-Bermejo, J. C., & Vicario, F. J., & Muñoz, A. (2006). Procesos metalúrgicos de aceros microaleados de alta resistencia 450 EMZ tipo II en la zona afectada térmicamente debido al efecto multipasada en soldaduras por arco sumergido. *Rev. Metal. Madrid*, *42*, 256–269.

González, R., & García, A. (2010). Innovación abierta: Un modelo preliminar desde la gestión del conocimiento. *Intangible Capital*, *7*(1), 82–115. doi:10.3926/ic.2011.v7n1.p82-115

Goosen, R., & Rudman, R. (2013). An integrated framework to implement it governance principles at a strategic and operational level for medium-to large-sized South African businesses. *International Business and Economics Research Journal, 12*(7).

Goossens, A. J. M., & Basten, R. J. I. (2015). Exploring maintenance policy selection using the Analytic Hierarchy Process; An application for naval ships. *Reliability Engineering & System Safety*, *142*, 31–41. doi:10.1016/j.ress.2015.04.014

Gosselin, S. R., & Fleming, K. N. (1997). Evaluation of pipe failure potential via degradation mechanism assessment. *Proceedings of the 5th International Conference on Nuclear Engineering (INCONE).*

Granderson, J., Piette, M. A., Ghatikar, G., & Price, P. (2009). *Building Energy Information Systems: State of the Technology and User Case Studies.* Retrieved from https://eta-publications.lbl.gov/sites/default/files/lbnl-2899e.pdf

Greenland, S. (1988). Probability versus Popper: an elaboration of the insufficiency of current popperian approaches for epidemiologic analysis. Chesnut Hill, MA: ERI.

Greer, R. W. (1960). *Records Make the Difference* (Vol. 65). Coal Age.

Griffiths, K. M., Tang, T. T., Hawking, D., & Christensen, H. (2005). Automated assessment of the quality of depression websites. *Journal of Medical Internet Research*, *7*(5), e59. doi:10.2196/jmir.7.5.e59 PMID:16403723

Grimaldi, R. P. (1989). *Discrete and Combinatorial Mathematics.* Addison-Wesley.

Grubb., Edmonds, J., ten Brink, P., & Morrison, M. (1993). The cost of limiting fossil-fuel CO2 emissions: A survey and analysis. Ann Rev Energy Environ, Ann Rev, California., 18(1), 397–478. doi:10.1146/annurev.eg.18.110193.002145

Grzenda, M. (2012). Consumer-oriented heat consumption prediction. *Control and Cybernetics*, *41*(1), 213–240.

Guerras, L., & Navas, J. (2015). *La dirección estratégica de la empresa. Teoría y aplicaciones.* Thomson Reuters.

Guha, S. K., Kundu, A., & Dattagupta, R. (2016). Energy Stability in Cloud for Web Page Ranking. In *Proceeding of the International Conference on Information Systems Design and Intelligent Applications.* Springer India. 10.1007/978-81-322-2752-6_61

Guha, S. K., Kundu, A., & Duttagupta, R. (2015). Energy Based Web Page Ranking. *International Journal of Artificial Life Research*, *5*(1), 27–40. doi:10.4018/IJALR.2015010102

Gutin, G. (2013). Section 4.6 traveling salesman problems. Handbook of Graph Theory, 336–359.

Gutin, G., Yeo, A., & Zverovich, A. (2002). *Traveling salesman should not be greedy: domination.* Academic Press.

Gutin, G., & Punnen, A. P. (Eds.). (2002). *The traveling salesman problem and its variations.* Kluwer Academic Publishers.

Gutin, G., & Punnen, A. P. (Eds.). (2002). *The Traveling Salesman Problem and Its Variations.* Kluwer Academic Publishers.

Gutin, G., & Yeo, A. (2002). Anti-matroids. *Operations Research Letters*, *30*(2), 97–99. doi:10.1016/S0167-6377(02)00117-7

Hamel, G., & Prahalad, C. (1989). Strategic intent. *Harvard Business Review*, 63–76. PMID:10303477

Hanks, H. (1961). Program for Maintenance of Mobile Equipment. *Min. Congress Joum.*, *47*(9), 35–38.

Hansen, M. T., Nohria, N., & Tierney, T. (1999, Mar.). What's your strategy for managing knowledge? Harvard Business Review, 106–116. PubMed

Hardy, M., & Bryman, A. (2004). *Handbook of data analysis.* Sage Publications., doi:10.4135/9781848608184.

Harrison, J. D. (1970, March). An Analysis of Data on non-Propagating Fatigue Cracks on a Fracture Mechanics Basis. *British Welding Journal*, *2*(3).

Hartung, P. (2018). *Link Between Usability Engineering and Risk Management*. https://www.regulatory-affairs.org/en/quality-management/news-page/link-between-usability-engineering-and-risk-management/

Haupt, R. (1989). A Survey of Priority-Rule Based Scheduling. *OR-Spektrum*, *11*(1), 3–16. doi:10.1007/BF01721162

Hedlud, O. (1994). A model of knowledge management and the N-form corporation. *Strategic Management Journal*, *15*(S2), 73–90. doi:10.1002/smj.4250151006

Heise, D., Strecker, S., & Frank, U. (2014). ControlML: A domain-specific modeling language in support of assessing internal controls and the internal control system. *The International Journal of Accounting*.

Hemmati, N., Galankashi, M. R., Imani, D. M., & Farughi, H. (2018). Maintenance policy selection: A fuzzy-ANP approach. *Journal of Manufacturing Technology Management*, *29*(7), 1253–1268. doi:10.1108/JMTM-06-2017-0109

Henriques de Gusmão, A. P., Mendonça Silva, M., Poleto, T., Camara e Silva, L., & Cabral Seixas Costa, A. P. (2018). Cybersecurity risk analysis model using fault tree analysis and fuzzy decision theory. *International Journal of Information Management*, *43*, 248–260. doi:10.1016/j.ijinfomgt.2018.08.008

Hernández, J., & Vizán, A. (2013). Lean Manufacturing Conceptos, Técnicas e implantación. Obtenido de https://www.eoi.es/es/savia/publicaciones/20730/lean-manufacturing-concepto-tecnicas-e-implantacion

Hiatt, B. (1999). *Best Practices Maintenance - A 13 Step Program in Establishing a World Class Maintenance Organization - USA*. Academic Press.

Hiles, A. (2002). *Enterprise Risk Assessment and Business Impact Analysis: Best Practices*. Rothstein Associates Inc.

Hlebnikov, A., Dementjeva, N., & Siirde, A. (2009). Optimization of Narva District Heating Network and Analysis of Competitiveness of Oil Shale CHP Building in Narva. *Oil Shale*, *26*(3), 269–282. doi:10.3176/oil.2009.3S.09

Holm, H., Korman, M., & Ekstedt, M. (2015). A bayesian network model for likelihood estimations of acquirement of critical software vulnerabilities and exploits. *Information and Software Technology*, *58*, 304–318. doi:10.1016/j.infsof.2014.07.001

Holthaus, O., & Rajendran, C. (1997). New Dispatching Rules for Scheduling in a Job Shop – An Experimental Study. *Advanced Manufacturing Technology*, *13*(2), 148–153. doi:10.1007/BF01225761

Hooff, B., & Huysman, M. (2009). Managing knowledge sharing: Emergent and engineering approaches. *Information & Management*, *46*(1), 1–8. doi:10.1016/j.im.2008.09.002

Horowitz, E., & Sahni, S. (1978). *Fundamentals of Computer Algorithms*. Computer Science Press.

Howson, C., & Urbach, P. (1991). Bayesian reasoning in science. *Nature*, *350*(6317), 371–374. doi:10.1038/350371a0

Huang, S., Chang, J., Leng, G., & Qiang, H. (2015). Integrated index for drought assessment based on variable fuzzy set theory: A case study in the Yellow River basin, China. *Journal of Hydrology (Amsterdam)*. Advance online publication. doi:10.1016/ j.jhydrol.2015.05.032

Huld, T., Müller, R., & Gambardella, A. (2012). A new solar radiation database for estimating PV performance in Europe and Africa. Solar Energy, 2012(86), 1803–1815. doi:10.1016/j.solener.2012.03.006

Hutagalung, A.O., & Hasibuan, S. (2019). Determining the priority of medical equipment maintenance with analytical hierarchy process. *International Journal of Online and Biomedical Engineering*, *15*(10), 107-120.

Hydraulic Institute (HI), Europump, and the US Department of Energy's Office of Industrial Technologies (OIT). (2001). *Pump life cycle costs: A guide to LCC analysis for pumping systems*. Author.

Idhammar, C. (1997, July). Maintenance management: moving from reactive to results-oriented. Journal Review Pima's Papermaker.

IEA. (n.d.a). Africa Energy Outlook. World Energy Outlook Special Report. OECD/IEA.

IEA. (n.d.b). Country Statistics. https://www.iea.org/statistics/statisticssearch/report/?country=CONGOREP&product=balances&year=2014

IEC 62366-1:2015 *Medical devices -- Part 1: Application of usability engineering to medical devices (1st ed.)*. International Electrotechnical Commission.

Ighravwe, D. E., & Oke, S. A. (2019). A multi-criteria decision-making framework for selecting a suitable maintenance strategy for public buildings using sustainability criteria. *Journal of Building Engineering*, *24*, 100753. doi:10.1016/j.jobe.2019.100753

Ilangkumaran, M., & Kumanan, S. (2009). Selection of maintenance policy for textile industry using hybrid multi-criteria decision making approach. *Journal of Manufacturing Technology Management*, *20*(7), 1009–1022. doi:10.1108/17410380910984258

International Copper Association (ICA). (2015). *Guía para la aplicación de la Norma NMX ISO 55001*. Naucalpán de Juarez.

Ishizaka, A., & Nemery, P. (2014). Assigning machines to incomparable maintenance strategies with ELECTRE-SORT. *Omega*, *47*, 45–59. doi:10.1016/j.omega.2014.03.006

ISO 14224. (2016). *Petroleum, petrochemical and natural gas industries — Collection and exchange of reliability and maintenance data for equipment.*

ISO 14971:2007 *Medical devices -- Application of risk management to medical devices (2nd ed.)*. International Organization for Standardization.

ISO 31000:2010. *Risk management. Principles and guidelines*. International Standard Organization

ISO 31010. (2019). *Risk Management — Risk assesment techniques.*

ISO 31010:2011. *Risk management. Risk assessment technics*. International Standard Organization ISO guide 73:2009. *Risk management. Vocabulary*. International Standard Organization

ISO 50001:2018 (en) Energy management systems – Requirements with guidance for use. (2018). Retrieved from https://www.iso.org/obp/ui/es/#iso:std:iso:50001:ed-2:v1:en

ISO 55000. (2014). *Asset Management — Overview, principles and terminology.*

ISO 55001. (2014). *Asset Management — Management systems — Requirements.*

ISO. (2014). International Organization for Standardization ISO 55000: Asset management - Principles and guidelines.

ISO. (2018). *(International Organization for Standardization) 31000*. Risk Management — Guidelines.

ISO. (2018). International Organization for Standardization ISO 31000: Risk management - Principles and guidelines.

ISO/IEC 27005:2008, "Information technology -- Security techniques-Information security risk management

ISO/IEC 27005:2018. *Information technology - Security techniques - Information security risk management*. International Standard Organization

Ittner, C., & Oyon, D. (2014). *The Internal Organization of Enterprise Risk Management*. Available at SSRN 2486588

Ivancevich, J. M., Lorenzi, P., Skinner, S. J., & Crosby, P. B. (1997). *Gestión calidad y competitividad*. McGraw-Hill.

Jain, A. S., & Meeran, S. (1999). Deterministic Job-Shop Scheduling: Past, Present, and Future. *European Journal of Operational Research*, *113*(2), 390–434. doi:10.1016/S0377-2217(98)00113-1

Jakoubi, S., & Tjoa, S. (2009) A reference model for risk-aware business process management. In Kalam, A., Y. Deswarte, & M. Mostafa (Eds.), *Risks and Security of Internet and Systems, 4th International Conference, CRiSIS'09, Toulouse, France, October 19-22, 2009, Proceedings*. IEEE Computer Society. 10.1109/CRISIS.2009.5411973

Jamsek, M., Dobersek, D., Goricanec, D., & Krope, J. (2010). Determination of Optimal District Heating Pipe Network Configuration. *WSEAS Transactions on Fluid Mechanics*, *3*(5), 167-174.

Jamshidi, A., Rahimi, S. A., Ait-kadi, D., & Ruiz, A. (2015). A comprehensive fuzzy risk-based maintenance framework for prioritization of medical devices. *Applied Soft Computing*, *32*, 322–334. doi:10.1016/j.asoc.2015.03.054

Johansson, C. (2010). *Towards Intelligent District Heating*. Retrieved from https://www.diva-portal.org/smash/get/diva2:833722/FULLTEXT01.pdf

Johnson, S. M. (1954). Optimal two-and-three-stage production schedules with set-up times included. *Naval Research Logistics*, *1*(1), 61–68. doi:10.1002/nav.3800010110

Jokipii, A. (2010). Determinants and consequences of internal control in firms: A contingency theory based analysis. *The Journal of Management and Governance*, *14*(2), 115–144. doi:10.100710997-009-9085-x

Jonsson, P. (1997). The status of maintenance management in Swedish manufacturing firms. *Journal of Quality in Maintenance Engineering*, *3-4*(4), 233–258. doi:10.1108/13552519710176863

Kahraman, C. (2008). *Fuzzy Multi-criteria decision making: Theori and applications with recent development*. Springer.

Kahraman, C. T., Ertay, T., & Büyüközkan, G. (2006). A fuzzy optimization model for QFD planning process using analytic network approach. *European Journal of Operational Research*, *171*(2), 390–411. doi:10.1016/j.ejor.2004.09.016

Kans, M. (2009). The advancement of maintenance information technology A literature review. *Journal of Quality in Maintenance Engineering*, *15*(1), 5–16. doi:10.1108/13552510910943859

Kar, A. K., & Pani, A. K. (2014). Exploring the importance of different supplier selection criteria. *Management Research Review*, *37*(1), 89–105. doi:10.1108/MRR-10-2012-0230

Karabacak, B., & Sogukpinar, I. (2005). ISRAM: Information security risk analysis method. *Computers & Security*, *24*(2), 147–159. doi:10.1016/j.cose.2004.07.004

Kelly, A. (2006). *Strategic Maintenance Planning*. Butterwoth-Heinemann.

Khamaksorn, A. (2016). Project Management Knowledge and Skills for the Construction Industry. *International Conference on Civil, Architecture and Sustainable Development (CASD-2016)*.

Khambhammettu, H., Boulares, S., Adi, K., & Logrippo, L. (2013). A framework for risk assessment in access control systems. *Computers and Security, 39*(A), 86–103.

Kirubakaran, B., & Ilangkumaran, M. (2016). Selection of optimum maintenance strategy based on FAHP integrated with GRA–TOPSIS. *Annals of Operations Research*, *245*(1), 285–313. doi:10.100710479-014-1775-3

Kitagawa, & Takahashi, S. H. (1976). Applicability of Fracture Mechanics to Very Small Cracks or the Cracks in the Early Stages. In *Proceedings of the Second International Conference on Mechanical Behaviour of Materials.* American Society for Metals.

Klir, G. J., & Yuan, B. (1995). *Fuzzy Sets and Fuzzy Logic: Theory and Applications.* Prentice Hall.

Klir, G., & Yuan, B. (1995). *Fuzzy sets and fuzzy logic* (Vol. 4). Prentice hall.

Kloot, L. (1997). Organizational learning and management control systems: Responding to environmental change. *Management Accounting Research, 8*(1), 47–73. doi:10.1006/mare.1996.0033

Knezevic, J. (1996). *Mantenimiento.* Isdefe.

Knoll, D., Reinhart, G., & Prüglmeier, M. (2019). Enabling value stream mapping for internal logistics using multidimensional process mining. *Expert Systems with Applications, 124*, 130–142. doi:10.1016/j.eswa.2019.01.026

Kobersi, I. S., Finaev, V. I., Almasani, S. A., & Ahmed Abdo, K. W. (2013). Control of the Heating System with Fuzzy Logic. *World Applied Sciences Journal, 23*(11), 1441–1447. doi:10.5829/idosi.wasj.2013.23.11.13156

Kodali, R., & Chandra, S. (2001). Analytical Hierarchy Process for justification of total productive maintenance. *Production Planning and Control, 12*(7), 695–705. doi:10.1080/09537280010024045

Kogut, B., & Zander, U. (1992). Knowledge of the firm, combinative capacities, and the replication of technology. *Organization Science, 7*(3), 502–517.

Komass, T., Sniders, A., & Laizans, A. (2012). Statistical Analysis of Municipal Heat Supply. *Proceedings of 11th International Scientific Conference, 11*, 1 – 8.

Kontu, K. (2013). *Forecasting district heating consumption based on customer measurements.* Report on the 36th Euroheat & Power Congress. Retrieved from http://www.lsta.lt/files/events/2013-05-27-29%20EHP%20kongresas%20Vienoje/09_130527_1600_1730_Pl_6_Kaisa+Kontu.pdf

Korte, B., & Lovasz, L. (1981). Mathematical structures underlying greedy algorithms. Lecture Notes in Computer Science, 117, 205–209. doi:10.1007/3-540-10854-8_22

Korte, B., & Lovasz, L. (1983). Structure properties of greedoids. *Combinatorica, 3*(3-4), 359–374. doi:10.1007/BF02579192

Korte, B., Lovasz, L., & Schrader, R. (1991). *Greedoids.* Springer-Verlag. doi:10.1007/978-3-642-58191-5

Korte, B., & Vygen, J. (2012). *Combinatorial optimization: theory and algorithms. Algorithms and Combinatorics* (Vol. 21). Springer-Verlag Berlin Heidelberg. doi:10.1007/978-3-642-24488-9

Kristin, D. (2012). Demand response in an isolated system with high wind integration. Power Systems. *IEEE Transactions on, 27*(1), 20–29.

Krupp, U., & Álvarez-Armas, I. (2014). Short Fatigue crack propagation during low-cycle, high cycle and very-high cycle fatigue of duplex steel a unified approach. *International Journal of Fatigue, 65*, 78–85. doi:10.1016/j.ijfatigue.2013.06.018

Kruskal, J. B. (1956). On the shortest spanning subtree and the traveling salesman problem. *Proceedings of the American Mathematical Society, 7*(1), 48–50. doi:10.1090/S0002-9939-1956-0078686-7

Kuhn, T. S. (1962). *The Structure of Scientific Revolutions* (3rd ed.). Chicago: University of Chicago Press.

Kujawski, D. (2013). Environment and Stress Interaction in Corrosion Fatigue and Stress Corrosion Cracking. *Metallurgical and Materials Transactions. A, Physical Metallurgy and Materials Science, 44A*(3), 1340–1352. doi:10.100711661-012-1507-8

Kujawski, D., & Sadananda, K. (2011). Effect of Crack-Tip Stresses on SCC Behaviour. *Metallurgical and Materials Transactions. A, Physical Metallurgy and Materials Science*, *42A*(2), 377–382. doi:10.100711661-010-0373-5

Kulkarni, U., Ravidran, S., & Freez, E. R. (2007). A knowledge management success model: Theoretical development and empirical validation. *Journal of Management Information Systems*, *23*(3), 309–347. doi:10.2753/MIS0742-1222230311

Kundu, A., Dutta, R., & Mukhopadhyay, D. (2006, December). An Alternate Way to Rank Hyper-linked Web-Pages. In *9th International Conference on Information Technology (ICIT'06)* (pp. 297-298). IEEE. 10.1109/ICIT.2006.26

Labib, W. A. (2004). Towards an Intelligent Holonic Maintenance System. *Maintenance Journal, 17*(2).

Labib, W. A., O'Connor, F. R., & Glyn, B. W. (1998). An effective maintenance system using the analytic hierarchy process. *Integrated Manufacturing Systems*, *9*(2), 87–98. doi:10.1108/09576069810202005

Lawler, E. L. (1985). *The Traveling Salesman Problem: A Guided Tour of Combinatorial Optimization*. John Wiley & Sons.

Lazakis, I., & Olcer, A. I. (2015). Selection of the best maintenance approach in the maritime industry under fuzzy multiple attribute group decision-making environment. *Proceedings of IMechE part M: Journal of Engineering for Maritime Environment*, 1-13. doi:10.1177/1475090215569819

Lazakis, I., Turan, O., & Olcer, A. I. (2012). Determination of the optimum ship maintenance strategy through multi attribute decision making. *11th International marine design conference*, 473-487.

Lazakis, I., Turan, O., & Aksu, S. (2010). Increasing ship operational reliability through the implementation of a holistic maintenance management strategy. *Journal of Ships and Offshore Structures*, *5*(4), 337–357. doi:10.1080/17445302.2010.480899

Lee, J., Holgado, M., Kao, H., & Macchi, M. (2014). New Thinking Paradigm for Maintenance Innovation design. *Proceedings of the 19th IFAC World Congress, IFAC Proceedings Volumes (IFAC-Papers Online)*, 24–29. 10.3182/20140824-6-ZA-1003.02519

Lee, J., Bagheri, B., & Kao, H. A. (2015). A Cyber-Physical Systems architecture for Industry 4.0-based manufacturing systems. *Manufacturing Letters*, *3*, 18–23. doi:10.1016/j.mfglet.2014.12.001

Leisinger, K. M., Blasco, J., Trigo, J., Petitbò, A., Álvarez, J. M., & Argandoña, A. P. (2010). *Hacia una nueva ética económica global. Innovación vs Status Quo*. Editorial Huygens.

Leiva, R., Donoso, J. R., Muehlich, U., & Labbe, F. (2004). Un enfoque de dos parámetros para el análisis del efecto del cordón de soldadura sobre el constreñimiento. *Rev. Metal Madrid*, *40*(1), 280–287. doi:10.3989/revmetalm.2004.v40.i1.237

Lei, Z., Hong, Y., Xie, J., Sun, C., & Zhao, A. (2012). Effects of inclusion size and location of very-high-cycle fatigue behaviour for high strength steels. *Materials Science and Engineering A*, *558*, 234–241. doi:10.1016/j.msea.2012.07.118

León, O. (1993). *Análisis de decisiones: técnicas y situaciones aplicables a directivos y profesionales*. McGraw-Hill.

Levary, R. R., & Han, D. (1995). Choosing a technological forecasting method. *Industrial Management (Des Plaines)*, *37*(1), 14–18.

Li, C., Lim, J., & Wang, Q. (2007). Internal and external influences on IT control governance. *International Journal of Accounting Information.*

Lima, F. R. Junior. (2013). Comparação entre os métodos Fuzzy TOPSIS e Fuzzy AHP no apoio à tomada de decisão para seleção de fornecedores (Mestrado em Engenharia de produção). Escola de Engenharia de São Carlos da Universidade de São Paulo, São Carlos.

Lin, Z., Zheng, J., & Wang, S. (2007). Study on Relation of Internal Control and Enterprise Value: An Empirical Analysis from Shanghai and Shenzhen Stock Markets. *Journal of Finance and Economics*.

Lin, H. F. (2007). A stage model of knowledge management: An empirical investigation of process and effectiveness. *Journal of Information Science*, *33*(6), 643–659. doi:10.1177/0165551506076395

Liu, W., Jin, Y., & Price, M. (2017). A new improved NEH heuristic for permutation flowshop scheduling problems. *International Journal of Production Economics*, *193*, 21–30. doi:10.1016/j.ijpe.2017.06.026

Liu, Y. B., Yanga, Z. G., Li, Y. D., Chena, S. M., Li, S. X., Hui, W. J., & … . (2010). Prediction of the S-N curve of high-strength steels in the very high cycle fatigue regime. *International Journal of Fatigue*, *32*(8), 1351–1357. doi:10.1016/j.ijfatigue.2010.02.006

Logan, L. H., Karlsson, E. M., Gall, H. E., Park, J., Emery, N., Owens, P., Niyogi, D., & Rao, P. S. C. (2013). Freshwater Wetlands: Balancing Food and Water Security with Resilience of Ecological and Social Systems. In Climate Vulnerability: Understanding and Addressing Threats to Essential Resources (pp. 105-116). Elsevier.

Loken, E. (2007). Use of multicriteria decision analysis methods for energy planning problems. *Renewable & Sustainable Energy Reviews*, *11*(7), 1584–1595. doi:10.1016/j.rser.2005.11.005

Longenecker, J. G., Moore, C. W., Petty, J. W., & Palich, L. E. (2013). *Administração de pequenas empresas*. Cencage Learning.

López Campos, M., Fumagalli, L., Gómez Fernández, J., Crespo Márquez, A., & Macchi, M. (2010). UML model for integration between RCM and CBM in an e-maintenance architecture. Proceedings of the First IFAC Workshop on Advanced Maintenance Engineering Services and Technology (A-MEST), 133-138. doi:10.3182/20100701-2-PT-4012.00020

Lopez, P., & Centeno, G. (2006). Integrated system to maximize efficiency in transit maintenance departments. *International Journal of Productivity and Performance Management*, *55*(8), 638–654. doi:10.1108/17410400610710189

Lujano-Rojas, J. M., Monteiro, C., Dufo-López, R., & Bernal-Agustín, J. L. (2012, June). Optimum residential load management strategy for real time pricing (RTP) demand response programs. *Energy Policy*, *45*, 671–679. doi:10.1016/j.enpol.2012.03.019

Lu, L. T., Zhnag, J. W., & Shiozawua, K. (2009). Influence of inclusion size on S-N curves characteristics of high-strength steels in very high cycle fatigue regime. *Fatigue & Fracture of Engineering Materials & Structures*, *32*(8), 647–655. doi:10.1111/j.1460-2695.2009.01370.x

Lund, H., Werner, S., Wiltshire, R., Svendsen, S., Thorsen, J. E., Hvelplund, F., & Mathiesen, B. V. (2014). 4th Generation District Heating (4GDH). *Energy*, *68*, 1–11. doi:10.1016/j.energy.2014.02.089

Lyons, L. (2019). *Digitalisation: Opportunities for heating and cooling*. doi:10.2760/00116

Mahfoud, H., Abdellah, E. B., & El Biyaali, A. (2018). Dependability-based maintenance optimization in healthcare domain. *Journal of Quality in Maintenance Engineering*, *24*(2), 200–223. doi:10.1108/JQME-07-2016-0029

Mahfoud, H., El Barkany, A., & El Biyaali, A. (2016). A Hybrid Decision-Making Model for Maintenance Prioritization in Health Care Systems. *American Journal of Applied Sciences*, *13*(4), 439–450. doi:10.3844/ajassp.2016.439.450

Make a game-changing impact – before the others do. The impact of digitalization in district heating. (2016). Retrieved from https://energia.fi/files/1345/2016_Impact_of_digitalization_in_district_heating_summary_ENG.pdf

Marathamuthu, S., & Jebaraj, S. (2015). The use of 5-WHYs technique to eliminate OEE's speed loss in a manufacturing firm. *Journal of Quality in Maintenance Engineering*, *21*(4), 419–435. doi:10.1108/JQME-09-2013-0062

Marconi, M. D. A., & Lakatos, E. M. (2010). *Fundamentos de metodologia científica. 7*. Atlas.

Martello, S., & Ries, B. (2015). Advances in Combinatorial Optimization. *Discrete Applied Mathematics*, *196*(C), 1–3. doi:10.1016/j.dam.2015.08.011

Martínez Marín, J. E., & Eguren Martí, M. L. (2013). Maritime Transport: A Theoretical Analysis Under a System's Approach. *Journal of Maritime Research*, 61-68.

Martinez, I., & Ruiz, J. (2002). Los procesos de creación del conocimiento: El aprendizaje y la espiral de creación del conocimiento. *XVI Congreso Nacional de AEDEM*. Alicante 2002.

Martin, J. (2008). Personal knowledge management: the basis of corporate and institutional knowledge management. In J. Martin & K. Wright (Eds.), *Managing Knowledge: Case Studies in Innovation*. Spotted Cow.

Martorell, S., Sánchez, A., Muñoz, A., Pitarch, J. L., Serradell, V., & Roldan, J. (1999). The use of maintenance indicators to evaluate the effects of maintenance programs on NPP performance and safety. *Reliability Engineering & System Safety*, *65*(2), 85–94. doi:10.1016/S0951-8320(98)00088-X

Martorell, S., Villanueva, J. F., Carlos, S., Nebot, Y., Sanchez, A., Pitarch, J. L., & Serradell, V. (2005). RAMS+C informed decision-making with application to multiobjective optimization of technical specifications and maintenance using genetic algorithms. *Reliability Engineering & System Safety*, *87*(1), 65–75. doi:10.1016/j.ress.2004.04.009

Maslow, A. (1943). A Theory of Human Motivation. Originally Published in Psychological Review, 50(4), 370–396. doi:10.1037/h0054346

Massa, F. H. D., Cazier, O., Tison, T., & Lallemand, B. (2015). Finite element analysis of frictionless contact problems using fuzzy control Poulinapproach. *Engineering Computations*, *32*(3), 585–606. doi:10.1108/EC-11-2013-0289

Matteson, D. (1985). Airline experience with reliability-centered maintenance. *Nuclear Engineering and Design*, *89*(2), 385–390. doi:10.1016/0029-5493(85)90076-7

McDonald, K.A., & Barnes. (1991). *Crack tip opening displacement fracture mechanics testing for project topsides*. Academic Press.

Medojevic, M., Villar, P. D., Cosic, I., Rikalovic, A., Sremcev, N., & Lazarevic, M. (2018). Energy Management in Industry 4.0 Ecosystem: a Review on Possibilities and Concerns. *Proceedings of the 29th DAAAM International Symposium*, 674 – 680. 10.2507/29th.daaam.proceedings.097

Mehta, K., Arora, N., & Singh, B. P. (2011). Low power efficient D flip flop circuit. In *International Symposium on Devices MEMS, Intelligent Systems & Communication (ISDMISC)* (pp. 16-19). IJCA.

Mell, P., & Grance, T. (2011). *The NIST definition of cloud computing*. NIST.

Mellor, P. (1966). A Review of Job Shop Scheduling. *Operational Research Quarterly*, *17*(2), 161–171. doi:10.1057/jors.1966.24

*Method for crack opening displacement (COD) testing Bs 5762*. (1979). British Standard Institution.

Methods for Crack Opening Displacement C.O.D. (Testing). (1979). *Methods for Crack Opening Displacement C.O.D. (Testing) DO: 19, (1.972)*.

Methods for Plane Strain Fracture Toughness KIC Testing. DD3. (1971). British Standard British standard 5762.

Miguel, P. A. C. (Ed.). (2012). *Metodologia de pesquisa em engenharia de produção e gestão de operações*. Elsevier.

Miles, M., & Huberman, A. (1994). *Qualitative Data Analysis: A sourcebook of new methods*. Sage Publications.

Mirabi, M. (2014). A novel hybrid genetic algorithm to solve the sequence-dependent permutation flow-shop scheduling problem. *International Journal of Advanced Manufacturing Technology, 71*(1-4), 429–437. doi:10.100700170-013-5489-5

Mitra, A. (2019). On investigating energy stability for Cellular Automata based PageRank validation model in Green Cloud. *International Journal of Cloud Applications and Computing, 9*(4), 66–85. doi:10.4018/IJCAC.2019100104

Mitra, A., & Kundu, A. (2015). Page Ranking Validation using Cellular Automata in Cloud. *International Journal of Cloud Applications and Computing, 5*(3), 1–19. doi:10.4018/IJCAC.2015070101

Mitra, A., & Kundu, A. (2017). Energy Efficient CA based Page Rank Validation Model: A Green Approach in Cloud. *International Journal of Green Computing, 8*(2), 59–76. doi:10.4018/IJGC.2017070104

Mobley, R. K. (2002). *An introduction to Predictive Maintenance.* Butterworth-Heinemann.

Mohammadi, M., Noorollahi, Y., Mohammadi-Ivatloo, B., Yousefi, H., & Jalilinasabady, S. (2017). Optimal Scheduling of Energy Hubs in the Presence of Uncertainty-A Review. *Journal of Energy Management and Technology, 1*, 1–17. doi:10.22109/JEMT.2017.49432

Möldner, A. K., Garza-Reyes, J. A., & Kumarc, V. (2018). Exploring lean manufacturing practices' influence on process innovation performance. *Journal of Business Research*, •••, 1–17. doi:10.1016/j.jbusres.2018.09.002

Montañez Naz, S. (2017). Del Falsifiabilityo a las inferencias bayesianas. *Blog: Educación y divulgación científica en la Comunidad de Madrid.* http://divulgamadrid.blogspot.com/

Moubray, J. (1991). *Reliability-Centered Maintenance.* Butterworth-Heinemann.

Muller, A., Crespo Márquez, A., & Iung, B. (2008). On the concept of e-maintenance: Review and current research. *Reliability Engineering & System Safety, 93*(8), 1165–1187. doi:10.1016/j.ress.2007.08.006

Munda, G. (2005). Multiple Criteria Decision analysis and sustainable development. In J. Figueira, S. Greco, & M. Ehrgott (Eds.), *Multicriteria Decision analysis. State of the Art Surveys* (pp. 953–988). Springer Verlag. doi:10.1007/0-387-23081-5_23

Muñoz, J. (1999). Sobre gestión del conocimiento, un intangible clave en la globalización. Economía Industrial, 330.

Murakami, Y. (2012). Material defects as the basis of fatigue design. *International Journal of Fatigue, 41*, 2–10. doi:10.1016/j.ijfatigue.2011.12.001

Murakami, Y., & Yamashita, Y. (2014). Prediction of life and scatter of fatigue failure originated at non-metallic inclusions. *Procedia Engineering, 74*, 6–11. doi:10.1016/j.proeng.2014.06.214

Musee, N., Lorenzen, L., & Aldrich, C. (2006). Decision support for evaluating energy demand in vinification processes using fuzzy sets theory. *Journal of Energy in Southern Africa, 17*(4), 4–18. doi:10.17159/2413-3051/2006/v17i4a3190

Musetti, T. F., & Escrivão Filho, E. (2016). Monitoramento do ambiente organizacional em pequenas empresas: Estudo de casos nos setores Metalmecânica e base tecnológica da região central do estado de São Paulo. *Gestão da Produção Operações e Sistemas, 11*(2), 139–154.

Nagano, M. S., & Moccellin, J. V. (2002). A high quality solution constructive heuristic for flow shop sequencing. *J Opl Res Soc, 53*(12), 1374–1379. doi:10.1057/palgrave.jors.2601466

Nakajima, S. (1988). *Introduction to TPM.* Productivity Press.

Nakajima, S. (1988). *TPM: Introduction to total productive maintenance.* Productivity Press Inc.

Nannapaneni, S., Mahadevan, S., & Rachuri, S. (2016). Performance evaluation of a manufacturing process under uncertaintyusing Bayesian networks. *Journal of Cleaner Production, 113*, 947–959. doi:10.1016/j.jclepro.2015.12.003

Nawaz, M., Enscore, E. E. Jr, & Ham, I. (1983). A heuristic algorithm for the m-machine, n-job flow-shop sequencing problem. *Omega International Journal of Management Science, 11*(1), 91–95. doi:10.1016/0305-0483(83)90088-9

Neapolitan, R. E., & Naimipour, K. (2003). *Foundations of Algorithms Using C++ Pseudo-code*. Jones & Bartlett Publishers.

Negrão, L., Godinho Filho, M., & Almeida Marodin, G. (2016). Lean practices and their effect on performance: A literature review. *Production Planning and Control, 28*(1), 33–56. doi:10.1080/09537287.2016.1231853

New Zealand Water and Wastes Association. (2008). *Visual assessment of utility assets*. Author.

Neyman, J. (1952). *Mathematical statistics and probability*. U.S. Department of Agriculture.

Niazi, M. A., & Hussain, A. (2011). A novel agent-based simulation framework for sensing in complex adaptive environments. *IEEE Sensors Journal, 11*(2), 404–412. doi:10.1109/JSEN.2010.2068044

Nishi, T., Ohno, A., Inuiguchi, M., Takahashi, S., & Ueda, K. (2017). A Combined column generation and heuristics for railway short-term rolling stock planning with regular inspection constraints. *Computers & Operations Research, 81*, 14–25. doi:10.1016/j.cor.2016.11.025

Nix, O. (2011). *Building Information Technology and Management*. Retrieved from https://www.researchgate.net/publication/265555488_Building_Information_Modelling_Adoption_An_Analysis_of_the_Barriers_to_Implementation

Nonaka, I. (1994). A dynamic theory of organizational knowledge creation. Organizatioll Saence, 5(1), 14–37. doi:10.1287/orsc.5.1.14

Nonaka, I. (1991, November). The knowledge-creating company. *Harvard Business Review*, •••, 96–104.

Nonaka, I., & Konno, N. (1998). The concept of "Ba": Building a foundation for knowledge creation. *California Management Review, 40*(3), 40–54. doi:10.2307/41165942

Nonaka, I., & Takeuchi, H. (1995). *The knowledge-Creating Company: How Japanese Companies Create the Dynamics of Innovation*. Oxford University Press.

Nonaka, I., & Takeuchi, H. (1999). *La Organización Creadora de Conocimiento*. Academic Press.

Noro, G. B., & Amarilho, A. (2012). *Gestão de Projetos Sustentáveis: o Caso América Latina Logística. Simpósio de excelência em Gestão e Tecnologia*. Evento SEGET.

Norton, D., & Kaplan, R. (1996). *The Balanced Scorecard: Translating Strategy into Action*. Harvard Business School Press.

NREL. (2009). *HOMER the Micro-Power Optimization Model; ver.2.68Beta*. NREL.

O'Dell, C., & Grayson, C. J. (1998). If only we knew what we know: Identification and transfer of internal best practices. *California Management Review, 40*(3), 154–170. doi:10.2307/41165948

Oakes, M. (1990). Statistical Inference. Chesnut Hill, MA: E.R.I.

OCDE. (1996). *The Knowledge-Based Economy*. OCDE.

OCDE. (2004). *Medición de la gestión de conocimientos en las empresas: primeros resultados*. Ed. OCDE.

OECD. (2015). Health at a Glance 2015. Obtenido de OECD ilibrary: https://www.oecd-ilibrary.org/social-issues-migration-health/health-at-a-glance-2015_health_glance-2015-en

Oiltech Analysis S.L. (1995). Mantenimiento Proactivo de sistemas mecánicos lubricados. Fluidos oleohidráulica neumática y automación, 24(208-209).

Olechowski, A., Oehmen, J., Seering, W., & Ben-Daya, M. (2016). The professionalization of risk management: What role can the ISO 31000 risk management principles play? *International Journal of Project Management*, *34*(8), 1568–1578. doi:10.1016/j.ijproman.2016.08.002

Online study material: What is automated decision making and profiling. (n.d.). Accessed from https://ico.org.uk/for-organisations/guide-to-data-protection/guide-to-the-general-data-protection-regulation-gdpr/automated-decision-making-and-profiling/what-is-automated-individual-decision-making-and-profiling/

Onyeji, I. (2014) Harnessing and Integrating Africa's Renewable Energy Resources. Renewable Energy Integration, 27-38.

Ordoñez de Pablos, P. (1999). *Gestión del conocimiento y medición del capital intelectual en la empresa internacional.* Proyecto de Doctorado, Universidad de Oviedo.

Ordoñez de Pablos, P. (2001). *Capital intelectual, gestión del conocimiento y sistemas de gestión de recursos humanos: Influencia sobre los resultados organizativos (Tesis Doctoral).* Universidad de Oviedo.

Ordoñez de Pablos, P., & Parreños, J. (2003). Aprendizaje organizativo en el contexto internacional: Implicaciones para la gestión del conocimiento. *Investigaciones Europeas de Dirección y Economía de la Empresa*, *9*(2), 205–216.

Ordoñez de Pablos, P., & Parreños, J. (2005). Aprendizaje organizativo y gestión del conocimiento: Un análisis dinámico del conocimiento en la empresa. *Investigaciones Europeas de Dirección y Economía de la Empresa*, *11*(1), 165–177.

Oskuii, S. T. (2004). *Comparative study on low-power high-performance flip-flops* (Unpublished Master of Technology Thesis). Linköping University. Retrieved May 15, 2017, from http://www.diva-portal.org/smash/get/diva2:19406/FULLTEXT01.pdf

Pace, S. (2004). A grounded theory of the flow experiences of Web users. *International Journal of Human-Computer Studies*, *60*(3), 327–363. doi:10.1016/j.ijhcs.2003.08.005

Page, E. S. (1961). An approach to the scheduling of jobs on machines. *Journal of the Royal Statistical Society. Series A (General)*, *23*, 484–492.

Palacios, D., & Garrigos, F. (2006). Propuesta de una escala de medida en la gestión del conocimiento en las industrias de biotecnología y telecomunicaciones. *Investigaciones Europeas de Dirección y Economía de la Empresa*, *12*(1), 207–224.

Palmer, D. S. (1965). Sequencing jobs through a multi-stage process in the minimum total time—A quick method of obtaining a near optimum. *Opns Res Quart*, *16*(1), 101–107. doi:10.1057/jors.1965.8

Panwalker, S. S., & Iskander, W. (1977). A Survey of Scheduling Rules. *Operations Research*, *25*(1), 45–61. doi:10.1287/opre.25.1.45

Papadimitriou, C. D., & Steiglitz, K. (1982). *Combinatorial Optimization: Algorithms and Complexity*. Prentice-Hall.

Parfenenko, Yu., Okopnyi, R., & Nenja, V. (2012). Creation of Control Instruments for Buildings' Heating. *The Journal of the National Technical University NTU KPI. Series Information Systems and Networks*, *34*, 93–97.

Parfenenko, Yu., Shendryk, V., Nenja, V., & Vashchenko, S. (2014). Information System for Monitoring and Forecast of Building Heat Consumption. *Communications in Computer and Information Science*, *465*, 1–11. doi:10.1007/978-3-319-11958-8_1

Paris, P. C. (1970). Testing for Very Slow Growth of Fatigue Cracks. *MTS Close Loop Magazine, 2*(5).

Paris, P. C., Bucci, R. J., Wessel, E. T., Clark, W. G., & Mager, T. R. (1972). *Extensive Study of Low Fatigue Crack Growth Rates in A533 and A508 Steels. ASTM STP513*. American Society for Testing and Materials.

Partington, D. (2000). Building grounded theories of management action. *British Journal of Management, 11*(2), 91–102. doi:10.1111/1467-8551.00153

Pasquinelli, M. (2009). Google's PageRank algorithm: A diagram of cognitive capitalism and the rentier of the common intellect. *Deep search: The politics of search beyond Google*, 152-162.

Paté-Cornell, M. E., Kuypers, M., Smith, M., & Keller, P. (2018). Cyber Risk Management for Critical Infrastructure: A Risk Analysis Model and Three Case Studies. *Risk Analysis, 38*(2), 226–241. doi:10.1111/risa.12844 PMID:28679022

Patel, Y. S., Mehrotra, N., & Soner, S. (2015, February). Green cloud computing: A review on Green IT areas for cloud computing environment. In *Futuristic Trends on Computational Analysis and Knowledge Management (ABLAZE), 2015 International Conference on* (pp. 327-332). IEEE. 10.1109/ABLAZE.2015.7155006

Patel, S. C., Graham, J. H., & Ralston, P. A. S. (2008). Quantitatively assessing the vulnerability of critical information systems: A new method for evaluating security enhancements. *International Journal of Information Management, 28*(6), 483–491. doi:10.1016/j.ijinfomgt.2008.01.009

Pauleen, D. (2009). Personal knowledge management: Putting the "person" back into the knowledge equation. *Online Information Review, 33*(2), 221–224. doi:10.1108/14684520910951177

Păunescu, C., Popescu, M. C., & Blid, L. (2018). Business impact analysis for business continuity: Evidence from Romanian enterprises on critical functions. *Management and Marketing, 13*(3), 1035–1050. doi:10.2478/mmcks-2018-0021

Pavez, A. S. (2000). *Modelo de implantación de Gestión del Conocimiento y Tecnologías de Información para la Generación de Ventajas Competitivas*. Universidad Técnica Federico Santa María.

Peluffo, M., & Catalán, E. E. (2002). *Introducción a la gestión del conocimiento y su aplicación al sector publico*. Ed. Instituto Latinoamericano y del Caribe de Planificación.

Peñalvo López, E. (2017a). Metodología de Evaluación y Optimización de Sistemas Renovables Híbridos para Electrificación de Zonas Aisladas de la Red (Doctoral dissertation). UPV.

Peñalvo López, E., Pérez-Navarro, A., Hurtado Pérez, E., & Cárcel Carrasco, J. (2019b). Comprehensive Methodology for Sustainable Power Supply in Emerging Countries. *Sustainability, 11*, 1–22. doi:10.3390u11195398

Peñalvo-López, E., & Cárcel-Carrasco, F. J. (2019a). An Analysis about Learning to Increase Women's Participation and Employment in Europe's Energy Transition: Evidence from the European Project MEnS. *Sustainability, 11*(16), 4345. doi:10.3390/su11164345

Peñalvo-López, E., Cárcel-Carrasco, F. J., Devece, C., & Morcillo, A. I. (2017b). A Methodology for Analysing Sustainability in Energy Scenarios. *Sustainability, 1590*(9), •••. doi:10.3390/su9091590

Peñalvo-López, E., Cárcel-Carrasco, F., Devece, C., & Morcillo, A. (2017). A methodology for analysing sustainability in energy scenarios. *Sustainability, 9*(9), 1590. doi:10.3390/su9091590

Peña, P. (2001). *To know or no to be. Conocimiento, el oro gris de las organizaciones*. DINTEL.

Pérez-Navarro, A. (2016). Experimental verification of hybrid renewable systems as feasible energy sources. *Renewable Energy, 86*, 384–391.

Peris, F., Loria, B., & Méndez, M. (2002). Creación de conocimiento y diseño de organizaciones: Equidad, confianza y objetivos compartidos como reto de la gestión del conocimiento. Cuadernos de CC.EE. y EE., 43, 41-56.

Peterson, D. L., Silsbee, D. G., & Schmoldt, D. L. (1994). A Case Study of Resources Management Planning with Multiple Objectives and Projects. *Environmental Management*, *18*(5), 729–742. doi:10.1007/BF02394636

Petrides, L. A., & Nodine, T. R. (2003). *Knowledge Management in Education: Defining the Landscape*. March Institute for the Study of Knowledge Management in Education.

Peulic, A., Dragicevic, S., Jovanovic, Z., & Krneta, R. (2013). Flexible GPS/GPRS Based System for Parameters Monitoring in the District Heating System. *International Journal of Computers, Communications & Control*, *8*(1), 105–110. doi:10.15837/ijccc.2013.1.174

Pfister, J. (2009). *Managing Organizational Culture for Effective Internal Control: From Practice to Theory*. Physica-Verlag. doi:10.1007/978-3-7908-2340-0

Phillips, J. G., Heidrick, T. R., & Potter, I. J. (2007). Technology futures analysis methodologies for sustainable energy technologies. *International Journal of Innovation and Technology Management*, *4*(02), 171–190. doi:10.1142/S0219877007001041

Pinedo, M., Zacharias, C., & Zhu, N. (2015). Scheduling in the service industries: An overview. *Journal of Systems Science and Systems Engineering*, *24*(1), 1–48. doi:10.100711518-015-5266-0

Pinson, E. (1995). The Job Shop Scheduling Problem: A Concise Survey and Some Recent Developments. In P. Chretienne, E. G. Coff-man Jr, J. K. Lenstra, & Z. Liu (Eds.), *Scheduling Theory and its Applications* (pp. 277–293). John Wiley & Sons.

Pintelon, L., & Gelders, L. (1992). Maintenance management decision-making. *European Journal of Operational Research*, *58*(3), 301–317. doi:10.1016/0377-2217(92)90062-E

Pohekar, S. D., & Ramachandran, M. (2004). Application of multi-criteria decision-making to sustainable energy planning–a review. *Renewable & Sustainable Energy Reviews*, *8*(4), 365–381. doi:10.1016/j.rser.2003.12.007

Polanyi, M. (1958). *Personal Knowledge: Towards a Post-Critical Philosophy*. University of Chicago Press.

Polanyi, M. (1966). *The tacit dimension*. Routledge & Kegan Paul.

Polatidis, H. (2006). Selecting an appropriate multi-criteria decision analysis technique for renewable energy planning. Energy Sources Part B Vol. doi:10.1080/009083190881607

Polikarpova, I., & Rosa, M. (2017). Energy reduction potential of the district heating company introducing energy management systems. *Energy Procedia*, *128*, 66–71. doi:10.1016/j.egypro.2017.09.016

Popper, K. R. (1983). *Conjeturas y refutaciones. El desarrollo del conocimiento científico*. Barcelona: Paidós, edición revisada y ampliada.

Popper, K. R. (1988). *Conocimiento objetivo. Un enfoque evolucionista*. Madrid: Tecnos, tercera edición.

Popper, K. R. (1959). *The Logic of Scientific Discovery*. Routledge.

Popper, K. R. (1972). *Conjeturas y refutaciones. El desarrollo del conocimiento científico*. Paidós.

Popper, K. R. (1985). *Realismo y el objetivo de la ciencia. Postscriptum a la lógica de la investigación científica* (Vol. 1). Tecnos.

Popper, K. R. (1994). *En busca de un mundo mejor*. Paidós.

Popper, K. R. (1997). *El mito del marco común. En defensa de la ciencia y la racionalidad.* Paidós.

Popper, K. R., & Miller, D. W. (1987). Why probabilistic support is not inductive. *Philosophical Transactions of the Royal Society of London, 321*, 569–591.

Porter, M., & M., K. (2011, Jan.). The Big Idea. Creating Shared Value. How to reinvent capitalism and unleash a wave of innovation and growth. *Harvard Business Review.*

Porter, M. (1985). *Competitive Advantage: Creating and Sustaining Superior Performance.* Free Press.

Porter, M., & Kramer, M. (2006). Strategy and Society: The link between competitive advantage and corporate social responsability. *Harvard Business Review.* PMID:17183795

Prajogo, D., Toy, J., Bhattacharya, A., Oke, A., & Cheng, T.C.E. (2018). The relationships between information management, process management and operational performance: Internal and external contexts. International Journal of Production Economics, 199, 95-103. doi: 10.1016 / j.ijpe.2018.02.019

Prawda, J. (1999). *Métodos y modelos de investigación de operaciones. Modelos estocásticos.* Limusa S.A.

Presidio Graduate School. (2006). *The dictionary of sustainable management.* Recuperado el 18 de 10 de 2015, de https://www.sustainabilitydictionary.com/

Priddle, E. K. (1982). *The Threshold Stress Intensity Factor for Fatigue Crack Growth in Mild Steel Plate and Weld Metal: Some Effects of Temperature and Environment. Fatigue Threshold-Fundamentals and Engineering Applications.* Engineering Materials Advisory Services Ltd.

Prieto, A. J., Macías, J. M., Chávez, M. J., & Alejandre, F. J. (2016). Expert system for predicting buildings service life under ISO 31000 standard. Application in architectural heritage. *Journal of Cultural Heritage, 18*, 209–218. doi:10.1016/j.culher.2015.10.006

Prim, R. C. (1957). Shortest connection networks and some generalizations. *The Bell System Technical Journal, 36*(6), 1389–1401. doi:10.1002/j.1538-7305.1957.tb01515.x

Project Management Institute. (2017). PMBOK 6th Edition. Project Management Body of Knowledge. PMI Standards Committee.

Pulido-Rojano, A. (2015). Methodological design for the prevention of risk in production Processes. *Dyna (Bilbao), 82*(193), 16–22. doi:10.15446/dyna.v82n193.42903

Pulleyblank, W. R. (Ed.). (2014). *Progress in combinatorial optimization.* Academic Press Canada.

Quan, G., Zou, C., & Hong, Y. (2001). Experimental and theoretical investigation of environmental media on very-High cycle fatigue behaviour for a structural steel. *Acta Materialia, 59*(4), 1321–1327. doi:10.1016/j.actamat.2010.10.064

Rainer, R. K. Jr, Snyder, C. A., & Carr, H. H. (1991). Risk analysis for information technology. *Journal of Management Information Systems, 8*(1), 129–147. doi:10.1080/07421222.1991.11517914

Rajbhandari, L. (2013). *Risk Analysis Using "Conflicting Incentives" as an alternative notion of Risk* (Ph.D. thesis). Gjøvik University College.

Rajendran, C. (1993). Heuristic algorithm for scheduling in a flowshop to minimize total flowtime. *International Journal of Production Economics, 29*(1), 65–73. doi:10.1016/0925-5273(93)90024-F

Rajendran, C. (1995). Heuristics for scheduling in flowshop with multiple objectives. *Eur J Opl Res, 82*(3), 540–555. doi:10.1016/0377-2217(93)E0212-G

Ramadhan, R. H., Hamad, I., Wahhab, A., & Duffuaa, S. O. (1999). The use of an analytical hierarchy process in pavement maintenance priority ranking. *Journal of Quality in Maintenance Engineering*, *5*(1), 25–39. doi:10.1108/13552519910257041

Rani, N. A. A., Baharum, M. R., Akbar, A. R. N., & Nawawi, A. H. (2015). Perception of Maintenance Management Strategy on Healthcare Facilities. *Procedia: Social and Behavioral Sciences*, *170*(27), 272–281. doi:10.1016/j.sbspro.2015.01.037

Rao, B. K. N. (1996). *Handbook of Condition Monitoring*. Elsevier.

Rastegari, A. (2017). *Condition Based Maintenance in the manufacturing industry from strategy to implementation* (Thesis). University of New South Wales, Stockholm, Sweden.

Rath, F. (2008). Tools for developing a quality management program: Proactive tools (process mapping, Value stream mapping, fault tree analysis and failure mode and effects analysis). *International Journal of Radiation Oncology, Biology, Physics*, *71*(1), 187–190. doi:10.1016/j.ijrobp.2007.07.2385 PubMed

Ravizza, A., Diaz Lantada, A., Ballesteros Sánchez, L. I., Sternini, F., & Bignardi, C. (2019). *Techniques for Usability Risk Assessment during Medical Device Design*. Conference BIOSTEC Biodevices., doi:10.5220/0007483102070214

Recommendations for implementing the strategic initiative INDUSTRIE 4.0. (2013). Final report of the Industrie 4.0 Working Group. Retrieved from https://en.acatech.de/publication/recommendations-for-implementing-the-strategic-initiative-industrie-4-0-final-report-of-the-industrie-4-0-working-group/

Reducing your ICT energy costs. (2015). Retrieved from https://www.jisc.ac.uk/guides/reducing-your-ict-energy-costs

Reiner, J., Koch, J., Krebs, I., Schnabel, S., & Siech, T. (2005). Knowledge management issues for maintenance of automated production systems. Proceedings of the IFIP International Conference on Human Aspects in Production Management, 160, 229—237. doi:10.1007/0-387-23078-5_18

Renovell, J. (2015). Estadísticas bayesianas vs. frecuentistas. *Blog: Tecnología para Negocio*. https://www.paradigma-digital.com/category/techbiz/

Ribstein, L. E. (2002). Market vs. Regulatory Responses to Corporate Fraud: A Critique of the Sarbanes-Oxley Act of 2002. *The Journal of Corporation Law*, *28*(1), 1–74. doi:10.2139srn.332681

Rivas, L., & Flores, B. (2007). La gestión de conocimiento en la industria automovilística. Estudios Gerenciales, 23(102), 83–100. doi:10.1016/S0123-5923(07)70003-7

Roberts, R., & Newton, C. (1984). Report on small-scale test correlations with $K_{IC}$ data. *WRC Bulletin, 299.*

Rodrigues, T. C. (2014). The MACBETH Approach to Health Value Measurement: Building a Population Health Index in Group Processes. *Procedia Technology*, *16*, 1361–1366. doi:10.1016/j.protcy.2014.10.153

Rodriguez, D. S. S., Costa, H. G., & Do Carmo, L. F. R. R. S. (2013). Métodos de auxílio multicritério à decisão aplicados a problemas de PCP: Mapeamento da produção em periódicos publicados no Brasil. Gestão Produção, São Carlos, 20(1), 134–146. doi:10.1590/S0104-530X2013000100010

Rodríguez, R. A. (2002). Popper y la tecnología. *Revista de Filosofía*, *27*(1), 135-159.

Rodríguez, D. (2006). Modelos para la creación y gestión del conocimiento: Una aproximación teórica. *Review of Education, 37*, 25–39.

Roiko, A., & Murakami, Y. (2012). A design approach for components in ultralong fatigue life with step loading. *International Journal of Fatigue*, *41*, 140–149. doi:10.1016/j.ijfatigue.2011.12.021

Rojas Lazo, O., Mavila Hinojoza, D., & Cortez Vasquez, A. (2014). Teoría de Popper y los riesgos en la ingeniería. *Industrial Data, 16*, 58. . doi:10.15381/idata.v16i1.6390

Roldán-Molina, G., Almache-Cueva, M., Silva-Rabadão, C., Yevseyeva, I., & Basto-Fernandes, V. (2017). A Comparison of Cybersecurity Risk Analysis Tools. *Procedia Computer Science, 121*, 568–575. doi:10.1016/j.procs.2017.11.075

Román, M., Padilla, D., & Uceda, K. (2018). Implementación del Plan Estratégico de Gestión de Activos en Red de Energía del Perú. Academic Press.

Romero, C. (1996). *Análisis de Decisiones Multicriterio*. Isdefe.

Roper, A. T., Cunningham, S. W., Porter, A. L., Mason, T. W., Rossini, F. A., & Banks, J. (2011). *Managing the Forecasting Project* (2nd ed.). Forecasting and Management of Technology. doi:10.1002/9781118047989

Rosés F., (2002). *Risk Management. A new way to assure business success*. ACV Editions.

Roweis, S. (n.d.). *Levenberg-Marquardt Optimization*. Retrieved September 24, 2019, from https://www.cs.nyu.edu/~roweis/notes/lm.pdf

Ruggles, R. (1998). The state of the notion: Knowledge management in practice. *California Management Review, 40*(3), 80–89. doi:10.2307/41165944

Ruiz, R., & Maroto, C. (2005). A comprehensive review and evaluation of permutation flowshop heuristics. *Eur J Opl Res, 165*(2), 479–494. doi:10.1016/j.ejor.2004.04.017

Rullo, P., Braccia, L., Luppi, P., Zumoffen, D., & Feroldi, D. (2019). Integration of sizing and energy management based on economic predictive control for standalone hybrid renewable energy systems. *Renewable Energy, 140*, 436–451. doi:10.1016/j.renene.2019.03.074

Runkler, T. (2012). Data Preprocessing. In T. Runkler (Ed.), *Data Analytics* (pp. 21–33). Springer Vieweg., doi:10.1007/978-3-8348-2589-6_3.

Saaty, R. W. (1987). The Analytic Hierarchy Process-what it is and how it is used. Mat/d Modelling, 9(3-5), 161-176. doi:10.1016/0270-0255(87)90473-8

Saaty, R. W. (1987). The analytic hierarchy process–what it is and how it is used. *Mathematical Modelling, 9*(3-5), 161–176. doi:10.1016/0270-0255(87)90473-8

Saaty, R. W. (1996). *Decision-making with dependence and feedback: the analytic network process*. RSW Publications.

Saaty, T. L. (1980). *The Analytic Hierarchy Process*. McGraw Hill.

Saaty, T. L. (1991). *Método de Análise Hierárquica. Tradução e Revisão por Wainer da Silveira e Silva*. McGraw-Hill.

Saaty, T. L. (2008). Decision making with the analytic hierarchy process. *International Journal of Services Sciences, 1*(1), 83–98. doi:10.1504/IJSSCI.2008.017590

Saboo, A., Garza-Reyes, J., Er, A., & Kumar, V. (2014). A VSM improvement-based approach for lean operations in an Indian manufacturing SME. Int. J. Lean Enterprise Research, 1(1), 41–58. doi:10.1504/IJLER.2014.062281

Sakai, T. (2009). Review and prospect for current studies on very high cycle fatigue of metallic materials for machine structural use. *J Solid Mech. Materials & Design, 3*, 425–439.

Salomon, V. A. P., Marins, F., & Duduch, M. (2009). Tomada de Decisões Múltiplas Aplicada à Seleção de Fornecedores de Equipamentos de uma Linha de Montagem em uma Fábrica de Autopeças. Pesquisa Operacional para o Desenvolvimento, 1(3), 208-217.

Sánchez, M., Chaminade, C., & Escobar, C. (1999). En busca de una teoría sobre medición y gestión de los intangibles en la empresa: Una aproximación metodológica. *Ekonomiaz*, *45*, 188–213.

Sander, M., Müler, T., & Lebahn, J. (2013). Influence of mean stress and variable amplitude loading on the fatigue behaviour of a high-strength steel in VHCF regime. *International Journal of Fatigue*, *62*, 10–20. doi:10.1016/j.ijfatigue.2013.04.015

Scandizzo, S. (2005). Risk mapping and key risk indicators in operational risk management. *Economic Notes*, *34*(2), 231–256. doi:10.1111/j.0391-5026.2005.00150.x

Schantl, S., & Wagenhofer, A. (2019). *On the Economics and Optimal Design of Internal Control Regulation*. SSRN.

Schenk, M., Wirth, S., & Müller, E. (2010). Factory Planning Manual. Springer. doi:10.1007/978-3-642-03635-4

Schmidt, R. A., & Paris, P. C. (1973). Threshold for Fatigue Crack Propagation and Effect of Load Ratio and Frequency. *ASTM Special Technical Publication*, 536.

Schmittling, R. A. M. (2010). Performing a Security Risk Assessment. *ISACA Journal*, *1*, 1–7.

Schneider. (n.d.). www.schneiderelectric.com

Schuller, R., Fitzka, M., Irrash, D., Tran, D., Pennings, B., & Mayer, H. (2015). VHCF properties of nitride 18 Ni of maraging steel thin sheets with different Co and Ti content. *Fatigue & Fracture of Engineering Materials & Structures*, *38*(5), 518–527. doi:10.1111/ffe.12251

Schweizer, C., Seifert, T., Nieweg, B., Von Hartrott, P., & Riedel, H. (2011). Mechanisms and modelling of fatigue crack growth under combined low and high cycle fatigue loading. *International Journal of Fatigue*, *33*(2), 194–202. doi:10.1016/j.ijfatigue.2010.08.008

Sellitto, M. A., & Walter, C. (2006). Avaliação do desempenho de uma manufatura de equipamentos eletrênicos segundo critérios de competição. *Produção*, *16*(1), 34–47.

Serratella, C. M., Wang, G., & Conachey, R. (2007). Risk-based strategies for the next generation of maintenance and inspection programs. *International Symposium on Maritime, Safety, Security and Environmental Protection (SSE)*, 20–21.

SESCAM. (2007). *Valoración de la calidad de la atención sanitaria pública en hospitales*. Memory of activities of SESCAM.

Shafer, S. M., & Byrd, T. A. (2000). A framework for measuring the effiency of organizational investments in information technology using data envelopment analysis. *Omega*, *28*(2), 125–141. doi:10.1016/S0305-0483(99)00039-0

Shafiee, M. (2015). Maintenance strategy selection problem: An MCDM overview. *Journal of Quality in Maintenance Engineering*, *21*(4), 378–402. doi:10.1108/JQME-09-2013-0063

Shafiee, M., Labib, A., Maiti, J., & Starr, A. (2019). Maintenance strategy selection for multi-component systems using a combined analytic network process and cost-risk criticality mode. Proceedings of the Institution of Mechanical Engineers, Part O. *Journal of Risk and Reliability*, *233*(2), 89–104.

Shahin, A., Shirouyehzad, H., & Pourjavad, E. (2012). Optimum maintenance strategy: A case study in the mining industry. *International Journal of Services and Operations Management*, *12*(3), 368–386. doi:10.1504/IJSOM.2012.047626

Sharma, R. K., Kumar, D., & Kumar, P. (2005). FLM to select suitable maintenance strategy in process industries using MISO model. *Journal of Quality in Maintenance Engineering*, *11*(4), 359–374. doi:10.1108/13552510510626981

Sheibani, K. (2004). An adaptive greedy genetic algorithm using fuzzy reasoning for the travelling salesman problem. *Proc. Hybrid Metaheuristics, 16th European Conf. on Artificial Intelligence*, 31-49.

Sheibani, K. (2005). *Fuzzy Greedy Evaluation in Search, Optimisation, and Learning* (PhD thesis). London Metropolitan University, London, UK.

Sheibani, K. (2008). Fuzzy Greedy Search in Combinatorial Optimisation. Tadbir Institute for Operational Research, Systems Design, and Financial Services.

Sheibani, K. (2013). A constructive heuristic for permutation flow-shop scheduling with the makespan criterion: in a particular case where the sum of the processing times of each job is the same on each machine. *Lecture Notes in Management Science, 5*, 18-24.

Sheibani, K. (2006). *Fuzzy greedy search and job-shop problem. Proc. the UK Planning & Scheduling Special Interest Group*.

Sheibani, K. (2010). A fuzzy greedy heuristic for permutation flow-shop scheduling. *The Journal of the Operational Research Society, 61*(5), 813–818. doi:10.1057/jors.2008.194

Sheibani, K. (2010). An evolutionary approach using fuzzy greedy initialization to permutation flow-shop scheduling with the makespan criterion. *Proc. IEEE Intl. Conf. on Industrial Engineering and Engineering Management*, 638-641. 10.1109/ICIEEM.2010.5646535

Sheibani, K. (2010). The fuzzy greedy search in combinatorial optimization with specific reference to the travelling salesman problem. *Proc. IEEE Intl. Conf. on Industrial Engineering and Engineering Management*, 1367-1370. 10.1109/IEEM.2010.5674393

Sheibani, K. (2011). A hybrid metaheuristic using fuzzy greedy search operator for combinatorial optimization with specific reference to the travelling salesman problem. *Iranian Journal of Operations Research, 2*(2), 63–71.

Sheibani, K. (2017). Fuzzy greedy search heuristic for combinatorial optimization with specific application to machine scheduling. *International Journal of Management Science and Engineering Management, 12*(2), 148–153. doi:10.1080/17509653.2016.1172993

Sheibani, K. (2018). A very fast heuristic for combinatorial optimization with specific application to priority rule sequencing in operations management: Fuzzy greedy search. *International Journal of Operations Research and Information Systems, 9*(3), 77–89. doi:10.4018/IJORIS.2018070104

Sheibani, K. (2019). An integrated heuristic for machine sequencing with specific reference to the permutation flow-shop scheduling problem. *International Journal of Strategic Engineering, 2*(2), 1–8. doi:10.4018/IJoSE.2019070101

Shendryk, V. V., & Vashchenko, S. M. (2011). The system of Data Collection, Allocation and Analysis. *Visnyk of the National University. Lviv Polytechnic, 715*, 343–353.

Shiozawa, K., Hasegawa, T., Kashiwagi, Y., & Lu, L. (2009). Very high cycle fatigue properties of bearing steel under axial loading condition. *International Journal of Fatigue, 31*(5), 880–888. doi:10.1016/j.ijfatigue.2008.11.001

Sikdar, P. (2017). Practitioner's Guide to Business Impact Analysis. Practitioner's Guide to Business Impact Analysis. doi:10.1201/9781315187884

Silverio, R. C., & Moura, T. (2003). Evaluation of cold crack susceptibility on HSLA steel welded joints. *Rev. Metal. Madrid, 39*(2), 83–89. doi:10.3989/revmetalm.2003.v39.i2.318

Simmons, M. (1997). *COSO based auditing*. Internal Auditor.

Singh, B., Garg, S., & Sharma, S. (2011). Value stream mapping: Literature review and implications for Indian Industry. *International Journal of Advanced Manufacturing Technology, 53*(5-8), 799–808. doi:10.1007/s00170-010-2860-7

Sinha, S. S. C., & Chandel, S. S. (2014, April 1). Review of software tools for hybrid renewable energy systems. *Renewable & Sustainable Energy Reviews*, *32*(Supplement C), 192–205. doi:10.1016/j.rser.2014.01.035

Sioud, A., & Gagné, C. (2018). Enhanced migrating birds optimization algorithm for the permutation flow shop problem with sequence dependent setup times. *European Journal of Operational Research*, *264*(1), 66–73. doi:10.1016/j.ejor.2017.06.027

Siru, N., Zhuang, Y., Jingjing, G., & Ying, H. (2016). A formal model and risk assessment method for security-critical real-time embedded systems. *Computers & Security*, *58*, 199–215. doi:10.1016/j.cose.2016.01.005

Skagestad, B., & Mildenstein, P. (n.d.). *District Heating and Cooling Connection Handbook*. Retrieved September 24, 2019, from https://www.districtheatingscotland.com/wp-content/uploads/2015/12/DistrictHeatingAndCoolingConnectionHandbook.pdf

Skolud, B., Krenczyk, D., & Davidrajuh, R. (2016). Solving Repetitive Production Planning Problems. An Approach Based on Activity-oriented Petri Nets. In International Conference on European Transnational Education. Springer International Publishing.

Smith, R. L., Bush, R. J., & Schmoldt, D. L. (1995). *A Hierarchical Analysis of Bridge Decision Makers*. USDA Forest Service, Timber Bridge Information Resource Center, in cooperation with the Center for Forest Products Marketing, Department of Wood Science and Forest Products, at Virginia Polytechnic Institute and State University.

Society for Risk Analysis (SRA). (2018). *Society for Risk Analysis Fundamental Principles*. Retrieved from https://www.sra.org/resources

Sokovic, M., Pavletic, D., & Kern Pipan, K. (2010). Quality Improvement Methodologies – PDCA Cycle, RADAR Matrix, DMAIC and DFSS. *Journal of Achievements in Materials and Manufacturing Engineering*, *43* (1), 476-483.

Soliman, M., Saurin, T. A., & Anzanello, M. J. (2018). The impacts of lean production on the complexity of socio-technical systems. *International Journal of Production Economics*, *197*, 342–357. doi:10.1016/j.ijpe.2018.01.024

Sols, A. (2000). Fiabilidad, Mantenibilidad, Efectividad, un enfoque sistémico. Comillas.

Soroudi, A., & Amraee, T. (2013). Decision making under uncertainty in energy systems: State of the art. *Renewable & Sustainable Energy Reviews*, *28*, 376–384. doi:10.1016/j.rser.2013.08.039

Souza, A. B., Bianchi, J. M. B. B., & Boente, A. N. P. B. (2015). Fuzzyahp: uma ferramenta de gestão Tecnológica para tomada de decisão Muticriterial. Revista científica digital da faetec, 8(1).

*Specification for weldable structural steels BS 4360*. (1990 September). British Standard Institution.

*Specification for weldable structural steels for fixed offshore structures Bs 7191*. (1989). British Standard Institution.

Spira, L. F., & Page, M. (2001). Risk management The reinvention of internal control and the changing role of internal audit. *Accounting, Auditing & Accountability Journal*, *16*(4), 640–661. doi:10.1108/09513570310492335

Sree, P. C. R. & Kujawski, D. (2014). A Deviatoric Stress-Strain Function for Fatigue Life Analysis. *Advanced Materials Research*, *891-892*, 789–802. doi: 10.4028/www.scientific.net/. /AMR.891-892.797

Srivastava, P., Khanduja, D., & Agrawal, V. (2017). A framework of fuzzy integrated MADM and GMA for maintenance strategy selection based on agile enabler attributes. *Mathematics in Industry Case Studies*, *8*(1), 5. doi:10.118640929-017-0014-y

Stafford, E. F. Jr, Tseng, F. T., & Gupta, J. N. D. (2005). Comparative evaluation of MILP flowshop models. *The Journal of the Operational Research Society*, *56*(1), 88–101. doi:10.1057/palgrave.jors.2601805

Standards from Australia / New Zealand. (2004). AS/NZS 4360:2004. *Risk management.*

Stecher, K., Brosowski, A., & Thrän, D. (n.d.). Biomass Potential in Africa. IRENA. Available at: http://www.irena.org/DocumentDownloads/Publications/IRENA-DBFZ_Biomass%20Potential%20in%20Africa.pdf

Sule, D. R. (1997). *Industrial Scheduling*. PWS Publishing Company.

Sun, C., Xie, J., Zhao, A., Lei, Z., & Hong, Y. (2012). Accumulation damage model for fatigue life estimation of high strength steels in high-cycle and very high-cycle fatigue regimes. *Fatigue & Fracture of Engineering Materials & Structures*, *35*(7), 638–647. doi:10.1111/j.1460-2695.2011.01658.x

Sun, L., Srivastava, R. P., & Mock, T. J. (2006). An information systems security risk assessment model under the Dempster-Shafer theory of belief functions. *Journal of Management Information Systems*, *22*(4), 109–142. doi:10.2753/MIS0742-1222220405

Šúri, M., Huld, T. A., Dunlop, E. D., & Ossenbrink, H. A. (2007). Potential of solar electricity generation in the European Union member states and candidate countries. Solar Energy, 2007(81), 1295–1305. doi:10.1016/j.solener.2006.12.007

Suriadi, S., Weiss, B., Winkelmann, A., Hofstede, A., Conforti, A. M., Fidge, R., La Rosa, C. M., Ouyang, C., Rosemann, M., Pika, A., & Wynn, M. (2014). Current research in risk-aware business process management: Overview, comparison, and gap analysis. *Communications of the Association for Information Systems*, *34*(1), 933–984. doi:10.17705/1CAIS.03452

Susser, M. (1973). *Causal thinking in the health sciences. Concepts and strategies of epidemiology*. Oxford University Press.

Taghipour, S., Banjevic, D., & Jardine, A. K. S. (2011). Prioritization of medical equipment for maintenance decisions. *The Journal of the Operational Research Society*, *62*(9), 1666–1687. doi:10.1057/jors.2010.106

Taha, H. A. (1998). *Investigación de Operaciones. Una introducción*. Pearson.

Taha, H. A. (2011). *Operations research. An introduction*. Prentice Hall.

Taillard, E. (1993). Benchmarks for Basic Scheduling Problems. *European Journal of Operational Research*, *64*(2), 278–285. doi:10.1016/0377-2217(93)90182-M

Taliotis, C., Bazilian, M., Welsch, D. G., & Howells, M. (2014). Grand Inga to power Africa: Hydropower development scenarios to 2035. *Energy Strategy Reviews*, *4*, 1–10.

Tanaka, K., & Akinawa, Y. (2002). Fatigue crack propagation behaviour derived from S-N data in very high cycle regime. *Fatigue & Fracture of Engineering Materials & Structures*, *25*(8-9), 775–784. doi:10.1046/j.1460-2695.2002.00547.x

Tari, J., & Garcia, M. (2009). Dimensiones de la gestión del conocimiento y la gestión de la calidad: Una revisión de la literatura. *Investigaciones Europeas de Dirección y Economía de la Empresa*, *15*(3), 135–148. doi:10.1016/S1135-2523(12)60105-1

Technology Futures Analysis Working Group, (2004). Technology futures analysis: toward integration of the field and new methods, *Technology Forecasting and Social Change*, *71*, 287-303.

Termis District Energy Management Software. (n.d.). Retrieved September 24, 2019, from http://7t.dk/products/termis/Download/product-brochures.aspx

*The Concise Oxford Dictionary*. (2012). Oxford University Press.

The Guidance for district heating companies. (n.d.). *The environmental benefits of district heating: using the new Ecoheat4cities label Guidance for district heating companies*. Retrieved September 24, 2019, from http://www.bre.co.uk/filelibrary/rpts/ecoheat4cities/Ecoheat4Cities_WP4_Guidance_for_companies.pdf

The Heat Monitoring Solution of ENMAT Software. (n.d.). Retrieved September 24, 2019, from http://www.en-mat.com/solutions/heat-monitoring

The Official Website of ePortal: Energy Insight Software. (n.d.). Retrieved September 24, 2019, from https://eportal.eu/en/

The Official Website of Landis+Gyr Company. (n.d.). Retrieved September 24, 2019, from https://www.landisgyr.eu/

The Official Website of Wattics Company. (n.d.). Retrieved September 24, 2019, from https://www.wattics.com/

Tiddens, W. W., Braaksma, A. J. J., & Tinga, T. (2015). The Adoption of Prognostic Technologies in Maintenance Decision Making: A Multiple Case Study. *Procedia CIRP*, *38*, 171-176.

Tiddens, W. W., Braaksma, A.J.J., & Tinga, T. (2018). Selecting Suitable Candidates for Predictive Maintenance. International Journal of Prognostics and Health Management, 9(1).

Tiddens, W. W., Tinga, T., Braaksma, A. J. J., & Brouwer, O. (2017). The business case for condition-based maintenance: A hybrid (Non-) financial approach. *Proceedings of the 27th European Safety and Reliability Conference*, 821-829.

Tiemessen, L., Lane, H. W., Crossan, M. M., & Yinkpen, A. C. (1997). Knowledge management in international joint ventures. In *North American perspectives. The Cooperative Strategies Series.* The New Lexington Press.

Tirpak, T. M. (2005). Five steps to effective knowledge management. *Research Technology Management*, *48*(3), 15–16. doi:10.1080/08956308.2005.11657310

Tjoa, S., Jakoubi, S., Goluch, G., Kitzler, G., Goluch, S., & Quirchmayr, G. (2011). A formal approach enabling risk-aware business process modeling and simulation. *IEEE Transactions on Services Computing*, *4*(2), 153–166. doi:10.1109/TSC.2010.17

Toyota. (2019). A production system based on the philosophy of achieving the complete elimination of all waste in pursuit of the most efficient methods. Obtenido de Toyota: https://global.toyota/en/company/vision-and-philosophy/production-system

Triantaphyllou, E., Lootsma, F. A., Pardalos, P. M., & Mann, S. H. (1994). On the evaluation and application of different scales for quantifying pairwise comparisons in Fuzzy Sets. *Journal of Multi-Criteria Decision Analysis*, *3*(3), 133–155. doi:10.1002/mcda.4020030302

Triantaphyllou, E., & Mann, S. H. (1995). Using the analytic hierarchy process for decision making In engineering applications: Some challenges. *International Journal of Industrial Engineering: Applications and Practice*, *2*(1), 35–44.

Tucker, A. (1995). *Applied Combinatorics.* John Wiley & Sons.

Tumenbayeva, O. S., & Zhaksybekova, G. N. (2016). Implementation of the Integrated System of Risk Management in the Banks of Kazakhstan. *Indian Journal of Science and Technology*, *9*(5), •••. doi:10.17485/ijst/2016/v9i5/87607

Turan, O., Lazakis, I., Judahb, S., & Incecika, A. (2011). Investigating the Reliability and Criticality of the Maintenance Characteristics of a Diving Support Vessel. *Quality and Reliability Engineering International*, *27*(7), 931–946. doi:10.1002/qre.1182

U.S. Securities and Exchange Comission. (1 de 10 de 2013). The Laws That Govern the Securities Industry. Recuperado el 18 de 10 de 2015, de Sarbanes-Oxley Act of 2002: http://www.sec.gov/about/laws.shtml#sox2002

USEPA GHD. (2006). *Asset Management Training Workshops.* Author.

Valdimarsson, P. (2014). *District Heat Distribution Networks.* Retrieved from http://www.os.is/gogn/unu-gtp-sc/UNU-GTP-SC-18-27.pdf

Van der Veeken, P., Van Schooten, S., Shinde, R., Dunnewind, M., & van den Berg, J. (2016, December). Applying Bayesian game theory to analyse cyber risks of bank transaction systems. In *2016 International Conference on Computing, Analytics and Security Trends (CAST)* (pp. 84-89). IEEE. 10.1109/CAST.2016.7914945

Van Laarhoven, P. J. M., & Pedrycz, W. (1983). A fuzzy extension of saaty's priority theory. *Fuzzy Sets and Systems, 11*(1-3), 229–241. doi:10.1016/S0165-0114(83)80082-7

Vatn, J. (1997). Maintenance optimisation from a decision theoretical point of view. *Reliability Engineering & System Safety, 58*(2), 119–126. doi:10.1016/S0951-8320(97)00025-2

Veldman, J., Klingenberg, W., & Wortmann, H. (2011). Managing condition based maintenance technology: A multiple case study in the process industry. *Journal of Quality in Maintenance Engineering, 17*(1), 40–62. doi:10.1108/13552511111116240

Velmurugan, R., & Dhingra, T. (2015). Maintenance strategy selection and its impact in maintenance function. *International Journal of Operations & Production Management, 35*(12), 1622–1661. doi:10.1108/IJOPM-01-2014-0028

Ventura, J. (1996). *Análisis dinámico de la estrategia empresarial: Un ensayo interdisciplinar. Servicio de publicaciones.* Universidad de Oviedo.

Ventura, J., & Ordóñez de Pablos, P. (2007). Análisis de estrategias de conocimiento en la industria manufacturera española: Evidencia empíricas. *Tribuna de Economía, 836*, 141–161.

Vince, A. (2002). A framework for the greedy algorithm. *Discrete Applied Mathematics, 121*(1-3), 247–260. doi:10.1016/S0166-218X(01)00362-6

Volkel, M., & Abecker, A. (2008). Cost-benefit analysis for the design of personal knowledge management systems. ICEIS 2008—International Conference on Enterprise Information Systems, 95–105.

Volkel, M., & Haller, H. (2009). Conceptual data structures for personal knowledge management. *Online Information Review, 33*(2), 298–315. doi:10.1108/14684520910951221

Von Bertalanffy, L. (1989). *Teoría General de los Sistemas* (7th ed.). George Braziller.

Wah, L. (1999). Behind the Buzz. *Management Review, 88*(4), 16–19.

Wang, B., Xiong, H., & Jiang, C. (2014). A Multicriteria Decision Making Approach Based on Fuzzy Theory and Credibility Mechanism for Logistics Center Location Selection. TheScientificWorldJournal, 2014, 9. doi:10.1155/2014/347619 PubMed

Wang, M. (2010). The comparison between MAUT and PROMETHEE. IEEE international conference on industrial engineering and engineering management (IEEM), 753–757. doi:10.1016/j.esr.2014.01.001

Wang, L., Chu, J., & Wu, J. (2007). Selection of optimum maintenance strategies based on a fuzzy analytic hierarchy process. *International Journal of Production Economics, 107*(1), 151–163. doi:10.1016/j.ijpe.2006.08.005

Wang, Z., & Klir, G. (2013). *Fuzzy measure theory*. Springer Science & Business Media.

Weather, M. E. T. E. O. B. L. U. E. (n.d.). Weather history Kinsasa. https://www.meteoblue.com/en/weather/archive/export/kinsasa_dr-congo_8325634

Weber, C., Marechal, F., & Favrat, D. (2006). Design and Optimization of District Energy Systems. *Proceedings of the 10th International Symposium on District Heating and Cooling*. Retrieved from http://www.lsta.lt/files/events/16_weber.pdf

Weimer, C. W., Miller, J. O., & Hill, R. R. (2016, December). Agent-based modeling: An introduction and primer. In *Winter Simulation Conference (WSC)*, (pp. 65-79). IEEE.

Weixiong, Z. (2011). Computer Science & Engineering Mckelvey School of Engineering. Obtenido de https://www.cse.wustl.edu: http://iasri.res.in/ebook/win_school_aa/notes/Data_Preprocessing.pdf

Wernerfelt, B. (1984). A resource based view of the firm. *Strategic Management Journal*, *5*(2), 171–180. doi:10.1002/smj.4250050207

Wernstedt, F. (2003). *Multi-Agent Systems for District Heating Management*. Retrieved from https://www.diva-portal.org/smash/get/diva2:837359/FULLTEXT01.pdf

Whitney, H. (1935). On the abstract properties of linear dependence. *American Journal of Mathematics*, *57*(3), 509–533. doi:10.2307/2371182

Wieland, J. (2005). Corporate governance, values management, and standards: A European perspective. *Business & Society*.

Wilson, W. K. (1973). *Elastic-Plastic Analysis of Blunt Notched CT Specimens and Applications*. Research Report 73-1 E7-MAGRR-P1. Westinghouse Research Laboratories.

Wilson, R. J. (1985). *Introduction to Graph Theory* (3rd ed.). Longman.

Woiler, S., & Mathias, W. F. (2010). *Projetos. Planejamento, elaboração, análise* (2nd ed.). Editora Atlas.

Wolf, M. (2008). Strategic management of medical equipment. Starting and operating a TSC. *Proceedings of the 20th Congress of the International Federation of Hospital Engineering*.

Womack, J., & Jones, T. D. (2003). *Lean Thinking*. Simon & Schuster, Inc.

Woo, D. S., & Yim, D. S. (1998). A heuristic algorithm for mean flowtime objective in flowshop scheduling. *Comput Opl Res*, *25*(3), 175–182. doi:10.1016/S0305-0548(97)00050-6

World Bank. (n.d.). Deforestation Trends in the Congo Basin: Reconciling Economic Growth and Forest Protection. Working Paper 5: Wood-based Biomass Energy. Author.

Xiomara, P. (2009). La gestión del conocimiento y las Tics en el siglo XXI. Revista Universitaria de Investigación y Diálogo Académico, 5(1).

Xu, L. D., & Duan, L. (2019). Big data for cyber physical systems in industry 4.0: A survey. *Enterprise Information Systems*, *13*(2), 148–169. doi:10.1080/17517575.2018.1442934

Yamakawa, E. K., Miguel, P. A. C., & Aoki, A. R. (2014). Aplicação de fuzzyqualityfunctiondeployment para seleção de indicadores de eficiência energética para utilização em um software de gestão de energia. *Science & Engineering Journal*, *23*(2), 21–31.

Yang, J. (2006). La estrategia de gestión del conocimiento y su efecto en el crecimiento corporativo. *Economía Industrial*, *362*, 123–133.

Yazar, Z. (2002). A qualitative risk analysis and management tool–CRAMM. *SANS InfoSec Reading Room White Paper*, *11*, 12-32.

Yin, Q., & Chen, Q. (2012, January). A Social Network Analysis Platform for Organizational Risk Analysis—ORA. In *2012 Second International Conference on Intelligent System Design and Engineering Application (ISDEA)*, (pp. 760-763). IEEE. 10.1109/ISdea.2012.546

Yin, R. K. (1995). *Case study research: design and methods*. Sage Publications.

Yin, R. K. (2015). *Estudo de Caso: Planejamento e Métodos* (5th ed.). The Bookman.

Yu, H., Yu, H., & Wilamowski, B. M. (2011). Levenberg–Marquardt Training. In The Industrial Electronic Handbook: Intelligent Systems. Taylor & Francis Group, CRC Press. doi:10.1201/b10604-15

Zadeh, L. A. (1965). Fuzzy sets. *Information and Control*, *8*(3), 338–353. doi:10.1016/S0019-9958(65)90241-X

Zahir, S. (1999). Geometry of decision making and the vector space formulation of the analytic hierarchy process. *European Journal of Operational Research*, *112*(2), 373–396. doi:10.1016/S0377-2217(98)00020-4

Zaim, S., Turkyílmaz, A., Acar, M. F., Al-Turki, U., & Demirel, O. F. (2012). Maintenance strategy selection using AHP and ANP algorithms: A case study. *Journal of Quality in Maintenance Engineering*, *18*(1), 16–29. doi:10.1108/13552511211226166

Zandieh, M., Joreir-Ahmadi, M., & Fadaei-Rafsanjani, A. (2017). Buffer allocation problem and preventive maintenance planningin non-homogenous unreliable production lines. *Springer London*, *91*(5-8), 201–211.

Zaphiropoulos, E. P., & Dialynas, E. N. (2007). Methodology for the optimal component selection of electronic devices under reliability and cost constraints. *Quality and Reliability Engineering International*, *23*(8), 885–897. doi:10.1002/qre.850

Zare, R., Nouri, J., Abdoli, M. A., Atabi, F., & Alavi, M. (2016). The Integrated Fuzzy AHP and Goal Programing Model Based on LCA Results for Industrial Waste Management by Using the Nearest Weighted approximation of FN: Aluminum Industry in Arak, Iran. Advances in Materials Science and Engineering, 2016, 13. doi:10.1155/2016/1359691

Zhao, B., Tang, T., & Ning, B. (2016). Applying Hybrid Decision-Making Method Based on Fuzzy AHP-WOWA Operator for Emergency Alternative Evaluation of Unattended Train Operation Metro System. Mathematical Problems in Engineering, 2016, 12. doi:10.1155/2016/4105079

Zheng, T., & Yamashiro, M. (2010). Solving flow shop scheduling problems by quantum differential evolutionary algorithm. *International Journal of Advanced Manufacturing Technology*, *49*(5-8), 643–662. doi:10.100700170-009-2438-4

Zimmermann, H. J. (2001). *Fuzzy Set Theory and its Applications* (4th ed.). Springer-Verlag. doi:10.1007/978-94-010-0646-0

Zio, E. (2018). The future of risk assessment. *Reliability Engineering & System Safety*, *177*, 176–190. doi:10.1016/j.ress.2018.04.020

# About the Contributors

**Vicente González-Prida** is PhD in Industrial Engineering by the University of Seville (Extraordinary Prize of Doctorate), and Executive MBA by the Chamber of Commerce (First Class Honours). He has been honoured also with the National Award for PhD Thesis by the Spanish Association for Quality and by the Spanish Association for Maintenance, and Best Nomination from Spain for the Excellence Master Thesis Award bestowed by EFNMS. He currently works for the company GDELS, sharing his professional performance with the development of research projects for the University and teaching activities for degree programs in Spain and LATAM. Dr. González-Prida has edited three books, (IGI-Global 2015 and 2017; Springer Verlag 2018), regarding Maintenance Modelling, RAMS, Asset Management and Decision Making, and authored two books (Confemetal 2018; Springer Verlag 2014) about ISO-55000 and After Sales Management. His research works have been presented in conferences like WCEAM, ESREL, IEEE, ESREDA, etc., and published in journals such as RESS, PP&C, IMA, Computers in Industry, among others. Finally, Dr. González-Prida is fellow of the Club of Rome (Spanish Chapter) among other institutions.

**María Carmen Carnero** has a Doctorate in Industrial Engineering from the University of Castilla-La Mancha (2001) and is a lecturer at the Higher Technical School of Industrial Engineering at the University of Castilla-La Mancha. She has published two books, 14 chapters in science books, and has published research articles in Omega, Decision Support System, European Journal of Operational Research, Reliability Engineering and System Safety, European Journal of Industrial Engineering, Journal of Manufacturing Systems, The International Journal of Advanced Manufacturing Technology, Mechanical Systems and Signal Processing, Production Planning and Control, etc. She has 78 contributions to conferences. She was the lead researcher in three projects financed by the regional government, and has been a part of 16 other European, national, regional and local projects, and of ten contracts.

* * *

**Olga Aleksenko** is an Associate Professor in the Department of Computer Science at the Sumy State University where she has been a faculty member since 2005. She completed her Ph.D. and undergraduate studies at Sumy State University in 2006. Her research interests lie in the area of modelling of information systems and simulation of processes with a focus on probability problems.

**Iryna Baranova** is an Associate Professor in the Department of Computer Science at the Sumy State University.

**F. Javier Cárcel-Carrasco** is an Industrial Engineer with a PhD in Industrial Engineering from the Universitat Politècnica de València (UPV), Spain. He holds a PhD in Economic and Business Sciences from UNED, Spain. He is currently Professor in teaching and research in the Department of Building Construction in the UPV. He is a member of the Institute of Energy Engineering of the UPV. His main research areas are renewable energy sources, energy efficiency and industrial maintenance engineering.

**Baris Cimen** received the B.Sc. and M.A. degrees in Management Information Systems from Bogazici University, Istanbul, Turkey in 2014 and 2017 respectively. He is currently a PhD Candidate in the Management Information Systems Department of Bogazici University. His research interests include cybersecurity and artificial intelligence for information systems.

**Diego Figueroa** is an Industrial Engineer from the University of Federico Santa María, his areas of interest are related to Operations Management.

**Elias Hurtado-Pérez** holds a doctorate in industrial engineering. She is a member of the Institute of Energy Engineering of the UPV. Her main research areas are renewable energy sources, energy efficiency.

**Esra Koçak** received her Bachelor's degree in Mathematics from Marmara University. She received her M.A. degree in Management Information Systems from Bogazici University in 2017. She is currently a Ph.D. candidate in the Department of Management Information Systems at Bogazici University. Her research interests include information systems and technologies, customer relationship management, social networking sites, and digital marketing.

**Vicente López-Mateu** is a doctor architect from the Universitat Politècnica de València (UPV), Spain. He is currently Professor in the Department of Building Construction in the UPV. His research area is in the area of construction, environment and sustainability. He has published numerous articles in research journals, as well as research books.

**María de Lourdes Eguren Martí** has a Ph.D. in Business from the Universitat de Barcelona (UB). Her academic background includes a Bachelor in Mathematics (Universidad Metropolitana- UNIMET), System's Engineer degree (UNIMET), Master in e-Business (Universidad de Cantabria with a full scholarship from Fundación Carolina) and a Master of Research in Business Administration (UB). She's expert on value and risk management, process reengineering and systemic thinking, reflected on several publications, congress papers, conferences and book chapters. She has more than 15 years of professional experience among multinational companies of different sectors, and currently acts as Global Risk Management Director of a pharma company. She works as Associate professor at the Tecnocampus (Universitat Pompeu Fabra).

**Maria Carolina Martins-Rodrigues,** International PhD in Management, University of Extremadura, Spain Post-graduate in Knowledge management, University of Belgrano, Buenos Aires, Argentina, and a research fellow of CIEO (Algarve University). Her research interests currently are Sustainability, innovation, Knowledge Management, Entrepreneurship, Social Responsibility and Business Management. Has co-authored over 20 articles and book chapters and published in several scientific journals. She has also organized Iberian Conference on Entrepreneurship and international Conferences of Sustainability and Innovation. Founder member and vice-president of the direction of Empreend - Portuguese Association for Entrepreneurship. Was Professor in University Aberta (UAb) and University Moderna. In TAX Administration was Tax Administration Technician Advisor acting as coordinator in the International Relations Services Directorate (DSRI). and was Head of the Tax Service.

**Bilgin Metin** received the B.Sc. degree in Electronics and Communication Engineering from Istanbul Technical University, Istanbul, Turkey in 1996 and the M.Sc. and Ph.D. degrees in Electrical and Electronics Engineering from Bogazici University, Istanbul, Turkey in 2001 and 2007 respectively. He is currently an Assoc. Professor in the Management Information Systems Department of Bogazici University. His research interests include cybersecurity and electronic circuit design for information systems. He was given the best student paper award of ELECO'2002 conference in Turkey. He has over 40 publications in SCI/SCIE indexed scientific journals and more than 50 papers in the international conference proceedings. He has been serving in the Organization Committee of ELECO Conference on Electrical and Electronics Engineering since 2011. He gave lectures at Budapest University of Technology and Economics in 2014 as a part of Joint International Master in Smart Systems Integration (SSI). Bilgin Metin was one of the Local Co-Chairs for ECIS 2016 (24th European Conference on Information Systems). He is also the manager of the Bogazici University Cyber Security Center.

**Arnab Mitra** has received B.E. (Information Technology) from University of North Bengal, India, M.Tech (Computer Science and Engineering) from West Bengal University of Technology, India, and Ph.D. (Engineering) from Jadavpur University, India. He was an Erasmus Mundus visiting doctoral researcher at "Ghoerghe Asachi" Technical University of Iasi, Romania. His research interests include Cellular Automata, Distributed Computing, Cloud Applications.

**Meltem Mutlutürk** received a BA degree in Econometrics in 2013 and a MA in Management Information Systems in 2016. She is currently a PhD candidate in the Department of Management Information Systems at Bogazici University where she is also a Research Assistant. Her research interests include cybersecurity, risk assessment, simulation as well as data mining.

**Yuliia Parfenenko** is a researcher of computer science at the Sumy State University, Sumy, Ukraine. She has a Bachelor of Computer science from Sumy State University in 2007. She received her Master's Degree of Information Technology of Design (Diploma with distinction) in 2008. She has a PhD in Information Technologies in 2016. Now she is s an Associate Professor in the Department of Computer Science at the Sumy State University. Research direction is creation information technologies for energy management in district heating.

**Elisa Peñalvo-López** holds a doctorate in industrial engineering. She is a member of the Institute of Energy Engineering of the UPV. Her main research areas are renewable energy sources, energy efficiency.

**Angel Pérez-Navarro** holds a doctorate in physical sciences. She is a member of the Institute of Energy Engineering of the UPV. Her main research areas are renewable energy sources, energy efficiency.

**Kaveh Sheibani** holds a PhD in Operational Research from London Metropolitan University, UK. His main research interests lie in exploring search methodologies for hard combinatorial optimisation problems and their efficiency across a wide variety of applications.

**Sergii Shendryk** is a PhD student in Computer Science Department at Sumy State University.

**Vira Shendryk** started her academic career since Sept, 2002 at the Sumy State University, Ukraine. She was a visiting scholar at McMaster University, Canada in 2012 and was a visiting research fellow at the Department of Computer Science of the Faculty of Technology and Society, the Malmo University, Sweden in 2013. Her research interest is focused on the field of Information Systems and Decision Science particularly in decision making under uncertainty. She has written over 100 journal articles and conference papers and presentations. She has also been a Member of the Editorial Review Board of international multi-disciplinary quarterly journal «Information Technology and Economics», which focuses on the intersection of Information Technology and Economics and was Member of Programme Committees the International Conference on Information and Software Technologies (ICIST 2014-2019, Lithuania) and Advanced Information Systems and Technologies (AIST 2012-2017, Ukraine).

**Pablo Viveros Gunckel** is a Researcher and Academic at Technical University Federico Santa María, Chile. He has been active in national and international research, both for important journals and conferences. Also he has developed projects in the Chilean industry. Consultant specialist in the area of Reliability, Asset Management, System Modeling and Evaluation of Engineering Projects. He is PhD in Industrial Engineer and Master in Asset Management and Maintenance.

**Jesús Zamora Bonilla** graduated in Philosophy (1986) and in Economics (1995) at Universidad Autónoma de Madrid, where he also attained a PhD in each of both areas (1993 and 2001). He taught Economic Methodology at Universidad Carlos III (1994-2002) before becoming professor of Philosophy of Science at UNED (Universidad Nacional de Educación a Distancia), where he currently is the dean of the Faculty of Philosophy. His research has focussed around scientific realism and truthlikeness, the rationality of science, philosophy of social sciences, and the application of economic models to social epistemology. He has published papers in most of the top journals in philosophy of science and philosophy of economics and the social sciences (Synthese, Philosophy of Science, Erkenntnis, Theoria, Episteme, Journal for General Philosophy of Science, Journal of Economic Methodology, Philosophy of the Social Sciences, Economics & Philosophy, etc.) and is also co-editor, with Ian Jarvie, of The SAGE Handbook of Philosophy of Social Science (2011), besides having also published numerous papers and monographs in Spanish. He has been member of the Executive Board of INEM (2006-2012), is founding member of the steering committee of the European Network for the Philosophy of the Social Sciences, was treasurer of the Spanish Society of Logic, Methodology and Philosophy of Science (1999-2007), is full member of the Académie Internationale de Philosophie des Sciences, and has served as associate editor of the European Journal of Philosophy of Science (2009-2017).

# Index

# G



# H



# I



# K



# L



# M



# O



# P

## V



## W

www.ingramcontent.com/pod-product-compliance
Lightning Source LLC
LaVergne TN
LVHW062111300926
849161LV00010B/2428